PENGUIN
REFERENCE BOOKS

THE NEW PENGUIN DICTIONARY OF
BIOLOGY

Michael Abercrombie took a degree in Zoology at Oxford, and then began research work at Cambridge. Later he taught in the Zoology Department, Birmingham University. He was Professor of Embryology and then Professor of Zoology at University College London, and from 1970 to 1979 was Director of the Strangeways Research Laboratory, Cambridge. He was elected a Fellow of the Royal Society in 1958, and awarded an honorary doctorate by the University of Uppsala in 1977. He died in May 1979 just after finishing work on the seventh edition of this dictionary.

Michael Hickman was born in Worcestershire in 1943. He gained a First Class Honours B.Sc. in Biology and Botany from the University of Western Ontario in 1966 and a Ph.D. in Phycology and Freshwater Ecology from the University of Bristol in England in 1970. He has been Professor in Botany at the University of Alberta from 1981, after having been both Assistant Professor and Associate Professor in the same department since 1970. He has had numerous papers published on his investigative researches into lakes and rivers and has acted as a consultant for many organizations. He is currently engaged in research concerning effects of climatic change upon the past history of selected Alberta lakes. He lives in Edmonton with his wife and two children.

M. L. Johnson taught zoolo sity for some years, and re invertebrates and on nerv researched on selection an dents at University College

disorders in cerebral palsied children at Guy's Hospital, and became Reader in Architectural Education at University College London. She then worked with the new clinical course in the Cambridge medical school. Publications include *Anatomy of Judgement, Perceptual and Visuo-Motor Disorders in Cerebral Palsy, Aims and Techniques of Group Teaching*, and (with P. M. Terry) *Talking to Learn*.

Michael Thain was born in Hampstead in 1946 and educated at University College School and Keble College, Oxford, graduating in Zoology and gaining a Diploma in Human Biology. After an introduction to History and Philosophy of Science at University College London, he graduated in Philosophy from Birkbeck College, London, in 1983, and is currently studying History of Technology at Imperial College, London. In 1969 he joined the staff of Harrow School, where he was Head of Biology for eleven years and is now Head of Special Studies and in charge of the school's conservation area, running a local WATCH group for young naturalists. He is currently collaborating on a *Dictionary of Zoology*, also for Penguin. He lives in Harrow with his wife and two children.

THE NEW PENGUIN
DICTIONARY OF
BIOLOGY

M. Abercrombie
M. Hickman
M. L. Johnson
M. Thain

EIGHTH EDITION

PENGUIN BOOKS

PENGUIN BOOKS

Published by the Penguin Group
Penguin Books Ltd, 27 Wrights Lane, London W8 5TZ, England
Viking Penguin, a division of Penguin Books USA Inc.
375 Hudson Street, New York, New York 10014, USA
Penguin Books Australia Ltd, Ringwood, Victoria, Australia
Penguin Books Canada Ltd, 2801 John Street, Markham, Ontario, Canada L3R 1B4
Penguin Books (NZ) Ltd, 182–190 Wairau Road, Auckland 10, New Zealand

Penguin Books Ltd, Registered Offices: Harmondsworth, Middlesex, England

First published as *The Penguin Dictionary of Biology*, 1951
Second edition 1954
Third edition 1957
Fourth edition 1961
Fifth edition 1966
Sixth edition 1973
Seventh edition 1980
Eighth edition, entitled *The New Penguin Dictionary of Biology*, 1990

Copyright © M. Abercrombie, C. J. Hickman and M. L. Johnson, 1951, 1954, 1957,
1961, 1966, 1973, 1980
Copyright © Michael Thain, the Estate of M. Abercrombie,
the Estate of C. J. Hickman and the Estate of
M. L. Johnson, 1990
All rights reserved

The acknowledgements on pages ix–x constitute an extension of this copyright page

Printed in England by Clays Ltd, St Ives plc
Filmset in 9 on 10½pt Monophoto Times

FOR KATEY AND AVRIL, AND MARGARET

PREFACE TO EIGHTH EDITION

In the first major revision of this work for some years, changes were overdue to both its form and content if we were to ensure its continuance as a valuable reference book for school and undergraduate work. While remaining within the tradition set by the previous authors, we provide more of the encyclopaedic type of entry, in particular for terms central to theory and for those describing a phenomenon or process central to much else in the discipline. These articles and the extensive cross-referencing of most entries will not please everyone, least of all those who consult this dictionary merely as a lexicon. For brevity's sake we have dispensed with many definite and indefinite articles, but have tried to keep the result grammatical.

It may be argued against us that some head words rarely turn up in the literature. In defence, terms such as *arms race* and *cost of meiosis* are included here not because they are particularly common in the literature but because they represent convenient headings under which to include important material that would have been difficult to place elsewhere without over-stretching another entry. In a few cases, indeed on the very first page, several entries with a mutual bearing on one another have been brought together under a single head word or phrase, we hope for interest as well as convenience. Where this occurs, all included subterms are separately listed, directing the reader to larger entries. Terms in small capitals indicate where the reader might choose to pursue related matters raised by an entry; for a cross-referencing role lies firmly in this dictionary's tradition. Some will argue that this detracts from what little aesthetic appeal the work may have; but cross-references frequently provide information which it was the authors' hope should be read, although by no means all terms with separate entries are given small capitals. Italicized terms either indicate subdivisions of an entry or those terms with a particular tendency to be found associated with the head word. Others have less explicit claim to emphasis, but were simply considered worth stressing.

We have endeavoured throughout to include genuinely informative material, in addition to stating the obvious. Advanced readers will share with us the common frustration of finding too little interesting information in dictionary entries, and we have sought to avoid that. Each entry could have been written in many ways, and although we have never knowingly side-stepped the thorny problems that many attempts at definition pose, we must often have oversimplified through ignorance. However, it did not seem possible to do justice to such terms as *gene*, *classification* and *species*, without including some of the philosophical issues they raise. Likewise, it seemed at times inappropriate to exclude a little historical information. To this extent we hope to have promoted the view that biology is not an isolated discipline, and that its natural links with other fields of inquiry deserve explicit recognition and study in their own right.

It is a pleasure to record here the many people who have given advice during the preparation of this dictionary, both in formulating entries and in reviewing sections of manuscript. Pupils at Harrow and students at the University of Alberta, Edmonton, have provided valuable criticism as consumers. Particular thanks for advice and comments also go to Richard Burden, Michael Etheridge, John Hoddinott, Stephanie Holliday, John Imeson, Bill Richmond, Chris Stringer, Dick Vane-Wright, Dale H. Vitt and two very helpful referees who chose to remain anonymous. Eileen Michie did splendid copy-editing work. It goes without saying that any errors which remain are the sole responsibility of the authors, who would welcome corrections and suggestions for improvement.

The dictionary has isolated us from our families, sometimes at rather critical periods, and it is a special pleasure to be able to recognize their patience and forbearance, particularly Katey and Avril, and Margaret, to whom the work is dedicated.

M.H.
M.T.
February 1990

ACKNOWLEDGEMENTS FOR FIGURES

Some of the original illustrations and/or labelling may have been slightly altered.

Alberts, B., et al.: *Molecular Biology of the Cell* (1st edn), Garland (1983). Figs. 3a, 3b, 6a, 6b, 8, 9, 14, 20, 29, 44, 60.

Austin, C. R., and Short, R. V.: *Reproduction in Mammals, Book 2: Embryonic and Fetal Development* (2nd edn), CUP (1982). Figs. 16, 55.

Barrington, E. J. W.: *Invertebrate Structure and Function* (2nd edn), Nelson (1979). Figs. 17a, 17b, 17c.

Chapman, R. F.: *The Insects* (2nd edn), Hodder & Stoughton (1971). Fig. 49.

Charig, A.: *A New Look at the Dinosaurs*, British Museum (Natural History) (1979). Fig. 63.

Cohen, J.: *Reproduction*, Butterworths (1977). Figs. 62a, 62b.

Freeman, W. H., and Bracegirdle, B.: *An Advanced Atlas of Histology*, Heinemann (1976). Fig. 24.

Frobisher, M., et al.: *Fundamentals of Microbiology* (9th edn), Harcourt Brace Jovanovich (1974). Figs. 3c, 30.

Goodenough, U.: *Genetics* (3rd edn), Holt-Saunders (1984). Fig. 38.

Grimstone, A. V., Harris, H., and Johnson, R. T.: *Prospects in Cell Biology*, The Company of Biologists Ltd, Cambridge (1986). Fig. 37.

Hartman, P. E., and Suskind, S. R.: *Gene Action*, Viking Penguin Inc. (1965). Fig. 10.

Hopkins, C. R.: *Structure and Function of Cells*, Harcourt Brace Jovanovich (1978). Figs. 4a, 4b, 5a, 5b, 5c, 5d.

Hughes, G. M.: *Comparative Physiology of Vertebrate Respiration*, Harvard University Press (1963). Figs. 23b, 23c.

Katz, B.: *Nerve, Muscle and Synapse*, McGraw-Hill, Inc. (1966). Fig. 35.

Kingley, J. S.: *Outlines of Comparative Anatomy of Vertebrates*, The Blakiston Company (McGraw-Hill Book Company) (1928), with permission of McGraw-Hill, Inc. Fig. 23a.

Landsborough Thompson, A. (ed.): *A New Dictionary of Birds*, British Ornithologists' Union (1964). Figs. 18a, 18b.

Lehninger, A. L.: *Biochemistry* (2nd edn), Worth (1975). Figs. 12a, 12b, 25b, 26, 28, 33, 51, 64.

Lewis, K. R., and John, B.: *The Matter of Mendelian Heredity* (2nd edn), Longman (1972). Fig. 42(i).

Mather, K.: *Genetical Structure of Populations*, Chapman and Hall (1973). Figs. 21a, 21b.

May, R. M. (ed.): *Theoretical Ecology* (2nd edn), Blackwell (1981). Table 5.

Roberts, M. B. V.: *Biology: A Functional Approach* (3rd edn), Nelson (1982). Figs. 54a, 54b.

Romer, A. S.: *The Vertebrate Body* (5th edn), Holt-Saunders (1977). Figs. 2, 52b, 53.

Slack, J. M. W.: *From Egg to Embryo*, CUP (1983). Fig. 22.

Staines, N., Brostoff, J., and James, K.: *Introducing Immunology*, Gower Medical (1985). Figs. 41, 43.

Szalay, F. S., and Delson, E.: *Evolutionary History of the Primates*, Academic Press (1979). Fig. 34.

Tortora, G. J., and Anagnostokos, N. P.: *Principles of Anatomy and Physiology* (4th edn), Harper and Row (1984). Figs. 25a, 27, 32, 40, 47, 50, 52a, 57, 61a.

Watson, J. D.: *Molecular Biology of the Gene* (3rd edn), Benjamin/Cummings (1976). Fig. 39.

A

A-BAND. See STRIATED MUSCLE.

ABAXIAL. (Of a leaf surface) facing away from stem. Compare AD-AXIAL.

ABDOMEN. (1) Vertebrate body region containing viscera (e.g. intestine, liver, kidneys) other than heart and lungs; bounded anteriorly in mammals but not other classes by a diaphragm. (2) Posterior arthropod trunk segments, exhibiting TAGMOSIS in insects, but not in crustaceans.

ABDUCENS NERVE. Sixth vertebrate CRANIAL NERVE. Mixed, but mainly motor, supplying external rectus eye-muscle.

ABERRANT CHROMOSOME BEHAVIOUR. Departures from normal mitotic and meiotic chromosome behaviour, often with a recognized genetic basis. Includes (1) *achiasmate meiosis*, where chiasmata fail to form (e.g. in *Drosophila* spermatogenesis; see SUPPRESSOR MU-TATION); (2) *amitosis*, where a dumb-bell-like constriction separates into two the apparently 'interphase-like', but often highly polyploid, ciliate macronucleus prior to fission of the cell; (3) *chromosome extrusion* or loss, as with X-chromosomes in egg maturation of some parthenogenetic aphids (see SEX DETERMINATION); and in *Drosophila* where gynandromorphs may result; but notably in some midges (e.g. *Miastor, Heteropeza*) where paedogenetic larvae produce embryos whose somatic cells contain far fewer chromosomes than GERM LINE cells, owing to selective elimination during cleavage (see WEISMANN). In some scale insects, males and females develop from fertilized eggs, but males are haploid because the entire paternal chromosome set is discarded at cleavage (see HET-EROCHROMATIN, PARASEXUALITY, GYNOGENESIS); (4) *meiotic drive*, where a mutation causes the chromosome on which it occurs to be represented disproportionately often in gametes produced by meiosis, as with the segregation distorter (SD) locus of *Drosophila*; mutants homozygous for the SD allele are effectively sterile; (5) *premeiotic chromosome doubling* (see AUTOMIXIS); (6) ENDO-MITOSIS, where chromosomes replicate and separate but the nucleus and cell do not divide; (7) POLYTENY, where DNA replication occurs but the strands remain together to form thick, giant chromosomes.

ABIOTIC. Environmental features, such as climatic and EDAPHIC factors, that do not derive directly from the presence of other organisms. See BIOTIC.

ABOMASSUM. The 'true' stomach of RUMINANTS.

ABSCISIC ACID (ABSCISIN, DORMIN). Inhibitory plant GROWTH SUB-
STANCE (a sesquiterpene). Present in a variety of plant organs –
leaves, buds, fruits, seeds and tubers. Promotes senescence and abscission of leaves; induces dormancy in buds and seeds. Antagonizes
influences of growth-promoting substances. Believed to act by inhibiting nucleic acid and protein synthesis.

ABSCISSION LAYER. Layer at base of leaf stalk in woody dicotyledons
and gymnophytes, in which the parenchyma cells become separated
from one another through dissolution of the middle lamella before
leaf-fall.

ABSORPTION SPECTRUM. Graph of light absorption versus wavelength
of incident light. Shows how much light (measured as quanta) is
absorbed by a pigment (e.g. plant pigments) at each wavelength.
Compare ACTION SPECTRUM.

ABYSSAL. Inhabiting deep water, roughly below 1000 metres.

ACANTHODII. Class of primitive, usually minnow-sized, fossil fish
abundant in early Devonian freshwater deposits. Earliest known
gnathostomes. Bony skeletal tissue. Fins supported by very stout
spine; several accessory pairs of fins common. Row of spines between
pectoral and pelvic fins. Heterocercal tail. Relationships with osteichthyan fishes uncertain, but probably not directly ancestral. See
PLACODERMI.

ACANTHOPTERYGII. Spiny-rayed fish. Largest superorder of (teleost)
fishes. Spiny rays in their fins consist of solid pieces of bone (and not
numerous small bony pieces); are unbranched and pointed at their
tips. Radial bones of each ray are sutured or fused, preventing relative
lateral movement. Often have short, deep bodies, and relatively large
fins, making these fish very manoeuvrable. See TELEOSTEI.

ACARI (ACARINA). Order of ARACHNIDA including mites and ticks.
External segmentation much reduced or absent. Larvae usually with
three pairs of legs, nymphs and adults with four pairs. Of considerable
economic and social importance as many are ectoparasites and vectors of pathogens.

ACCESSORY BUD. A bud generally situated above or on either side of
main axillary bud.

ACCESSORY CHROMOSOME. See SUPERNUMERARY CHROMOSOME.

ACCESSORY NERVE. Eleventh cranial nerve of tetrapod vertebrates,
unusual in originating from both brain stem and spinal cord. A
mixed nerve, whose major motor output is to muscles of throat, neck
and viscera.

ACCESSORY PIGMENT. Pigment that captures light energy and transfers it to chlorophyll *a*, e.g. chlorophyll *b*, carotenoids, phycobiliproteins.

ACCOMMODATION. Changing the focus of the eye. In man and a few other mammals occurs by changing the curvature of the lens; at rest lens is focused for distant objects and is focused for near objects by becoming more convex with contraction of the ciliary muscles in the CILIARY BODY. See EYE, OCULOMOTOR NERVE.

ACELLULAR. Term sometimes applied to organisms or their parts in which no nucleus has sole charge of a specialized part of the cytoplasm, as in unicellular organisms. Applicable to coenocytic organisms (e.g. many fungi), and to tissues forming a SYNCYTIUM. Sometimes preferred to 'unicellular'. See MULTICELLULARITY.

ACENTRIC. (Of chromosomes) chromatids or their fragments lacking any CENTROMERES.

ACETABULUM. Cup-like hollow on each side of hip girdle into which head of femur (thigh bone) fits, forming hip joint in tetrapod vertebrates. See PELVIC GIRDLE.

ACETIFICATION. See FERMENTATION.

ACETYLCHOLINE (ACh). NEUROTRANSMITTER of many interneural, neuromuscular and other *cholinergic* effector synapses. Relays the electrical signal in chemical form, with transduction back to an electrical signal at the postsynaptic membrane. Initiates depolarization of those postsynaptic membranes to which it binds, but hyperpolarizes vertebrate cardiac muscle membranes (slowing heart rate). Stored in SYNAPTIC VESICLES inside axon terminals and released there in quantal fashion in response to calcium ion uptake on arrival of an ACTION POTENTIAL. It diffuses across the synaptic cleft and binds to receptor sites on the postsynaptic membrane, whereupon these ion channels open and allow appropriately sized positive ions to enter cell, initiating membrane depolarization. Hydrolysis to choline and acetate by cholinesterase attached to the postsynaptic membrane ensures that the chemical signal is appropriately brief. Vertebrate ACh postsynaptic receptors are distinguished as nicotinic or muscarinic on the results of alkaloid administration. Nicotinic receptors (ganglia, neuromuscular junctions and possibly some brain and spinal cord regions) are blocked by curare, muscarinic (peripheral autonomic interneural synapses) by atropine. ACh is found in some protozoans. Compare ADRENERGIC.

ACETYLCHOLINESTERASE. See CHOLINESTERASE.

ACETYL COENZYME A. See COENZYME A.

ACHENE. Simple, dry, one-seeded fruit formed from a single carpel, without any special method of opening to liberate seed; seed coat is not adherent to the pericarp; may be smooth-walled (e.g. buttercup), feathery (e.g. traveller's joy), spiny (e.g. corn buttercup), or winged (when termed a *samara*) as in sycamore and maple.

ACHIASMATE. Of meioses lacking chiasmata. One form of ABERRANT CHROMOSOME BEHAVIOUR. See SUPPRESSOR MUTATION.

ACHLAMYDEOUS. (Of flowers) lacking petals and sepals; e.g. willow.

ACID DYES. Dyes consisting of an acidic organic compound (anion) which is the actively staining part, combined with an inorganic cation, e.g. eosin. Stain particularly cytoplasm and collagen. See BASIC DYES.

ACID HYDROLASE. Any hydrolytic enzyme whose optimum pH of activity is in the acidic range. Many different examples occur in LYSOSOMES. Pepsin is an acid protease.

ACID PHOSPHATASE. One of several acid hydrolases located in LYSOSOMES and concentrated in the trans-most cisternae of the GOLGI APPARATUS.

ACID RAIN. Rainfall (precipitation) with a pH less than 5.6. Rain dissolves carbon dioxide, forming carbonic acid, giving it a normal pH of 5.6, but lower pHs result as it dissolves atmospheric pollutants such as oxides of nitrogen and sulphur dioxide. Some acid rain results from effects of atmospheric ozone production, some natural and some attributable to human activity. Its most serious consequence is the release of cations from the soil resulting in leaching. In the case of Mg^{++} ions this leads to chlorosis of leaves and poor plant growth, even death.

ACINAR CELLS. See ACINI.

ACINAR GLAND. (Zool.) Multicellular gland (e.g. seminal vesicle) with flask-like secretory portions.

ACINI. (Zool.) Cells lining tubules of pancreas and secreting digestive juices. Their secretory vesicles (*zymogen granules*) concentrate the enzymes and fuse with the apical portion of the plasmalemma under the stimulus of ACETYLCHOLINE or CHOLECYSTOKININ, releasing their contents into the lumen of the duct. Much used in the study of secretion.

ACOELOMATE. Having no COELOM. Refers to some lower animal phyla, e.g. coelenterates, platyhelminths, nemerteans and nematodes.

ACOUSTIC NERVE. See VESTIBULOCOCHLEAR NERVE.

ACOUSTICO-LATERALIS SYSTEM. See LATERAL LINE SYSTEM.

ACQUIRED CHARACTERISTICS, INHERITANCE OF. See LAMARCKISM.

ACQUIRED IMMUNE RESPONSE. Secondary antibody response to presence of antigen and differing from the initial response (which may precede it by a matter of years) in that it appears more quickly, achieves a higher antibody titre (concentration) in the blood and in that the principal IMMUNOGLOBULIN species present is IgG rather than IgM. See ANTIBODY, IMMUNITY.

ACRANIA. See CEPHALOCHORDATA.

ACRASIOMYCOTINA (ACRASIALES). Cellular slime moulds. Those MYXOMYCOTA which may exist as separate amoebae (myxamoebae), and retain their original identities within the pseudoplasmodium (slug) formed by swarming.

ACROCENTRIC. Of chromosomes and chromatids in which the CENTROMERE is close to one end.

ACROMION. Point of attachment of clavicle to scapula in mammals and mammal-like reptiles. A bone process.

ACROPETAL. (Bot.) Development of organs in succession towards apex, the oldest at base, youngest at tip (e.g. leaves on a shoot). Also used in reference to direction of transport of substances within a plant, i.e. towards the apex. Compare BASIPETAL.

ACROSOME. Specialized penetrating vesicular organelle, formed from GOLGI APPARATUS and part of the nuclear envelope at the tip of a spermatozoon. It contains HYALURONIDASE, several lytic enzymes and acid hydrolases released when the sperm cell membrane fuses at several points with the acrosome during the *acrosome reaction*, dissolving the jelly around the egg so that the sperm can penetrate it. Some sperm discharge an *acrosomal process* composed of rapidly polymerizing ACTIN which punctures the egg membranes prior to fusion with ovum (e.g. in some echinoderms).

ACTH (ADRENOCORTICOTROPIC HORMONE, CORTICOTROPIN). A polypeptide of 39 amino acids secreted by anterior lobe of the pituitary, involved in the growth and secretory activity of adrenal cortex. Has a minor positive effect on aldosterone secretion, but an important role in glucocorticoid secretion. Both stress and low blood glucocorticoid levels cause release from the hypothalamus of *corticotropin releasing factor* (*CRF*) which initiates ACTH release. See ADRENAL GLAND, CORTISOL.

ACTIN. Diagnostic eukaryotic protein, absent from prokaryotes. Filamentous actin (*F-actin*) is composed of globular protein monomers (*G-actin* molecules) polymerized to form long fibrous molecules, two of which coil round one other in the thin actin filaments of muscle and other eukaryotic cells, where they are termed *microfilaments*. Each *G-actin* molecule binds one calcium ion and one ATP or ADP molecule, when it polymerizes to form *F-actin* with ATP hydrolysis.

Like MICROTUBULES, the opposite ends of actin filaments grow and depolymerize at different rates and play a vital role in CYTO-SKELETON structure. Stress fibres are bundles of actin filaments and other proteins at the lower surfaces of cells in culture dishes and will contract if exposed to ATP in vitro. Microfilaments are involved in the building of FILOPODIA, microspikes and MICROVILLI where, as in stress fibres, they form paracrystalline bundles. Filaments of actin and MYOSIN are capable of contracting together as ACTOMY-OSIN in both muscle and non-muscle cells, e.g. in the contractile ring of dividing cells, in belt DESMOSOMES and in CYTOPLASMIC STREAMING.

ACTINOMORPHIC. (Of flowers) regular; capable of bisection vertically in two (or more) planes into similar halves, e.g. buttercup. Such flowers are also said to exhibit RADIAL SYMMETRY.

ACTINOMYCETE. Member of an order (Actinomycotales) of Gram-positive bacteria with cells arranged in hypha-like filaments. Mostly saprotrophs, some parasites. Source of streptomycin.

ACTINOMYCIN D. Antibiotic derived from species of the bacterial genus *Streptomyces.* Binds to DNA between two G–C base pairs and prevents movement of RNA polymerase, so preventing transcription in both prokaryotes and eukaryotes. Penetrates into intact cells. See ANTI-BIOTIC.

ACTINOPTERYGII. Ray-finned fishes. Generally regarded as subclass of Osteichthyes, and includes all common fish except sharks, skates and rays. Earliest forms (chondrosteans) represented in the Devonian by the palaeoniscoids and today by e.g. *Polypterus*; later forms (holosteans) were predominantly Mesozoic fishes but represented today by e.g. *Lepisosteus* (gar pikes); teleosts are the dominant fish of the modern world and represent the subclass in almost every part of the globe accessible to fish. Internal nostrils absent; SCALES ganoid in primitive forms, but reduced or even absent in teleosts. The paired fins are webs of skin braced by horny rays (like ribs of a fan), each a row of slender scales, there being no fleshy fin lobes except in the most primitive forms. A swim bladder is present and the skeleton is bony. Internal groupings given here probably represent GRADES rather than CLADES. See TELEOSTEI, ACANTHOPTERYGII.

ACTINOZOA (ANTHOZOA). Sea anemones, corals, sea pens, etc. A class of Coelenterata (Subphylum Cnidaria). The body is a polyp, there being no medusoid stage in the life cycle. Polyp more complexly organized than that of other coelenterates; coelenteron divided by vertical mesenteries. May have external calcareous skeleton as in well-known corals, but some forms have internal skeleton of spicules in meso-gloea.

ACTION POTENTIAL. Localized reversal and then restoration of electrical potential between the inside and outside of a nerve or muscle cell (or fibre) which marks the position of an impulse as it travels along the cell. See IMPULSE, ACTIVATION.

ACTION SPECTRUM. Plot of the quanta of different wavelengths required for a photochemical response against the wavelength of light used. Its reciprocal indicates photochemical efficiency.

ACTIVATED SLUDGE. Material consisting largely of bacteria and protozoa, used in and produced by one method of sewage disposal. Sewage is mixed with some activated sludge and agitated with air; organisms of the sludge multiply and purify the sewage, and when it is allowed to settle they separate out as a greatly increased amount of activated sludge. Part of this is added to new sewage and part disposed of.

ACTIVATION. (Of eggs). When the membrane of the sperm ACROSOME fuses with the egg plasma membrane, an *activation reaction* passes over the egg surface involving an ACTION POTENTIAL of longer duration than in nerve or muscle. It signifies the onset of embryological development and may be achieved merely by pricking of some eggs (e.g. frog).

ACTIVATION ENERGY. Free energy of activation is the amount of energy needed to bring all the molecules in 1 mole of a substance at a given temperature to the transition state (when there is high probability that a chemical bond will be made or broken) at the top of an activation barrier. Its biological significance is that enzymes accelerate reactions by lowering their energies of activation, the principal factor permitting such complicated chemistry to occur at relatively low temperatures.

ACTIVE SITE. Part of an enzyme molecule in its natural hydrated state which, by its three-dimensional conformation and charge distribution, confers upon the enzyme its substrate specificity. It binds to a substrate molecule, forming a transient enzyme–substrate complex. Enzymes may have more than one active site and so catalyse more than one reaction. Competitive inhibitors of an enzyme reaction bind reversibly to the active site and reduce its availability for normal substrate. Active sites may only take on their appropriate conformation after the enzyme has combined at some other site with an appropriate modulator molecule. Some active sites require metal ions as prosthetic groups (e.g. human carboxypeptidase requires a zinc atom). See ENZYME.

ACTIVE TRANSPORT. The energy-dependent carriage of a substance across a cell membrane, accumulating it on the other side in opposition to chemical or electrochemical gradients (i.e. 'uphill'). The process involves 'pumps' composed of protein molecules in the membrane (often traversing it) which carry out the transport. Requires an appropriate energy supply, commonly ATP, or a gradient of protons across the membrane itself usually generated by redox, photochemical or ATP-hydrolysing reactions. Collapse of this gradient drives proton-linked symports or antiports (see TRANSPORT PROTEINS). Alternatively, a membrane potential arising from ion asymmetry across the membrane may drive specific ions through special transport systems. Probably all cells engage in active transport. See SODIUM PUMP, ELECTRON TRANSPORT SYSTEM, FACILITATED DIFFUSION.

ACTOMYOSIN. Complex formed when the pure proteins ACTIN and MYOSIN are mixed, resulting in increased viscosity of the solution. Actomyosin undergoes dissociation in the presence of ATP and magnesium ions (Mg^{++}), when ATP hydrolysis occurs. Completion of this hydrolysis results in reaggregation of the two proteins. Live muscle cells have an absolute requirement for calcium ions (Ca^{++}) before myosin and actin filaments will interact, and when Ca^{++} is removed the actin and myosin dissociate. Such interactions form the basis of many biological force-generating events, notably during MUSCLE CONTRACTION, CYTOPLASMIC STREAMING, CELL LOCOMOTION and blood clot contraction.

ADAPTATION. (1) Evolutionary. Some property of an organism is normally regarded as an adaptation (i.e. fits the organism in its environment) if (a) it occurs commonly in the population, and (b) the cause of its commonness was NATURAL SELECTION in its favour. Adaptations are not, therefore, fortuitous benefits, the implication being that they have a genetic basis, since selection operates only upon genetic differences between individuals. Alternatively, we often in practice identify an adaptation by its effects rather than its causes. Learned abilities which improve an individual's FITNESS or inclusive fitness, but without clear genetic causation, are cases in point. See TELEOLOGY. (2) Physiological. A change in an organism, resulting from exposure to certain environmental conditions, allowing it to respond more effectively to them. (3) Sensory. A change in excitability of a sense organ through continuous stimulation, increasingly intense stimuli being required to produce the same response.

ADAPTIVE ENZYME. Inducible enzyme. See ENZYME.

ADAPTIVE IMMUNE RESPONSE. Response, ultimately by B-CELLS, to the presence of foreign antigen, in which large quantities of antigen-specific antibody appear in the blood while MEMORY CELLS with antigen-specific binding sites persist with capability of rapid clonal expansion on subsequent triggering by the antigen.

ADAPTIVE RADIATION. Evolutionary diversification from a single ancestral (prototype) population of descendant populations into more and more numerous ADAPTIVE ZONES and ecological NICHES. May involve both ANAGENESIS and CLADOGENESIS.

ADAPTIVE ZONE. A more or less distinctive set of ecological niches established and occupied by an evolutionary lineage with time.

ADAXIAL. (Of a leaf surface) facing the stem. Compare ABAXIAL.

ADENINE. A purine base of DNA, RNA, some nucleotides and their derivatives.

ADENOHYPOPHYSIS. See PITUITARY GLAND.

ADENOSINE DIPHOSPHATE. See ADP.

ADENOSINE MONOPHOSPHATE. See AMP.

ADENOSINE TRIPHOSPHATE. See ATP.

ADENOVIRUS. One kind of DNA tumour virus of animals. See VIRUS.

ADENYL CYCLASE (ADENYLATE CYCLASE). A plasma membrane-bound enzyme converting ATP to cyclic AMP (see AMP). Many peptide hormones and local chemical signals operate through activation of this enzyme.

ADH. See ANTIDIURETIC HORMONE.

ADHESION. Cells of a multicellular animal must be able to recognize and adhere to each other in order to group together as tissues. It is not yet clear how this happens, but INTERCELLULAR JUNCTIONS are implicated. Involved in MORPHOGENESIS and MULTICEL-LULARITY.

ADIPOSE TISSUE. A connective tissue. (1) Brown adipose tissue (brown fat) comprises cells whose granular cytoplasm is due to high concentration of cytochromes and whose function appears to be release of heat in the neonatal mammal. Distributed around neck and between scapulae in these and hibernating mammals but not otherwise extensively in adults. Richly innervated and vascularized. (2) White adipose tissue is distributed widely in animal bodies, comprising large cells (fat cells) each with single large fat droplet inside a thin rim of cytoplasm. This depot fat is composed largely of triglyceride. AD-RENALINE, GLUCAGON, GROWTH HORMONE and ACTH all stimulate release of fatty acids and glycerol via activation of intrinsic lipases, probably via cyclic AMP (see AMP, SECOND MESSENGER). Its nerve supply is less than that of brown adipose tissue.

ADP (ADENOSINE DIPHOSPHATE). A nucleoside diphosphate found universally inside cells. Phosphorylated to ATP during energy-yielding catabolic reactions and produced in turn when ATP itself is hydrolysed.

ADRENAL GLAND (SUPRARENAL G., EPINEPHRIC G.). Endocrine gland of most tetrapod vertebrates lying paired on either side of the mid-line, one atop each kidney. Each is a composite of an outer cortex derived from coelomic mesoderm, making up the bulk of the gland, and an inner medulla derived from neural crest cells of the ectoderm. Rarely found as a composite gland in fish. Cortex comprises three zones, the outermost secreting aldosterone which promotes water retention by kidneys by increasing renal potassium excretion and sodium retention; the middle zone secretes cortisol and other glucocorticoids under ACTH control and enhancing GLUCONEOGENESIS; the innermost zone secretes sex (mainly male) hormones. The medulla comprises *chromaffin cells*, effectively postganglionic sympathetic nerve cells that have lost their axons and secrete adrenaline (epinephrine) and a little noradrenaline (norepinephrine) into the large surrounding blood-filled sinuses. These mimic effects of the sympathetic nervous system (see AUTONOMIC NERVOUS SYSTEM), release being under hypothalamic control via the splanchnic nerve. They promote liver and muscle glycogenolysis via cyclic AMP (see AMP), lipolysis in ADIPOSE TISSUE, vasodilation in skeletal and heart muscle and brain, and vasoconstriction in skin and gut. They relax bronchi and bronchioles and increase rate and power of heart beat, raising blood pressure. All adrenal hormones are known as 'stress' hormones, those of the cortex responding to internal physiological stress such as low blood temperature or volume, while medullary hormones are released in response to stress situations (often auditory or visual) outside the body. See L-DOPA.

ADRENALINE. (In USA, EPINEPHRINE.) Hormone derivative of amino acid tyrosine secreted by chromaffin cells of ADRENAL GLAND and to a lesser extent by sympathetic nerve endings.

ADRENERGIC. Of a motor nerve fibre secreting at its end noradrenaline (norepinephrine) or, less commonly, adrenaline. Characteristic of postganglionic sympathetic nerve endings. Compare CHOLINERGIC.

ADRENOCORTICOTROPIC HORMONE (ACTH). See ACTH, ADRENAL GLAND.

ADVENTITIOUS. Arising in an abnormal position; of roots, developing from part of the plant other than roots (e.g. from stem or leaf cutting); of buds, developing from part of the plant other than a leaf axil (e.g. from a root).

AECIOSPORE. Binucleate spore of rust fungi produced in a cup-shaped structure, the aecium (pl. aecia).

AERENCHYMA. Secondary spongy tissue of some aquatic plants, with intercellular air spaces formed by the activity of a CORK cambium or phellogen. May develop in a lesser way from the lenticels of land plants such as willow (*Salix*), and poplar (*Populus*) if partially submerged. Seems to function mainly in a flotation capacity rather than as a respiratory aid.

Aerobic. Requiring free (gaseous or dissolved) oxygen. In most cases the oxygen is utilized in aerobic respiration, but a few enzymes (oxygenases) insert oxygen atoms directly into organic substrates. See RESPIRATION.

Aerobic respiration. See AEROBIC, RESPIRATION.

Aestivation. (Bot.) Arrangement of parts in a flower-bud. (Zool.) DORMANCY during summer or dry season as e.g., in lungfish (dipnoans). See HIBERNATION.

Aethelium. A sessile, rounded or pillow-shaped fruitification formed by a massing of the whole plasmodium in the Myxomycota.

Afferent. Leading towards, as of arteries leading to vertebrate gills or of nerve fibres (sensory) conducting an input towards the central nervous system. Opposite of EFFERENT.

A-form helix. Less common right-handed double helical form of DNA (compare B-FORM and Z-FORM HELICES), and, under some conditions, the most stable form of double-stranded DNA.

After-ripening. Dormancy exhibited by certain seeds (e.g. hawthorn, apple) which, although embryo is apparently fully developed, will not germinate immediately seed is formed. Even when removed from seed coat and provided with favourable conditions, the embryo has to undergo certain chemical and physical changes before it can grow. Possibly associated with delay in production of required growth substances, or with gradual breakdown of growth inhibitors. See DORMANT.

Agamospecies. See SPECIES.

Agamospermy. Any plant APOMIXIS in which embryos and seeds are formed but without prior sexual fusion. Excludes vegetative reproduction (vegetative apomixis). Occurs widely in higher plants, both ferns and flowering plants. Unknown in gymnophytes. See PSEUDOGAMY.

Agamospory. Asexual formation of an embryo and the subsequent development of a seed.

Agar. Mucilage obtained from cell walls of certain red algae. Mixture of polysaccharides, some sulphated, forming gel with water and melting at a higher temperature than that at which it solidifies. Used as a solidifying base for culture media in microbiology.

Agarose. Polysaccharide used as gel in column chromatography and in electrophoresis. See SOUTHERN BLOT TECHNIQUE.

Ageing (senescence). Progressive deterioration in function of cells, tissues, organs, etc., related to the period of time since that function commenced. By dividing indefinitely, bacteria and many protozoans avoid ageing; higher plants often seem capable of unlimited vegetative propagation. Regeneration and renewal in many simple invertebrates seem to permit escape from senescence. GERM LINES of sexual metazoa are potentially immortal (see WEISMANN). Expressed as disintegration of somatic tissue, ageing may be due to gradual accumulation of somatic mutations or to late expression of genes not subject to strong selection. Some evidence suggests loss of DNA METHYLATION may be involved. In the population context, it may be due to inbreeding or to some other factor reducing genetic variation.

AGGLUTINATION. Sticking together or clumping; as of bacteria (an effect of antibodies), or through mismatch of AGGLUTINOGENS of red blood cells and plasma AGGLUTININS in blood transfusions. See LECTIN.

AGGLUTININS (ISOANTIBODIES). Plasma and cell-surface proteins that by interacting with AGGLUTINOGENS (antigens) on foreign cells can cause cell clumping (AGGLUTINATION). Commonly LECTINS.

AGGLUTINOGEN. Proteins acting as cell-surface antigens of red blood cells and interacting with AGGLUTININS to cause red cell clumping and possible blockage of blood vessels. Genetically determined, and the basis of BLOOD GROUPS.

AGGREGATE FRUIT. Fruit which develops from several separate carpels of a single flower (e.g. magnolia, raspberry, strawberry).

AGNATHA. Class of Subphylum Vertebrata (sometimes also a superclass, other vertebrates forming Superclass Gnathostomata). Modern forms (cyclostomes) include lampreys (Subclass Monorhina) and hagfishes (Subclass Diplorhina), but fossil forms included anaspids, osteostracans and heterostracans. Jawless vertebrates. Buccal chamber acts as muscular pump sucking water in, serving for filter-feeding in lamprey larvae as well as ventilating gills – an advance over ciliary mechanisms. Paired appendages almost unknown. Earliest forms (heterostracans) appear in the late Cambrian.

AGONISTIC BEHAVIOUR. Intraspecific behaviour normally interpreted as attacking, threatening, submissive or fleeing. Actual physical injury tends to be rare in most apparently aggressive encounters.

AGROBACTERIUM. Bacterial genus noted for crown gall tumour-inducing ability. Oncogenic strains are host to a tumour-inducing (Ti) PLASMID which can be transmitted between species. A segment (T) of the Ti plasmid is transmitted to the plant host cell and is the immediate agent of tumour induction. See ONCOGENE.

AHNFELTAN. A complex phycocolloid substance occurring in the cell walls of some red algae (Rhodophyta).

AIR BLADDER. See GAS BLADDER.

AIR SACS. (1) Expanded bronchi in abdomen and thorax of birds, initially in five pairs but one or more pairs fusing to form thin-walled passive sacs with limited vascularization. Ramify throughout the body and within bones. Connected to lung by small tubes whose relative diameters are probably crucially important in establishing a unidirectional passage of air from lung to sacs and back to lung. The avian ventilation system lacks a tidal rhythm characteristic of mammals. (2) Expansions of insect tracheae into thin-walled diverticulae whose compression and expansion assist VENTILATION.

AKINETE. Vegetative cell which becomes transformed into a thick-walled, resistant spore. Formed by certain Cyanobacteria and some algae (e.g. some Chlorophyta).

ALBINISM. Failure to develop pigment, particularly melanin, in skin, hair and iris. Resulting *albinos* light-skinned with white hair and 'pink' eyes due to reflection from choroid capillaries behind retina. In mammals, including humans, usually due to homozygous autosomal recessive gene resulting in failure to produce enzyme tyrosine 3-monooxygenase.

ALBUMEN. Egg-white of birds and some reptiles comprising mostly solution of ALBUMIN with other proteins and fibres of the glyco-protein *ovomucoid*. Contains the dense rope-like CHALAZA and with yolk supplies protein and vitamins to embryo, but is also major source of water and minerals.

ALBUMIN. Group of several small proteins produced by the liver, forming up to half of human plasma protein content, with major responsibility for transport of free fatty acids, for blood viscosity and OSMOTIC POTENTIAL. If present in low concentration oedema may result, as in kwashiorkor.

ALBUMINOUS CELLS. Ray and parenchyma cells in gymnophyte phloem, closely associated morphologically and physiologically with sieve cells.

ALCYONARIA. Order of coelenterates within the Class Actinozoa. Sea pens, soft corals, etc. Have eight pinnate tentacles and eight mes-enteries. Polyps colonial, with continuity of body wall and enteron. Skeleton, often of calcareous spicules, within mesogloea and occasionally externally.

ALDOSTERONE. Hormone of ADRENAL cortex. See OSMOREGULA-TION.

ALEURONE GRAINS. Membrane-bound granules of storage protein occurring in the outermost cell layer of the endosperm of wheat and other grains.

ALEURONE LAYER. Metabolically active cells of outer cereal endosperm (in contrast to metabolically inactive cells of most of the endosperm) containing *aleurone grains*, several hydrolytic enzymes and reserves of *phytin* (releasing inorganic phosphate and inositol on digestion by phytase). During germination, aleurone cells secrete α-amylase into the endosperm, initiating its digestion. Recent work suggests that the synthesis of enzymes by aleurone cells may not be as specifically induced by gibberellins from the embryo axis as was once thought, although these growth substances are certainly implicated in the control of endosperm digestion.

ALEUROPLAST. Colourless plastid (leucoplast) storing protein; found in many seeds, e.g. brazil nuts.

ALGAE. Informal term covering many simple photosynthetic plants, including prokaryotic forms (CYANOBACTERIA, PROCHLORO-PHYTA), although the majority are eukaryotic. The algal plant body (THALLUS) may be unicellular or multicellular, filamentous, or flattened and ribbon-like, with relatively complex internal organization in the higher forms, e.g. some of the brown algae (Phaeophyta). Algae are either aquatic (marine or freshwater) or of damp situations, such as damp walls, rock faces, tree trunks, moss hummocks, or soil.

Algal sexual reproduction differs from that of other chlorophyllous plants; when unicellular, the entire organism may function as a gamete; when multicellular, gametes may be formed in unicellular or multicellular gametangia, each cell of the latter being fertile and producing a gamete. These characteristics distinguish algae from higher plants.

The formal taxon 'algae' has been abandoned in recent classifications, component groups being considered sufficiently distinctive to merit divisional status, dependent upon similarities and differences between pigments, assimilatory products, flagella, cell wall chemistry and aspects of cell ultrastructure. Eukaryotic algae include the following divisions: Bacillariophyta, Chlorophyta, Charophyta, Euglenophyta, Chrysophyta, Xanthophyta, Prymnesiophyta, Pyrrophyta (Dinophyta), Eustigmatophyta, Cryptophyta, Rhodophyta and Phaeophyta.

ALGIN. A complex phycocolloid occurring in the cell walls and intercellular spaces of brown algae (Phaeophyta), and commercially marketed.

ALIMENTARY CANAL. The gut; a hollow sac with one opening (an enteron) or a tube (said to be 'entire' since it opens at both mouth and anus) in whose lumen food is digested, and across whose walls the digestion products are absorbed. The epithelium lining the lumen is endodermal in origin, but the bulk of the organ system in higher forms is mesodermal, and is muscularized and vascularized. There are usually many associated glands.

ALKALINE PHOSPHATASE. Broad specificity enzyme, hydrolysing many phosphoric acid esters with an optimum activity in the basic pH range. Breaks down pyrophosphate in vertebrate blood plasma, enabling bone mineralization.

ALKALOIDS. Group of clinically important basic nitrogenous organic compounds produced by a few families of dicotyledonous plants, e.g. Solonaceae, Papaveraceae; possibly end-products of nitrogen metabolism, e.g. atropine, caffeine, cocaine, morphine, nicotine, quinine, strychnine.

ALKYLATING AGENT. A substance introducing alkyl groups (e.g. $-CH_3$, $-C_2H_5$, etc.) into either hydrocarbon chains or aromatic rings. Alkylation of DNA residues important in regulating transcription. See DNA REPAIR MECHANISMS, DNA METHYLATION.

ALLANTOIS. Stalk of endoderm and mesoderm which grows out ventrally from the posterior end of embryonic gut in AMNIOTES, expanding in reptiles and birds into a large sac underlying and for much of its surface attached to the CHORION. May represent precocious development of ancestral amphibian bladder. One of the three EXTRAEMBRYONIC MEMBRANES. A richly vascularized organ of gaseous exchange within cleidoic eggs, also functioning as a bladder to store embryo's nitrogenous waste. In higher primates and rodents, persists into later life as the urinary bladder.

ALLELES (ALLELOMORPHS). Representatives of the same gene LOCUS, and as such said to be alleles of (allelomorphic to) one another, a relational property dependent upon the prior concept of gene locus. Identical and non-identical alleles occur, being represented singly in haploid cells. Classically, alleles were ascribed to the same gene on the basis of two criteria: (i) failure to recombine with one another at meiosis, as if occupying the same locus, and (ii) failure, when mutant, to exhibit COMPLEMENTATION when present together in a diploid. Alleles of the same gene differ by MUTATION at one or more nucleotide sites within the same length of DNA, and back-mutation from one to another may occur. There may be many alleles of a gene in a population, but normally only two in the same diploid cell. See MULTIPLE ALLELISM.

ALLELIC COMPLEMENTATION. Interaction between individually defective mutant alleles of the same gene to give a phenotype more functional than either could produce by itself. Due to interaction (hybridization) of protein products. A source of confusion in the delineation of CISTRONS. See COMPLEMENTATION.

ALLELOPATHY. Inhibition of one species of plant by chemicals produced by another plant (e.g. by *Salvia leucophylla* – purple sage).

ALLEN'S RULE. States that the extremities (tail, ears, feet, bill) of

ENDOTHERMIC animals tend to be relatively smaller in cooler regions of a species range. See BERGMANN'S RULE.

ALLERGIC REACTION. Release of histamine and other mediators of ANAPHYLAXIS, producing symptoms of asthma, hay fever and hives. Membrane receptors of mast cells and basophilic leucocytes bind Ig antibodies which in turn bind antigen (allergen) and trigger histamine release. Often controllable by antihistamines. Disposition to allergic reaction is termed an allergy.

ALLOANTIBODY. Antibody introduced into an individual but produced in a different member of the same species.

ALLOANTIGEN (ISOANTIGEN). Antigen stimulating antibody response in genetically different members of the same species.

ALLOCHRONIC. Of species or species populations that are either sympatric at different times of the year or otherwise have non-overlapping breeding seasons (e.g. different flowering seasons in anthophytes). See ALLOPATRY, SYMPATRIC.

ALLOCHTHONOUS. Originating somewhere other than where found.

ALLOGAMY. (Bot.) Cross-fertilization.

ALLOGENEIC (ALLOGENIC). With different genetic constitutions. Often refers to intraspecific genetic variations. See INFRASPECIFIC VARIATION.

ALLOGRAFT (HOMOGRAFT). Graft between individuals of the same species but of different genotypes (allogeneic). See AUTOGRAFT, ISOGRAFT, XENOGRAFT.

ALLOGROOMING. Grooming of one individual by another of the same species (a conspecific).

ALLOMETRY. Study of relationships between size and shape. Organisms do not grow isometrically; rather proportions change as size changes. Thus juvenile mammals have relatively large heads, while limb proportions of arthropods alter in successive moults. Summarized by the exponential equation $y = bx^a$, where y = size of structure at some stage, b = a constant for the structure, x = body size at the stage considered and a = allometric constant (unity for isometric growth). The analysis is open to multivariate generalization. See HETEROCHRONY.

ALLOPATRIC. Geographical distribution of different species, or subspecies or populations within a species, in which they do not occur together but have mutually exclusive distributions. Populations occupying different vertical zones in the same geographical area may still be fully allopatric. See ALLOCHRONY, SYMPATRIC.

ALLOPOLYPLOID. Typically, a TETRAPLOID organism derived by chromosome doubling from a hybrid between diploid species whose chromosomes have diverged so much that little or no synapsis occurs between them at meiosis, so that only bivalents are formed (e.g. New World cottons, *Gossypium* spp.). This clearly distinguishes the term from AUTOPOLYPLOID, but some polyploids do not fall readily into either category. Allopolyploids may back-cross with one or other diploid parent stock; hence allotetraploids, which are generally themselves fully fertile (since they form bivalents at meiosis), behave in effect as new reproductively isolated species. However, if the original diploid progenitors were closely related species, or even ecotypes of the same species, then MULTIVALENTS may arise in meioses, which then resemble meioses in typical autopolyploids. Nevertheless, as a result of their greater fertility classical allopolyploids have been more significant in evolution than have classical autopolyploids. Many new plant species have arisen this way. Cultivated wheat (*Triticum aestivum*) is an allohexaploid, combining doubling in a triploid hybrid between an allotetraploid and a diploid.

ALL-OR-NONE RESPONSE. Ability of certain excitable tissues, under standardized conditions, to respond to stimuli of whatever intensity in just two ways: (a) no response (stimulus sub-threshold), or (b) a full-size response (stimulus at or above threshold). ACTION POTENTIALS of nerve and muscle membranes are characterized by all-or-none behaviour. Where thresholds of different units in a response differ, as in the many motor fibres of the sciatic nerve, or the various MOTOR UNITS of an entire muscle, an increase in stimulus intensity may bring progressively more units to respond. In muscle, this constitutes spatial SUMMATION. Nerve signals cannot use such amplitude variations.

ALLOSTERIC. Of those molecules (typically proteins) whose three-dimensional configurations alter in response to their environmental situation, normally registered by a change in molecule function. Often the key to regulation of critical biochemical pathways, serving as a feedback monitoring device in cybernetic circuits both inside and outside cells (see REGULATORY ENZYMES). At least as significant is allosteric control of GENE EXPRESSION by regulatory proteins. Among non-enzyme proteins, the haemoglobin molecule is allosteric under different blood pH values, with marked effects upon its oxygen saturation curve (see BOHR EFFECT). For allosteric inhibition and induced fit of enzymes, see ENZYME.

ALLOTETRAPLOID. An ALLOPOLYPLOID derived by doubling the set of chromosomes resulting from fusion between haploid gametes from more or less distantly related parental species. In classical cases, there is no meiotic SYNAPSIS between the chromosomes of different origin, and more or less complete fertility is achieved. Far more common in plants than animals, probably through comparative rarity of vegetative habit and/or parthenogenesis in the latter, in which it is difficult to rule out autopolyploidy as the source. See POLYPLOIDY.

ALLOTOPIC. Of closely related sympatric populations, whose distributions are such that both occupy the same geographical range, but each occurs in a different habitat within that range.

ALLOTYPE. Genetic variant within a LOCUS of a given species population, such as allelic forms within a BLOOD GROUP SYSTEM or variants of heavy chain constant regions of ANTIBODY molecules. See IDIOTYPE, ISOTYPE.

ALLOZYMES. Forms of an enzyme that are encoded by different allelic genes.

ALPHA-ACTININ (α-ACTININ). An accessory protein of muscle, anchoring actin filaments at the Z-disc and cross-linking adjacent sarcomeres; also cross-links actin in many other cells to contribute to the CYTOSKELETON.

ALPHA BLOCKER. Drug-blocking ADRENERGIC alpha receptors, preventing activity of the sympathetic neurotransmitter NORADRENALINE.

ALPHA HELIX. (Of proteins) a common secondary structure, in which the chain of amino acids is coiled around its long axis. Not all proteins adopt this conformation, it depending upon the molecule's primary structure. When adopted there are about 3.6 amino acids per turn (corresponding to 0.54 nm along the axis), amino acid R-groups pointing outwards. Hydrogen bonds between successive turns stabilize the helix. The α-helix may alternate with other secondary structures of the molecule such as β-sheets or 'random' sections. See PROTEIN.

ALPHA RECEPTOR. ADRENERGIC membrane receptor site binding NORADRENALINE in preference to ADRENALINE. May be excitatory or inhibitory, depending on the tissue. As with beta receptors, effects are mediated through an adenylate cyclase molecule adjacent in the membrane. The commonest receptors on postsynaptic membranes of postganglionic cells of sympathetic system. See CHOLINERGIC, AUTONOMIC NERVOUS SYSTEM, ALPHA BLOCKER.

ALPHA-RICHNESS. Number of species present in a small, local, homogeneous area. See DIVERSITY.

ALTERNATION OF GENERATIONS. Either (1) *metagenesis*, a life cycle alternating between a generation reproducing sexually and another reproducing asexually, the two often differing morphologically; or (2) the alternation within a life cycle of two distinct cytological generations, one being haploid and the other diploid. See LIFE CYCLE.

Metagenesis occurs in a few animals, e.g. CNIDARIA and parasitic flatworms, where both generations are normally diploid. The alternation of distinct cytological generations is clearest in plants such as ferns and some algae, where the two generations (gametophyte and sporophyte) are independent and either identical in appearance (alternation of *isomorphic* generations) or quite dissimilar (alternation of *heteromorphic* generations). In mosses and liverworts the dominant (vegetative) plant is the gametophyte while the sporophyte (the capsule) is more or less nutritionally dependent on the gametophyte. In flowering plants, the male (micro-) and female (macro-) gametophytes are reduced to microscopic proportions, the male gametophyte being shed as the pollen grain and the female gametophyte (embryo sac) being retained on the sporophyte in the ovule. A clearcut alternation of physically distinct plants is avoided here, although alternating cytological phases are still discernible. In vascular plants generally, the sporophyte generation is the vegetative plant itself, be it a fern, herb, shrub or tree.

ALTRICIAL. Animals born naked, blind and immobile (e.g. rat and mouse pups, many young birds). See NIDICOLOUS.

ALTRUISM. Behaviour benefiting another individual at the expense of the agent. Widespread and apparently at odds with Darwinian theory, which predicts that any genetic component of such behaviour should be selected against. Theories of altruism in biology tend to be concerned with cost-benefit analysis, as dictated by the logic of natural selection. One component of Darwinian FITNESS may be the care a parent bestows upon its offspring, although this is not usually considered altruism. HAMILTON'S RULE indicates the scope for evolutionary spread of genetic determinants of altruistic character traits, compatibly with Darwinian theory, and explains the evolution of parental care, while showing that *reciprocal altruism* can evolve even in the absence of relatedness between participants (e.g. members of different species). MULTICELLULARITY may afford opportunities for sacrifice of somatic cells (e.g. leucocytes) for a genetically related germ line harbouring the potentially immortal UNITS OF SELECTION. See ARMS RACE.

ALVEOLUS. (1) Minute air-filled sac, grouped together as *alveolar sacs* to form the termini of bronchioles in vertebrate lungs. Their thin walls are composed of squamous epithelial and surfactant-producing cells. A rich capillary network attached to the alveoli supplies blood for gaseous exchange across the huge total alveolar surface. A surfactant (lecithin) layer reduces surface tension, keeping alveoli open from birth onwards, and provides an aqueous medium to dissolve gases. Macrophages in the alveolar walls remove dust and debris. (2) Expanded sac of secretory epithelium forming internal termini of ducts of many glands, e.g. mammary glands. (3) Bony sockets into which teeth fit in mandibles and maxillae of jawed vertebrates, lying in the *alveolar processes* of the jaws. (4) An elongated chamber on the cell wall of some diatoms (Bacillariophyta) from the central axis to the margin, and opening to the inside of the cell wall.

AMACRINE CELL. One of three classes of neurone in mid-layer of vertebrate retina. Conducts signals laterally without firing action potentials.

AMASTIGOMYCOTA. Division of fungi that lack a motile stage and are not usually adapted to aquatic habitats. Includes Subdivisions ZYGOMYCOTINA, ASCOMYCOTINA, BASIDIOMYCOTINA and DEUTEROMYCOTINA.

AMBER MUTATION. One of three mRNA CODONS not recognized by transfer RNAs commonly present in cells, and bringing about normal polypeptide chain termination. Its triplet base sequence is UAG. Any mutation producing this sequence within a reading frame results in termination of the TRANSLATION process and release of incomplete polypeptide. *Missense* or *stop mutation*. See OCHRE and OPAL MUTATIONS, GENETIC CODE.

AMENSALISM. Interaction in which one animal is harmed and the other unaffected. See SYMBIOSIS.

AMES TEST. Test assessing mutagenic potential of chemicals. Strains of the bacterium *Salmonella typhimurium* having qualities such as permeability to chemicals, inability to repair DNA damage, or ability to convert DNA damage into heritable mutations, are made AUXOTROPHIC for histidine. After mixing with potential mutagen prior to plating, increase in normal (PROTOTROPHIC) colonies indicates mutagenicity.

AMETABOLA. Primitively wingless insects (APTERYGOTA).

AMINO ACID. Amphoteric organic compounds of general structual formula

$$\mathrm{{}^{H}_{H}\!\!>\!\!N-\underset{R}{\overset{H}{C}}-C\!\!\underset{OH}{\overset{O}{\diagup}}}$$ (where R may be one of 20 atomic groupings)

occurring freely within organisms, and polymerized to form dipeptides, oligopeptides and polypeptides. Amino acids differ in their R-groups and the amino acid sequence in a protein molecule determines not only its charge sequence but also its configuration in solution. Relative molecular masses of the common forms vary from 75 (glycine) to 204 (tryptophan). Only three commonly contain sulphurous R-groups: methionine, cysteine and cystine (formed from two oxidized cysteines, providing 'sulphur bridges'). During PROTEIN SYNTHESIS the carboxy- and amino-terminal ends of adjacent amino acids condense to form peptide bonds, leaving only the N-terminal and C-terminal ends of the protein and some R-groups ionizable. About 20 amino acid radicals occur commonly in proteins, encoded by the GENETIC CODE. Their modification after attachment to a transfer RNA molecule may result in rare non-encoded amino acids occurring in proteins. Some amino acids (e.g. ornithine) never occur in proteins. Most naturally-occurring amino acids except proline have a free carboxyl and a free amino group on the α-carbon atom (alpha amino acids). *Essential amino acids* are required by an organism from its environment, due to inability to synthesize them from precursors (see VITAMINS, which they are not); there are about 10 such for humans.

AMITOSIS. See ABERRANT CHROMOSOME BEHAVIOUR (2).

AMMOCOETE. Filter-feeding larva of lamprey, capable of attaining lengths of over 10 cm if conditions for metamorphosis do not prevail.

AMMONIFICATION. Decomposition of amino acids and other nitrogenous organic compounds; results in production of ammonia (NH_3) and ammonium ions (NH_4^+). Bacteria involved are *ammonifying bacteria*. See NITROGEN CYCLE.

AMMONITES. Group of extinct cephalopod molluscs (Subclass Ammonoidea, Order Ammonitida) dominating the Mesozoic cephalopod fauna. Had coiled shells, with protoconch (calcareous chamber) at origin of the shell spiral. Of great stratigraphic value.

AMMONOTELIC. (Of animals) whose principal nitrogenous excretory material is ammonia. Characterizes aquatic, especially freshwater, forms. See UREOTELIC, URICOTELIC.

AMNION. Fluid-filled sac in which AMNIOTE embryo develops. An EXTRAEMBRYONIC MEMBRANE (Fig. 16) formed in reptiles, birds and some mammals by extraembryonic ectoderm and mesoderm growing up and over embryo, the (amniotic) folds overarching and fusing to form the amnion surrounding the embryo, and the CHORION surrounding the amnion, ALLANTOIS and YOLK SAC. The amnion usually expands to meet the chorion. In humans and many other mammals the amnion originates by rolling up of some of the cells of the INNER CELL MASS during GASTRULATION. Amniotic fluid (amounting to about one dm^3 at birth in humans) is circulated in placental mammals by foetal swallowing, enabling wastes to pass to the placenta for removal. Provides a buffering cushion against mechanical damage, helps stabilize temperature and dilate the cervix during birth. In *amniocentesis*, amniotic fluid containing cells from the foetus is withdrawn surgically for signs of abnormal development.

AMNIOTE. Reptile, bird or mammal. Distinguished from anamniotes by presence of EXTRAEMBRYONIC MEMBRANES in development.

AMNIOTIC EGG. Egg type characteristic of reptiles, birds and PROTOTHERIA (much modified in placental mammals). Shell leathery or calcified; ALBUMEN and yolk typically present. EXTRAEMBRYONIC

MEMBRANES occur within it during development. See CLEIDOIC EGG.

AMOEBA. Genus of sarcodine protozoans. Single-celled animals of irregular and protean shape, moving and feeding by use of PSEUDO-PODIA. Some slime mould cells are also loosely termed 'amoebae', while any CELL LOCOMOTION resembling an amoeba's is termed 'amoeboid'.

AMOEBOCYTE. Cell (haemocyte) capable of active amoeboid locomotion found in blood and other body fluids of invertebrates; in sponges, an amoeboid cell type implicated in mobilization of food from the feeding CHOANOCYTES and its conveyancing to non-feeding cells in absence of true vascular system.

AMOEBOID. Describing cells resembling those of the genus AMOEBA.

AMP. Adenosine monophosphate. Nucleotide component of DNA and RNA (in deoxyribosyl and ribosyl forms respectively), and hydrolytic product of ADP and ATP. Converted to cyclic AMP (cAMP) by ADENYLATE CYCLASE, intracellular concentrations of cAMP rising rapidly in response to extracellular (esp. hormonal) signals and falling rapidly due to activity of intracellular phosphodiesterase. Its level dictates rates of many biochemical pathways, depending upon cell type. See CASCADE, SECOND MESSENGER, G-PROTEIN, GTP.

AMPHIBIA. Class of tetrapod vertebrate, its first fossil representatives being Devonian ichthyostegids and its probable ancestors rhipidistian crossopterygian fishes. A POLYPHYLETIC origin has not been ruled out. Many early forms had scaly skins, almost entirely lost in the one modern Subclass (Lissamphibia) of three orders: Apoda, legless caecilians; Urodela, salamanders and newts; Anura, toads and frogs. Compared with their mainly aquatic ancestors, the more terrestrialized amphibians have: vertebrae with larger, more articulating neural arches and larger intercentra (see VERTEBRAL COLUMN); greater freedom of the PECTORAL GIRDLE from the skull, allowing some lateral head movement; PELVIC GIRDLE composed of three paired bones (pubis, ischium and ilium) with some fusion to form the rigid PUBIC SYMPHYSIS; eardrums (homology with part of the spiracular gill pouch of fish) and a single middle ear ossicle, the columella, homologous with the hyomandibular bone of fish. Fertilization is internal or external (but intromittant organs are lacking). Most return to water to lay anamniote eggs, although some are viviparous. The skin is glandular for gaseous exchange. Modern forms specialized and not representative of the Carboniferous amphibian radiation.

AMPHICRIBRAL. (Bot.) Type of vascular arrangement where phloem surrounds the xylem. Compare AMPHIPHLOIC.

AMPHIDIPLOID. See ALLOTETRAPLOID.

AMPHIMIXIS. Normal sexual reproduction, involving meiosis and fusion of haploid nuclei, usually borne by gametes. See AUTOMIXIS, APOMIXIS, PARTHENOGENESIS.

AMPHINEURA. Minor Class of MOLLUSCA, including the chitons. Marine, mostly on rock surfaces; head reduced and lacking eyes and tentacles; mantle all round head and foot; commonly eight calcareous shell plates over visceral hump; nervous system primitive, lacking definite ganglia.

AMPHIOXUS. Lancelets (Subphylum CEPHALOCHORDATA). Widely distributed marine filter-feeding burrowers up to 5 cm long. Two genera (*Branchiostoma*, *Asymmetron*). Giant larva resulting from prolonged pelagic life once given separate genus (*Amphioxides*) and develops premature gonads, providing support for the evolutionary origin of vertebrates by PROGENESIS.

AMPHIPHLOIC. (Bot.) Type of vascular arrangement where phloem is on both sides of the xylem. Compare AMPHICRIBRAL.

AMPHIPODA. Order of Crustacea (Subclass Malacostraca). Lack carapace; body laterally flattened. Marine and freshwater forms; about 3600 species. Very important detritus feeders and scavengers. Includes gammarids.

AMPHISTYLIC. Method of upper jaw suspension in a few sharks, in which there is support for the jaw both from the hyomandibular and the brain case. See AUTOSTYLIC, HYOSTYLIC.

AMPULLA. (Of inner ear) see VESTIBULAR APPARATUS.

AMYGDALA (AMYGDALOID BODIES or NUCLEI). Basal ganglia of the subcortical region of the most ancient part of the vertebrate CEREBRAL HEMISPHERES, gathering olfactory and visceral information. They appear to be involved in the generation of emotions. Removal in humans increases sexual activity.

AMYLASES (DIASTASES). Group of enzymes hydrolysing starches or glycogen variously to dextrins, maltose and/or glucose; α-amylase (in saliva and pancreatic juice) yields maltose and glucose; β-amylase (in malt) yields maltose alone. Present in germinating cereal seeds (see ALEURONE LAYER), where only α-anylase can digest intact starch grains, and produced by some microorganisms.

AMYLOPECTIN. Highly branched polysaccharide component of the plant storage carbohydrate STARCH. Consists of homopolymer of α[1,4]-linked glucose units, with α[1,6]-linked branches every 30 or so glucose radicals. Like GLYCOGEN it gives a red-violet colour with iodine/KI solutions. See AMYLASES.

AMYLOPLAST. Colourless plastid (leucoplast) storing STARCH; e.g. found in cotyledons, endosperm and storage organs such as potato tubers.

Amylose. Straight-chain polysaccharide component of STARCH. Comprises α[1,4]-linked glucose units. Forms hydrated micelles in water, giving the impression of solubility. Gives a blue colour with iodine/KI solutions. Hydrolysed by AMYLASES to maltose and/or glucose.

Anabolism. Enzymatic synthesis (build-up) of more complex molecules from more simple ones. Anabolic processes include multi-stage photosynthesis, nucleic acid, protein and polysaccharide syntheses. ATP or an equivalent needs to be available and utilized for the reaction(s) to proceed. See CATABOLISM, GROWTH HORMONE, METABOLISM.

Anadromous. Animals (e.g. lampreys, salmon) which must ascend rivers and streams from the sea in order to breed. See OSMO-REGULATION.

Anaerobic. (Of organisms) ability to live *anoxically* i.e. in the absence of free (gaseous or dissolved) oxygen. (Of processes) occurring in the absence of such oxygen. *Anaerobic respiration* is the enzyme-mediated process by which cells (or organisms) liberate energy by oxidation of substances but without involving molecular oxygen. This involves less complete oxidation of substrates, with less energy released per g of substrate used, enabling *anaerobes* to exploit environments unavailable to obligate aerobes. *Facultative anaerobes* can switch metabolism from aerobic to anaerobic under anoxic conditions, as required of many internal parasites of animals, some yeasts and other microorganisms. GLYCOLYSIS is anaerobic but may require aerobic removal of its products to proceed. Relatively anoxic environments include animal intestines, rumens, gaps between teeth, sewage treatment plants, polluted water, pond mud, some estuarine sediments and infected wounds. See OXYGEN DEBT, RESPIRATION.

Anagenesis. (1) Process by which characters change during evolution within species, by NATURAL SELECTION or GENETIC DRIFT. (2) Any non-branching speciation in which species originate along a single line of descent yet only one species represents the lineage after any speciation event (contrast CLADOGENESIS). Gradual anagenetic speciation is not possible within the *biological species concept*, for reproductive isolation is never completed between ancestral and descendant species. CLADISTICS excludes anagenetic speciation by definition, but the term is retained in the context of characters. See SPECIES.

Analogous. A structure present in one evolutionary lineage is said to be analogous to a structure, often performing a similar function, within the same or another evolutionary lineage if their phyletic and/or developmental origins were independent of one another; i.e. if there is HOMOPLASY. Tendrils of peas and vines and eyes of squids

and vertebrates are pairs of analogous structures. See CONVERG-
ENCE, HOMOLOGY, PARALLEL EVOLUTION.

ANAMNIOTE. (Of vertebrates) more primitive than the AMNIOTE grade.
Includes agnathans, all fish, and amphibians.

ANANDROUS. (Of flowers) lacking stamens.

ANAPHASE. Stage of mitosis and meiosis during which either bivalents
(meiosis I) or sister chromatids (mitosis, meiosis II) separate and
move to opposite poles of the cell. See SPINDLE.

ANAPHYLAXIS. A type of hypersensitivity to antigen (allergen) in
which IgE antibodies attach to mast cells and basophils. May result
in circulatory shock and asphyxia. See ALLERGIC REACTION.

ANATROPOUS. (Of ovule) inverted through 180°, micropyle pointing
towards placenta. Compare ORTHOTROPOUS, CAMPYLOTROPOUS.

ANDRODIOECIOUS. Having male and hermaphrodite flowers on separate plants. Compare ANDROMONOECIOUS.

ANDROECIUM. A collective term referring to the stamens of a flower.
Compare GYNOECIUM.

ANDROGEN. Term denoting any substance with male sex hormone
activity in vertebrates, but typically steroids produced by vertebrate
testis and to a much lesser extent by adrenal cortex. See TESTOSTER-
ONE.

ANDROMONOECIOUS. Having male and hermaphrodite flowers on the
same plant. Compare ANDRODIOECIOUS.

ANEMOPHILY. The pollination of flowers by the wind. Compare ENTO-
MOPHILY.

ANEUPLOID (HETEROPLOID). Of nuclei, cells or organisms having more or
less than an integral multiple of the typical haploid chromosome
number. See EUPLOID, MONOSOMIC, TRISOMIC, NULLISOMIC.

ANGIOSPERM. Literally, a seed borne in a vessel (carpel); thus one of a
group of plants (the flowering plants) whose seeds are borne within a
mature ovary (fruit). See ANTHOPHYTA, which replaces Angiospermae.

ANGIOTENSINS. *Angiotensin I* is a decapeptide produced by action of
kidney enzyme, renin, on the plasma protein angiotensinogen when
blood pressure drops. It is in turn converted by a plasma enzyme in
the lung to the octapeptide *angiotensin II*, an extremely powerful
vasoconstrictor which raises blood pressure and also results in sodium
retention and potassium excretion by kidney. See OSMOREGULA-
TION.

ÅNGSTRÖM UNIT (Å). Unit of length, 10^{-10} metres (0.1nm); 10^{-4}
microns. Not an SI unit.

ANIMALIA. Animals. Kingdom containing those eukaryotes combining a lack of cell wall material with heterotrophic nutrition (although endosymbiotic photosynthetic cells may occur in some tissues). Commonly divided into subkingdoms PROTOZOA, PARAZOA and METAZOA. Classifications recognizing Kingdom Protista or Kingdom Protoctista would exclude protozoans from the Kingdom Animalia. In these systems, Kingdom Animalia includes all heterotrophic eukaryotes lacking cell wall material and having a blastula stage in their development.

ANIMAL POLE. Point on surface of an animal egg nearest to nucleus, or extended to include adjacent region of cell. Often marks one end of a graded distribution of cytoplasmic substances. See POLARITY.

ANISOGAMY. Condition in which gametes which fuse differ in size and/ or motility. In OOGAMY, gametes differ in both properties. Significantly, the sperm often contributes the sole centriole for the resulting zygote. See FERTILIZATION, ISOGAMY, PARTHENOGENESIS.

ANNELIDA (ANNULATA). Soft-bodied, metamerically segmented coelomate worms with, typically, a closed blood system; excretion by NEPHRIDIA; a central nervous system of paired (joined) nerve cords ventral to the gut, and a brain comprising paired ganglia above the oesophagus, linked by commissures to a pair below it. Cuticle collagenous, not chitinous. Chitin present in CHAETAE, which may be quite long, bristle-like and associated laterally with fleshy parapodia (e.g. ragworms, Class Polychaeta) or shorter and not housed in parapodia (e.g. earthworms, Class Oligochaeta). Leeches (Class HIRUDINEA) have 34 segments, confused by surface annulations. CLITELLUM present in both oligochaetes and leeches. Septa between segments often locally or entirely lost. The coelom acts as a hydrostatic skeleton against which longitudinal and circular muscle syncytia (and diagonal muscles in leeches) contract. Cephalization most pronounced in polychaetes (largely marine); eyes and mandibles often well developed but oligochaetes lack specialized head structures. Gametes leave via COELOMODUCTS. Oligochaetes and leeches are typically hermaphrodite, polychaetes frequently dioecious.

ANNUAL. Plant completing its life cycle, from seed germination to seed production followed by death, within a single season. Compare BIENNIAL, EPHEMERAL, PERENNIAL. See DESERT, r-SELECTION.

ANNUAL RING. Annual increment of secondary wood (xylem) in stems and roots of woody plants of temperate climates. Because of sharp contrast in size between small wood elements formed in late summer and large elements formed in spring the limits of successive annual rings appear in a cross-section of stem as a series of concentric lines.

ANNULAR THICKENING. In protoxylem, internal thickening of a xylem vessel or tracheid wall, in rings at intervals along its length. Provides mechanical support, permitting longitudinal stretching as neighbouring cells grow.

ANNULUS. (1) Ring of tissue surrounding the stalk (stipe) of fruit bodies of certain Basidiomycotina (e.g. mushrooms); (2) line of specialized cells involved in opening moss capsules and fern sporangia to liberate spores.

ANOESTRUS. Period between breeding seasons in mammals, when OESTROUS CYCLES are absent. See OESTRUS.

ANOPLURA. See SIPHUNCULATA.

ANOXIA. Deficiency or absence of free (gaseous or dissolved) oxygen.

ANTAGONISM. Opposition of two or more processes or systems. (1) Of organisms, one interfering with or inhibiting growth or presence of another; (2) of drugs, hormones, etc., producing opposite physiological effects; (3) of muscles, producing opposite movements so that contraction of one must be accompanied by relaxation of the other. The normal way by which muscles regain their relaxed shape after contraction is by being extended by antagonistic muscle contraction. Contrast SYNERGISM.

ANTENNA. Paired, preoral, tactile and olfactory sense organs developing from second or third embryonic somites of all arthropod classes other than Onychophora and Arachnida. Usually much jointed and mobile. In some crustaceans locomotory or for attachment, a pair of ANTENNULES (often regarded as antennae) typically occurring on the segment anterior to that with antennae. ONYCHOPHORA have pair of cylindrical preantennae on first somite. See TENTACLES.

ANTENNA COMPLEX. Clusters of several hundred chlorophyll molecules fixed to the thylakoid membranes of chloroplasts by proteins in such a way as to harvest light energy falling on them, and relaying it to a special chlorophyll molecule in an associated PHOTOSYSTEM. See PHOTOSYNTHESIS and Fig. 14.

ANTENNAPEDIA COMPLEX. Complex of HOMOEOTIC and segmentation loci in *Drosophila* which, when homozygously mutant, may result in conversion of antennal parts into leg structures. Intensely studied in contexts of MORPHOGENESIS, and POSITIONAL INFORMATION. Some loci in the complex appear to be expressed only in specific embryonic COMPARTMENTS. See HOMOEOBOX.

ANTENNULE. Paired and most anterior head appendages of crustaceans; uniramous, whereas antennae like most appendages in the class are biramous.

ANTHER. Terminal portion of a STAMEN, containing pollen in *pollen sacs*.

ANTHERIDIOPHORE. In some liverworts, a stalk that bears the antheridia.

ANTHERIDIUM. 'Male' sex organ (gametangium) of fungi, and of plants other than seed plants (e.g. algae, bryophytes, lycophytes, sphenophytes and pterophytes).

ANTHEROZOID. Synonym of SPERMATOZOID.

ANTHESIS. Flowering.

ANTHOCEROTOPSIDA. Hornworts. Class of BRYOPHYTA. Small, widely distributed group, especially in tropical and warm temperate regions, growing in moist, shaded habitats. Plant a thin, lobed, dorsiventral THALLUS, anchored by rhizoids. Each cell usually has a single large chloroplast rather than the many small discoid ones found in cells of other bryophytes and vascular plants; and each chloroplast possesses a PYRENOID, all features suggesting algal affinities. Some (e.g. *Anthoceros*) contain Cyanobacteria (e.g. *Nostoc* spp.), supplying fixed nitrogen to their host plants.

ANTHOCYANINS. Group of water-soluble, flavonoid pigments (glycosides) occurring in solution in vacuoles in flowers, fruits, stems and leaves. Change colour, depending on acidity of solution. Responsible for most red, purple and blue colours of plants, especially in flowers; contribute to autumn (fall) colouring of leaves and tint of young shoots and buds in spring. Colours may be modified by other pigments, e.g. yellow flavonoids.

ANTHOPHYTA. Flowering plants (formerly Angiospermae). Division of plant kingdom. Seed plants whose ovules are enclosed in a carpel, and with seeds borne within fruits. Vegetatively diverse; characterized by FLOWERS; pollination basically by insects, but other modes (e.g. ANEMOPHILY) have evolved in a number of lines. Gametophytes much reduced; male gametophyte, initiated by pollen grain (microspore), comprising two non-motile gamete nuclei and a tube cell nucleus each associated with a little cytoplasm in the pollen tube; female gametophyte developing entirely within wall of megaspore which at maturity is a large cell containing eight nuclei, the EMBRYO SAC. Characteristic DOUBLE FERTILIZATION.

Two classes: *Monocotyledonae* (monocots, about 65 000 spp.), with flower parts usually in threes, leaf venation usually parallel, primary vascular bundles in the stem scattered, true secondary growth absent, and just a single cotyledon present; *Dicotyledonae* (dicots, about 170 000 spp.), with flower parts usually in fours or fives, leaf venation usually net-like, primary vascular bundles in the stem forming a ring, often with true secondary growth and vascular cambium, and two cotyledons present.

ANTHOZOA. See ACTINOZOA.

ANTHROPOID APES. Members of Family Pongidae (Order PRIMATES). Include orang-utan, chimpanzee and gorilla. Common ancestor of pongids and hominids ('men') probably Miocene in age. Gibbons (Family Hylobatidae) are in same suborder (Anthropoidea) as 'great apes' (pongids) and occasionally included in the term 'anthropoid ape'. Much ape anatomy stems from a brachiating mode of progression. Fundamentally quadrupedal; tendency to bipedal gait limited. Markedly prognathous, with diastemas. All are Old World forms.

ANTHROPOIDEA. Suborder of PRIMATES. Three living superfamilies: Ceboidea (New World monkeys); Cercopithecoidea (Old World monkeys); Hominoidea (gibbons, great apes and man). Eyes large and towards front of face; brain expansion associated with relative expansions of frontal, parietal and occipital bones of skull; great manual dexterity.

ANTIAUXINS. Chemicals which can prevent the action of AUXINS in plants, e.g. 2,6-dichlorophenoxyacetic acid; 2,3,5-triiodobenzoic acid.

ANTIBIOTIC RESISTANCE ELEMENT. Genetic element, composed of DNA and often borne on a TRANSPOSON, conferring bacterial resistance to an antibiotic. Often with INSERTION SEQUENCES at either end, when capable of moving between PLASMID, viral and bacterial DNA and selecting insertion sites, sometimes turning off expression of genes it inserts into or next to. Able to spread rapidly across species and other taxonomic boundaries, making design of new antibiotic drugs even more urgent. Many common pathogenic bacterial strains are now resistant to some of the best-known drugs. Non-homologous recombination between plasmids can give rise to multiple-resistance plasmids, bacterial plasmid R1 conferring resistance to chloramphenicol, kanamycin, streptomycin, sulphonamide and ampicillin. See PLASMID.

ANTIBIOTICS. Diverse group of generally low molecular mass organic compounds (in the category of non-essential 'secondary metabolites'). Characteristically produced by spore-forming soil microorganisms during or just prior to sporulation they tend to inhibit growth of potential competitors either reversibly (when bacteristatic), or irreversibly (bactericidal), generally by blocking one or more enzyme reactions in the affected cell. *Streptomycin* affects the syntheses of DNA, RNA and proteins, and alters the cell membrane and respiratory activity of sensitive cells; *penicillin* prevents cross-linking of glycan chains of the peptidoglycans of the bacterial cell walls, resulting in wall-less or wall-deficient cells; ACTINOMYCIN prevents TRANSCRIPTION; *puromycins* prevent TRANSLATION; *anthracyclines* block DNA replication and RNA transcription. In most cases these are achieved by the antibiotic forming complexes with, or otherwise inserting itself into, a nucleic acid. Antibiotics have been widely used as clinical drugs. Their use has generated new selection pressures for the target microorganisms

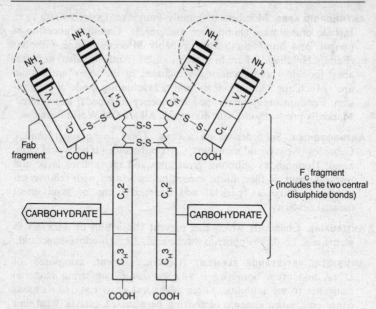

Fig. 1. *Diagram of IgG structure; rectangles are protein subunits and hypervariable regions are shown as dark lines. Antigen-binding sites are within dotted lines; other domains exist for complement fixation, for binding to F_c receptors, neutrophils and K cells.*

(see DEUTEROMYCOTINA), resulting in the spread of ANTIBIOTIC RESISTANCE ELEMENTS in what resembles an ARMS RACE.

ANTIBODY (IMMUNOGLOBULIN). Class of glycoprotein produced by vertebrate white blood cells (B-CELLS), after maturation into *plasma cells*. Their main function is to bind highly selectively to foreign molecules (antigens), which then clump together (agglutinate) so that phagocytic white cells can engulf them.

Five major classes differ principally in their type of heavy protein chain, and the degree to which the molecule is a polymer of immunoglobulin 'monomers'. Each immunoglobulin unit comprises two identical H- (heavy) and two identical L- (light) polypeptide chains forming mirror images of each other and joined by a flexible hinge region involving disulphide bridges. They bind to antigen at specific antigen-binding regions provided uniquely by the combination of H- and L-chain amino-terminal portions (see Fig. 1), which are extremely variable in their amino acid sequences between different antibodies, in contrast to constant regions at their carboxy-terminal portions. Only about 20–30 amino acids of the variable regions of H- and L-chains contribute to the antigen-binding site, these being located in three short hypervariable regions of each variable region. These lie themselves within relatively invariant 'framework regions' of the variable regions. The other biological properties of the molecule are determined by the constant domains of the heavy chains.

Digestion of antibody with papain produces two identical Fab (antigen-binding) fragments and one Fc (crystallizing) fragment. The latter region in the intact Ig (immunoglobulin) molecule is responsible for determining which component of the immune system the antibody will bind to. The Fc region of IgG may bind phagocytes and the first component of COMPLEMENT. Only the IgG antibody can cross the mammalian placenta. IgM is the major Ig type secreted in a primary immune response, but IgG dominates in secondary immune responses (see B-CELL).

Transformation of B-cells into differentiated antibody-producing plasma cells generally requires both antigen-presenting cells and a signal from a helper T-cell (see T-CELL). Because B-cells have only a a few days' life in culture they are not suitable for commercial antibody production: however, if an antibody-producing B-cell from an appropriately immunized mouse is fused to an appropriate mutant tumour B-cell, the hybrid cell formed may continue dividing and producing the particular antibody required. The resulting HY-BRIDOMA can be sub-cloned indefinitely, giving large amounts of antibody. Initial isolation of the appropriate B-cell follows discovery of the required antibody in the growing medium. The purity of the resulting *monoclonal antibody* and its production in response to what is possibly a minor component of an impure antigen mixture are both desirable features of the technique. See ANTIBODY DIVERSITY, ANTIGEN–ANTIBODY REACTION, IgA–IgM.

ANTIBODY DIVERSITY (A. VARIATION). Production of different ANTIBODY molecules by different B-CELLS. (see Fig. 1 for symbols). Light and heavy chains are encoded by different gene clusters. In humans, light chain genes lie on chromosomes 2 and 22, heavy chain genes on chromosome 14, and the light chains of a particular immunoglobulin molecule are encoded either by chromosome 2 or chromosome 22, not both. Any particular B-cell assembles in a line all the heavy chain genes needed to make its own unique antibody type, joining first the genes for variable (V), hypervariable (HV) and joining (J) regions of the molecule, then linking this combination to the genes for the constant (C) regions of the molecule, with different constant regions for different immunoglobulin classes (see IM-MUNOGLOBULIN references). The enzymes bringing together genes from different parts of a chromosome are performing a form of RECOMBINATION. Diversity arises from the randomness with which particular genes from heavy and light chain clusters are brought together. In addition, extra short nucleotide sections (N segments) get inserted, probably in some rule-following way, into the DNA encoding the antigen-binding regions of the molecule, and this together with variation in RNA PROCESSING of the hnRNA transcript increases still further the total antibody diversity, often classified as follows: (1) Allotypic: variation in the C_H1, C_H3 and C_L antibody regions caused by allelic differences between individuals at one or more loci for a subclass of immunoglobulin chains; (2) Idiotypic: variation in the V_L and V_H regions (especially in the hypervariable regions) that are generally characteristic of a particular antibody clone, and therefore not present in all members of a population; (3) Isotypic: variation in the C_L, and in the C_H1-3 antibody regions, determined by loci whose representative alleles are shared by all healthy members of a population.

ANTICLINAL. (Bot.) Alignment of the plane of cell division approximately at right angles to the outer surface of the plant part. Compare PERICLINAL.

ANTICOAGULANT. Any substance preventing blood clotting. Blood naturally contains such substances: fibrin and antithrombin III absorb much of the thrombin formed in the clotting process and HEPARIN inhibits conversion of prothrombin to thrombin. Blood-sucking animals (leeches, insects, bats, etc.) frequently produce anticoagulants in their saliva. Artificial anticoagulants (e.g. dicumarol) are either helpful to patients, or prevent blood samples from clotting in blood banks (e.g. EDTA). The rat poison warfarin is an anticoagulant. See BLOOD CLOTTING.

ANTICODON. The triplet sequence of tRNA nucleotides capable of base-pairing with a codon triplet of an mRNA molecule. See PROTEIN SYNTHESIS.

ANTIDIURETIC HORMONE (ADH, VASOPRESSIN). Ring-structure octapeptide hormone produced by hypothalamic neurosecretory cells and released into posterior pituitary circulation if blood water potential drops below the homeostatic norm. Has marked vasoconstrictor effects on arterioles, raising blood pressure, and increases water permeability of collecting ducts and distal convoluted tubules to the 10–20% of the initial glomerular filtrate still remaining (see KIDNEY), resulting in water retention. See OSMOREGULATION, OXYTOCIN.

ANTIGEN. Molecule (often protein or glycoprotein) which induces production of specific ANTIBODY, and to which the latter binds at a specific configurational domain of the antigen molecule called the antigenic determinant, or epitope. See ANTIGEN–ANTIBODY REACTION.

ANTIGEN–ANTIBODY REACTION. Non-covalent bonding between antigenic determinant of ANTIGEN and antigen-combining site on an immunoglobulin molecule (see ANTIBODY). Several such bonds form simultaneously. The reactions show high specificity but cross-reactivity may result if some determinants of one antigen are shared by another. Antibodies seem to recognize the three-dimensional configuration and charge distribution of an antigen rather than its chemical make-up as such. Such reactions form the basis of humoral and of many cell-mediated immune responses. See AGGLUTININ, COMPLEMENT, PRECIPITIN, IMMUNITY.

ANTIGENIC VARIATION. Ability of some pathogens, notably viruses, bacteria and protozoa, to change their coat antigens during infection. Trypanosomes and some stages in the malarial life-cycle achieve it, making the search for vaccines to some devastating human diseases very difficult.

ANTIGEN-PRESENTING CELL (APC). Few antigens bind directly to antigen-sensitive T-CELLS or B-CELLS but are generally 'presented' to these lymphocytes on the surfaces of other cells, the antigen-presenting cells. Dendritic cells with a large resultant surface area for antigen-attachment are widely distributed in the human body and trap antigens. Those in spleen and lymph nodes trap lymph- and blood-borne antigen and present it to lymphocytes there. Other antigen-bearing dendritic cells migrate from non-lymphoid regions to lymph nodes. Clonal expansion of appropriately stimulated B-lymphocyte classes occurs, resulting in specific antibody production. See IMMUNITY.

ANTIGIBBERELLINS. Organic compounds of varied structure causing plants to grow with short, thick stems or with appearance opposite to that obtained with GIBBERELLIN, which can reverse the action of most of these compounds. Of agricultural importance, they include phosphon and maleic hydrazide (retarding growth of grass, reducing frequency of cutting).

ANTIPODALS. Three (sometimes more) cells of the mature EMBRYO SAC, located at the end opposite the micropyle.

ANTIPORT. See TRANSPORT PROTEINS.

ANTISEPTIC. Substance used on a living surface (e.g. skin) to destroy microorganisms and sterilize it. Ethyl and isopropyl alcohol, diluted 70% with sterile water, kill vegetative bacteria and some viruses, but not spores of bacteria or fungi. Iodine (dissolved with potassium iodide in 90% ethanol) is rapidly bactericidal, killing both vegetative cells and spores. It does, however, stain. See DISINFECTANT, AUTO-CLAVE.

ANTISERUM. SERUM containing antibodies with affinity for a specific antigenic determinant (see ANTIGEN) to which they bind. May result in cross-reactivity (see ANTIGEN–ANTIBODY REACTION) within recipient.

ANTLER. Bony projection from skull of deer. Unlike HORN (which is matted hair) they are often branched, are shed annually, and are confined to males (except in reindeer).

ANURA (SALIENTIA). Frogs and toads. An order of the Class AMPHIBIA. Hind legs modified for jumping and swimming; no tail; often vocal.

ANUS. The opening of the alimentary canal to the exterior through which egested material, some excretory material and water may exit. When present, the gut is said to be entire. Absent from coelenterates and platyhelminths. See PROCTODAEUM.

AORTA. Term applied to some major vertebrate arteries. See AORTA, DORSAL; AORTA, VENTRAL; AORTIC ARCH.

AORTA, DORSAL. Major vertebrate (and cephalochordate) artery through which blood passes to much of body, supplying arteries to most major organs. In sharks a single dorsal aorta collects oxygenated blood from the gills, but in bony fish paired dorsal aortae on either side in the head region perform this task before uniting as a single median vessel. Oxygenated blood then passes backwards to the body; but in fish too blood flowing up through the third AORTIC ARCH tends to pass anteriorly through the aorta(e) rather than posteriorly (see CAROTID ARTERY). In adult tetrapods, those parts of the single or paired dorsal aortae between the third and fourth aortic arches tend to disappear, blood from the fourth (systemic) arch(es) passing back within two uniting dorsal aortae (terrestrial salamanders, lizards) or within a single dorsal aorta (most reptiles, birds and mammals) derived from the right arch (reptiles and birds) and from the left arch (mammals). Protected throughout in vertebrates by proximity of bone above (typically vertebrae).

AORTA, VENTRAL. Large median artery of fish and embryonic amniotes leading anteriorly from ventricle of heart, either giving off branches to gills or running uninterrupted as AORTIC ARCHES to dorsal aorta(e). In lungfish, branches differ in this respect. In living amphibians it has disappeared, while in other tetrapods it serves merely as a channel supplying blood to aortic arches III, IV and VI.

AORTIC ARCHES. Paired arteries (usually 6, but up to 15 in hagfishes) of vertebrate embryos connecting ventral aorta with dorsal aorta(e) by running up between gill slits or gill pouches on each side, one in each VISCERAL ARCH. The study of their comparative anatomy in embryos and, where they persist, in adults provides striking support for macroevolutionary change. Each is given a Roman numeral, beginning anteriorly. Arches I and II do not persist in post-embryonic tetrapods, but arch II at least is present in sharks, some bony fish and lungfish. Arch III usually serves (with parts of the dorsal aortae) as the tetrapod carotid arteries, but in fish is usually interrupted by gills; arch IV is separated from the anterior arches in most tetrapods and becomes the systemic arch (see AORTA, DORSAL); arch V is absent from adult tetrapods other than urodeles, but serves as the DUCTUS ARTERIOSUS in development prior to lung function; arch VI then shifts to supply the lungs.

APE. General term for HOMINOID primates of families Hylobatidae (gibbons, siamangs) and Pongidae ('great apes'). See ANTHROPOID APE.

APETALOUS. Lacking petals, e.g. flower of wood anemone.

APHANIPTERA. See SIPHONAPTERA (fleas).

APHID. Green fly or black fly. Homopteran insect (Superfamily Aphidoidea) notorious for sucking plant juices, for transmitting plant viral diseases, and for phenomenal powers of increase by viviparous PARTHENOGENESIS.

APHYLLOUS. Leafless.

APICAL DOMINANCE. (Bot.) Influence exerted by a terminal bud in suppressing growth of lateral buds. See AUXINS.

APICAL MERISTEM. Growing point (zone of cell division) at tip of root and stem in vascular plants, having its origin in a single cell (*initial*), e.g. Pterophyta, or in a group of cells (initials), e.g. Anthophyta. In the latter, the growing point apex (*promeristem*) consists of actively dividing cells. Behind this, division continues and differentiation begins, becoming progressively greater towards mature tissues. One (older) concept of growing point organization in flowering plants recognizes differentiation into three regions (*histogens*): *dermatogen*, a superficial cell layer giving rise to the epidermis; *plerome*, a central core of tissue giving rise to the vascular cylinder and pith; and *periblem*, tissue lying between dermatogen and plerome, that gives rise to cortex. It is now evident that respective roles assigned to these histogens are by no means universal; nor can periblem and plerome always be distinguished, especially in the shoot apex. Becoming widely accepted is the *tunica-corpus* concept, an interpretation of the shoot apex recognizing two tissue zones in the promeristem: *tunica*, consisting of one or more peripheral layers, in which the planes of cell division are predominantly anticlinal, enclosing the *corpus* or central tissue of irregularly arranged cells in which the planes of cell division vary. No relation is implied between cells of these two regions and differentiated tissue behind the apex as in the histogen concept. Although epidermis arises from the outermost tunica layer, underlying tissue may originate in tunica or in corpus, or in both, in different plant species.

Besides providing for growth in length of main axis, apical meristem of stem is the site of origin of leaf and bud primordia. In roots, two types of apical meristem occur, one in which vascular cylinder, cortex and root cap can be traced to distinct layers of cells in the promeristem, and a second type in which all tissues have a common origin in one group of promeristem cells. In contrast to those of stems, apical meristems of roots provide only for growth in length, lateral roots originating some distance from apex and, endogenously, from pericycle.

APOCARPOUS. (Of the gynoecium of flowering plants) having separate carpels, e.g. buttercup. See FLOWER.

APOCRINE GLAND. Type of gland in which only the apical part of the cell from which the secretion is released breaks down during secretion, e.g. mammary gland. Compare HOLOCRINE GLAND, MEROCRINE GLAND.

APODA (GYMNOPHIONA). Caecilians. Order of limbless burrowing amphibians with small eyes and, sometimes, a few scales buried in the dermis of the skin, and a pair of tentacle-like structures in grooves above the maxillae.

APOENZYME. The protein component of a holoenzyme (enzyme-cofactor complex) when the COFACTOR is removed. It is catalytically inactive by itself.

APOGAMY. See APOMIXIS.

APOGEOTROPIC. Growth of roots away from the earth and from the force of gravity (i.e. into the air).

APOMICT. Plant produced by APOMIXIS.

APOMIXIS. Most common in botanical contexts. (1) AGAMOSPERMY, reproduction which has the superficial appearance of ordinary sexual reproduction (amphimixis) but occurs without fertilization and/or meiosis. Affords the advantages of the seed habit (dispersal, and survival through unfavourable conditions) without risks in achieving pollination. Often genetically equivalent to asexual reproduction. See PARTHENOGENESIS. (2) Vegetative apomixis; ASEXUAL methods of propagation such as by rhizomes, stolons, runners and bulbils.

APOMORPHOUS. In evolution, of a character derived as a novelty from pre-existing (plesiomorphous) character. The two form a homologous pair of characters, termed an *evolutionary transformation series* in CLADISTICS. See SYNAPOMORPHY.

APOPLAST. The cell wall continuum of a plant or organ; movement of substances in the cell walls is termed apoplastic movement or transport. Compare SYMPLAST.

APOSEMATIC. Colour, sound, behaviour or other quality advertising noxious or otherwise potentially harmful qualities of an animal. See MIMICRY.

APOSPORY. See APOMIXIS, PARTHENOGENESIS.

APOTHECIUM. Cup- or saucer-shaped fruit body of certain Ascomycotina and lichens; lined with a hymenium of asci and paraphyses. Sessile or stalked, often brightly coloured; varying from a few mm to more than 40 cm across.

APPENDAGE. A functional projection from an animal surface; termed *paired appendages* if bilaterally symmetrical. Two such pairs (e.g. limbs, fins) generally occur in gnathostome vertebrates. Primitively one pair per segment in arthropods (walking legs, mouthparts, antennae) and polychaetes (parapodia).

APPENDIX, VERMIFORM. Small diverticulum of human caecum, of many other primates, and of rodents, containing lymphoid tissue. Not a vestigial structure, contrary to common belief.

APPETITIVE BEHAVIOUR. Behaviour (e.g. locomotory activity) variable with circumstances, increasing the chances of an animal satisfying some need (e.g. for food, nesting material) usually through a more stereotyped CONSUMMATORY ACT, such as eating. To this extent it is *goal-oriented*.

APPOSITION. (Bot.) Growth in thickness of cell walls by successive deposition of material, layer upon layer. Compare INTUSSUSCEPTION.

APTEROUS. Wingless; either of insects which are polymorphic for winged and wingless forms, e.g. aphids, many social insects; or of insects which have discarded wings, as do some ants and termites; or of primitively wingless (apterygotan) insects.

APTERYGOTA (AMETABOLA). Subclass of primitively wingless insects. Comprises orders Thysanura (bristletails, silverfish), Collembola (springtails), Protura and Diplura. Probably a polyphyletic assemblage. Some abdominal segments in members of all four orders have small paired lateral appendages, another primitive characteristic. Metamorphosis slight or absent. See PTERYGOTA.

AQUEOUS HUMOUR. Fluid filling the space between cornea and VITREOUS HUMOUR of vertebrate EYE. The iris and lens lie in it. Much like cerebrospinal fluid in composition. Continuously secreted by ciliary body, and drained by canal of Schlemm into blood. Much less viscous than vitreous humour. Links circulatory system to lens and cornea, neither having blood vessels for optical reasons; also maintains intraocular pressure.

ARACHNIDA. Class of chelicerate arthropods. Most living forms terrestrial, using lung books (scorpions), lung books and tracheae (spiders), tracheae alone (e.g. pseudoscorpions, larger mites), or just the body surface (smaller mites) for gaseous exchange. Usually there is TAGMOSIS into a *prosoma* of eight adult segments anteriorly and an *opisthosoma* of 13 segments posteriorly. No head/thorax distinction. Prosoma lacks antennae and mandibles; first pair of appendages clawed and prehensile chelicerae; second pair (pedipalps) may be prehensile and sensory, copulatory or stridulatory devices. Remaining four pairs of prosomal appendages are legs. Bases (gnathobases) of second and subsequent pairs of appendages are often modified for crushing and 'chewing' (in absence of true jaws). Includes orders: Acari (mites and ticks), Araneae (spiders), Scorpiones (scorpions), Pseudoscorpiones (false scorpions), Palpigrada (palpigrades), Solifugae (solfugids) and Opiliones (harvestmen). Xiphosura (king crabs), and the predatory and extinct Eurypterida are usually placed as subclasses of the MEROSTOMATA.

ARACHNOID MEMBRANE. One of the MENINGES around vertebrate spinal cord and brain.

ARANEAE (ARANEIDA). Order of ARACHNIDA. Spiders. Abdomen (opisthosoma) almost always without any trace of segmentation and joined to prosoma (cephalothorax) by 'waist'; silk produced from two to four spinning glands (spinnerets); pedipalps in male modified as intromittant organs for copulation; ends of chelicerae modified as poisonous fangs.

ARCHAEAN (ARCHAEOZOIC). Geological division preceding PROTEROZOIC; earlier than about 2600 Myr BP.

ARCHAEBACTERIA. Ancient lineage of bacteria distinct from other bacteria (eubacteria) and from eukaryotes. Many live in hot acidic conditions (i.e. they are thermophilic and acidophilic), growing best at temperatures approaching 100°C. Formerly in two groups, either aerobic (Sulfolobales) or anaerobic (Thermoproteales), facultative anaerobic forms are now known. Many unusual biochemical characteristics including possession of a novel 16S-like ribosomal RNA component in the small ribosome subunit, which with their peculiar membrane composition indicates that there may be a deep divide among prokaryotes between archaebacteria and eubacteria. Halophiles, methanogens and sulphur-dependent thermophiles occur.

ARCHAEOPTERYX. Most ancient recognized fossil bird (late Jurassic, 150–145 Myr BP). Exhibits mixture of reptilian and bird-like characters, having feathered wings and tail (impressions clear in limestone) and furcula (fused clavicles and interclavicles); but with teeth, bony tail, and claws on three digits of fore-limbs. Only known representative of Subclass Archaeornithes of Class AVES.

ARCHEGONIOPHORE. In some liverworts, a stalk bearing archegonia.

ARCHEGONIUM. 'Female' sex organ of liverworts, mosses, ferns and related plants, and of most gymnophytes. Multicellular, with neck composed of one or more tiers of cells, and swollen base (venter) containing egg-cell.

ARCHENTERON. Cavity within early embryo (at gastrula stage) of many animals, communicating with exterior by BLASTOPORE. Formed by invagination of mesoderm and endoderm cells at gastrulation; becomes the gut cavity.

ARCHESPORIUM. Cells or cell from which spores are ultimately derived, e.g. in developing pollen sac, fern sporangium.

ARCHOSAURS. 'Ruling reptiles'; the Subclass Archosauria. Originating with thecodonts in the Triassic, it includes the bipedal carnivorous dinosaurs (sauriscians) and the bird-like dinosaurs (ornithiscians). Crocodiles and alligators are living representatives. Birds are descendants. See DINOSAUR.

ARGINASE. Enzyme catalysing hydrolysis of arginine to ornithine and urea in urea cycle (see UREA); in mammals occurs in liver cell cytosol.

ARIL. Accessory seed covering, often formed from an outgrowth at the base of the OVULE (e.g. yew); often brightly coloured, aiding dispersal by attracting animals that eat it and carry seed away from the parent plant.

ARISTA. See AWN.

ARMS RACE. Term sometimes used to express the dialectical changes in selection pressure that occur when regular, often unavoidable, conflicts of interest between two or more 'ways of life' favour an adaptation for one party which creates a fresh 'challenge' for the other to respond by adapting to. Such conflicts are common: predator/prey; parasite/host; parent/offspring and male/female. It has been argued that selection will be the stronger where one party has more to lose by 'not evolving' and minimizing the probability of losing the conflict. Consequence to a prey organism of losing a predator/prey conflict is probably more serious than to a predator on any occasion. Much depends on how likely such conflict encounters are as to whether selection will favour whatever 'costs' may be involved in evolving a ploy to avoid or win the conflict. Conflicts are best generalized as conflicts of 'ways of life', or strategies, rather than as conflicts between individuals *per se*. Some conflicts of interest may resolve in favour of one of the parties through inability of the genetic system to 'represent' the other in the arms race. See ALTRUISM, COEVOLUTION.

AROUSAL. General causal term (and factor) invoked to account for the fact that animals are variably alert and responsive to potential stimuli. There may be a general 'sleeping/waking' difference; but it is less clear that there is a continuum of levels of awareness or responsiveness during either of these states. The phenomenology of arousal may be correlated with neural activity in the RETICULAR FORMATION of the medulla, hypothalamus and cortex of the vertebrate brain. Physiological processes which facilitate certain behaviours include hormone release and endogenous rhythms. Both may then be said to be arousal mechanisms, or to affect motivation.

ARRHENOTOKY. See MALE HAPLOIDY.

ARTERY. Any relatively large blood vessel carrying blood (not necessarily oxygenated) from the heart towards the tissues. Vertebrate arteries have thick elastic walls of smooth muscle and connective tissue (larger ones have capillaries in them), damping blood pressure changes. Their innermost layer is endothelium, as with all vertebrate blood vessels. They divide repeatedly to form arterioles.

ARTHROPODA. The largest phylum in the animal kingdom in terms of both number of taxa and (protozoans probably notwithstanding) biomass. Bilaterally symmetrical and metamerically segmented coelomates, with appendages on some or all segments (somites). A chitinous cuticle provides the exoskeleton, flexible to provide joints. Haemocoele is the main body cavity (coelom reduced). They lack true nephridia and cilia (onychophorans have the latter); with an annelid-like central nervous system and one or more pairs of coelomoducts acting as gonoducts or excretory ducts. Taxonomy varies. Thirteen classes are widely recognized, including: Onychophora (peripatids), Myriapoda (centipedes and millipedes), Insecta (insects), Trilobita (trilobites, extinct), Merostomata (king, or horseshoe crabs and extinct eurypterids), Arachnida (scorpions, spiders, harvestmen, solfugids, mites and ticks), Crustacea (crabs, prawns and shrimps, water-fleas). The extent and patterning of TAGMOSIS reflects the locomotory method, while appendages have proved marvellously adaptable and account in large measure for the success of the group. There appear to be three major evolutionary lineages: the *Onychophora-Myriapoda-Insecta* group, the *Merostomata-Arachnida-Trilobita* group, and the *Crustacea*. The phylum many be regarded as a GRADE, a polyphyletic origin not yet discounted.

ARTICULAR CARTILAGE. Cartilage providing the articulating surfaces of vertebrate joints.

ARTIFICIAL INSEMINATION. Artificial injection of semen into female reproductive tract. Much used in animal breeding.

ARTIFICIAL KEY. Any IDENTIFICATION KEY not based upon evolutionary relationships but rather upon any convenient distinguishing characters. See NATURAL CLASSIFICATION.

ARTIFICIAL SELECTION. Directional selection imposed by humans, deliberately or otherwise, upon wild or domesticated organisms. Crop plants originated in many cases from such deliberate crosses, sometimes involving one or more polyploid stocks. Procedures employed in harvesting these crops commonly involve unintentional (but still artificial) selection upon plants growing with the crops, favouring weed properties (see WEEDS). The phenomenon was well known to DARWIN and examples of conscious human selection provided an analogy through which his readers could grasp the theory of NATURAL SELECTION.

ARTIODACTYLA. Order of eutherian mammals. Ungulates (hoofed), with even number of toes upon which they walk. Includes pigs and hippopotamuses (four toes) and camels, deer, antelopes, etc. (two toes – the cloven-hoof). See PERISSODACTYLA, RUMINANT.

ASCHELMINTHES. A phylum of pseudocoelomate animals, probably representing a grade rather than a clade. The body cavity is a persistent blastocoele. Many are worm-like, and most either aquatic, soil-dwelling or parasitic. They are bilaterally symmetrical, unsegmented, and frequently minute, composed of remarkably few cells in view of their anatomical complexity. The classes (often regarded as phyla) include: NEMATODA, ROTIFERA, KINORHYNCHA, GASTROTRICHA and NEMATOMORPHA. See PSEUDOCOELOM.

ASCIDIAN. Sea squirt, or tunicate; member of Class Ascidiacea of Subphylum UROCHORDATA.

ASCOCARP. Fruiting body of the ASCOMYCOTINA (Eumycota); usually an open cup (apothecium), a vessel (perithecium, locule) or closed sphere on or in which asci are borne.

ASCOGENOUS HYPHAE. HYPHAE containing paired haploid 'male' and 'female' nuclei; develop from ascogonia which are the oogonia or 'female' gametangia of the Ascomycotina; eventually give rise to asci.

ASCOGONIUM. Oogonium or 'female' gametangium of the ASCOMYCOTINA.

ASCOMYCOTINA. Formerly Ascomycetes. Group of fungi (Eumycota) in which sexual spores are produced within asci (see ASCUS). Typically either filamentous moulds or non-filamentous yeasts. Some, such as *Candida*, can grow in either form. See ASCOCARP.

ASCORBIC ACID (VITAMIN C). A water-soluble sugar acid; a VITAMIN for a few arthropods and vertebrates (including humans), readily undergoing oxidation without loss of vitamin activity. Especially abundant in citrus fruits and tomatoes. Hydrolysis of its lactone ring destroys its vitamin activity, as often occurs in cooking (e.g. of potatoes). Mild dietary deficiency affects connective tissue as it appears to be required for collagen formation; its complete dietary absence in humans results in symptoms of scurvy, with poor wound-healing.

ASCOSPORE. Fungal spore of ASCOMYCOTINA, produced within an ascus.

ASCUS. In some fungi (Ascomycotina) a spherical, cylindrical or club-shaped cell in which fusion of haploid nuclei occurs during sexual reproduction, followed by meiosis and formation of, usually, eight haploid ascospores.

ASEPTATE. Those algal filaments and fungal hyphae devoid of cross-walls (septa).

ASEXUAL. Of reproduction (or organisms) lacking all the following: meiosis, gamete production, fertilization (leading to genome or nuclear union), transfer of genetic material between individuals, and PARTHENOGENESIS. Includes, or is synonymous with, VEGETATIVE REPRODUCTION. Often employed (with parthenogenesis) as a means of rapidly increasing progeny output during a favourable period (these having practically uniform genotype); hence common in internal parasites (see POLYEMBRYONY). The basis of natural cloning (artificially imposed in the propagation of plants by cuttings). May alternate with sexual phase in LIFE CYCLE (see ALTERNATION OF GENERATIONS). Some organisms (e.g. *Amoeba*, trypanosomes) are obligately asexual, and this raises questions about the evolutionary and ecological significance of SEX.

A-SITE. (Of ribosome) binding site on ribosome for charged (amino-acyl, hence A for acyl) tRNA molecule in PROTEIN SYNTHESIS. See P-SITE.

ASSIMILATION. Absorption of simple substances by an organism (i.e. across cell membranes) and their conversion into more complicated molecules which then become its constituents.

ASSOCIATION. (Of plants) climax plant community dominated by particular species and named according to them; e.g. oak–beech association of deciduous forest. Sometimes applied to very small natural units of vegetation. See CONSOCIATION.

ASSORTATIVE MATING (A. BREEDING). Non-random mating, involving selection of breeding partner, usually based on some aspect of its phenotype. This 'choice' (consciousness not implied) may be performed by either sex, and may be positively assortative (choice like self in some respect) or negatively assortative (disassortative, choice unlike self in some respect). Likely to have consequences for degree of inbreeding and maintenance of POLYMORPHISM. Some INCOMPATIBILITY mechanisms in plants are analogies. See SEXUAL SELECTION.

ASTER. A system of cytoplasmic striations radiating from the centriole and consisting of MICROTUBULES; often conspicuous during cleavage of egg, or during fusion of nuclei at fertilization. Also probably present in many other animal cells during division. Absent from higher plants.

ASTEROIDEA. Class of ECHINODERMATA. Starfishes. Star-shaped; arms, containing projections of gut, not sharply marked off from central part of body; mouth downwards; suckered TUBE FEET; spines and pedicillariae. Carnivorous (some notoriously on oysters or corals).

ASTHMA. See ALLERGIC REACTION.

ASTROCYTE. One type of GLIAL CELL of central nervous system. Star-shaped, with numerous processes, they provide mechanical support for transmitting cells by twining round them and attaching them to their blood vessels. Different types are found in white and grey matter of the CNS.

ATHEROSCLEROSIS. See SCLEROSIS.

ATLAS. First VERTEBRA, modified for articulation with skull. Modified further in amniotes, which have freer head movement, than in amphibians. Consists of simple bony ring, while a peg (odontoid process) of the next vertebra (the AXIS) projects forward into the ring (through which the spinal cord also runs). This peg represents part of the atlas (its centrum) which has become detached and fused to the axis. Nodding the head takes place at the skull–atlas joint; rotation of head at atlas–axis joint.

ATP. Adenosine triphosphate. Adenyl nucleotide diphosphate. The common 'energy currency' of all cells, whose hydrolysis accompanies and powers most cellular activity, be it mechanical, osmotic or chemical. Its two terminal phosphate groups have a more negative STANDARD FREE ENERGY of hydrolysis than phosphate compounds below it on the thermodynamic scale (e.g. sugar phosphates), and less negative than those higher (e.g. phosphocreatine, phosphoenolpyruvate), but this varies with intracellular concentrations of ATP, ADP and free phosphate as well as pH. A HIGH-ENERGY PHOSPHATE (see Fig. 33), it tends to lose its terminal phosphate to substances lower on the scale, provided an appropriate enzyme is present, and its mid-position on the scale enables it to serve as a common intermediate in the bulk of enzyme-mediated phosphate-group transfers in cells. Its relationship with ADP and AMP may be summarized:

$$ATP + H_2O \rightleftharpoons AMP + PP_i \text{ (pyrophosphate)} - 10 \text{ kcal mol}^{-1}$$

$$ATP + AMP \rightleftharpoons ADP + ADP$$

$$ATP \rightleftharpoons ADP + P_i \text{ (inorganic phosphate)} - 7.3 \text{ kcal mol}^{-1}$$

The energy values are STANDARD FREE ENERGY changes
at pH 7, standard temperature and pressure, at 25°C.

Cells normally contain about ten times as much ATP as ADP and AMP, but when metabolically active the drop in the ATP/(ADP+ AMP) ratio results in acceleration of GLYCOLYSIS and aerobic respiration (see RESPIRATION), the signal being detected by ALLOSTERIC enzymes in these pathways whose modulators (see ENZYME) are ATP, ADP or AMP. ATP is not a reservoir of chemical energy in the cell but rather a transmitter or carrier of it. The bulk of ATP in eukaryotic cells is provided by mitochondria, where these are present. Some extra-mitochondrial ATP is produced anaerobically in the cytosol, and chloroplasts produce it but do not export it. ATP hydrolysis is used to transfer energy when work is done in cells.

ATPase activity is found in MYOSIN (e.g. MUSCLE CONTRAC-TION) and DYNEIN (e.g. ciliary/flagellar beating). Membrane ion pumps (e.g. sodium and calcium pumps) and macromolecular syntheses of all kinds involve ATPase activity. Ultimately the energy source for ATP formation in the biosphere is solar energy trapped by autotrophs in photosynthesis – plus some lithotrophy. All heterotrophs depend upon respiratory oxidation of these organic compounds to power their own ATP synthesis. ATP is, like the other common nucleoside triphosphates in cells (CTP, GTP, TTP, UTP), a substrate in nucleic acid synthesis, and its hydrolysis provides the energy needed to build the resulting AMP monomer into the growing polynucleotide chain. These other triphosphates may participate in some other energy transfers; but ATP has by far the major role. See AMP, ADP, PHOSPHAGEN, BACTERIORHODOPSIN.

ATPase. Enzyme bringing about either (i) orthophosphate (P_i) cleavage of ATP yielding ADP and inorganic phosphate, or (ii) pyrophosphate ($2P_i$) cleavage of ATP to yield AMP and pyrophosphate. The latter provides a greater decrease in free energy and is involved where a 'boost' is needed for an enzyme reaction. ATPase activity is found in myosin and dynein molecules, chloroplast thylakoids and inner mitochondrial membranes (as ATP synthetase). See MITOCHONDRION, CHLOROPLAST, BACTERIORHODOPSIN.

ATRIUM. (1) Chamber, closed except for a small pore, surrounding gill slits of Amphioxus and urochordates. (2) A type of heart chamber of vertebrate chordates synonymous with 'auricle'; receives blood from major vein and passes it to ventricle. Walls not as muscular as those of ventricle. Fishes have single atrium, but tetrapods, breathing mainly or entirely by lungs, have two: one (the left) receives oxygenated blood from lungs, the other (the right) receives deoxygenated blood from the body. Much of the blood flow though the atria is passive (see HEART CYCLE). Non-chordates may have an atrial component of the heart, e.g. some polychaete worms and most molluscs, in which the term 'auricle' is sometimes preferred. (3) A space or cavity in some invertebrates (e.g. platyhelminths, some molluscs) known as the genital atrium, which houses the penis and/or opening of the vagina, and into which these may open.

ATROPHY. Diminution in size of a structure, or in the amount of tissue of part of the body. Generally involves destruction of cells, and may be under genetic and hormonal control, as is frequently the case in metamorphosis. May also result from starvation. Compare HYPERTROPHY.

ATROPINE. ALKALOID inhibiting action of acetylcholine and parasympatheticomimetic drugs; its application may double heart rate in man, and generally blocks the effects of MUSCARINE on effectors of the vagus nerve.

ATTENUATION. (Of pathogenic microorganisms) loss of virulence. May be achieved by heat treatment. See VACCINE.

AUDITORY (OTIC) CAPSULE. Part of vertebrate skull, enclosing auditory organ.

AUDITORY NERVE. See VESTIBULOCOCHLEAR NERVE.

AUDITORY ORGAN. Sense organ detecting pressure waves in air ('sound'), in vertebrates represented by the cochlea of the inner ear, but the term often intended to include VESTIBULAR APPARATUS detecting positional and vectorial information as well. See LATERAL LINE SYSTEM.

AUERBACH'S PLEXUS. That part of the autonomic nervous system in vertebrates (mostly from the vagus nerve) lying between the two main muscle layers of the gut and controlling its peristaltic movements.

AURICLE. (Zool.) (1) Often used synonymously with atrial heart chamber (see ATRIUM). (2) External ear of vertebrates. (Bot.) Small ear- or claw-like appendage occurring one on each side at the bases of leaf-blades in certain plants.

AUSTRALIAN REGION. ZOOGEOGRAPHICAL REGION consisting mainly of Australia, New Guinea and the Celebes; demarcated from south-east Asia by WALLACE'S LINE.

AUSTRALOPITHECINE. Member of genus *Australopithecus*, now extinct; of the primate family Hominidae (see HOMINID). It appears to have been a long extant genus (roughly from 4–1 Myr BP), with perhaps as many as five African species. *A. afarensis* (4.0–3.0 Myr BP) appears to have been a conservative species near the common ancestry of later forms; *A. aethiopicus* and *A. africanus* were later contemporaries (2.1–3 Myr BP) and possibly sister species with *A. afarensis* as common ancestor; *A. robustus* and *A. boisei* were later still (approx. 1.2–2 Myr BP) and shared several derived features (synapomorphies). *A. aethiopicus*, *A. robustus* and *A. boisei* were all 'robust' australopithecines, with heavy skulls and facial features; *A. africanus* had more 'gracile' features and was possibly ancestral to HOMO. Cranial capacities of typical australopithecines were from 400–600 cm^3 (modern man averages 1400 cm^3), and the animals had bipedal posture and gait. More fossil material and analysis are needed for definitive assessment of genus relationships. All the fossil material comes from Africa.

AUTECOLOGY. Ecology of individual species, as opposed to communities (synecology).

AUTOANTIGEN. Molecular component of organism, normally regarded as 'self' by its immune system, but here recognized as 'non-self' and eliciting an AUTOIMMUNE REACTION.

AUTOCATALYST. Any molecule catalysing its own production. The more produced, the more catalyst there is for further production. Most likely, some such process was involved in the origin of pre-biological systems which, once enclosed in a membrane, might be called 'living'. The current process whereby nucleic acid codes for enzymes that decode and replicate it is, in a strong sense, autocatalytic. See ORIGIN OF LIFE.

AUTOCHTHONOUS. (1) Of soil microorganisms whose metabolism is relatively unaffected by increase in organic content of soil. See ZYMOGENOUS. (2) The earliest inhabitant or product of a region (aboriginal). In this sense contrasted with *allochthonous* (not native to a region).

AUTOCLAVE. Widely-used equipment for heat-sterilization. Commonly air is either pumped out prior to introduction of steam, or simply replaced by steam as the apparatus is heated under pressure. Material being sterilized is usually heated at 121°C and 138–172 kNm pressure for 12 minutes, which destroys vegetative bacteria, all bacterial spores and viruses; but these figures will vary with the size and nature of material.

AUTOECIOUS. Of rust fungi (Basidiomycotina) having different spore forms of the life cycle all produced on one host species, as in mint rust. Compare HETEROECIOUS.

AUTOGAMY. Fusion of nuclei derived from the same zygote but from different meioses. Includes all forms of self-fertilization. See AUTO-MIXIS.

AUTOGRAFT. Tissue grafted back onto the original donor. See ALLO-GRAFT, ISOGRAFT, XENOGRAFT.

AUTOIMMUNE REACTION. Response by an organism's immune system to molecules normally regarded as 'self' but which act as antigens. Quite often the thyroid gland, adrenal cortex or joints become damaged by the action there of antibodies or sensitized lymphocytes. See B-CELL, T-CELL.

AUTOLYSIS. (1) Self-dissolution that tissues undergo after death of their cells, or during metamorphosis or atrophy. Involves LYSOSOME activity within cell. (2) Prokaryotic self-digestion, dependent upon enzymes of cell envelope.

AUTOMIXIS. Fusion of nuclei derived both from the same zygote and from the same meiosis. See AUTOGAMY, PARTHENOGENESIS.

AUTONOMIC NERVOUS SYSTEM (ANS). Term sometimes referring to the entire vertebrate visceral nervous system, but more often restricted to the efferent (motor) part of it (the visceral motor system), supplying smooth muscles and glands. Neither anatomically nor physiologically autonomous from the central nervous system. Sometimes termed the 'involuntary' nervous system; but here again its effects (in humans) can largely be brought under conscious control with training. For convenience the ANS can be subdivided into two components: the *parasympathetic* and *sympathetic* systems.

Parasympathetic nerve fibres are CHOLINERGIC and in mammals are found as motor components of CRANIAL NERVES III, VII, IX and X, as well as of three spinal nerves in sacral segments 2–4. Most of its effects are brought about by its distribution in the vagus (CN X), serving the gut (see AUERBACH'S PLEXUS), liver and heart among other organs.

Fibres of the sympathetic system originate within the spinal cord of the thoracic and lumbar segments, but beyond the vertebrae each departs from the cord and turns ventrally in a short white ramus (rami communicantes) to enter a sympathetic ganglion, a chain of which lies on either side of the mid-line. In the sympathetic ganglia many of the preganglionic fibres relay with postganglionic fibres that innervate target organs (e.g. mesenteries and gut); others pass straight through as splanchnic nerves and meet in plexi (collectively termed the *solar plexus*) beneath the lumbar vertebrae. From here postganglionic fibres innervate much of the gut, liver, kidneys and adrenals. Postganglionic fibres usually liberate catecholamines, particularly noradrenaline. Preganglionic parasympathetic fibres are relatively long and usually synapse in a ganglion on or near the effector, postganglionic fibres being short. In general ANS preganglionic fibres are myelinated, postganglionic unmyelinated and usually (there are exceptions) where the sympathetic stimulates, the parasympathetic inhibits, and vice versa; but organs are not always innervated by both. Both are, however, under central control, notably via the hypothalamus. Together they afford homeostatic nervous control of the internal organs, often reflexly.

AUTOPHAGY. Process whereby some secondary LYSOSOMES come to contain and digest organelles of the cell in which they occur.

AUTOPOLYPLOID. In classical cases, a POLYPLOID (commonly a tetraploid) in which all the chromosomes are derived from the same species, frequently the same individual. Compare ALLOPOLYPLOID.

AUTORADIOGRAPHY. Method using the energy of radioactive particles taken up by cells, tissues, etc., from an artificially enriched medium and localized inside them, to expose a plate sensitive to the emissions, thus indicating where radioactive atoms lie. Much used in tracing pathways within cells. See LABELLING.

AUTOSOME. Chromosome that is not a SEX CHROMOSOME.

AUTOSTYLIC JAW SUSPENSION. The method of upper jaw suspension of modern chimaeras and lungfishes; presumed to have been that employed by earliest jawed fishes, in which the hyomandibular bone has no role in the suspension. The upper jaw (palatoquadrate) attaches directly to the cranium. See HYOSTYLIC, and AM-PHISTYLIC JAW SUSPENSION.

AUTOTOMY. Self-amputation of part of the body. Some lizards can break off part of the tail when seized by a predator, muscular action snapping a vertebra. Both here and in many polychaete worms which can shed damaged parts of the body, REGENERATION restores the autotomized part.

AUTOTROPHIC. Of those organisms independent of external sources of organic carbon (compounds) for provision of their own organic constituents, which they can manufacture entirely from inorganic material. Autotrophs may be phototrophic or chemotrophic as regards energy supply, using solar or chemical energies respectively. Most chlorophyll-containing plants are autotrophic, manufacturing organic material from water, carbon dioxide and mineral ions (nitrates, phosphates, sulphates, metal ions, etc.) and using solar energy phototrophically (see PHOTOSYNTHESIS). Some bacteria and the blue-green algae are also phototrophic; other bacteria are chemotrophic, using energy released from inorganic oxidations (e.g. of hydrogen sulphide by sulphur bacteria, or of hydrogen by hydrogen bacteria). The term *lithotroph* is applied to those autotrophs which employ inorganic oxidants to oxidize inorganic reductants (e.g. the nitrifying bacteria *Nitrosomonas* and *Nitrobacter*, of the NITROGEN CYCLE). All other organisms, *heterotrophs* in one form or another, depend upon the synthetic activities of autotrophic organisms for their energy and nutrients. See ECOSYSTEM.

AUXINS. Group of plant GROWTH SUBSTANCES, produced by many regions of active cell division and enlargement (e.g. growing tips of stems and young leaves), that regulate many aspects of plant growth. Promote growth by increasing rate of cell elongation, apparently by activating a proton pump in the plasmalemma, pumping H^+ out of the cell, acidifying the cell wall and loosening bonds within it, so promoting cell expansion through turgor. Auxins also affect GENE EXPRESSION, increasing production of cell wall material in the longer term (apparently independently of the proton pump). Transported basipetally in shoots at a rate of about 1 cm/hour, they act synergistically with GIBBERELLIN in stem cell elongation, and with CYTOKININ in control of buds behind the apical bud (apical dominance). Effects of auxins on cell growth include curvature responses, such as GEOTROPISM and PHOTOTROPISM. Auxins may also have mitogenic effects, as in initiation of cambial activity in association with cytokinins, and in adventitious root formation in cuttings. They are also implicated in flower initiation, sex determination, fruit growth, and delays in leaf fall and fruit drop.

Naturally-occurring auxins include indole-3-acetic acid (IAA) and indole-3-acetonitrile (IAN). IAA has been isolated from such diverse sources as corn endosperm, fungi, bacteria, human saliva and, the richest natural source, human urine. In addition to naturally occurring auxins, many substances with plant growth regulatory activity (*synthetic auxins*) have been produced. Some are used on a very large scale for regulating growth of agriculturally and horticulturally important plants, as in inhibition of sprouting in potato tubers, prevention of fruit drop in orchards, achievement of synchronous flowering (and hence fruiting) in pineapple; also parthenocarpic fruit production, as in tomato, avoiding risks of poor pollination. At increased, though still relatively low, concentration, auxins inhibit growth, sometimes resulting in death. Some synthetic auxins have differential toxicity in different plants: toxicity of 2,4-dichlorophenoxyacetic acid (2,4–D) to dicotyledonous and non-toxicity to monocotyledonous plants is perhaps best known, being exploited successfully in control of WEEDS in cereal crops and lawns.

AUXOTROPH. Mutant strain of bacterium, fungus or alga requiring nutritional supplement to the MINIMAL MEDIUM upon which the wild-type strain can grow. See PROTOTROPH.

AVES. Birds. A class of vertebrates. Feathered ARCHOSAURS whose earliest known fossil, *Archaeopteryx*, was of upper Jurassic date and the sole known representative of the Subclass Archaeornithes. All other known birds (including fossils) belong to the Subclass Neornithes. The two superorders with living representatives are the Palaeognathae (ratites) and Neognathae (20 major orders; about half the 2900 living species, including songbirds, belonging to the Order Passeriformes, or 'perching' birds). Distinctive features include: FEATHERS; furcula (WISHBONE); forelimbs developed as wings. Bipedal and homeothermic, laying cleidoic eggs and (excluding *Archaeopteryx* and two other fossil genera) lacking teeth, but with the skin of the jaw margins cornified to form a beak (bill), whose diversity of form is in large part responsible for the Cretaceous, and subsequent, adaptive radiation of the group.

AWN (ARISTA). Stiff, bristle-like appendage occurring frequently on the flowering glumes of grasses and cereals.

AXENIC. (Of cultures of organisms) a pure culture.

AXIAL SYSTEM. In secondary xylem and secondary phloem, collective term for those cells originating from fusiform cambial initials. Long axes of these cells are orientated parallel with the main axis of the root or stem.

AXIL. (Of a leaf) the angle between its upper side and the stem on which it is borne; the normal position for lateral (axillary) buds.

AXILLARY. Term used to describe buds, etc., occurring in the AXIL of a leaf.

AXIS. (1) Embryonic axis of animals. There are generally three such: antero-posterior, dorso-ventral and medio-lateral, established very early in development, and sometimes by the POLARITY of the egg. The genetics of early development is under active study. See HOMOEOTIC, COMPARTMENT. (2) Second amniote VERTEBRA, modified for supporting the head. See ATLAS.

AXON. The long process which grows out from the cell bodies of some neurones towards a specific target with which it connects and carries impulses away from the cell body. See NEURONE, NERVE FIBRE.

AXONEME. Complex microtubular core of CILIUM and flagellum.

AXOPOD. PSEUDOPODIUM of some sarcomastigophoran protozoans in which there is a thin skeletal rod of siliceous material upon which streaming of the cytoplasm occurs. They may bend to enclose prey items.

B

BACILLARIOPHYTA. Diatoms. Division of the ALGAE. Microscopic unicellular plants, occurring singly or grouped in colonies. In addition to chlorophylls *a* and *c*, chloroplasts contain α, β and ε carotenes, and xanthophylls, including fucoxanthin. Cells surrounded by rigid, siliceous and finely sculptured cell wall (*frustule*) or two parts (*valves*). Asexual reproduction is by cell division; sexual reproduction isogamous or anisogamous, resulting in the characteristic *auxospore*. Abundant in marine and fresh waters, both plankton and benthos. Past deposition of countless numbers of silicified cell walls has formed siliceous or diatomaceous earths, while oil stores of past diatoms have contributed to the petroleum supplies of today. Extremely important microfossils in palaeolimnology, enabling interpretations of past lake histories.

BACILLUS. General term for any rod-shaped bacterium. Also a genus of bacteria: *Bacillus*.

BACKBONE. See VERTEBRAL COLUMN.

BACKCROSS. Cross (mating) between a parent and one of its offspring. Employed in CHROMOSOME MAPPING, when the parent is homozygous and recessive for at least two character traits, and to ascertain genotype of offspring (i.e. whether homozygous or heterozygous for a character), the parent used being the homozygous recessive. Where the organism of known genotype in the cross is not a parent of the other, the term *testcross* is often used.

BACTERIA. Unicellular, filamentous and mycelial PROKARYOTES, of the Kingdom Monera. Among the simplest of all known organisms (see CELL for diagram). Opinions differ on whether to include the blue-green algae within bacteria (see CYANOBACTERIA). Work on ARCHAEBACTERIA suggests that these form a distinctive side-branch. The description which follows relates to 'true' bacteria, or EUBACTERIALES, which vary greatly in shape, being rod-like (bacilli), spherical (cocci), more or less spiral (spirilli), filamentous and occasionally mycelial. Multiplication is by simple fission; other forms of asexual reproduction, e.g. production of aerially dispersed spores or flagellated swarmers, occur in some bacteria. As prokaryotes, they have no meiosis or syngamy, but genetic recombination occurs in many of them (see F FACTOR, PILI, PLASMID, RECOMBINATION).

Bacteria are ubiquitous, occurring in a large variety of habitats. 1 g of soil may contain from a few thousand to several million; 1 cm³ of sour milk, many millions. Most are saprotrophs or parasites; but a few are autotrophic, obtaining energy either by oxidation processes or from light (in the presence of bacteriochlorophyll). In soil, their activities are of the utmost importance in the decay of dead organic matter and return of minerals for higher plant growth (see DE-COMPOSERS, CARBON CYCLE, NITROGEN CYCLE). Some are sources of ANTIBIOTICS (e.g. *Streptomyces griseus* produces streptomycin). As agents of plant disease bacteria are less important than fungi; but they cause many diseases in animals and humans (e.g. diphtheria, tuberculosis, typhoid, some forms of pneumonia). See ANTISEPTIC, BACTERICIDAL, BACTERIORHODOPSIN, ESCHERICHIA COLI, GRAM'S STAIN, MYCOPLASMAS, SPIRO-CHAETES.

BACTERICIDAL. Substances which kill bacteria. Includes many ANTI-BIOTICS. For *colicins*, see PLASMID. Compare BACTERIOSTATIC.

BACTERIOPHAGE (PHAGE). A VIRUS parasitizing bacteria. The genetic material is always housed in the centre of the phage particle and may be DNA or RNA – the former either double- or single-stranded, the latter double-stranded. RNA phages are very simple; the more complex T-even DNA phages have a head region, collar, tail and tail fibres; some (e.g. φX174) are entirely spherical; others (e.g. M13) are filamentous. Some (e.g. PM2) contain a lipid bilayer between an outer and inner protein shell. The complex phage T4 has a polyhedral head about 70 nm across containing double-stranded DNA, enclosed by a coat (capsid) of protein subunits (capsomeres); a short collar or 'neck' region; a cylindrical tail (or tail sheath) region and six tail fibres. The particle (viron) of T4 phage is quite large – about 300 nm in length. It initiates infection by attachment of tail fibres to the bacterial cell wall at specific receptor sites. This is followed by localized lysis of the wall by previously inert viral enzymes in the tail and by contraction of the viral tail sheath, forcing the hollow tail core through the host cell wall to inject the phage DNA. *Virulent phages* (such as T4, T7, φX174) engage in a subsequent *lytic cycle* inside the host cell. Their circular genomes are transcribed and translated, causing arrest of host macromolecule synthesis and production of virion DNA and coat proteins. Lysis of the cell releases the virions. *Filamentous phages* (e.g. MI3, fl, fd) do not lyse the host cell, but even permit it to multiply. Eventually they get extruded through the cell wall. Some *temperate phages* (e.g. λ) can insert their genome into their host's and be replicated with it; others (e.g. P1) replicate within a bacterial plasmid. This non-lethal infective relationship does not involve transcription of phage genome, and is termed *lysogeny*, the phage being termed *prophage*. Lysogenic bacteria (those so infected) can produce infectious phage, but are immune to lytic infection by the same or closely related phage (superinfection immunity). Conversion from lysogenic state to lytic cycle (induction) may be enhanced by UV light and other mutagens. A replication cycle from adsorption to release of new phage takes $\simeq 15$–20 mins. *Transduction* occurs when temperate phage from one lysogenic culture infects a second bacterial culture, taking with it a small amount of closely-linked DNA which remains as a stable feature of the recipient cell. Antibiotic resistance can be transferred this way (see ANTI-BIOTIC RESISTANCE ELEMENT). See PHAGE CONVERSION, PHAGE RESTRICTION.

BACTERIORHODOPSIN. Conjugated protein of the 'purple membrane' of the photosynthetic bacterium *Halobacterium halobium*, forming a proton channel whose prosthetic group (retinal) is light-absorbing and identical to that of vertebrate rod cells. Allosteric change on illumination of the pigment results in proton ejection from the cell, the resulting proton gradient being used to power ATP synthesis. Its integration into LIPOSOMES along with mitochondrial ATP synthetase showed that the latter is also a proton channel, driven by a proton gradient. If chlorophyll-based photosynthesis evolved after a rhodopsin-based variety then absence of green-absorption in chlorophyll's action spectrum may be due to selection in favour of a pigment avoiding competition with abundant rhodopsin-based forms. See CELL MEMBRANES, CHLOROPLAST, ELECTRON TRANSPORT SYSTEM, MITOCHONDRION, ROD CELL.

BACTERIOSTATIC. Inhibiting growth of bacteria, but not killing them. See BACTERICIDAL.

BALANCED LETHAL SYSTEM. Genetic system operating when the two homozygotes at a locus represented by two alternative alleles each produces a lethal phenotype, yet the two alleles persist in the population through survival of heterozygotes, which thus effectively breed true. Compare HETEROZYGOUS ADVANTAGE.

BALANCE OF NATURE. Phrase glossing observations that in natural ecosystems, communities and the biosphere at large, herbivores do not generally overgraze, predators do not generally over-predate nor parasites decimate host populations; that populations of all appear to be held roughly in equilibrium, and that drastic (sometimes trivial) disturbance of this harmony between organisms and the physical environment will have inevitable and generally unfavourable consequences for mankind. Causal processes involved in the complex systems with which ecologists deal are increasingly amenable to computer simulation. See DENSITY DEPENDENCE, NEGATIVE FEEDBACK.

BALANOGLOSSUS (GLOSSOBALANUS). Genus of acorn worms. Worm-like members of the HEMICHORDATA, with vertebrate affinities.

BALBIANI RING. See PUFF.

BALDWIN EFFECT. Reinforcement or replacement of individually acquired adaptive responses to altered environmental conditions, through selection (artificial or natural) for genetically determined characters with similar functions. Not regarded as evidence of Lamarckism. See GENETIC ASSIMILATION, LAMARCK.

BALEEN. 'Whalebone'; thin sheets of cornified skin hanging from the roof of the mouth of the largest (whalebone) whales, with which food is filtered. Such whales are toothless.

BARBS. Side-branches in a row on each side of the rachis of a contour FEATHER making up the expanded pennaceous part (vane), and giving off barbules.

BARBULES. Minute filaments lying in two rows, one proximal and one distal to each barb of a contour FEATHER, those distal bearing hooks which slot into grooves on the proximal barbules and linking barbs together, as occurs during preening.

BARK. Protective corky tissue of dead cells, present on the outside of older stems and roots of woody plants, e.g. tree trunks. Produced by activity of cork cambium. Bark may consist of cork only or, when other layers of cork are formed at successively deeper levels, of alternating layers of cork and dead cortex or phloem tissue (when it is known as *rhytidome*). Popularly regarded as everything outside the wood.

BARORECEPTOR (PRESSORECEPTOR). Receptor for hydrostatic pressure of blood. In man and most tetrapods, located in carotid sinuses, aortic arch and wall of the right atrium. Basically a kind of stretch receptor. When stimulated, those in the first two locations activate the cardio-inhibitory centre and inhibit the CARDIO-ACCELERATORY CENTRE, while those in the atrium stimulate the cardio-acceleratory centre, helping to regulate blood pressure.

BARR BODY. Heterochromatic X-chromosome occurring in female placental mammals. Paternally- or maternally-derived X-chromosomes may behave in this way, often a different X-chromosome in different cell lineages. See HETEROCHROMATIN.

BASAL BODY. Structure indistinguishable from centriole, acting as an organizing centre (nucleation site) for eukaryotic cilia and flagella, unlike which its 'axoneme' is a ring of nine triplet microtubules, each comprising one complete microtubule fused to two incomplete ones. It is a permanent feature at the base of each such flagellum or CILIUM. See CENTRIOLE for details.

BASAL LAMINA. Thin layer of several proteins, notably collagen and the glycoprotein laminin, about 50–80 nm thick, varying in composition from tissue to tissue. Underlying and secreted by sheets of animal epithelial cells and tubes they may form (e.g. many glands, and endothelial linings of blood vessels). In kidney glomerulus and lung alveolus, may be an important selective filter of molecules between cell sheets. See BASEMENT MEMBRANE.

BASAL METABOLIC RATE (BMR). The respiratory rate of a resting animal, normally measured by oxygen demand. The 'background' respiration rate, as required for unavoidable muscle contractions (e.g. heart), growth, repair, temperature maintenance, etc. See THYROID HORMONES.

BASAL PLACENTATION. (Bot.) Condition in which ovules are attached to the bottom of the locule in the ovary.

BASE. Either a substance releasing hydroxyl ions (OH⁻) upon dissociation, with a pH in solution greater than 7, or a substance capable of acting as a proton acceptor. In this latter sense, the nitrogenous cyclic or heterocyclic groups combined with ribose to form nucleosides are termed bases. See PURINE, PYRIMIDINE.

BASEMENT MEMBRANE. Combination of BASAL LAMINA with underlying reticular fibres and additional glycoproteins, situated between many animal epithelia and connective tissue.

BASE PAIRING. Hydrogen bonding between appropriate purine and pyrimidine bases of (antiparallel) nucleic acid sequences, as during DNA synthesis, mRNA transcription and during translation (see PROTEIN SYNTHESIS). If two DNA strands align then adenine in one strand pairs with a thymine and a guanine with a cytosine (i.e. A:T, G:C; but if a DNA strand aligns with an RNA strand, then adenines in the DNA pair align with uracils in the RNA (i.e. A:U, G:C). Without these 'rules' there could be no GENETIC CODE or heredity as we know it. See DNA HYBRIDIZATION.

BASE RATIO. The $(A + T):(G + C)$ ratio in double-stranded (duplex) DNA. It varies widely between different sources (i.e. from different species). The amount of adenine equals the amount of thymine; the amount of guanine equals the amount of cytosine. See BASE PAIRING.

BASIC DYES. Dyes consisting of a basic organic grouping of atoms (cation) which is the actively staining part, usually combined with an inorganic acid. Nucleic acids, hence nuclei, are stained by them and are consequently referred to as BASOPHILIC.

BASIDIOMYCOTINA. Subdivision of the fungi (EUMYCOTA), known informally as 'basidiomycetes'. Contains a large variety of species (e.g. jelly fungi, bracket fungi, smuts, rusts, mushrooms, toadstools, puffballs). Most of these common names refer to the visible part of the fungus, the conspicuous reproductive or 'fruiting' body (*basidiocarp*) supported nutritionally by an extensive assimilative mycelium that penetrates the plant or soil and derives nutrients. Primarily terrestrial, with perforated septa (cross-walls) in their hyphae. Complete septa cut off the reproductive bodies. Chitin predominates in the hyphal walls; sexual reproduction involves formation of basidia.

BASIDIOSPORE. Characteristic spore type of Basidiomycotina, produced within the BASIDIUM by meiosis.

BASIDIUM. Specialized reproductive cell of Basidiomycotina; often club-shaped, cylindrical, or divided into four cells. Nuclear fusion and meiosis occur within it, resulting in formation of four *basidiospores* borne externally on minute stalks called *sterigmata*.

BASIPETAL. (Bot.) (Of organs.) Developing in succession towards the base, oldest at the apex, youngest at the base. Also used of the direction of transport of substances within a plant: away from apex. Compare ACROPETAL.

BASOPHIL. Blood POLYMORPH. Very similar to MAST CELL in structure and probably function, but with peroxidase rather than acid and alkaline phosphatase activity in its granules.

BASOPHILIC. Staining strongly with basic dye. Expecially characteristic of nucleic acids, and hence of nucleus (during division of which the condensed chromosomes are strongly basophilic), and of cytoplasm when actively synthesizing proteins (due to rRNA and mRNA).

BATRACHIA. Rarely used term for AMPHIBIA.

B-CELL (B LYMPHOCYTE). A LEUCOCYTE, derived from a LYMPHOID TISSUE stem cell which has migrated from foetal liver to bone marrow and has not entered the thymus but has settled either in a lymph node or in the spleen. B-cells express a specific immunoglobulin (Ig, or ANTIBODY) on their plasma membranes. This can bind appropriate antigen, when the cell becomes activated to divide repeatedly by mitosis and produce a clone of specific antibody-coated cells. This primary immune response (see IMMUNITY) is also characterized by their secretion of specific IgM antibody into the blood, the combined effect being to remove antigen. Some B-cells do not greatly participate in Ig production, but circulate in the body and may persist for years as memory cells, capable of clonal expansion and rapid Ig secretion (a secondary immune response) if activated by the initial antigen. This provides immunological memory. Still other B-cells mature into plasma cells (the major Ig-producer in a secondary response) after multiplication, and although these progeny cells may have Igs of more than one class, they all have the same antigen specificity. A fully mature plasma cell will have little surface Ig but will be secreting an Ig of one class and of one antigen specificity. See IgA-IgM, MYELOID TISSUE, T-CELL.

B-CHROMOSOME. See SUPERNUMERARY CHROMOSOME.

BELT DESMOSOME. See DESMOSOME.

BENEDICT'S TEST. A modification of Fehling's test for sugar using just one solution. Contains sodium citrate, sodium carbonate and copper sulphate dissolved in water in the ratio $1.7:1.0:0.17$ g: 30 cm³ water. Five drops of test solution are added to 2 cm³ Benedict's solution. If a REDUCING SUGAR is present then a rust-brown cuprous oxide precipitate forms on boiling.

BENNETTITALES (CYCADEOIDALES). Extinct fossil gymnophytes, present from the Triassic to the Cretaceous. Resembled cyads (Cycadales), with leaves entire or pinnate. Epidermal cells differed from cycads in having sinuose cell walls, the guard cells and subsidiary cells originating from the same mother cell. Further, in most forms, the cones were bisexual, with both micro- and megasporophylls. They may have self-pollinated.

BENTHOS. General term referring to those animals and plants living on the bottom of the sea, lake or river (crawling or burrowing there, or attached as with sea weeds and sessile animals), from high water mark down to deepest levels. There have been many schemes for subdividing the benthos. With respect to aquatic higher plants and algae the benthos can be subdivided as follows: (1) *Rhizobenthos* – vegetation rooted in the sediments, e.g. *Chara*, submersed and emergent aquatic higher plants. (2) *Haptobenthos* – plants attached to solid surfaces. (3) *Herpobenthos* – the community living on, or moving through, sediments, e.g. ENDOPELON, EPIPELON. (4) *Endobenthos* – the community living and often boring into solid substrata, e.g. endolithon living inside rock. Organisms feeding primarily upon the benthos are termed *benthophagous*. See PELAGIC, NEKTON, PLANKTON.

BERGMANN'S RULE. States that in geographically variable species of HOMOIOTHERMIC animals, body size tends to be larger in cooler regions of a species range. See ALLEN'S RULE.

BERRY. Many-seeded succulent fruit, in which the wall (*pericarp*) consists of an outer skin (*epicarp*) enclosing a thick fleshy *mesocarp* and inner membranous *endocarp*, as in gooseberry, currant, tomato. Compare DRUPE.

BETA-BLOCKER. Substance, such as the drug propanolol, selectively blocking BETA RECEPTORS. Clinical use is to slow heart rate and lower blood pressure. See OPIATES.

BETA CELLS. See PANCREAS.

BETA-GLOBULINS. A class of vertebrate plasma proteins including certain lipoproteins, TRANSFERRIN and plasminogen (precursor of fibrinolysin, see FIBRINOLYSIS).

BETA RECEPTOR. ADRENERGIC receptor sites (associated with adenylate cyclase in appropriate postsynaptic membranes, but not identical with it) binding preferentially to adrenaline rather than noradrenaline. Heart muscle has beta receptors predominantly which result in increased heart rate and blood pressure when stimulated; other beta effects include dilation of arterioles supplying skeletal muscle, bronchial relaxation and relaxation of the uterus. All these effects are mimicked by the synthetic drug isoprenaline. See ALPHA RECEPTOR, BETA BLOCKER, AUTONOMIC NERVOUS SYSTEM.

BETA-SHEETS (BETA-CONFORMATION). One type of protein secondary structure. See PROTEIN, AMINO ACID.

B-FORM HELIX. Paracrystalline form of DNA (DNA-B) adopted in aqueous media, as in cells. See A-FORM and Z-FORM HELIX.

BICOID GENE. (*bcd*). A pattern-specifying MATERNAL GENE in *Drosophila*, whose mRNA transcript passes from maternal ovary nurse cells to anterior poles of developing oocytes where it becomes localized, apparently trapped by the oocyte cytoskeleton. Its eventual protein product is a transcription activator (with a DNA-binding homoeodomain), is located in nuclei of the syncytial blastoderm, and is translocated half-way along the embryo from the anterior pole. With a similar translocation of OSKAR GENE product in the opposite direction, two opposing gradients of maternally encoded proteins provide quantitative POSITIONAL INFORMATION which the embryo genome converts into qualitative phenotypic differences. Genes involved in this conversion process, i.e. *Krüppel* (*Kr*), *hunchback* (*hb*) and *knirps* (*kni*), belong to the GAP GENE class, and all three encode proteins with DNA-binding FINGER DOMAINS. Bicoid product is responsible for production of a normal head and thorax, structures absent in embryos from females lacking functional *bcd*, and both it and the oskar product are examples of MORPHOGENS.

By repressing transcription of *Kr* in anterior and posterior embryo regions, bicoid and oskar morphogens only permit *Kr* expression in mid-embryo. High levels of *bcd* product activate *hb* transcription, while low levels permit *Kr* transcription. However, their abilities to elicit such zygotic (embryo) gene expression depend upon the affinities of their respective DNA-binding sites for them, and these are known to vary. The *hb* and *Kr* products are mutual repressors of each others' transcription, so that their 'domains' of product expression become stable and restricted. These *gap gene domains* delimit the domains of HOMOEOTIC GENE expression but also enable position-specific regulation of the PAIR-RULE GENES, whose expression leads to metameric segmentation of the embryo.

BICOLLATERAL BUNDLE. See VASCULAR BUNDLE.

BIENNIAL. Plant requiring two years to complete life cycle, from seed germination to seed production and death. In the first season, biennials store up food which is used in the second season to produce flowers and seed. Examples include carrot and cabbage. Compare ANNUAL, EPHEMERAL.

BILATERAL CLEAVAGE. Old term for radial cleavage. See CLEAVAGE.

BILATERAL SYMMETRY. Property of most metazoans, having just one plane in which they can be separated into two halves, approximately mirror images of one another. This is usually the antero-posterior and dorso-ventral plane, separating left and right halves. Major metazoan phyla excluded are coelenterates and echinoderms (having radial symmetry as adults). In flowers, the condition is termed zygomorphy.

BILE. Fluid produced by vertebrate liver cells (hepatocytes), containing both secretory and excretory products, and passed through bile duct to duodenum. Contains (i) BILE SALTS (e.g. sodium taurocholate and glycocholate) which emulsify fat, increasing its surface area for lipolytic activity. Also form micelles for transport of sterols and unsaturated fatty acids towards intestinal villi (see CHYLOMICRON); (ii) bile pigments, breakdown products of haemoglobin such as bilirubin, which are true excretory wastes and colour faeces; (iii) LECITHIN and CHOLESTEROL as excretory products. Bile is aqueous and alkaline due to $NaHCO_3$, providing a suitable pH for pancreatic and subsequent enzymes. Stored in gall bladder. See MICELLE.

BILE DUCT. Duct from liver to duodenum of vertebrates, conveying BILE. See GALL BLADDER.

BILE SALTS. Conjugated compounds of bile acids (derivatives of CHOLESTEROL and taurine and glycine), forming up to two thirds of dry mass of hepatic BILE.

BILHARZIA. Schistosomiasis. See SCHISTOSOMA.

BINARY FISSION. Asexual reproduction occurring when a single cell divides into two equal parts. Compare MULTIPLE FISSION.

BINOCULAR VISION. Type of vision occurring in primates and many other active, predatory vertebrates; eyeballs can be so directed that an image of an object falls on both retinas. Extent to which eyes converge to bring images on to the foveas of each retina gives proprioceptive information needed in judging distance of object. Stereoscopic vision (perception of shape in depth) depends on two slightly different images of an object being received in binocular vision, the eyes viewing from different angles.

BINOMIAL NOMENCLATURE. The existing method of naming organisms scientifically and the lasting contribution to taxonomy of LINNAEUS. Each newly described organism (usually in a paper published in a recognized scientific journal) is placed in a genus, which gives it the first of its two (italicized) latin names – its generic name – and is always given a capital first letter. Thus the genus *Canis* would probably be given to any new placental wolf discovered. Within this genus there may or may not be other species already described; in any case, the new species will receive a second (specific) name to follow its generic name, this time with a lower case first letter. The wolf found in parts of Europe, Asia and North America belongs to the species *Canis lupus*; jackals, coyotes and true dogs (including the domesticated dog) also belong to the genus *Canis*, but each species is further identified by a different specific name: the domestic dog, in all its varieties, is *C. familiaris* (often only the first letter of the genus is given, if it is contextually clear what the genus is). If a previously described species is subsequently moved into another genus, it takes the new generic name, but carries the old specific name. The author who published the original description of a species often receives credit in the form of an abbreviation of their name after the initial mention of the species in a paper or article. Thus one might see *Canis lupus* Linn., since Linnaeus first described this species. The specimen upon which the initial published description was based is termed the type specimen, or 'type', and is probably housed in a museum for comparison with other specimens. If sufficient variation exists within a species range for SUBSPECIES or varieties to be recognized, then a trinomial is employed to identify the subspecies or variety. The British wren rejoices under the trinomial *Troglodytes troglodytes troglodytes*.

BIOASSAY. Quantitative estimation of biologically active substances by measurement of their activity in standardized conditions on living organisms or their parts. A 'standard curve' is first produced, relating the response of the tissue or organism to known quantities of the substance. From this the amount giving a particular experimental response can be read.

BIOCHEMICAL OXYGEN DEMAND (BOD). Amount of dissolved oxygen (in mg/dm^3 water) which disappears from a water sample in a given time at a certain temperature, through decomposition of organic matter by microorganisms. Used as an index of organic pollution, especially sewage.

BIOCHEMISTRY. The study of the chemical changes within, and produced by, living organisms. Includes molecular biology, or molecular genetics.

BIOGENESIS. The theory that all living organisms arise from pre-existing life forms. The works of Redi (1688) for macroorganisms, and of Spallanzani (1765) and PASTEUR (1860) in particular for microorganisms, stand as landmarks in the overthrow of the theory of SPONTANEOUS GENERATION. Since their time it has become generally accepted that every individual organism has a genetic ancestry involving prior organisms. The first appearance of living systems on the Earth (see ORIGIN OF LIFE) is still problematic, but there is no reason in principle to dispense with faith in natural and terrestrial causation, and every reason to pursue that line of inquiry.

BIOGENETIC LAW. Notorious view propounded by Ernst Haeckel in about 1860 (a more explicit formulation of his mentor Muller's view) that during an animal's development it passes through ancestral adult stages ('ontogenesis is a brief and rapid recapitulation of phylogenesis'). Much of the evidence for this derived from the work of embryologist Karl von Baer. It is now accepted that embryos often pass through stages resembling related *embryonic*, rather than adult, forms. Such comparative embryology provides important evidence for EVOLUTION. See PAEDOMORPHOSIS.

BIOGEOGRAPHY. The study and interpretation of geographical distributions of organisms, both living and extinct. One approach (*dispersal biogeography*) stresses the role of dispersal of organisms from a point of origin across pre-existing barriers; another approach (*vicariance biogeography*) takes barriers which occur within a pre-existing continuous distribution to be more important. See GONDWANALAND, LAURASIA, ZOOGEOGRAPHY.

BIOLOGICAL CLASSIFICATION. See CLASSIFICATION.

BIOLOGICAL CLOCK. In its widest sense, any form of biological timekeeping, such as heart beat or ventilatory movements; more often used in contexts of PHOTOPERIODISM, DORMANCY and DIAPAUSE; but most usually associated with physiological, behavioural, etc., rhythms relating to environmental cycles, notably CIRCADIAN, tidal, lunar-monthly and annual. Mechanisms 'counting' the number of cell divisions that have elapsed since some signal would also be included. See PINEAL GLAND.

BIOLOGICAL CONTROL. Artificial control of PESTS and parasites by use of organisms or their products; e.g. of mosquitoes by fish and aquatic insectivorous plants which feed on their larvae; or of the prickly pear (*Opuntia*) in Australia by the moth *Cactoblastis cactorum*. There is increasing use of PHEROMONES in attracting pest insects, which may then be killed or occasionally sterilized and released. Sometimes attempts are made to encourage spread of genetically harmful factors in the pest population's gene pool, although this needs great care. Success depends on a thorough grasp of relevant ecology.

BIOLOGY. Term coined by LAMARCK in 1802. The branch of science dealing with properties and interactions of physico-chemical systems of sufficient complexity for the term 'living' (or 'dead') to be applied. These are usually cellular or acellular in organization; but since viruses share some of the same polymers (nucleic acid and protein) as cells, and moreover are parasitic, they are regarded as biological systems but not usually as organisms. See LIFE.

BIOLUMINESCENCE. Production of light by 'photogenic' organisms, e.g. fireflies, glow-worms, crustaceans, coelenterates, fungi, some dinoflagellates and bacteria – the last often in symbiotic association with animals, e.g. fish, which thereby become luminous (see PHOTO-PHORE). Some animals are self-luminous, having special photogenic organs containing photocytes. In the coelenterate *Obelia* these are scattered through the endoderm. The functional photoprotein is often luciferin, which is reduced by ATP and the enzyme luciferase but oxidized in the presence of free oxygen, light being released as luciferin returns to its stable ground state. Flashing that occurs in luminous organs of many animals (often serving in mate or prey attraction) is often under nervous control, the organs having a rich supply of nerve terminals. See PHOSPHORESCENCE.

BIOMASS. The total quantity of matter (the non-aqueous component frequently being expressed as dry mass) in organisms, commonly of those forming a trophic level or population, or inhabiting a given region. See STANDING CROP, PYRAMID OF BIOMASS.

BIOME. Major regional ecological complex of COMMUNITIES extending over large natural areas and characterized by distinctive vegetation and climate; e.g. TROPICAL RAIN FOREST, grasslands, coral reef. Plants of land biomes comprise the *formations* of plant ecologists.

BIOMETRY. Branch of biology dealing specifically with application of mathematical techniques to the quantitative study of varying characteristics of organisms, populations, etc.

BIOPHYSICS. Fields of biological inquiry in which physical properties of biological systems are of overriding interest. Biophysics departments tend to work in such areas as crystal structures of macromolecules and neuromuscular physiology.

BIOPOIESIS. The generation of living from non-living material. See ORIGIN OF LIFE.

BIOSPHERE. That part of Earth and its atmosphere inhabited by living organisms.

BIOSYSTEMATICS. Branch of SYSTEMATICS concerned with variation and evolution of taxa. More or less equivalent to experimental taxonomy, using any techniques that might yield relevant information. See CLASSIFICATION.

BIOTECHNOLOGY. Application of discoveries in biology to large-scale production of useful organisms and their products. Centres on development of enzyme technology in industry and medicine, and of GENE MANIPULATION, often in the service of plant and animal breeding. Together, these constitute biomolecular engineering. Branches include fermentation technology, waste technology (e.g. recovery of metals from mining waste) and renewable resources technology, such as the use of LIGNOCELLULOSE to generate more usable energy sources. Organisms involved tend to be microorganisms, and their traditional involvement in the brewing, baking and cheese/yoghurt industries is also affected by the new technology.

BIOTIC FACTORS. Those features of the environments of organisms arising from the activities of other living organisms; as distinct from such abiotic factors as climatic and edaphic influences.

BIOTIN. VITAMIN of the B-complex made by intestinal bacteria, so difficult to prevent its uptake and assimilation. However avidin, a component of raw egg white, binds tightly to biotin and prevents its uptake from the gut lumen, resulting in biotin deficiency in those eating raw eggs too avidly. A modified biotin when bound to the enzyme propionyl-CoA carboxylase acts as a coenzyme in the conversion of pyruvic acid to oxaloacetic acid (see KREBS CYCLE) and in the synthesis of fatty acids and purines, again facilitating carboxylation and decarboxylation reactions.

BIOTYPE. (1) Naturally occurring group of individuals of the same genotype. See INFRASPECIFIC VARIATION. (2) See PHYSIOLOGIC SPECIALIZATION.

BIRAMOUS APPENDAGE. Paired crustacean appendage branching distally from a basal region (protopodite) to form two rami, the exopodite and endopodite, the exopodite often being more slender and flexible. Subject to considerable adaptive radiation, serving varied functions, such as locomotion, feeding and gaseous exchange. Trilobite limbs also had a biramous structure, but the origin of the second ramus is different. *Stenopodia* has been the term used for crustacean appendages in which one or more processes, epipodites, lie on the outer sides of protopodites; *phyllopodia* are broader and flatter than most stenopodia and may be unjointed, bearing lobes known as endites and exites. See UNIRAMOUS APPENDAGE.

BISACCATE. Pollen grains possessing two air sacs or bladders; mainly coniferous pollen, e.g. pine pollen.

BISEXUAL. See HERMAPHRODITE.

BISPORANGIATE CONE. (Bot.) Cone containing both megaspores and microspores (e.g. in the lycopsid *Selaginella*).

BISPORIC EMBRYO SAC. Embryo sac developing from the inner spore of a diad of spores produced by incomplete meiosis (as in the onion, *Allium*).

BIURET REACTION. Reaction often used as a test for presence of proteins and peptides (as a result of peptide bonds; but also works for any compound containing two carbonyl groups linked via a nitrogen or a carbon atom). A few drops of 1% copper sulphate solution and sodium hydroxide solution are added to the test sample, when a purple Cu^{2+}-complex is produced if positive; solution stays blue otherwise.

BIVALENT. Pair of homologous chromosomes during pairing (synapsis) at the first meiotic prophase and metaphase. See MEIOSIS.

BIVALVE. Broadly speaking, any animal with a shell in two parts hinged together, e.g. bivalve molluscs (with which the term is often equated) and BRACHIOPODA. Occasionally ostracod crustaceans are said to have a bivalve carapace, although this is not a shell. See BIVALVIA.

BIVALVIA (LAMELLIBRANCHIA, PELECYPODA). Class of MOLLUSCA, with freshwater, brackish and marine forms with greatly reduced head (no eyes, tentacles or radula), and body laterally compressed and typically bilaterally symmetrical. Shell composed of two hinged valves (both lateral) under which lie two large gills (*ctenidia*) used for gaseous exchange and generally for filter-feeding. Sexes nearly always separate; trochosphere and veliger larvae, but occasionally a parasitic glochidium. Considerable radiation, including fixed forms (e.g. mussels, clams) and burrowers (e.g. shipworm, razorshell). See BRACHIOPODA.

BLADDER. (1) Urinary bladder. Part of the anterior ALLANTOIS of embryonic amniotes, which persists into adult life and receives the openings of the ureters either directly (mammals) or via a short part of the cloaca (lower tetrapods). In fish the urinary ducts themselves may enlarge as 'bladders', or fuse for part of their length forming a single sac, receiving also the oviducts in females, as a urogenital sinus. In all cases the bladder serves for urine retention, and in tetrapods is distensible with a thick smooth-muscle wall. In lower tetrapods it opens into the cloaca, but in mammals its contents leave via the urethra. (2) Other sac-like fluid-filled structures in animals termed bladders include GALL BLADDER and GAS BLADDER.

BLADDERWORM. The CYSTICERCUS stage of some tapeworms.

BLASTEMA. Mass of undifferentiated tissue which forms, often at a site of injury or amputation, and from which regenerating parts regrow and differentiate. See REGENERATION.

BLASTOCOELE. Primary body cavity of metazoans, arising as fluid-filled cavity in the BLASTULA and persisting as blood and tissue fluids where these occur; otherwise obliterated during GASTRULATION. Much expanded in most arthropods as the HAEMOCOELE. Compare COELOM, PSEUDOCOELOM.

BLASTOCYST. Stage of mammalian development at which implantation into the uterine wall occurs, the INNER CELL MASS spreading inside the blastocoele as a flat disc.

BLASTODERM. Sheet of cells, usually one or just a few cells thick, surrounding the blastocoele in non-yolky eggs, or covering it in yolky ones. Consists usually of small, tightly packed cells.

BLASTOMERE. Cell produced by cleavage of an animal egg, up to the late blastula stage. In yolky eggs especially this results in smaller *micromeres* at the animal pole, with larger yolky *macromeres* at the vegetal pole where the rate of division is slower. See CLEAVAGE.

BLASTOPORE. Transitory opening on surface of gastrula by which the internal cavity (archenteron) communicates with the exterior; produced by invagination of superficial cells during GASTRULATION. It becomes the mouth in PROTOSTOMES and the anus in DEUTERO-STOMES. The *dorsal lip* of amphibian blastopores is famous for INDUCTION of overlying tissue in gastrulation. Transplant experiments indicate that a region up to about 60° around the original blastopore material is the source of a dorsoventral gradient responsible for regional subdivision of embryonic mesoderm. See POSITIONAL INFORMATION, ORGANIZER.

BLASTULA. Stage of animal development at or near the end of cleavage and immediately preceding gastrulation. May consist of a hollow ball of cells (*blastomeres*), especially in non-yolky embryos.

BLEPHAROPLAST. Rare term for CENTRIOLE.

BLIND SPOT. Region of vertebrate retina at which optic nerve leaves; devoid of rods and cones and hence 'blind'.

BLOOD. Major fluid transport medium of many animal groups, including nemerteans, annelids, arthropods, molluscs, brachiopods, phoronids and chordates. Derived from the BLASTOCOELE. Comprises an aqueous mixture of substances in solution (e.g. nutrients, wastes, hormones, gases and osmotically active compounds) in which are suspended cells (haemocytes) functioning either in defence (e.g. phagocytes) or oxygen transport (e.g. RED BLOOD CELLS). Blood is moved by muscle contraction in some of the vessels it passes through. Hearts are such specialized vessels, the hydrostatic pressure generated being employed in filtration processes (in capillaries generally, and kidney glomeruli in particular), in locomotion (e.g. many bivalve molluscs) and other activities besides solute translocation. Arthropod blood (haemolymph) is hardly confined to vessels (open circulation), insects and onychophorans having least vascularization, the haemocoele being much expanded. When blood is confined to vessels, the circulation is said to be closed, as it is throughout annelid and vertebrate bodies, although expanded blood sinuses are a feature of lower vertebrates. Respiratory pigment (absent in most insects) is in simple solution in invertebrate blood, but confined to red blood cells (erythrocytes) in vertebrates. See TISSUE FLUID.

BLOOD CLOTTING (B. COAGULATION). Adaptive response to haemorrhage involving local conversion of liquid blood to a gel, which plugs a wound. In vertebrates, blood does not clot in normal passage through vessels since the smooth endothelial lining prevents damage to platelets. The altered surface of damaged vessel walls plus turbulence of blood flow results in ADP release from platelets, and the exposed collagen in vessel walls causes adherence of platelets, which release Ca^{++} and cause more aggregation. Vasoconstriction, stemming blood flow, results from SEROTONIN released when platelets fragment. A temporary plug of platelets is made permanent by a CASCADE of enzymic reactions caused by release of phospholipid and the protein thromboplastin (thrombokinase) by damaged cells and platelets. Ten clotting factors in addition to the above have been isolated; but the main sequence may be summarized as: thromboplastin (Factor III) converts prothrombin (Factor II) to thrombin, which converts fibrinogen (Factor I) to an insoluble fibrin meshwork, trapping erythrocytes, platelets and plasma to form clot, stabilized by Factor XIII in presence of Ca^{++} (Factor IV). Fibrin absorbs 90% of the thrombin formed and prevents the clot from spreading away from the damaged area. Several genetic disorders cause poor clotting. Factor VIII, one of the substances required for thromboplastin formation, is absent in classical X-linked haemophilia. See ANTICOAGULANT, FIBRINOLYSIS.

BLOOD GROUP. Either a group of people bearing the same antigen(s) on their red blood cells within a particular BLOOD GROUP SYSTEM; or the blood characteristic used to distinguish groups of individuals within a blood group system. Main clinical significance lies in blood transfusion and Rhesus incompatibility between mother and foetus. Without due matching of donor and recipient blood, death or severe illness of a recipient may result from a single transfusion (in ABO system) or after repeated tranfusion (in Rhesus system). Causes of death include haemolysis (red cell rupture), agglutination (red cell clumping) with blockage of capillaries etc., and tissue damage. Danger in transfusion comes when donor antigens meet antibodies to them in recipient plasma and elsewhere resulting in INNATE IMMUNE RESPONSE. Antibodies to antigens of the ABO system occur naturally in plasma; those to Rhesus antigens occur in plasma only as a result of immunization, during pregnancy or transfusion. People with an antigen of the ABO system on their red cells automatically lack the antibody to it in their plasma, whereas lack of an ABO antigen on the red cells is coupled to presence of antibody to it in plasma (Table 1a).

Success or failure of transfusions is determined by the data outlined in

	Blood Group				Donor Blood Group			
	A	B	AB	O		A	B	AB	O

	A	B	AB	O
Antigen(s) on red cells	A	B	A & B	—
Antibody (ies) in plasma	anti-B	anti-A	—	anti-A anti-B

Table 1a. Relationship of antigen and antibody distributions in the ABO blood group system.

		Donor Blood Group			
		A	B	AB	O
Recipient Blood Group	A	√	×	×	√
	B	×	√	√	×
	AB	√	√	√	√
	O	×	×	×	√

Table 1b. Success [√] or failure [×] of blood transfusions from specific donor blood groups to specific recipient blood groups.

Blood Group (phenotype)	Possible Genotypes
A	$I^A I^A$ or $I^A I^O$
B	$I^B I^B$ or $I^B I^O$
AB	$I^A I^B$
O	$I^O I^O$

Table 2a. Genetics of ABO system.

P₁ phenotypes		A	×	O
		Case 1	*Case 2*	
P₁ genotypes		$I^A I^A$	$I^A I^O$ × $I^O I^O$	
Gametes		I^A	I^A & I^O × I^O	
F₁ genotypes		$I^A I^O$	$I^A I^O$: $I^O I^O$	
F₁ phenotypes		all A	A:O	

Table 2b. Example showing expected offspring from marriage between group A and group O people.

Table 1b. People with O-type are universal donors; those with AB-type blood are universal recipients. The Rhesus antibody may be found in plasma of Rhesus negative (Rh−) women who have been pregnant with Rhesus positive (Rh+) babies, who have Rhesus antigen on their red cells. This is because mothers may become sensitized to the Rhesus antigen if foetal blood leaks across the placenta during birth of a Rh+ baby. Later Rh+ foetuses are at risk if anti-Rh antibody crosses the placenta from the mother, who had produced it as an immune response to the earlier leaked foetal blood. Foetal haemolysis can be averted by injecting anti-Rh antibody (anti-D) to a Rh− mother prior to delivery of any Rh+ offspring; immediate blood transfusion of babies ('blue babies') born with haemolytic anaemia is an alternative. Blood groups are a classic instance of POLYMORPHISM in man. The genetics involved is often complex (there are at least three loci responsible for the Rh system), but the ABO system is a simple multiple allelism, as shown in Table 2.

The ratios of different blood groups within a blood group system differ geographically and indicate either NATURAL SELECTION or GENETIC DRIFT among different populations. Their genetic basis may be used in cases of contested parentage. See MENDELIAN HEREDITY.

BLOOD GROUP SYSTEM. A person's blood groups are genetically determined, each person belongs to several, and most are determined by loci situated on autosomes. The alleles at these loci determine antigens on a person's red blood cell membranes. A blood group system refers to the range of red cell antigens determined by the alleles at one such locus, or sometimes a group of closely linked loci. Some non-human primates share certain human blood group systems. In humans there are at least 14 such systems (ABO, Rhesus, MNS, P, Kell, Lewis, Lutheran, Duffy, Kidd, Diego, Yt, I, Dombrock and Xg). Only the Xg locus is sex-linked. No linkage between any of these systems is apparent, other than between Lewis and Lutheran.

BLOOD PLASMA. Clear yellowish fluid of vertebrates, clotting as easily as whole blood and obtained from it by separating out suspended cells by centrifugation. An aqueous mixture of substances, including plasma proteins. See BLOOD.

BLOOD PLATELETS. See PLATELETS, BLOOD CLOTTING.

BLOOD PRESSURE. Usually refers to pressure in main arteries; in humans, normally between 120 mm Hg at SYSTOLE and 80 mm Hg at DIASTOLE (i.e. 120/80). Pressure drops most rapidly in arterioles, falling further in capillaries and more slowly in venules and veins. In the venae cavae it is 2 mm Hg, and 0 mm Hg in the right atrium. Homeostatically controlled (see BARORECEPTOR).

BLOOD SERUM. Fluid expressed from clotted blood or from clotted blood plasma. Roughly, plasma deprived of fibrinogen and other clotting proteins.

BLOOD SUGAR. Glucose dissolved in blood. Homeostatically regulated in humans at about 90 mg glucose/100 cm^3 blood. Hormones affecting level include: INSULIN, GLUCAGON, ADRENALINE, GROWTH HORMONE and THYROXINE.

BLOOD SYSTEM (BLOOD VASCULAR SYSTEM). The system of blood vessels (in sequence: arteries, arterioles, metarterioles, capillaries, venules and veins) and/or spaces (often sinuses) through which blood flows in most animal bodies. See BLOOD for other details.

BLOOM. A term used to describe the dense growth of planktonic algae which imparts a distinct colour to the water. Commonly, blue-green algae (Cyanobacteria) form such blooms in eutrophic lakes.

BLUBBER. Thick layer of fatty tissue below dermis of skin, probably serving for thermal insulation, of many aquatic mammals, e.g. whales, seals, etc.

BLUE-GREEN ALGAE. See CYANOBACTERIA.

BOD. See BIOCHEMICAL OXYGEN DEMAND.

BODY CAVITY. (1) Primary body cavity; see BLASTOCOELE. (2) Secondary body cavity; see COELOM.

BODY CELL. (Bot.) The cell of the microgametophyte or pollen grain of gymnophytes which divides mitotically to form two sperm. (Zool.) See SOMATIC CELL.

BOHR EFFECT. The effect of dissolved carbon dioxide on the oxygen equilibria curves or respiratory pigments (see HAEMO-GLOBIN), whereby increased CO_2 concentration (pCO_2) shifts the curve to the right, i.e. decreases the percentage O_2-saturation of the pigment for a given pO_2, increasing the rate of oxygen unloading in regions of high respiratory activity (the tissues) and loading of oxygen in regions of low pCO_2 (e.g. lungs, gills). Brought about by ALLOSTERIC effect of pH on haemoglobin molecule.

BONE. Vertebrate connective tissue laid down by specialized meso-dermal cells (osteocytes) lying in LACUNAE within a calcified matrix which they secrete, containing about 65% by weight inorganic salts (mainly hydroxyapatite, providing hardness); remainder largely organic and comprising mostly COLLAGEN fibres providing tensile strength. *Compact bone* forms outer cylinder of shafts of long bones of limbs, and is typified by HAVERSIAN SYSTEMS; *spongy bone* forms vertebrae, most 'flat' bones (e.g. skull), and ends of long bones (epiphyses), typified by presence of TRABECULAE. Bone is a living tissue, supplied with nerves and blood vessels. Its constitution changes under hormone influence (see CALCITONIN, PARATHYROID HORMONE). Besides its skeletal role in providing levers for movement and support for soft parts of the body, it protects many delicate tissues and organs. See OSSIFICATION, HAEMOPOIESIS, CARTILAGE.

BONY FISH. See OSTEICHTHYES.

BORAX CARMINE. See STAINING.

BOTANY. The branch of biological sciences concerned with plants. Compare ZOOLOGY.

BOUTON. Knob-like terminal of nerve axon, containing synaptic vesicles. See SYNAPSE.

BOWMAN'S CAPSULE. Cup-like receptacle composed of two epithelial layers, formed by invagination of a single layer, surrounding a glomerulus and forming part of a vertebrate KIDNEY nephron.

BRACHIAL. Relating to the arm; e.g. brachial plexus, the complex of interconnections of spinal nerves V–IX supplying the tetrapod forelimb.

BRACHIATION. The arm-over-arm locomotion adopted by many arboreal primates.

BRACHIOPODA. Lamp shells. A small phylum of marine coelomate bivalve invertebrates, valves being dorsal and ventral (see BIVALVIA); with teeth and sockets along the hinge in articulate, but absent from inarticulate forms. Ciliated LOPHOPHORE serves for feeding and gaseous exchange. Of enormous value in dating rock strata, appearing in Lower Cambrian, with major extinctions in Devonian and Permian and expansions in the Ordovician, Carboniferous and Jurassic. One of the few surviving genera, *Lingula* (inarticulate) is almost unchanged since the Ordovician.

BRACT. Small leaf with relatively undeveloped blade in axil of which arises a flower or a branch or an inflorescence.

BRACTEOLE. Small BRACT.

BRADYCARDIA. Slowing of heart (hence pulse) rate. Compare TACHY-CARDIA.

BRADYKININ. Nonapeptide hormone of submaxillary salivary gland, inducing vasodilation and increased capillary permeability of tissues. Released by parasympathetic stimulation. Implicated in ALLERGIC REACTIONS.

BRAIN. Enlargement of the central nervous system of most bilaterally symmetrical animals with an antero-posterior axis. Its development anteriorly is a major component of the process of CEPHALIZATION.

(1) In vertebrates, an anterior enlargement of the hollow neural tube, sharing features with the spinal cord, such as relative positioning of white matter (myelinated axons) externally and grey matter (un-myelinated neurones) internally – although in higher forms there is migration of grey matter cell bodies above the white matter to form a third layer, reaching its zenith in the human cerebrum. Three dilations of the neural tube give rise to forebrain, midbrain and hindbrain.

The *forebrain* comprises the diencephalon anteriorly and the telencephalon posteriorly. Primitively, this is olfactory but in higher vertebrates the diencephalon roof is expanded to form the paired cerebral hemispheres and dominates the rest of the brain in its sensory and motor function. The telencephalon retains its association with smell, forming a pair of olfactory lobes.

The *midbrain* is primitively an optic centre, the pair of optic nerves entering after DECUSSATION at the optic chiasma. Although terminations of these fibres enter the visual cortex of the cerebrum in higher vertebrates, the midbrain still retains integrative functions (see BRAINSTEM) and is the origin of some CRANIAL NERVES.

The *hindbrain* generally has its anterior roof enlarged to form a pair of cerebellar hemispheres associated with proprioceptive coordination of muscle activity in posture and locomotion. Its floor is thickened to form the pons anteriorly and the medulla oblongata posteriorly (see BRAINSTEM). The central canal of the spinal cord expands to form the brain ventricles, and the whole is surrounded by MENINGES.

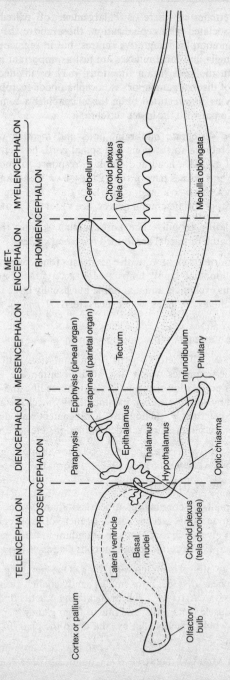

Fig. 2. Diagram showing median section of typical mature vertebrate brain, illustrating the principal divisions and structures.

(2) In invertebrates, there is enlargement of paired anterior GANGLIA associated with cephalization, those above the gut often fusing to form complex integrative centres; but in segmented forms, segmental ganglia and local reflexes are just as important in nervous integration. In molluscs 'brain functions' may be divided between several pairs of distinct ganglia or, as in cephalopods (octopus, squid, etc.), as many as 30 integrated brain lobes, enabling a complexity of behaviour on a par with the lower vertebrates.

BRAINSTEM. The vertebrate midbrain, pons and medulla oblongata. It links the cerebrum to the pons and spinal cord, has reflex centres for control of eyeballs, head and trunk in response to auditory and other stimuli, and houses part of the RETICULAR FORMATION. The respiratory, VASOMOTOR and CARDIAC CENTRES are here, associated with appropriate cranial nerves. See VENTILATION.

BRANCHIAL. Relating to gills. The aortic arches serving the gills in fish (the third arch onwards) are termed *branchial arches*.

BRANCHIOPODA. CRUSTACEA with at least four fairly uniform pairs of phyllopodial trunk limbs (see BIRAMOUS APPENDAGE) used in gaseous exchange, and a variable but usually large number of somites. Primitive. Includes water fleas (e.g. *Daphnia*) and brine shrimps (e.g. *Artemia*).

BREATHING. See VENTILATION.

BREEDING SYSTEM. All factors, apart from mutation, affecting the degree to which gametes which fuse at fertilization are genetically alike. Includes population size, and levels of inbreeding and outbreeding controlled by variable selfing, incompatibility mechanisms, assortative mating, heterostyly, dichogamy, arrangement and distribution of sex organs, mechanism of sex-determination, etc. See GENETIC VARIATION.

BRISTLETAILS. See APTERYGOTA.

BRITTLE STARS. See ECHINODERMATA.

BRONCHIOLE. Small air-conducting tube (less than 1 mm in diameter) of tetrapod lung, arising as a branch of a bronchus and terminating in alveoli. Smooth muscle abundant in walls, controlling lumen diameter. Lacks cartilage and mucous glands of bronchi.

BRONCHUS. Large air tube of tetrapod lung. One per lung, connecting it to trachea; divided into smaller and smaller bronchi, and finally into BRONCHIOLES. Each bronchus has cartilaginous plates and smooth muscle in its walls, and mucus from glands traps bacteria and dust, the whole being beaten by cilia up to the pharynx for swallowing.

BROWN FAT. See ADIPOSE TISSUE.

BRUSH BORDER. Animal cell surface (often apical), often a whole epithelial surface, covered in MICROVILLI and serving for absorption and/or DIGESTION.

BRYOPHYTA. Division of plant kingdom comprising HEPATICOPSIDA (liverworts), BRYOPSIDA (mosses) and ANTHOCEROTOPSIDA (hornworts). Small group of plants with wide distribution. Habitats various, e.g. wet banks, on soil, rock surfaces; some epiphytic, others aquatic. Small plants, flat, prostrate or with a central stem up to 30 cm in length, bearing leaves. Vascular tissue poorly developed; attached to substratum by RHIZOIDS. Reproducing sexually by fusion of macro- and microgametes produced in multicellular sex organs; antherida liberate microgametes, motile by flagella; archegonia contain a single macrogamete (egg cell). Sexual reproduction is followed by development of capsule containing spores giving rise via PROTONEMATA to new plants. A well-marked ALTERNATION OF GENERATIONS: the leafy vegetative plant is the gametophyte generation; the capsule (and seta) comprise the sporophyte generation, partially dependent nutritionally upon the gametophyte.

BRYOPSIDA (MUSCI). Mosses. Class of BRYOPHYTA, having fairly conspicuous PROTONEMATA and multicellular RHIZOIDS. The sporophyte develops from an apical cell, and the capsule has a complex opening mechanism. ELATERS are absent. Cosmopolitan, occurring in damp habitats (e.g. moist humus, peat, wet cliff faces, dry boulders, etc.) and as epiphytes on branches and tree trunks. Sex organs (antheridia and archegonia) borne either terminally or in lateral bud-like structures (perigonia and perichaetia), either on separate or the same plants. Fertilization is followed by development of the sporophyte, the capsule elevated on a seta and covered when young by a gametophytic CALYPTRA. A filamentous or thalloid protonema gives rise to new moss plants from lateral buds.

BRYOZOA. See ECTOPROCTA.

BUCCAL CAVITY. The mouth cavity, lined by ectoderm of stomodaeum, leading into the PHARYNX.

BUD. (Bot.) (1) Compact embryonic plant shoot comprising a short stem bearing crowded overlapping immature leaves. (2) Vegetative outgrowth of yeasts and some bacteria serving for vegetative reproduction. See BUDDING. (Zool.) Outgrowth of organism that may detach, as in BUDDING; or a morphogenic feature of a growing region, as in vertebrate limb buds (see REGENERATION).

BUDDING. (Bot.) (1) Grafting in which grafted part is a bud. (2) Asexual reproduction in which a new cell is formed as an outgrowth of a parent cell, e.g. in yeast. Compare BINARY FISSION. (Zool.) ASEXUAL method of reproduction common in many invertebrates such as sponges, coelenterates (e.g. *Hydra*), ectoproctans and urochordates, but also a feature of the HYDATID CYSTS of tape-worms. Rarely known as gemmation. See POLYEMBRYONY.

BUFFER. A solution resisting pH change on addition of acid (i.e. H^+) or alkali (OH^-), absorbing protons from acids and releasing them on addition of alkali. Usually consists of a mixture of a weak acid and its conjugate base, or vice versa. Intracellular and extracellular buffers may differ; thus the commonest intracellular buffer is the acid-base pair $H_2PO_4^-$–HPO_4^{2-} and such organic phosphates as ATP, but the bicarbonate buffer system (H_2CO_3– HCO_3^-) is common extracellu-larly, as in vertebrate blood plasma, where plasma proteins are also a major buffer system. HAEMOGLOBIN acts as a buffer during the CHLORIDE SHIFT. If pH varies, PROTEIN shape and function may be affected. See KIDNEY.

BUFFON, G.L.L. de. French naturalist (1707–1788), with comparable influence to LINNAEUS on contemporaries. His great work is *Histoire Naturelle* (1749–1788), a natural history in 44 volumes. Espoused NOMINALISM with regard to species and other taxa, but had noted that species seemed to 'breed true'. He held that the environment had important effects on animal characteristics, amounting in time to a sort of degeneration from original types rather than being in any way creative. He favoured the theory of the GREAT CHAIN OF BEING, indicating his distance from the later DARWINISM.

BUG. See HEMIPTERA.

BULB. Organ of vegetative reproduction; modified shoot consisting of very much shortened stem enclosed by fleshy scale-like leaves (e.g. tulip) or leaf-bases (e.g. onion). See BULBIL. Compare CORM.

BULBIL. Dwarf shoot occurring in place of a flower (e.g. *Saxifraga, Festuca, Allium*), borne either in lower part of the inflorescence or in axils of leaves (lesser celandine) and serving as an organ of vegetative reproduction. See APOMIXIS.

BULK FLOW. Overall movement of water or some other liquid induced by gravity or pressure or the interplay of both.

BULLA, AUDITORY. Bony protection of middle ear cavity in most placental, but not most marsupial, mammals. Absent also in earliest mammals.

BUNDLE SCAR. The scar from a vascular bundle remaining on a leaf scar after ABSCISSION.

BUNDLE SHEATH. Layer of cells which surrounds a vascular bundle, comprising cells of PARENCHYMA, SCLERENCHYMA, or both. See KRANZ ANATOMY.

BURSA OF FABRICIUS. Thymus-like LYMPHOID organ found in birds, but not in mammals, developing dorsally from CLOACA. Like the mammalian THYMUS, it appears to cause differentiation of circulating stem cells into immunoglobulin-producing cells; hence bursa-derived cells (B-CELLS). Contains ANTIGEN-PRESENTING CELLS. See T-CELL.

BUTTERFLY. See LEPIDOPTERA.

BUTTRESS ROOT. An ADVENTITIOUS root on a stem, functioning in support; found mainly in monocotyledonous plants.

C

CADUCOUS. (Bot.) Not persistent. Of sepals, falling off as flower opens (e.g. poppy); of stipules, falling off as leaves unfold (e.g. lime).

CAECUM. Blind-ending diverticulum, commonly of gut. One or two may be present at junction of vertebrate ileum and colon, housing cellulose-digesting bacteria. Thin-walled, sometimes with spiral valve for increased surface area, terminating in vermiform APPENDIX. Mesenteric (midgut) caeca in some annelids (e.g. leeches) and many arthropods are secretory and absorptive, and may be generally 'liver-like' en masse.

CAINOZOIC. See CENOZOIC.

CALCIFEROL. See VITAMIN D.

CALCITONIN (CT). Polypeptide hormone of parafollicular cells (*C-cells*) of thyroid gland. Lowers plasma calcium and phosphate levels by inhibiting bone degradation and stimulating their uptake by bone. May have evolved alongside conquest of land by vertebrates, given its role in regulating plasma ion levels. Antagonized by PARATHYROID HORMONE. See OSSIFICATION.

CALCIUM PUMP. An ATP-driven TRANSPORT PROTEIN. Calcium ions (Ca^{++}) act as SECOND MESSENGERS in the cell cytosol and their changing concentration there, particularly in eukaryotes, is significant. Although total cell Ca^{++} concentration approximates to that of the environment, it is unevenly distributed and Ca^{++} pumps in the plasma membrane expel Ca^{++} when it enters. Much is accumulated by pumps in MITOCHONDRIA and other organelles, causing a thousandfold drop in Ca^{++} concentration across the plasma membrane, a gradient down which the ion moves. The calcium pump in the sarcoplasmic reticulum of striated muscle accumulates Ca^{++} from the cytosol, enabling muscle relaxation. See MUSCLE CONTRACTION, CALMODULIN.

CALLOSE. A complex branched polysaccharide associated with the sieve areas of SIEVE ELEMENTS. May form in reaction to injury of these and parenchyma cells and be deposited so that their activity is impaired or finished, permanently or seasonally.

CALLUS. (Bot.) Superficial tissue developing in woody plants, usually through cambial activity, in response to wounding, protecting the injured surface. Often used in tissue culture, when the effects of hormones upon cell differentiation can be studied. (Zool.) Fibrocartilage produced at bone fracture, developing into bone as blood vessels grow into it and pressure and tension are applied.

CALMODULIN. Small multiply-allosteric protein required for Ca^{++}-dependent activities of many cellular (esp. membrane-bound) enzymes. Said to be activated when bound to Ca^{++}. Ubiquitous cellular component related to troponin C (see MUSCLE CONCENTRATION) which, once activated can in turn bind to several cell proteins (e.g. adenylate cyclase, some ATPases and membrane pumps) and regulate their activities. Has been implicated in geotropic response of roots. Its importance is being rapidly appreciated.

CALVIN CYCLE. A series of enzymic photosynthetic reactions in which carbon dioxide is reduced to 3-phosphoglyceraldehyde, while the carbon dioxide acceptor ribulose-1,5, bisphosphate is regenerated. For every six molecules of carbon dioxide that enter the cycle, a net gain of two molecules of glyceraldehyde-3,phosphate results. See PHOTOSYNTHESIS.

CALYPTRA. Hood-like covering of moss and liverwort capsules, developing from the archegonial wall.

CALYPTROGEN. Layer of actively dividing cells formed over apex of growing part of roots in many plants, e.g. grasses, giving rise to ROOT CAP.

CALYX. Outermost part of a flower, consisting usually of green, leaf-like members (sepals) that in the bud stage enclose and protect other flower parts. See FLOWER.

CAM. See CRASSULACEAN ACID METABOLISM.

CAMBIUM. A MERISTEM. A layer of actively dividing cells lying between xylem and phloem; forms additional xylem and phloem elements during SECONDARY THICKENING. The cambium of a vascular bundle is *fascicular* cambium; that formed from the parenchyma between vascular bundles, linking the fascicular cambium to form a complete ring, is *interfascicular* cambium. See CORK.

CAMBRIAN. The earliest period of the Palaeozoic era, and hence the start of the Phanerozoic Age (evident animals). Although Precambrian fossils are known, the earliest structural fossils are found in the Cambrian. Its fauna included trilobites, crustaceans, king crabs and eurypterids (see ARTHROPODA), annelids and brachiopods. Extended from 600–500 Myr BP.

CAMPYLOTROPOUS. (Of ovule) curved over so that funicle appears to be attached to the side, between chalaza and micropyle. Compare ANATROPOUS, ORTHOPTEROUS.

CANCER CELL. Cell which has escaped normal controls regulating its growth and division, producing a clone of dividing daughter cells which invade adjacent tissues and may interfere with their activities. Despite a normal oxygen uptake, cancer cells tend to use several times the normal glucose requirement. In vertebrates they produce lactic acid under aerobic conditions (termed *aerobic glycolysis*). This places a burden on the liver, which must use ATP to get rid of lactate. Cancer cells that proliferate but stay together form benign tumours; those that not only proliferate but also shed cells, e.g. via the blood or lymphatic system (metastasis) to form colonies elsewhere form malignant tumours, and cancer generally refers to a disease resulting from either. Among these, *carcinomas* are malignant tumours of epithelial cells. *Teratocarcinomas* are carcinomas that can be cultured in vitro and serially grafted to other hosts (see TERATOMA); *sarcomas* are cancers of connective tissue; *myelomas* are malignant tumours of bone marrow. There are some viral cancers (see VIRUS, ONCOGENE); others are caused by somatic mutations, perhaps affecting genes controlling cell division and growth. Some cancers may be regarded as genetically deviant cell lines that create selection forces within the multicellular organism. A notable property of cancer cells in culture is their loss of CONTACT INHIBITION. Some cancers are now thought to arise through a cell's lack of suppressor genes inhibiting growth. See CARCINOGEN.

CANINE TOOTH. Dog- or eye-tooth of mammals; usually conical and pointed, one on each side of upper and lower jaws between incisors and premolars. Missing or reduced in many rodents and ungulates, they are used for puncturing flesh, threat, etc. Sometimes enlarged as tusks (e.g. in wild boar).

CAPACITATION. Final stage in maturation of mammalian spermatozoa, without which they cannot engage in fertilization. Generally occurs in female tract (sometimes *in vitro*) where substances, perhaps secreted by the ovary or by the uterine lining, must be encountered for the sperm to undergo the acrosome reaction (see ACROSOME). Sperm do 'wait' at specific points on the uterus wall, and may be capacitated then.

CAPILLARY. (1) (Of blood system) an endothelial tube, one cell thick and 5–20 μm in internal diameter, on a BASAL LAMINA, and linking a narrow metarteriole to a venule. Permits exchange of water and solutes between blood plasma and tissue fluid (hence called exchange vessels). Their walls lack smooth muscle and connective tissue, and their permeability depends on the junctions between the endothelial cells. Three main types: (i) *continuous capillaries* (e.g. in muscle), where just one endothelial cell with overlapping ends forms the whole tightly sealed structure; (ii) *fenestrated capillaries* (as in intestine, endocrine glands), where pores through the cell are closed by just a cell membrane diaphragm, offering little resistance to small solute molecules; or (e.g. in glomeruli) pores occur between the adjacent cells, the basal lamina alone restricting solute passage; (iii) *sinusoids*, discontinuous capillaries (as in liver and spleen), where complete gaps occur between fenestrated endothelial cells. These incomplete capillaries have the highest permeability and largest diameters of the three, proteins passing through, although few blood cells do. *Precapillary sphincters* at the junctions of capillaries and metarterioles can slow or shut off blood flow in response to pH, carbon dioxide, oxygen, temperature, dilator and constrictor agents (e.g. ADRENALINE, NORADRENALINE). Blood pressure squeezes water and solutes from plasma across capillaries, forming the tissue fluid bathing body cells. Blood cells and plasma proteins are retained on grounds of size, the latter causing the relatively low WATER POTENTIAL of plasma at the venule end of a capillary, returning water to the blood. Solutes diffusing across the endothelium include oxygen, glucose, amino acids and salts (all outwards), carbon dioxide and metabolic wastes (e.g. urea in liver) inwards. Capillaries are absent from animals with open blood systems. (2) (Of the LYMPHATIC SYSTEM) structurally similar to blood capillaries, but blind-ending and with non-return valves, draining off surplus water from the tissue fluid. See BLOOD, INFLAMMATION.

CAPILLITIUM. (Bot.) (1) Tubular protoplasmic threads in fruiting bodies of Myxomycota (slime moulds), assisting discharge of spores in some species by their movements in response to changes in humidity. See ELATERS. (2) Sterile hyphae in the fruiting bodies of certain fungi, e.g. puff-balls.

CAPITULUM. (Bot.) (1) In flowering plants, inflorescence composed of dense aggregation of sessile flowers. (2) In the Sphagnidae (BRYOPSIDA), a dense tuft of branches at the apex of the gametophyte.

CAPPING. (1) See RNA CAPPING. (2) *Cell capping*. Process by which antibodies or other membrane components are attached by cross-linking ligands (see LECTIN) to cell-surface antigens and then swept along the surface to one end (cap) of a motile cell (commonly the rear) where they may be ingested by endocytosis. Unlinked membrane components diffuse fast enough in the membrane to avoid being swept back.

CAPSID. Coat of virus particle, composed of one or a few protein species whose molecules (capsomeres) are arranged in a highly ordered fashion. See VIRUS, BACTERIOPHAGE.

CAPSOMERE. See CAPSID.

CAPSULE. (Bot.) (1) In flowering plants, dry indehiscent fruit developed from a compound ovary; opening to liberate seeds in various ways, e.g. by longitudinal splitting from apex to base, separated parts being known as *valves* (e.g. iris); by formation of pores near top of fruit (e.g. snapdragon) or in the *pyxidium*, by detachment of a lid following equatorial dehiscence (e.g. scarlet pimpernel). (2) In liverworts and mosses, organ within which spores are formed. (3) In some kinds of bacteria, a gelatinous envelope surrounding the cell wall. (Zool.) Connective tissue coat of an organ, providing mechanical support.

CARAPACE. (1) Bony plates, often fused, beneath the horny scutes of the chelonian dorsal skin (turtles, tortoises). See PLASTRON. (2) Dorsal skin fold of many crustaceans arising from posterior border of head and reaching to varying extent over trunk somites. May enclose whole body (ostracods), the thorax (malacostracans), or be absent altogether (e.g. copepods). May enclose chamber in which gills are housed, embryos protected, etc.

CARBOHYDRATE. The class of organic compounds with the approximate empirical formula $C_x(H_2O)_y$, (i.e. literally 'hydrated carbon'), where $y = x$ (monosaccharides) or $y = x - [n - 1]$ (di- oligo- and polysaccharides) where n is the number of monomer units in the molecule. Of enormous biological importance both structurally and as energy stores. Sometimes atoms of nitrogen and other elements are also present (e.g. acetylglucosamine). They include CELLULOSE, CHITIN, GLUCOSE, GLYCOGEN, RIBOSE, STARCH and SUCROSE, but also occur as components of GLYCOLIPIDS and GLYCOPROTEINS.

CARBON CYCLE. The constant recycling of carbon atoms between inorganic (CO_2, carbonates) and organic sources. Both abiotic factors (e.g. rock weathering) and biotic factors are involved. The major carbon-fixing process is PHOTOSYNTHESIS, during which autotrophs incorporate CO_2 enzymatically into organic compounds. Heterotrophs (herbivores, secondary consumers, decay organisms) use organic products of photosynthesis for their own metabolic processes; but respiration by almost all organisms releases CO_2 for re-fixation. Fossil fuels (coal, oil) are being burnt at an increasing rate, contributing to the level of atmospheric CO_2. See FOOD WEB, ECOSYSTEM, GREENHOUSE EFFECT.

CARBONIC ANHYDRASE. Vertebrate enzyme of red blood cells, brush borders of kidney proximal convoluted tubule, and other body cells. Essential in catalysing the reaction: $CO_2 + H_2O \leftrightarrow H^+ + HCO_3^-$ (reversibly under different blood pH conditions in lungs and tissues), speeding CO_2 transport. Also involved in blood pH regulation by kidney.

CARBONIFEROUS. A Palaeozoic period, lasting from 345–280 Myr BP. Notable for its coal measures, with *Lepidodendron* and *Calamites* among the dominant plants; and for thick limestone deposits rich in brachiopods. The present-day continents were in greater contact than today, but PANGAEA had yet to form. Amphibians radiated during it, and reptiles appeared in its lower deposits.

CARBOXYLASE. Enzyme fixing carbon dioxide or transferring COO^- groups. Important carboxylases occur in both respiration and photosynthesis.

CARBOXYPEPTIDASE (CARBOXYPOLYPEPTIDASE). Pancreatic enzyme, digesting peptides to amino acids.

CARBOXYSOME. Structure in some bacteria (e.g. chemoautotrophic *Thiobacillus*), housing the CO_2-fixing enzyme ribulose-1,5, bisphosphate carboxylase.

CARCINOGEN. Any factor resulting in transformation of a normal cell into a CANCER CELL. The AMES TEST assesses carcinogenicity of a substance. Gut bacteria and other fermenting organisms often produce carcinogens as by-products, many of them glycosides (sugar-containing). When bacterial glycosidase cleaves the sugar group, they become mutagenic and potentially carcinogenic. Red wine and tea appear more carcinogenic than white wine and coffee. Ultraviolet and X-radiation and mustard gas are classic carcinogens, their effects usually being attributable to mutation. See MUTAGEN.

CARCINOMA. See CANCER CELL.

CARDIAC. Of the HEART; hence cardiac cycle (see HEART CYCLE), CARDIAC MUSCLE, cardiac sphincter (at the junction of oesophagus and stomach, near the heart).

CARDIAC CENTRES. See CARDIO-ACCELERATORY CENTRE.

CARDIAC MUSCLE. One of three vertebrate MUSCLE types; restricted to the heart walls. Striated, and normally involuntary. Myogenic (see PACEMAKER). Most obvious structural distinctions from skeletal muscle are its anastomosing (branching and rejoining) fibres, and the periodic irregularly thickened sarcolemma, forming *intercalated discs* which appear dark in most stained light microscope preparations. Unlike skeletal muscle, cardiac muscle tissue is not a multinucleate syncytium; each fibre is uninucleate and limited by its sarcolemma. Cardiac muscle tissue has a longer REFRACTORY PERIOD than skeletal muscle (see MUSCLE CONTRACTION). Both pacemaker and accompanying Purkinje fibres are modified cardiac tissue, but with neurone-like properties.

CARDINAL VEINS. Paired veins dorsal to the gut of fish and tetrapod embryos, taking blood toward the heart from the head/front limb region (anterior cardinals) to join posterior cardinals (from trunk), forming a common cardinal (Cuvierian) duct which enters the sinus venosus. Replaced by venae cavae in adult tetrapods.

CARDIO-ACCELERATORY AND CARDIO-INHIBITORY CENTRES (CARDIAC CENTRES). Association centres in medulla oblongata (see BRAIN-STEM) of homeothermic vertebrates, with reciprocal effects on heart rate. The former employs sympathetic nerves, the latter parasympathetic (vagus). Regulated by hypothalamus and cerebrum. Adjustments of heart rate involve BARORECEPTORS. See CAROTID SINUS, PACEMAKER, VASOMOTOR CENTRE.

CARINATE. Of those birds (the majority apart from RATITES) with a keel (*carina*) on the sternum. The group so formed is not now regarded as more than a GRADE.

CARNASSIAL TEETH. Modified last premolars in each upper half-jaw and corresponding first lower molars of carnivorous mammals. Between them they shear and slice, e.g. tendons and bones, when jaw closes.

CARNIVORA. Order including all living carnivorous mammals. Fossil CREODONTS, also carnivorous, form a separate order. Two suborders: FISSIPEDIA (dogs, cats, weasels, civets), and PINNIPEDIA (seals, walrus). Canine and carnassial teeth and retractile claws usually present.

CARNIVORE. Any meat-eater. Sometimes indicates a member of the CARNIVORA.

CAROTENE. Yellow or orange pigments found in the chloroplasts of plant cells. See CAROTENOIDS.

CAROTENOIDS. Group of yellow, orange and red plant pigments located in chloroplasts and plastids in parts of plant where chlorophyll is absent (e.g. carrot roots, many flowers); also in photosynthetic lamellae of blue-green algae (Cyanobacteria), in some bacteria and some fungi. Increase in concentration in many ripening fruits, e.g. tomato. Not essential to photosynthesis, but serve as accessory pigments, absorbing photons of different energy and passing this energy on to chlorophyll. Protect cells from photochemical damage in several photosynthetic and non-photosynthetic bacteria. Long-chain compounds (tetraterpenes), they include CAROTENES (α, β, etc.) and xanthophylls (lutein, flavoxanthin, fucoxanthin, zeaxanthin, etc.). Dietary carotene is converted to VITAMIN A in the vertebrate liver.

CAROTID ARTERY. Major paired vertebrate artery, one on each side of neck, supplying oxygenated blood to head from heart. Derived from third AORTIC ARCH. Each common carotid branches into internal and external carotids; their origins from the aorta vary.

CAROTID BODY. Small neurovascular structure near branch of internal and external carotids (near carotid sinus); supplied by vagus and glossopharyngeal nerves (see CRANIAL NERVES). Sensitive to oxygen content of blood, assisting in homeostatic reflex control of ventilation. May also assist carotid sinus reflexes.

CAROTID SINUS. Small swelling in internal carotid artery (therefore paired) in whose walls lie BARORECEPTORS innervated by the glosso-pharyngeal nerve (cranial nerve IX). Increase in arterial pressure and sinus stimulation causes reflex homeostatic drop in heart rate and vasodilation, involving the cardio-inhibitory and vasomotor centres of the BRAINSTEM.

CARPAL BONES. Bones of proximal part of hand (roughly the wrist) of vertebrates. Compact group of primitively 10–12 bones, reduced to 8 in man. Articulate with radius and ulna on proximal side, and with metacarpals on distal side. See PENTADACTYL LIMB.

CARPEL. Female reproductive organ (megasporophyll) of flowering plants. Consists of ovary containing one or more OVULES (which become seeds after fertilization), and a STIGMA, a receptive surface for pollen grains. Often borne at apex of a stalk, the STYLE. See FLOWER.

CARPELLATE. See PISTILLATE.

CARPOGONIUM. Female sex organ of red algae (Rhodophyta). Consists of swollen basal portion containing the egg, and an elongated ter-minal projection (trichogyne) receiving the microgamete.

CARPOSPORE. In red algae (Rhodophyta), the single diploid protoplast found within a containing cell (the carposporangium). Formed after fertilization and borne at the end of an outgrowth of the mature car-pogonium.

CARPUS. Region of vertebrate fore-limb containing carpal bones. Approximates to wrist in man.

CARRAGEENAN. Complex phycocolloid occurring in cell walls of some red algae (Rhodophyta).

CARRIER. (1) An individual HETEROZYGOUS for a recessive character and who does not therefore express it, but half of whose gametes would normally contain the allele for the character (sex linkage excepted). (2) An individual infected with a transmissible pathogen and who may or may not suffer from the disease.

CARRIER MOLECULE. See PERMEASE.

CARTILAGE. With BONE, the most important vertebrate skeletal con-nective tissue. Cells (chondroblasts) derive from mesenchyme and become *chondrocytes* when surrounded within lacunae by the ground substance they secrete. This amorphous matrix (chondrin) contains glycoproteins, basophilic chondroitin and fine collagen fibres, varying proportions of which determine whether it is hyaline (gristle), elastic or fibrocartilage. The surface of cartilage is surrounded by irregular connective tissue forming the perichondrium. Growth may be inter-stitial (endogenous) resulting from chondrocyte division and matrix deposition within existing cartilage; or appositional (exogenous) resulting from activity of deeper cells of the perichondrium. Lacks blood vessels or nerves. Cartilage is more compressible than bone and in the form of intercostal cartilage absorbs stresses generated throughout the vertebral column during locomotion, lifting, etc.; costal cartilage caps the articulating bone surfaces of JOINTS. The trachea is kept open by rings of hyaline cartilage; the pinnae of ears and auditory tubes contain elastic cartilage. In some kinds of OSSIFICATION cartilage is destroyed and replaced by bone. The CHONDRICHTHYES have entirely cartilaginous skeletons.

CARTILAGE BONE. See OSSIFICATION.

CARUNCLE. Warty outgrowth on seeds of a few flowering plants, e.g. castor oil; obscures MICROPYLE.

CARYOPSIS. A simple, dry, single-seeded indehiscent fruit. An ACHENE with ovary wall (pericarp) firmly united with seed coat (testa). Characteristic of grasses (Fam. Poaceae).

CASCADE. Biological process by which progressive amplification of a signal via a sequence of biochemical/physiological events results in a very localized response. Such a sequence might involve a hormone or other ligand binding to a membrane receptor site, activation of membrane adenylate cyclase producing many cAMP molecules, each activating many kinase molecules, which in turn activate many enzyme molecules, each producing quantities of product. Activation of COMPLEMENT, BLOOD CLOTTING, FIBRINOLYSIS, RHODOPSIN activity and embryonic acquisition of POSITIONAL INFORMATION all result from cascades. SERINE PROTEASES are often involved.

CASEIN. Conjugated milk protein. A phosphate ester of serine residues. RENNIN and calcium precipitate it to produce curd; also a major component of cheese.

CASTE. In EUSOCIAL insects, a structurally and functionally specialized individual: a MORPH. Caste determination may depend upon the state of ploidy (e.g. haploid bees, ants and wasps are male), or a combination of ploidy (the number of haploid chromosome sets) and environmental factors (e.g. worker and queen bees are diploid and female, but only queens are fed royal jelly as larvae); in lower termites at least it appears to be non-genetic, pheromones produced by king and queen controlling differentiation of caste. Hymenopteran castes are: queen, worker (some ant species having soldier and non-soldier sub-castes of worker) and drone; termite castes include: primary reproductives (king, queen), supplementary reproductives, workers and soldiers. See POLYMORPHISM.

CATABOLISM. The sum of enzymatic breakdown processes, such as digestion and respiration in an organism. Opposite of ANABOLISM.

CATABOLITE. Metabolite broken down enzymatically.

CATABOLITE REPRESSION. Suppression by a fuel molecule, or one of its breakdown products, of synthesis of inducible enzymes which would make use of alternative fuel molecules in the cell. Glucose represses production of galactosidase and some respiratory enzymes (the glucose effect) in bacteria. This involves GENE REPRESSION, a glucose breakdown product combining with the cell's cyclic AMP (cAMP), reducing the amount available for transcription of the operon which includes the galactosides gene. See GENE REGULATION, PASTEUR EFFECT.

CATALASE. Haem enzyme of PEROXISOMES of many eukaryotic cells. Converts hydrogen peroxide, produced by certain dehydrogenases and oxidases, to water and oxygen. Used commercially in converting latex to foam rubber and in removing hydrogen peroxide from food.

CATALYST. Substance speeding up a chemical reaction and altering its equilibrium point. Biological catalysts are ENZYMES.

CATAPHYLL. Small scale-like leaf in flowering plants, often serving for protection.

CATARRHINE. Old World monkeys and apes, and all humans; i.e. all cercopithecoids and hominoids. The Anthropoidea other than PLATYRRHINES. Characterized by narrow nasal septum and by menstrual cycle. Never have prehensile tail.

CATECHOLAMINES. Tyrosine derivatives; biologically important ones include the hormones and NEUROTRANSMITTERS adrenaline, noradrenaline and dopamine.

CATERPILLAR. Larval stage of Lepidoptera, Mecoptera and some Hymenoptera, bearing abdominal prolegs in addition to thoracic legs. Generally poorly sclerotized and inactive, living close to food.

CATHEPSINS. A group of proteolytic enzymes occurring in LYSOSOMES.

CATHETER. Tube, often plastic, inserted into gut, blood vessels, etc. for withdrawal/introduction of material. Balloon catheters have an inflatable tip and may be used to dilate blocked vessels (e.g. the coronary artery).

CATION. A positively charged ion.

CAUDAL. (Of the tail) caudal vertebrae are tail vertebrae, the caudal fin of a fish is its tail fin.

CAULINE. Belonging to the stem, or arising from it.

CAULINE BUNDLE. A VASCULAR BUNDLE forming part of the stem tissue.

CAVITATION. Occurrence of air pockets and/or bubbles in xylem vessels when tension exerted on the water column exceeds that enabling cohesion. It may occur during water stress. An alternative route for the TRANSPIRATION stream would be needed, bypassing the blockage; however, air pockets so formed may be squeezed out again by ROOT PRESSURE.

cDNA (COMPLEMENTARY DNA). DNA complementary to RNA and produced by REVERSE TRANSCRIPTASE activity. Initially single-stranded, can be converted to double-stranded cDNA by DNA POLY-MERASE activity. cDNA complementary to mRNA lacks INTRON sequences, useful when cloning functional DNA.

CELL. Mass of protoplasm made discrete by an enveloping plasma membrane (plasmalemma). Any cell wall material is, strictly speaking, extracellular (e.g. in most plants and fungi); but distinctions between intracellular and extracellular may be arbitrary (see GLYCOCALYX).

The two basic types of cell architecture are those of PROKARYOTES and EUKARYOTES. See Fig. 3. In the former, cells consist entirely of cytoplasm (lacking nuclei); in the latter, cells have (or had) in addition one or more nuclei. Eukaryotic cells have greater variety of organelles, many enclosed in one or more membranes (see CELL MEMBRANES). They are further distinguished by the presence of distinctive proteins, particularly ACTIN, MYOSIN, TUBULIN and HISTONE, that have very significant uses and are entirely absent from prokaryotic cells. Actin is paramount in the structure of the eukaryotic CYTO-SKELETON; tubulin is fundamental in cilium and flagellum structure, and in mitotic and meiotic spindles – none of these being found in prokaryotes, whose flagella are rigid and of a completely different structure. These and other features indicate how similar even such apparently dissimilar cells as those of plants and animals are when compared with those of prokaryotes (bacteria and blue-green algae). Basic eukaryotic cell architecture is elaborated upon in many ways, notably by fungi, where true cells are commonly absent in much of the vegetative body, organization being COENOCYTIC. A similar multinucleate situation, without intervening cell membranes, arises where eukaryotic cells fuse to form a SYNCYTIUM. Both may be termed ACELLULAR. The plasmodesmata uniting many plant cells may be regarded as producing an intermediate condition. See MULTI-CELLULARITY, ORIGIN OF LIFE.

CELL BODY (PERIKARYON). Region of a neurone containing the nucleus and its surrounding cytoplasm. Generally swollen compared with rest of cell. Some ganglia consist of aggregations of cell bodies.

CELL CAPPING. See CAPPING.

CELL CENTRE (CENTROSOME). Region of animal cell cytoplasm adjacent to nucleus at which interphase microtubules appear to terminate. In its centre lies a CENTRIOLE pair.

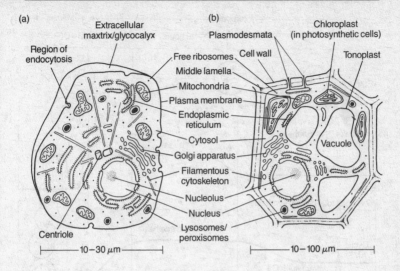

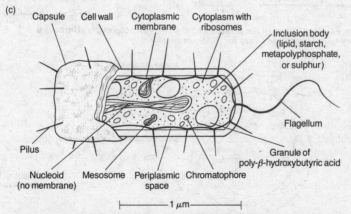

Fig. 3a & b. *Generalized eukaryotic structure (a) of animal cell and (b) of plant cell as seen in low-power electron microscopy. Small dots are ribosomes, both free in the cytosol and attached to endoplasmic reticulum. (c) Generalized prokaryotic (bacterial) cell structure.*

CELL CYCLE. Events occurring between one mitotic cell division (cytokinesis) and the next division of one (or both) of its daughter cells.

CELL DIVISION. Process by which a cell divides into two. (a) *Prokaryotic*: one event achieves separation of both DNA and cytoplasm into daughter cells. Since the two sister chromosomes are attached separately to the cell membrane they can become separated by the cleavage furrow formed between them as the cell membrane invaginates.

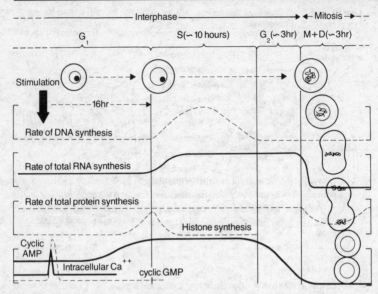

Fig. 4a. *Breakdown of the phases of the eukaryotic cell cycle. Times are only approximate and vary for different systems.*

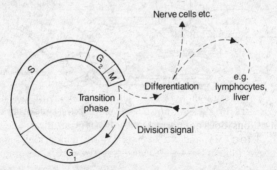

Fig. 4b. *Indication of the cell cycle in relation to differentiation. A few cell types can avoid the differentiation and ageing pathway by re-entering the division cycle; undifferentiated cells in embryos often do so.*

Fission occurs as membrane intuckings fuse. No microtubules occur in prokaryotes, so there is no mitosis or meiosis; (b) *eukaryotic*: nuclear and cytoplasmic divisions are achieved by separate mechanisms. In higher eukaryotes the nuclear membrane breaks down and chromosomes attach to microtubules of the spindle by their KINETOCHORES; cytoplasmic division (*cytokinesis*) usually starts in mitotic anaphase and proceeds by a furrowing of the plasmalemma in the plane of the metaphase plate, achieved by a contractile ring of ACTIN filaments. Fusion of the invaginating plasmalemma then occurs. In plant cells

Low Speed (1000 G, 10 min)	Medium Speed (20 000 G, 20 min)	High Speed (80 000 G, 1 h)	Very High Speed (150 000 G, 3 h)
nuclei	mitochondria	microsomes	ribosomes
whole cells	lysosomes	(rough & smooth ER)	viruses
cytoskeletons	peroxisomes	small vesicles	large macro-molecules

Table 3. Centrifugation of cell components.

with walls the new wall is built upon a CELL PLATE. Golgi vesicles travelling towards it on microtubules deposit their wall precursor molecules, extending the plate to the cell membranes and pinching off the cell into two. See MITOSIS and MEIOSIS for nuclear division.

CELL FRACTIONATION. Process whereby cells are first appropriately buffered (often in sucrose solution) and then disrupted (by osmotic shock, sonic vibration, maceration or grinding with fine glass, sand, etc.); the cell fragments are then spun in a refrigerated centrifuge. Different cell components descend to the bottom of the centrifuge tube at different speeds, and these can be increased progressively. Forces generated may be 500 000 times that of gravity (G). The G-forces and times required to spin down different cell constituents are shown in Table 3.

CELL FUSION. Process involving fusion of plasma membranes of two cells to form one resultant cell. Naturally occurring cell fusion may or may not result in hybridization (unity of genomes). Fusion of MYOBLASTS in skeletal muscle development, and other syncytial organizations, does not normally involve hybridization. The processes of PLASMOGAMY and KARYOGAMY are temporally separated in those fungal life cycles where a DIKARYON occurs at some stage. In fertilization, separation of plasmogamy and karyogamy is usually brief. Artificial cell fusion is often achieved by treatment with inactivated viruses, or a glycol. The heterokaryon, with its separate nuclei intact, may then divide, in which case all chromosomes may end up within a single nuclear membrane. Irregular chromosome loss may permit CHROMOSOME MAPPING in tissue culture, as with mouse-human hybrid cells. Techniques resulting in fusion and hybridization of normal and tumour B-CELLS have yielded HYBRIDOMAS capable of generating monoclonal antibodies on a commercial or clinical scale. Protoplasts resulting from enzymic digestion of plant cell walls can be encouraged to fuse, and may generate heterokaryons or even fusion hybrids. Appropriate horticulture can generate somatic hybrid plants between species that would not normally hybridize. As with mouse-human somatic hybrids, chromosome loss often prevents a genetically stable product.

CELL HYBRIDIZATION. See under CELL FUSION.

CELL LINEAGE. The cellular ancestry of any cell in a multicellular organism; in most cases starts with the zygote from which all an organism's cells are derived by cell division but often implies a more or less restricted CLONE of cells forming a functional subset of the whole cell population. See WEISMANN.

CELL LOCOMOTION. There are various methods by which cells move, those of prokaryotes having apparently little in common with those of eukaryotes. For the latter, most mechanisms seem to involve protein tubules or filaments sliding past one another and generating force. The details of how force is transmitted to the substratum remain largely unknown.

(1) *Bacterial:* H^+ gradients across the inner cell membrane provide the motive force for rotation of the FLAGELLUM, whose fixed helix of protein subunits permits clockwise and counter-clockwise rotations, like a corkscrew. This involves an extraordinary 'wheel-like' rotor in the inner membrane, and cylindrical fixed bearing in the outer membrane. Reversal of flagellar rotation alters the behaviour of the cell.

(2) *Eukaryotic.* (*a*) *Ciliary/flagellar:* see CILIUM for structure. Paired outer microtubules slide over adjacent pairs in response to forces generated by dynein arms coupled to their ATPase activity. Radial spokes and the inner sheath apparently convert this sliding to bending of the organelle. The axoneme can beat without the cell membrane sheath around it. The dynein arms probably act in an equivalent fashion to myosin heads during MUSCLE CONTRACTION and make contact with adjacent microtubule pairs during their power stroke. Control of ciliary/flagellar beat appears to be independent of Ca^{++} flux, but may be dependent upon signal relay via proteins of the actual structure. However, reversal of ciliary beat in some ciliates is associated with membrane voltage change brought about by Ca^{++} influx. It is still uncertain how waves of ciliary beating in cell surfaces are coordinated. (*b*) *'Fibroblastic' crawling:* the leading edge of a cell engaged in this method of locomotion, characteristic of fibroblasts, extends forwards and, after attachment to the substratum, pulls the rest of the cell forward by contraction of actin microfilaments under influence of myosin. Typical features associated with this method are lamellipodia and microspikes (see CELL MEMBRANES), which both pass backwards in waves along the upper cell surface ('ruffling'), typically when the anterior of the cell has failed to attach to the substratum. Molecular mechanisms involved are not clear, but it seems that random endocytosis of plasmalemma and its subsequent restricted exocytosis (incorporating the membrane pieces) at the

anterior of the cell generates a circulation of membrane akin to movement of tank caterpillar tracks. The protein FIBRONECTIN is involved in fibroblast crawling. (c) '*Amoeboid*' (*pseudopodial*): the cells outermost layer is gel-like (plasmagel) while the core is a fluid sol (plasmasol). It is possible that contraction of the thick cortical plasmagel squeezes the plasmasol and generates pseudopodial extensions of the cell, at the tips of which sol-to-gel transformation occurs. Gel-to-sol changes accompany this elsewhere in the cell, e.g. as a pseudopod retracts. Just how these CYTOPLASMIC STREAMING events are coupled to locomotion is not clear, but motive force must act against regions where the cell adheres to its substratum (see FIBRONECTIN). Apparently, surfaces of large amoebae are relatively permanent, undergoing folding and unfolding to accommodate pseudopod extension and retraction. ACTIN is implicated in the process. Characteristic of amoebae, macrophages. See CAPPING, DESMID.

CELL-MEDIATED IMMUNITY. See IMMUNITY.

CELL MEMBRANE (PLASMA MEMBRANE, PLASMALEMMA). The membrane surrounding any cell. See CELL MEMBRANES.

CELL MEMBRANES. Cells may have a wide variety of membranes (often called 'unit' membranes) varying from 5–10 nm in thickness; but all have a plasma membrane (plasmalemma), the outer limit of the cell proper, which is generally quite distinct from any cell wall material present (which is extracellular).

Major membrane functions include: restriction and control of movements of molecules (e.g. holding the cell together) enabling scarce metabolites to reach local concentrations sufficient to enhance enzyme-substrate interactions; to act as platforms for the spatial organization of enzymes and their cofactors, holding otherwise scattered molecules in functional contact; and separation and localization of incompatible reactions. Many eukaryotic organelles have one or two membranes around them, chloroplasts having yet a third system within. The currently accepted structure of most cell membranes is that proposed in the *fluid mosaic model*, the evidence coming from X-ray crystallography, freeze-fracture and freeze-etching electron microscopy (see MICROSCOPE), radiolabelling, electron spin resonance spectroscopy and fluorescence depolarization. The last two involve insertion of *molecular probes* with particular spectroscopic features adding peaks or troughs to the lipid spectrum.

In this model an outer and an inner phospholipid monolayer (major components phosphatidyl ethanolamine and lecithin) lie with their polar phosphate heads in the direction of the water which the bilayer thus separates. Specific, and different, proteins lie in one or other layer or traverse the bilayer, making the membrane *asymmetric*.

The whole structure has fluid properties resulting from rapid lateral movement of most of its molecular components through

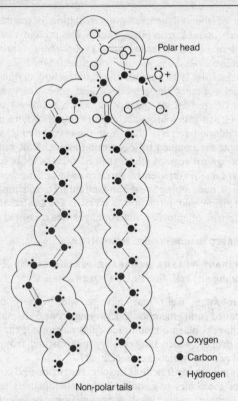

Polar head

○ Oxygen
● Carbon
· Hydrogen

Non-polar tails

Fig. 5a. A phospholipid molecule (phosphatidylserine).

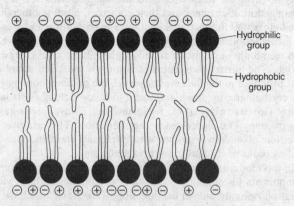

Hydrophilic group

Hydrophobic group

Fig. 5b. A phospholipid bilayer.

thermal agitation (1 μm.s^{-1} for lipids, 10 μm.min^{-1} for proteins). Thus fused mouse and human cells, each with differently labelled

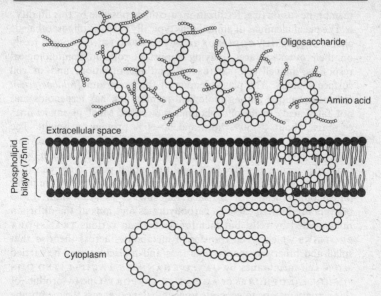

Fig. 5c. *Diagram showing the position of a glycophorin molecule (a glycoprotein) which penetrates the membrane.*

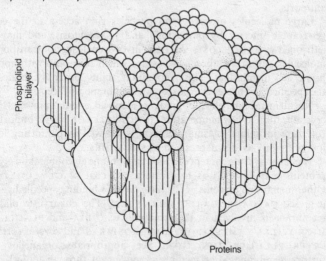

Fig. 5d. *Diagram indicating the fluid-mosaic model of cell membranes.*

membrane proteins, exhibit rapid mixing of labels over the entire cell surface. The ionosphere gramicidin functions only when the two halves of the molecule, one in each half of the bilayer, come together – which they do in a quantized way, indicating membrane fluidity. Endocytosis, exocytosis and other processes involving membrane fusion (e.g. fertilization) are made possible by this fluidity.

The plasmalemmas of animal cells typically have the oligosaccharide chains of their GLYCOLIPIDS and GLYCOPROTEINS exposed freely on their outer surfaces, playing important roles in immunological responses, in cell–cell ADHESION and identification, and in cell surface changes. The plasmalemma of the bacterium *Halobacterium halobium*, unique among biological membranes, has terpenoids and not fatty acids in its phospholipid molecules. Most plasmalemmas comprise about 40–50% lipid and 50–60% protein by weight. The phospholipid bilayer has a non-polar hydrophobic interior, preventing passage of most polar and all charged molecules. Small non-polar molecules readily dissolve in it, and uncharged polar molecules (e.g. H_2O) can also diffuse rapidly across it, possibly assisted by the polar phospholipid heads, or by such ionospheres as gramicidin. Lipid bilayers are impermeable to carbohydrates and ions at the diffusion rates needed by cells; but membranes contain various TRANSPORT PROTEINS which speed transfer of metabolites across them so that small and otherwise inaccessible ions and molecules may be carried across cell membranes by GATED CHANNELS, FACILITATED DIF-FUSION, IONOPHORES or ACTIVE TRANSPORT. One by-product of this activity may be to generate ionic imbalances across the membrane which may be used to power ATP synthesis, or drive symports and antiports.

Large molecules or even solid particles gain access to the cell's geographic interior by pinocytosis and phagocytosis, and may be jettisoned by EXOCYTOSIS. All these involve enclosure of transported molecules within membranous vesicles which fuse only with appropriate cell membranes. This recognition ability probably resides in the specificity of proteins exposed at a membrane surface.

Not all membrane phospholipids are identical, and this prevents their crystallization at low temperatures (CHOLESTEROL has an important role here in animal plasma membranes) as well as permitting local loss of fluidity as at synapses and DESMOSOMES.

The carbohydrate content of plasma membrane glycolipids and glyco-proteins may be such as to create a cell coat or GLYCOCALYX. Other membrane proteins act as *receptor sites* binding specific ligands (e.g. see CASCADE, GATED CHANNELS). The eukaryotic plasma membrane is involved in the structures of CILIA and FLAGELLA, MICROVILLI, LAMELLIPODIA, MICROSPIKES and several sorts of INTERCELLULAR JUNCTION. See appropriate organelles for further membranes. For membrane movement through the cell, see GOLGI APPARATUS, LYSOSOME.

CELL PLATE. (Bot.) 'Plate' of differentially staining material which appears at telophase in the PHRAGMOPLAST across the equatorial plane of the spindle. Believed to be forerunner of MIDDLE LAMELLA. See CELL WALL.

CELL THEORY. The theory, first proposed by Schwann in 1839, that organic structure originates through formation and differentiation of units, the cells, by whose divisions and associations the complex bodies of organisms are formed. Much of the original theory is now untenable. Schleiden's name is also associated with the theory. See VIRCHOW.

CELLULAR IMMUNITY AND CELLULAR RESPONSE. See IMMUNITY.

CELLULOSE. The most abundant organic polymer. A polysaccharide, occurring as the major structural cell wall material in the plant kingdom. Some fungi have it as a component of their hyphal walls, and it may occur in animal cell coats (see GLYCOCALYX). A long-chain polysaccharide of repeating *cellobiose* units, it may also be considered as a long chain of β[1, 4]-linked glucose units. Hydrogen bonding both within each molecule as well as between parallel molecules (producing crystalline *microfibrils*) gives cellulose its great tensile strength; but microfibrils can be loosened by lowered pH (an effect of AUXINS on the cell) allowing for wall extension in cell growth, when more cellulose may be laid down between existing microfibrils. With LIGNIN, it forms *lignocellulose*. The fibrous texture of cellulose is responsible for its use in textile industries (cotton, linen, artificial silk). See CELL WALL for cellulose distribution.

CELL WALL. Permeable extracellular coat of bacterial and plant cells, secreted by the protoplasm and closely investing it. The bacterial wall is a component of its *envelope* and contains either a thick layer of PEPTIDOGLYCAN, or rather little (see GRAM'S STAIN). MUCINS may also be present. Comparatively rigid, both these and the chitinous walls of fungal cells and hyphae provide mechanical support.

Walls of newly formed plant cells are at first very thin, thickening as cells assume their permanent characters. At plant cell division, PECTIC COMPOUNDS are laid down in the CELL PLATE across the equatorial plane of the division's spindle forming the MIDDLE LAMELLA, intercellular material cementing adjacent cells together. Each new cell lays down a *primary wall* consisting of CELLULOSE, HEMICELLULOSES and negatively charged pectins. *Hemicellulose* molecules bind by hydrogen bonds to cellulose microfibrils, cross-linking them. *Pectin* molecules, being negatively charged, bind cations such as calcium (Ca^{++}) and in so doing form a gel-like matrix filling the interstices between the cellulose microfibrils, holding them together. Glycoprotein molecules probably attach to pectins (see Fig. 6a). At full size, a cell may remain with just its primary wall (e.g. some forms of PARENCHYMA); in others, after cell growth is complete, a *secondary wall* is laid down inside the primary wall (see Fig. 6b). During deposition of these layers, certain small areas remain largely unthickened, forming PITS. Pits of ajacent walls usually

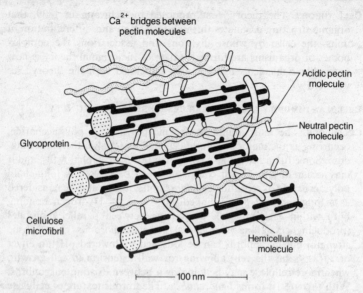

Fig. 6a. The relative arrangements of molecule types in a primary cell wall.

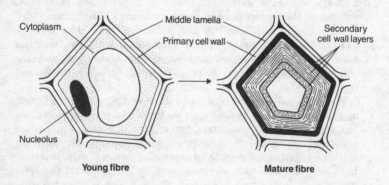

Fig. 6b. Secondary cell wall deposition by a phloem fibre cell, to show different wall layers.

coincide, so that in these areas protoplasts are separated only by the *pit membrane* on each side. Through pit membranes pass the majority of plasmodesmata, fine protoplasmic connections which are elements of the SYMPLAST. Some walls undergo further modification, as in cuticularization of epidermal cells and suberization of cork cells, making them impermeable to water. Lignification of fibres, vessels and tracheids gives them more strength and rigidity.

The plant cell wall limits cell growth (see AUXINS), is a barrier to digestion (especially when toughened by aromatic polymers, see TANNINS), and glues adjacent cells together, playing an important role in plant morphogenesis. Its stretch-resistance is a major contributory factor to a plant cell's WATER POTENTIAL. Cell walls can contain enzymes that can incompletely digest its polysaccharides, releasing oligosaccharides that can act like growth substances and serve in cell-cell signalling. See CELLULOSE, CHITIN.

CEMENT (CEMENTUM). Modified bone surrounding roots of vertebrate teeth (i.e. below gum), binding them to periodontal ligament by which tooth is attached to jaw. In some herbivorous mammals, occurs between folds of the tooth, forming part of the grinding surface.

CENOZOIC (CAINOZOIC). The present geological era; extends from about 65 Myr BP to the present. The 'age of mammals'. Its two periods, the Tertiary and the Quarternary, are sometimes regarded as eras in their own rights.

CENTRAL DOGMA. Proposal by F. H. C. Crick in 1958 that the flow of molecular information in biological systems is from DNA to RNA and then to protein. RNA viruses (e.g. RNA tumour viruses) have since been shown to transcribe single-stranded DNA from RNA templates by means of the enzyme reverse transcriptase, providing exceptions to the generalization. See cDNA.

CENTRAL MOTHER CELLS. Relatively large vacuolated cells in a subsurface position in the apical meristem of a plant shoot.

CENTRAL NERVOUS SYSTEM (CNS). A body of nervous tissue integrating animal sensory and motor functions and providing through-conduction pathways to transmit impulses rapidly, usually medially, along the body. In vertebrates it comprises the BRAIN and SPINAL CORD; in annelids and arthropods a pair of solid ventral nerve chains, each with segmental ganglia, and a pair of dorsal ganglia anteriorly serving as a 'brain', united to the nerve chains by commisures. Impulses travel to and from the CNS via peripheral nerves (vertebrate spinal nerves), while local reflex arcs (vertebrate spinal reflexes) produce adaptive responses to stimuli independently of higher centres (the brain), although these centres initiate and coordinate actions and store memory. See NERVOUS SYSTEM, SPINAL CORD.

CENTRIC DIATOM. A diatom which is radially symmetrical in VALVE VIEW.

CENTRIOLE. Organelle (probably of endosymbiotic origin) found in cells of those eukaryotic organisms which have cilia or flagella at some stage in their life cycle; hence absent from higher plants. Each comprises a hollow cylinder composed of nine sets of triplet microtubules held together by accessory proteins. Each is 300–500 nm long and 150 nm in diameter. Often functionally interconvertible with BASAL BODY. They occur at right angles to each other near the nucleus, separating at cell division amd organizing the spindle microtubules (which arise from material surrounding the centriole, but possibly in turn organized by it). Centrioles generally arise at right angles to existing centrioles. Normally an animal obtains its centrioles from the sperm cell at fertilization; rarely, an egg may form its own (see PARTHENOGENESIS). Centrioles possess their own DNA and appear to be self-replicating, and there may be a link between centriole replication and nuclear DNA replication. Similar or identical structures (BASAL BODIES), possibly functionally interconvertible, occur at the bases of cilia or flagella in cells which have these.

CENTROLECITHAL. Of eggs (typically insect) where yolk occupies centre of egg as a yolky core. See TELOLECITHAL.

CENTROMERE (SPINDLE ATTACHMENT). A chromosome region holding sister chromatids together until mitotic or second meiotic anaphase. The position of a centromere defines the ratio between the lengths of the two chromosome arms. Centromeres may be associated with REPETITIVE DNA sequences (not in the yeast *Saccharomyces cerevisae*) and centromeric DNA may be late-replicating. They either include or correspond to KINETOCHORES, which attach to the spindle fibres and by replicating at late metaphase allow the forces pulling sister chromatids apart to operate only if chromosomes are properly aligned. Normally one per chromosome; but chromasomes with 'diffuse' centromeres (e.g. those of many lepidopterans) permit spindle fibre attachment along the whole chromosome length. See ACENTRIC, ACROCENTRIC, METACENTRIC, TELOCENTRIC.

CENTROSOME. See CELL CENTRE.

CENTRUM. Bulky part of a vertebra, lying ventral to spinal cord. In function, as in development, replaces the notochord. Each is firm but flexible, attached to adjacent centra by collagen fibres.

CEPHALASPIDA (OSTEOSTRACI). Extinct group of monorhine vertebrates. See AGNATHA.

CEPHALIC INDEX. Measure of skull shape introduced by anatomist Retzius. It relates breadth as a percentage of length $((B/L) \times 100)$.

CEPHALIZATION. The tendency, during evolution of animals with an antero-posterior axis, for sense organs, feeding apparatus and nerve tissue to proliferate and enlarge at the anterior end, forming a head.

CEPHALOCHORDATA (ACRANIA). Subphylum of marine chordates characterized by persistence of notochord in adult, extending (unlike in vertebrates) to the tip of the snout. Metameric segmentation, dorsal hollow nerve cord, gill slits and post-anal tail also present. Amphioxus (*Branchiostoma*) is typical. Compare UROCHORDATA.

CEPHALOPODA. Most advanced class of the phylum MOLLUSCA.

All are aquatic, and most marine, possessing a well-developed head surrounded by a ring of prehensile tentacles; and a muscular siphon derived from the foot through which water is forced from the mantle cavity during locomotion. Primitively (e.g. *Nautilus* and extinct AM-MONITES) the animal inhabits the last chamber of an external spiral shell which also serves for buoyancy; in the cuttlefish *Sepia* the shell is internal, while in squids it is much reduced, and absent altogether in *Octopus*. The complexity of cephalopod eyes rivals that of vertebrates (and provides an example of convergent evolution), while the large brain enables powers of learning and shape recognition on a par with simple vertebrates. Much has to be learnt about cephalopod communication; some believe that cuttlefish employ their phenomenal powers of colour and pattern change to this effect.

CEPHALOTHORAX. Term indicating either fusion of, or indistinctness between, head and some or all anterior trunk (thoracic) segments in crustacean and arachnid arthropods.

CERCARIA. The last larval stage of flukes (Order Digenea); produced asexually by POLYEMBRYONY within preceding redia larva inside secondary host, often a snail, from which it emerges and swims with its tail to penetrate skin of primary host (e.g. man in *Schistosoma* causing bilharzia) or to encyst as a metacercaria awaiting ingestion by primary host.

CERCI. A pair of appendages, often sensory, at the end of the abdomen of some insects. Long in mayflies, short in cockroaches and earwigs (where they are curved).

CERCOPITHECOIDEA. Old World monkeys. See ANTHROPOIDEA.

CEREAL. Flowering plant of the family Graminae, whose seeds are used as human food, e.g. wheat, oats, barley, rye, maize, rice, sorghum.

CEREBELLUM. Enlargement of the hindbrain of vertebrates, anterior to the medulla oblongata. Coordinates posture (balance) during rest and activity through reflexes initiated by inputs mainly from the VESTIBULAR APPARATUS fed via acoustic regions of the medulla (the lower vertebrate ACOUSTICO-LATERALIS SYSTEM), and from the PROPRIOCEPTORS in muscles and tendons. In mammals, covered in a cortex of grey matter.

CEREBRAL CORTEX (PALLIUM). Layer of GREY MATTER rich in synapses lying atop white matter, covering cerebral hemispheres of amniote and some anamniote vertebrates. In advanced reptiles and all mammals a new association centre, the neopallium, appears in the cortex receiving sensory inputs from the brainstem and initiating actions via motor bundles of the pyramidal tract. Its evolving dominance in mammalian brain involves its reception of increasingly wide ranges of sensory information via the thalamus and the emergence of higher neural (i.e. mental) activities based upon these data. Folding of the cortical surface in mammals provides a large surface area for synaptic association.

CEREBRAL HEMISPHERES (CEREBRUM). Paired outpushings of vertebrate forebrain, originally olfactory in function, whose evolution has involved progressive movement of grey matter to its surface and an increasing role as an association and motor control centre. The CEREBRAL CORTEX dominates the mammalian brain both physically and functionally.

CEREBROSIDE. SPHINGOLIPIDS of the myelin sheaths of nerves, the commonest being *galactocerebrosides* with a polar head group containing D-galactose. Other tissues contain small amounts of glucose-containing cerebrosides.

CEREBROSPINAL FLUID (CSF). Fluid filling the hollow neural tube and subarachnoid space of vertebrates. Secreted continuously into ventricles of the brain by the choroid plexuses and reabsorbed by veins. Clear and colourless fluid, with some white blood cells, supplying nutrients. Serves as shock absorber for the central nervous system. About 125 cm^3 present in humans. See MENINGES.

CEREBRUM. See CEREBRAL HEMISPHERES.

CERVICAL. (Adj.) Of the neck; or CERVIX. Cervical vertebrae have reduced or absent ribs; almost all mammals (including giraffe) have seven.

CERVIX. Cylindrical neck of mammalian uterus, leading into vagina. Glands secrete mucus into vagina.

CESTODA. Tapeworms. Class of endoparasitic PLATYHELMINTHES lacking gut and absorbing digested food from host gut lumen across microtriches, minute folds of the surface epithelial cell membranes similar to microvilli. Tapeworms are unsegmented, but body sections (proglottides) budded off from head region (scolex) give segmented appearance. Sequentially hermaphrodite, young proglottides male but become female with age. Self-fertile. Life cycle involves primary and secondary hosts. Larva a six-hooked onchosphere egested in proglottis with faeces of primary host. Sense organs reduced.

CETACEA. Whales. Order of placental (eutherian) mammals. Entirely aquatic. Doubtful credont ancestry. Morphology convergent with ichthyosaurs, with a dorsal fin, forelimbs developed as flippers, and tail a powerful fluked swimming organ. Traces only of pelvic girdle. Subcutaneous fat (blubber) for thermal insulation. Dorsal blowhole connects with lungs. Includes Odontoceti (toothed whales, including porpoises and dolphins) and Mysticeti (whalebone whales). Earliest fossils from Eocene.

CHAETA. Chitinous bristle characteristic of oligochaete (where few) and polychaete (where many) annelid worms. In polychaetes they are borne on parapodia. Assist in contact with substratum during locomotion. See SETA.

CHAETOGNATHA. Arrow-worms. Small phylum of marine coelomate invertebrates, abundant in plankton. Hermaphrodite.

CHALAZA. (Bot.) Basal region of ovule, where the stalk (funiculus) unites with the INTEGUMENTS and the NUCELLUS. (Zool.) Of a bird's egg, the twisted strand of fibrous ALBUMEN; two are attached to the vitelline membrane, one each at opposite poles of the yolk, lying in the long axis of the egg. They stabilize the position of the yolk and early embryo in the albumen.

CHALONE. Substances (possibly glycoproteins) difficult to extract, but alleged to occur in mammalian tissues and having anti-mitotic effects of a self-regulatory kind dependent upon thickness of tissue producing it.

CHAMAEROPHYTES. Class of RAUNKIAER'S LIFE FORMS.

CHARACTER DISPLACEMENT. Evolutionary phenomenon whereby, it is believed, interspecific competition causes two closely related species to become more different in regions where their ranges overlap than in regions where they do not. Such differences are often anatomical, but may involve any aspect of phenotype. Few rigorously documented instances exist where such differences have been shown to be due to competition.

CHAROPHYTA. Stoneworts (from their characteristic incrustations with calcium carbonate). Division of the ALGAE. Occur in ponds and lakes, where they often form extensive underwater growths. Possess both chlorophylls *a* and *b* and store food as starch. Thallus multicellular, attached to substratum by rhizoids and bearing lateral branches in whorls. Cell walls composed of cellulose. Multicellular sex organs (antheridia and oogonia).

CHELA. The last joint of an arthropod limb, if it can be opposed to the joint preceding it so that the appendage is adapted for grasping, as in pincers of lobster and some CHELICERAE. Such a limb is termed *chelate*.

CHELICERAE. Paired, prehensile first appendages of CHELICERATA, contrasting with antennae of other groups. Often form CHELAE (when said to be *chelate*).

CHELICERATA. Probably natural assemblage containing those arthropods with chelicerae. Includes MEROSTOMATA and ARACHNIDA. No true head, but an anterior tagma termed the PROSOMA. Mandibles absent. Probably closely related to trilobites. See ARTHROPODA, MOUTHPARTS.

CHELONIA (TESTUDINES). Tortoises and turtles. Anapsid reptile order, with bony plates enclosing body or covered by epidermal horny plates. Shoulder and pelvic girdles uniquely within rib cage. Teeth absent.

CHEMIOSMOTIC THEORY (CHEMIOSMOTIC-COUPLING HYPOTHESIS). Hypothesis of P. Mitchell, now generally accepted, that chloroplasts and mitochondria require their appropriate membrane to be intact so that a proton gradient created across it by integral membrane pumps can be coupled to ATP synthesis as protons return across the membrane down their electrochemical gradient. See CHLOROPLAST, ELECTRON TRANSPORT SYSTEM, MITOCHONDRION.

CHEMOAUTOTROPHIC (CHEMOSYNTHETIC). Organism obtaining energy from a simple inorganic reaction, the nature of which varies according to the species; e.g. oxidation of hydrogen sulphide to sulphur by *Thiobacillus*. Several autotrophic bacteria are chemoautotrophs. See CHEMOTROPHIC.

CHEMOHETEROTROPHIC. See CHEMOTROPHIC.

CHEMORECEPTOR. RECEPTOR responding to chemical aspects of internal or external environment. Taste and olfaction are chemosenses. See CAROTID BODY.

CHEMOSYNTHETIC. See CHEMOAUTOTROPHIC.

CHEMOTAXIS. TAXIS, along a chemical gradient.

CHEMOTROPHIC. Of organisms obtaining energy by chemical reactions independent of light. Reductants obtained from the environment may be inorganic (CHEMOAUTOTROPHIC), or organic (chemoheterotrophic). See AUTOTROPHIC, HETEROTROPHIC, PHOTOTROPHIC.

CHEMOTROPISM. (Bot.) TROPISM in which stimulus is a gradient of chemical concentration, e.g. downward growth of pollen tubes into stigma due to presence of sugars. (Zool.) Rarely used as a synonym of CHEMOTAXIS.

CHIASMA (pl. CHIASMATA). (1) The visible effects of the process of genetic CROSSING-OVER between chromosomes which have paired up (i.e. between bivalents) in appropriately stained meiotic cells, and hence indicators of homologous (non-random) RECOMBINATION. Each chiasma may involve either of the two chromatids of each chromosome. Appreciation that chiasmata result from breakage and reciprocal refusion between chromatids during the first meiotic prophase was a major achievement of classical cytogenetics and is due largely to Jannsens and Darlington (their *chiasmatype theory*). The molecular mechanism involved may incorporate enzymes that were formerly part of a DNA REPAIR MECHANISM. Several chiasmata may occur per bivalent, longer bivalents having more on average. Their frequency and distribution are not entirely random and are sometimes under genetic control. See SUPPRESSOR MUTATION. (2) See OPTIC CHIASMA.

CHILOPODA. Centipedes and their allies. Class (or Subclass) of ARTHROPODA. See MYRIAPODA.

CHIMAERA. (1) An organism with tissues of two or more different genotypes, often a result of mutation, grafting, or the introduction into a very early embryo of cells from a genetically distinct individual. See MOSAIC. (2) Genus of holocephalan fish.

CHIROPTERA. Bats. Order of eutherian (placental) mammals; characterized by membranous wing spread between arms, legs, and sometimes tail, generally supported by greatly elongated fingers. Use of echolocation for avoidance of objects and food capture during commonly nocturnal insectivorous feeding. Some are plant pollinators. Diurnal fruit-eating bats may have had different evolutionary origin from other bats.

CHI-SQUARED (χ^2) TEST. Statistical test for assessing the significance of the deviation of a set of whole numbers (e.g. from scoring the classes of phenotypes from a genetic cross) from those expected from hypothesis. The formula used is

$$\chi^2 = \sum \frac{(n_{obs} - n_{exp})^2}{n_{exp}}$$

The value obtained has to be assessed in relation to the number of *degrees of freedom*, which is the number of classes minus 1, and a χ^2 table will then give the probability (P) of finding as poor a fit with the expected results owing to random sampling error. If, for instance, $P < 0.05$, the data are said to be significantly different from expectation at the 5 per cent level. The χ^2 test becomes seriously inaccurate if any of the expected numbers is less than 5.

CHITIN. Nitrogenous polysaccharide found in many arthropod exoskeletons and hyphal walls of many fungi. Comprises repeated N-acetylglucosamine units (β[1,4]-linked). Strictly a PROTEOGLYCAN, owing to peptide chains attached to its acetamido groups. Of considerable mechanical strength, hydrogen bonding between adjacent molecules stacked together forming fibres giving structural rigidity; also resistant to chemicals. With lignocellulose, among the most abundant of biological products.

CHLAMYDOSPORE. Thick-walled fungal spore capable of surviving conditions unfavourable to growth of the fungus as a whole; asexually produced from a cell or portion of a hypha.

CHLORAMPHENICOL. Antibiotic, formed originally by *Streptomyces* bacteria, inhibiting translation of mRNA on prokaryotic ribosomes, eukaryotic translation being unaffected. Its use can thus distinguish proteins synthesized by mitochondrial/chloroplast ribosomes from those manufactured in the rest of eukaryotic cell. See CYCLO-HEXIMIDE.

CHLORENCHYMA. Parenchymatous tissue containing chloroplasts.

CHLORIDE SHIFT. Entry/exit of chloride ions across red blood cell membranes to balance respective exit/entry of hydrogen carbonate ions resulting from CARBONIC ANHYDRASE activity. See BOHR EFFECT.

CHLOROCRUORIN. Respiratory pigment (green, fluorescing red) dissolved in plasma of certain polychaete worms. Conjugated iron-porphyrin protein resembling haemoglobin.

CHLOROPHYLL. Green pigment found in all algae and higher plants except a few saprotrophs and parasites. Responsible for light capture in PHOTOSYNTHESIS. Located in CHLOROPLASTS, except in CYANOBACTERIA (blue-green algae) where borne on numerous photosynthetic membranes (thylakoids) dispersed in the cytoplasm at the periphery of the cell. Each molecule comprises a magnesium-containing *porphyrin* group, related to the prosthetic groups of haemoglobin and the cytochromes, ester-linked to a long *phytol* side chain. Several chlorophylls exist (*a*, *b*, *c*, *d* and *e*), with minor differences in chemical structure. Chlorophyll *a* is the only one common to all plants (and the only one found in blue-green algae). In photosynthetic bacteria, other kinds of chlorophyll (*bacteriochlorophylls*) occur. Can be extracted from plants with alcohol or acetone and separated and purified by chromatography. See ACCESSORY PIGMENTS, ANTENNA COMPLEX.

CHLOROPHYTA. Green algae. Division of ALGAE, possessing chlorophylls *a* and *b*, *β*-carotene and xanthophylls, and storing starch. Cell wall of cellulose. Largest group of algae; extremely diverse. Primitive forms microscopic, unicellular and either flagellate (base number = 2) or non-motile, occurring singly or grouped together in colonies (*coenobial* or *palmelloid*). Higher forms are multicellular with filamentous (branched or unbranched) or flattened thallus. Asexual reproduction occurs by cell division, fragmentation, aplanospores, zoospores or akinetes. Sexual reproduction can be isogamous, anisogamous or oogamous. Widely distributed; mainly aquatic (freshwater and marine), but aerial algal floras flourish given sufficient moisture (e.g. on the barks of trees, attached to mosses).

CHLOROPLAST. Chlorophyll-containing plastid; the organelle within which both light and dark phases of plant PHOTOSYNTHESIS occur. Present in nearly all plants, but not usually in all their cells. Where present, there may be one to many per cell. Usually disc-shaped (about 2×5 μm, sometimes larger) in higher plants, arranged in a single layer in the cytoplasm but changing shape and position in relation to light intensity (see CYCLOSIS). In algae, either cup-shaped, spiral, stellate forming a network, often accompanied by PYRENOIDS.

Mature chloroplasts of some algae, bryophytes and lycopods can multiply by division; there is little evidence for mature chloroplast division in higher plants, continuity being through growth and division of *proplastids* in meristematic regions. Mature chloroplast typically comprises two outer membranes enclosing a homogeneous *stroma*, where reactions of the dark phase occur. In the stroma are embedded a number of *grana*, each comprising a stack of *thylakoids*; flattened, discoid, membranous vesicles. Thylakoid membranes house the photosynthetic pigments and ELECTRON TRANSPORT SYSTEM involved in the light-dependent phase of photosynthesis. Grana are generally linked by pigment-free *intergranal lamellae*. See Fig. 14b.

Typical mesophyll chloroplasts of C_4 plants (see PHOTOSYNTHESIS) have grana and few, small, starch grains (as in C_3 plants). Bundle sheath chloroplasts are larger, with prominent starch grains (in light) and thylakoid lamellae which run parallel from end to end, without grana. Algal chloroplasts resemble those of bundle sheath cells of C_4 plants. Photosynthetic prokaryotes lack chloroplasts, the numerous thylakoids lying free in the cytoplasm and varying in arrangement and shape in different forms. Chloroplasts contain circular DNA (see cpDNA) and protein-synthesizing machinery, including RIBOSOMES of a prokaryotic type. The evolutionary origin of chloroplasts is currently explained in terms of ENDOSYMBIOSIS.

CHLOROSIS. Disease of green plants characterized by yellow (chlorotic) condition of parts that are normally green; caused by conditions preventing chlorophyll formation e.g. lack of light or of appropriate soil nutrients.

CHOANAE (INTERNAL NARES). Paired connections between nasal and oral cavities of typical crossopterygian (lobe-finned) fish, some teleosts, lungfish and higher vertebrates; probably evolved independently in different fish groups. Not used for respiratory purposes in any living jawed fish, but providing a passage for ventilation of lungs in tetrapods. Situated near front of roof of mouth, unless false palate (see PALATE) present, when they are at the back. See CHOANICHTHYES, NARES.

CHOANICHTHYES (SARCOPTERYGII). A probably unnatural subclass (GRADE) of the OSTEICHTHYES, including the DIPNOI (lungfishes) and CROSSOPTERYGII (lobe-finned fishes). Bony skeleton; paired fins with central axes and fleshy lobes; external and typically internal NARES (see CHOANAE); COSMOID SCALES, or their derivatives.

CHOANOCYTE (COLLAR CELL). Cell with single flagellum generating currents by which sponges (PORIFERA) draw water through their ostia and catch food particles which stick to the outside of cylindrical protoplasmic collar around base of flagellum. Affinities of sponges with the protozoan choanoflagellates problematical.

CHOLECYSTOKININ (PANCREOZYMIN). Hormone of mucosa of small intestine, released in response to presence of CHYME. Causes pancreas to release enzymatic juice and gall bladder to eject bile. Promotes intestinal secretion but inhibits gastric secretion. Role in brain uncertain, but may assist control of feeding. See SECRETIN.

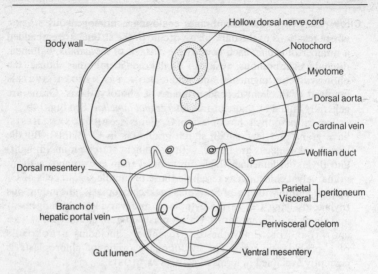

Fig. 7. Transverse section through embryonic vertebrate, indicating the layout of the trunk region prior to origin of the skeleton.

CHOLESTEROL. Sterol lipid derived from squalene, forming a major component of animal CELL MEMBRANES where it affects membrane fluidity. Absent from higher plants and most bacteria. Precursor of several potent steroid hormones (e.g. corticosteroids, sex hormones) which are in turn converted back to it in liver. Synthesis by liver suppressed by dietary cholesterol. Most plasma cholesterol is transported esterified to long-chain fatty acids within a micellar lipoprotein complex. These structures, *low-density lipoproteins* (LDL), are about 22 nm in diameter and adhere to plasma membrane receptor sites produced and found on COATED PITS when a cell needs to make more membranes using the cholesterol in the LDL. Cholesterol is excreted in BILE, both in native form (as micelles) and conjugated with taurine or glycine as bile salts.

CHOLINE. An organic base (formula $OHC_2H_4.N[CH_3]_3OH$), a vitamin for some animals, and a component of some lipids (e.g. lecithin), and ACETYLCHOLINE.

CHOLINERGIC. Of nerve fibres which secrete ACETYLCHOLINE. In vertebrates, motor fibres to striated muscle, parasympathetic fibres to smooth muscle, and fibres connecting CNS to sympathetic ganglia are cholingergic, as are some invertebrate neurones.

CHOLINESTERASE. Hydrolytic enzyme anchored to BASAL LAMINA between synapsing membranes of most (especially vertebrate) neuromuscular junctions and of cholinergic synapses. Degrades ACETYL-CHOLINE to choline and acetate.

CHONDRICHTHYES. Vertebrate class containing cartilaginous fish, first appearing in the Devonian. Includes HOLOCEPHALI (e.g. ratfish, *Chimaera*) and ELASMOBRANCHII (sharks, skates and rays). Cartilaginous skeleton; PLACOID SCALES (denticles), modified to form replaceable teeth; intromittant organs (claspers) formed from male pelvic fins. No GAS BLADDER. See OSTEICHTHYES.

CHONDROBLAST, CHONDROCYTE. See CARTILAGE.

CHONDROCRANIUM. Part of the skull first formed in vertebrate embryos as cartilaginous protection of brain and inner ear. Usually ossified during development to form membrane bones. See OSSIFICATION.

CHONDROITIN. Sulphated GLYCOSAMINOGLYCAN composed largely of D-glucuronic acid and N-acetylgalactosamine. Found in cartilage, cornea, bone, skin and arteries.

CHONDROSTEI. Group (often considered a superorder) of the AC-TINOPTERYGII. Includes the primitive Palaeozoic palaeoniscoids represented today by the bichirs (*Polypterus*), paddlefishes and sturgeons. Ganoid scales of bichirs are lost altogether in paddlefish, sturgeons having rows of bony plates lacking ganoine. Ancestral bony internal skeleton largely substituted by cartilage. Primitive heterocercal tail present in sturgeons and paddlefish. Bichirs have lungs, sturgeons a gas bladder.

CHORDATA. Animal phylum, characterized by presence at some stage in development of a NOTOCHORD, by the dorsal hollow nerve cord, pharyngeal gill slits and a post-anal tail. Includes the invertebrate subphyla UROCHORDATA and CEPHALOCHORDATA, and vertebrates (Subphylum VERTEBRATA).

CHORION. (1) One of three EXTRAEMBRYONIC MEMBRANES of amniotes. Comprises the TROPHOBLAST with an inner lining of mesoderm, coming to enclose almost the entire complement of embryonic structures. In reptiles and birds it forms with the ALLANTOIS a surface for gaseous exchange within the egg. In most mammals it combines with the allantois to form the PLACENTA. (2) Egg shell of insects, secreted by follicle cells of ovary, and often sculptured externally.

CHOROID. Mesodermal layer of vertebrate eyeball between outer sclera and retina within. Soft and richly vascularized (supplying nutrition for retina); generally pigmented to prevent internal reflection of light, but reflecting crystals of TAPETUM, part of the choroid, increase retinal stimulation in many nocturnal/deepwater vertebrates. Becomes the CILIARY BODY anteriorly.

CHOROID PLEXUSES. Numerous projections of non-nervous epithelium into ventricles of brain, secreting CEREBROSPINAL FLUID from capillary networks. One plexus occurs in the roof of each of the four ventricles in man.

CHROMAFFIN CELL, C. TISSUE. Cells derived from NEURAL CREST tissue, which having migrated along visceral nerves during development come to lie in clumps in various parts of the vertebrate body (e.g. the adrenal medulla). They are really postganglionic neurones of the sympathetic nervous system, which have 'lost' their axons and secrete the catecholamines ADRENALINE and NORADRENALINE into the blood, the former more abundantly. Stain readily with some biometric salts (hence name).

CHROMATID. One of the two strands of CHROMATIN, together forming one CHROMOSOME, which are held together after DNA replication during the cell cycle by one or more CENTROMERES prior to separation at either mitotic anaphase or second meiotic anaphase. In mitosis the strands are genetically identical (barring mutation), but in meiosis crossing-over increases the likelihood of dissimilarity.

CHROMATIN (NUCLEOHISTONE). The material of which eukaryotic CHROMOSOMES are composed. Consists of DNA and proteins, the bulk of them HISTONES, organized into nucleosomes. See EU-CHROMATIN, HETEROCHROMATIN.

CHROMATOGRAPHY. Techniques involving separation of components of a mixture in solution through their differential solubilities in a moving solvent (*mobile phase*) and absorptions on, or solubilities in, a *stationary phase* (often gel or paper). In *gel filtration*, mixture to be separated (often proteins) is poured into column containing beads of inert gel and then washed through with solvent. Speed of passage depends on relative solubilities in solvent and on ability to pass through the pores in the gel, a function of relative molecular size. Components may then be identified. See ELECTROPHORESIS.

CHROMATOPHORE. (Zool.) Animal cell lying superficially (e.g. in skin), with permanent radiating processes containing pigment that can be concentrated or dispersed within the cell under nervous and/or hormonal stimulation, effecting colour changes. When dispersed, the pigment of groups of such cells is noticeable; when condensed in centre of cells the region may appear pale. Three common types occur in vertebrates: *melanophores*, containing the dark brown pigment melanin; *lipophores*, with red to yellow carotenoid pigments; *guanophores*, containing guanine crystals whose light reflection may lighten the region when other chromatophores have their pigments condensed. MELANOCYTE-STIMULATING HORMONE disperses melanin, while melatonin (see PINEAL GLAND) and adrenaline concentrate it. (Bot.) (1) See CHROMOPLAST. (2) In prokaryotes (bacteria, blue-green algae), membrane-bounded vesicles (thylakoids) bearing photosynthetic pigments. See PROCHLOROPHYTA.

CHROMATOSOME. A NUCLEOSOME core particle plus a number of adjacent DNA base pairs on either side. Obtained by moderate nuclease digestion of a polynucleosome fibre.

CHROMOCENTRE. Region of constitutive HETEROCHROMATIN which aggregates in interphase nucleus. In *Drosophila* all four chromosome pairs become fused at their centromere regions in POLYTENE nuclei to form a large chromocentre.

CHROMOMERES. Darkly staining (heterochromatic) bands visible at intervals along chromosomes in a pattern characteristic for each chromosome. Especially visible in mitotic and meiotic prophases, and at bases of loops of LAMPBRUSH CHROMOSOMES. Probably reflects tight clustering of groups of chromosome loops (see CHROMOSOME). Dark bands of polytene chromosomes are probably due to multiple parallel chromomeres.

CHROMONEMA. Term usually used for chromosome thread while extended and dispersed throughout nucleus during interphase.

CHROMOPLAST (CHROMATOPHORE). Pigmented plant cell PLASTID. May be red, orange or yellow, e.g. tomato fruits, carrot roots (containing carotenoid pigments), or green (chloroplasts), containing chlorophyll. Former are common in fruits and flowers and develop from *leucoplasts* or chloroplasts. See CHLOROPLAST.

CHROMOSOME. Literally, a coloured (i.e. stainable) body; originally observed as threads within eukaryote nuclei during mitosis and meiosis. Composed of nucleic acid, most commonly DNA, usually in conjunction with various attendant proteins, in which form the genetic material of all cells is organized. Chromosomes are linear sequences of GENES, plus additional non-genetic (i.e. apparently nonfunctional) nucleic acid sequences. Gene sequence is probably never random, being the result of selection for particular LINKAGE groups (but see TRANSPOSABLE ELEMENT). Prokaryotes and eukaryotes differ in the amount of genetic material which needs to be packaged, and in resulting complexities of their chromosomes. Thus the absence to date from prokaryotic chromosomes of the DNA-binding proteins, histones, has some taxonomic value (see CHROMATIN). Non-histone proteins (e.g. *protamines*) form part of the structure of all chromosomes, however, and their roles, for example as activators of transcription, are being increasingly elucidated. The DNA of a normal individual chromosome or chromatid is probably just one highly folded molecule.

The *prokaryotic chromosome* (usually one main chromosome per cell) is just over 1 mm in length, contains about 4×10^6 base pairs of DNA, is circular and is attached to the cell membrane, at least during DNA replication. It lacks the nucleosome infrastructure of eukaryotic chromatin. Additionally, there may be one or more PLASMIDS, some of which (*megaplasmids*) may constitute more than 2% of the cell's DNA. There is no nucleus to contain the chromosome, but the term 'nucleoid' may be used to indicate this region of the cell. The DNA appears to be packaged in a series of loops (see later). *Eukaryotic*

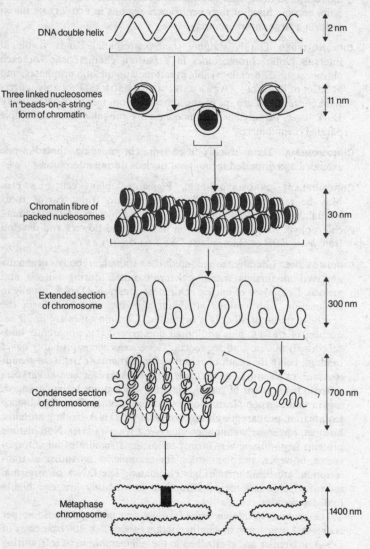

DNA double helix — 2 nm

Three linked nucleosomes
in 'beads-on-a-string'
form of chromatin — 11 nm

Chromatin fibre of
packed nucleosomes — 30 nm

Extended section
of chromosome — 300 nm

Condensed section
of chromosome — 700 nm

Metaphase
chromosome — 1400 nm

Fig. 8. *Possible progressive packing arrangement of a DNA duplex with histones to form nucleosomes and then subsequent packing of these, ultimately to form the structures visible in light microscopy.*

chromosomes are made of chromatin, containing DNA and five different histone species roughly equal in total weight to the DNA; plus various attendant proteins. The fundamental organizational unit is the NUCLEOSOME, a polynucleosome giving rise in turn during nuclease digestion to mononucleosomes (200 DNA base pairs), chromatosomes (165 DNA base pairs) and nucleosome core particles (145 DNA base pairs).

The polynucleosome filament has a diameter of about 10 nm, but adopts a tight 30 nm helix under physiological ion concentrations. This reduces the DNA length 50-fold and may be the normal interphase state of chromatin. Further looping along a single axis forms a fibre 0.3 μm in width which may in turn form a helix of radially arranged loops about 0.7 μm in diameter, possibly the metaphase chromatin condition. Bands seen in stained mitotic chromosomes probably reflect tight clustering of groups of loops, which stain more densely. Polytene chromosome bands (see POLYTENY) would result from lateral amplification of these tightly clustered loops. The higher orders of chromatin packing are features of HETEROCHROMATIN such as chromocentres, centromeres, and pericentric regions. The virus genome may be of either DNA or RNA, in single-stranded or double-stranded form, the protein coat of the virus often being instrumental in chromosome packing.

CHROMOSOME MAP. Linear map (circular in bacteria, plasmids, etc.) of the sequence of genes (cistrons or loci) on a chromosome as defined by CHROMOSOME MAPPING techniques. The MAP DISTANCE between two genes does not accurately reflect their physical separation but only their probability of RECOMBINATION.

CHROMOSOME MAPPING. Several techniques involved in constructing a model (or map) of the linear gene sequence of a chromosome (or chromosome pair). Traditionally this amounts to a map of the sequence of loci along a chromosome pair; but the rise of molecular genetics, particularly the use of microorganisms, has provided details of chromosome base sequences, and a shift from the locus to the CISTRON or OPERON as the functional unit of chromosome structure. For most eukaryotes, LINKAGE between two or more loci is normally detected by first obtaining a generation (normally an F1) heterozygous for the two loci concerned (i.e. doubly heterozygous). This normally achieved by first crossing two stocks, each purebreeding for *one* of the two mutant phenotypes involved. The F1 stock is then crossed to a doubly wild-type stock and the resulting offspring scored for phenotypes. If all four possible phenotypes (assuming complete dominance of wild-type over the mutant phenotype) are present in equal ratio, linkage is not probable; but if there is a departure in the ratios from those expected on the null hypothesis of no linkage, then this departure can be tested for its significance (using CHI-SQUARED TEST). Where the ratio is obviously non-Mendelian (i.e. departs obviously from $1:1:1:1$), with the parental classes outnumbering the recombinants, then a CROSS-OVER VALUE can be determined giving a map distance between the two loci.

When we wish to know whether the loci bearing the alleles for black body and vestigial wing (both recessive characters) in *Drosophila* are linked, then using the symbols

b = black body
+ = wild-type body
v = vestigial wing
+ = wild-type wing

first pure-breeding black body/wild-type wing flies ($bb++$) are mated with pure-breeding wild-type body/vestigial wing flies ($++$ vv). F1 offspring are then mated with a double recessive stock (i.e. pure-breeding black body/vestigial wing, $bbvv$) as a TEST CROSS. If all four resulting offspring phenotypes ($++$, $+v$, $b+$, bv) occur in equal ratio then, given adequate sample size, linkage is unlikely. If two phenotypic classes (the parental classes, $b+$, $+v$) outnumber the other two (the two recombinant classes, bv, $++$) then linkage is likely and a provisional map distance can be calculated, equal to the frequency of the recombinant offspring as a percentage of the total number of offspring. (The example is actually more complex, for only when male flies are used as the double recessive in the backcross do four phenotypic classes appear in the F2 generation. This is because in male *Drosophila* there is no crossing-over during meiosis (see SUPPRESSOR MUTATION) so the males cited only produce two gamete types, giving only two F2 phenotypes.) Sex-linked loci would give a different result, suitably modified to take account of the chromosome arrangement of the heterogametic sex. When testing for linkage between mutations for dominant characters, the recessive characters in the method employed above would be wild-type characters.

Chromosome mapping in bacteria can employ transformation, TRANSDUCTION or interrupted mating. In the latter, progress of the donor bacterial chromosome into the recipient cell during conjugation is interrupted, as by shaking (see F FACTOR, for *Hfr* strain). The map of cistrons on the incoming chromosome will be a function of the time allowed for conjugation before interruption, and is deduced from recipient cell phenotypes. The CIS/TRANS TEST may be used to determine whether two mutations lie within the same cistron. In *deletion mapping*, gene sequences can be ascertained by noting whether or not wild-type recombinants occur in appropriate crosses between mutant strains: they will not do so if the part of the chromosome needed for recombination is missing, so that fine mapping of such recombinants can indicate the limits of a deletion and the genes involved in it. Plasmid and viral chromosome maps may be constructed using *restriction fragment mapping* techniques in which different RESTRICTION ENDONUCLEASES digest the chromosome, and electrophoretic patterns of resulting fragments are used to reconstruct the complete nucleotide sequences of the chromosomes. New electrophoretic techniques with infrequently cutting restriction endonucleases now permit restriction fragment mapping of even entire mammalian chromosomes. See CELL FUSION.

CHROMOSOME PUFF. See PUFF.

CHRYSALIS. The PUPA of lepidopterans (butterflies and moths).

CHRYSOLAMINARIN (LEUCOSIN). Polysaccharide reserve food material present in algae belonging to divisions Chrysophyta and Xanthophyta.

CHRYSOPHYTA. Golden brown algae. Division of Algae, whose colour is due to abundance of carotenoid pigments, including β-carotene, fucoxanthin and other xanthophylls present within the chloroplast, with chlorophyll *a*. Reserve foods stored as oils and as the polysaccharide *chrysolaminarin* (= leucosin). Unicellular, many lacking a cell wall but where present this comprises pectic substances, with cellulose in some forms; may also bear superficial species-specific microscopic scales (calcified or silicified). Species producing silica scales are useful palaeolimnological indicators of environmental change. Flagella (one or two) variable in length and type. A diverse group, with interesting phylogenetic links to other simple organisms (e.g. protozoa and fungi). Colonial, filamentous and amoeboid forms occur. Many are planktonic.

CHYLE. The milky suspension of fat droplets within LACTEALS and THORACIC DUCTS of vertebrates after absorption of a meal.

CHYLOMICRON. Protein-bounded vesicle, up to 100 nm in diameter, containing reconstituted triglycerides, phospholipids and CHOLESTEROL produced by the epithelial cells of intestinal villi after long-chain fatty acids and monoglycerides have diffused across the microvilli. Also act as transport vehicles for dietary lipids within the LACTEALS, LYMPHATIC SYSTEM and BLOOD PLASMA, being absorbed ultimately by the liver. See FAT, LIPOPROTEIN.

CHYME. Partially digested food as it leaves the vertebrate stomach. See CHOLECYSTOKININ, SECRETIN.

CHYMOTRYPSIN. Proteolytic enzyme secreted as inactive chymotrypsinogen by vertebrate pancreas. An exopeptidase, it converts proteins to peptides and is activated by the enzyme enterokinase.

CILIARY BODY. Anterior part of the fused RETINA and CHOROID of the eyes of vertebrates and cephalopod molluscs; containing ciliary processes secreting the aqueous humour, and ciliary muscles (circular smooth muscle) which may permit ACCOMMODATION of the eye either by altering the focal length of the lens (amniotes), or by

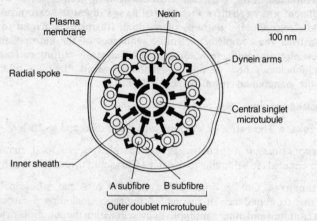

Fig. 9. *Diagram of a cilium or flagellum in cross-section, as viewed by light microscopy. The microtubule apparatus is termed the AXONEME.*

moving the lens to and fro (cephalopods, sharks and amphibians). In mammals the lens is suspended from it by ligaments, and the IRIS arises from the same region.

CILIARY FEEDING. Variety of feeding mechanism (MICROPHAGY) by which many soft-bodied aquatic invertebrates draw minute water-borne food particles through e.g. gills or the pharyngeal region of the gut, when the particles are frequently trapped in mucus and moved either towards the gut (often by further cilia) or further along it (by peristalsis).

CILIATA. Class of Protozoa (Subphylum Ciliophora) containing the most complex cells in the phylum. Covered typically in cilia, with meganucleus, micronucleus, and a cytostome (at the end of a depression, or 'mouth') at which food vacuoles form. Includes familiar *Paramecium* and *Vorticella*, and voracious predatory suctorians. CONJUGATION and BINARY FISSION both occur, as may AUTOGAMY (see PARTHENOGENESIS).

CILIATED EPITHELIUM. Layer of columnar cells with apices covered in cilia whose coordinated beating enables CILIARY FEEDING, movement of mucus in the respiratory tract, etc.

CILIOPHORA. Protozoan Subphylum containing the solitary Class CILIATA.

CILIUM. Organelle of some eukaryotic cells. See Fig. 9. Tubular extension of the cell membrane, within which a characteristic 9 + 2 apparatus of MICROTUBULES and associated proteins occurs (nine paired outer tubules and a lone central pair). Used either for CELL LOCOMOTION (see for details of ciliary action) or for movement of material past a CILIATED EPITHELIUM; but frequently sensory, especially elongated cilia known as FLAGELLA. Cilia may beat in an organized METACHRONAL RHYTHM, for which KINETODESMATA are probably responsible. Such rows of beating cilia may fuse to form *undulating membranes*; or several cilia may mat together to beat as one, as in the conical *cirri* of some ciliates used for 'walking'. For *stereocilium*, see HAIR CELL.

CIRCADIAN RHYTHM (DIURNAL RHYTHM). Endogenous (intrinsic) rhythmic changes occurring in an organism with a periodicity of approximately 24 h; even persisting for some days in the experimental absence of the daily rhythm of environmental cycles (e.g. light/dark) to which circadian rhythm is usually entrained. Widely distributed, including leaf movements, growth movements, sleep rhythms and running activity. In animals, rhythms of hormone secretion have been implicated in some circadian rhythms, these in turn requiring explanation. Their existence indicates a BIOLOGICAL CLOCK, but the detailed chemistry is usually unknown.

CIRCINATE VERNATION. Coiled arrangement of leaves and leaflets in the bud; gradually uncoils as leaf develops further, as in ferns. See PHYLLOTAXY.

CIRCULATORY SYSTEM. System of vessels and/or spaces through which blood and/or lymph flows in an animal. See BLOOD SYSTEM, LYMPHATIC SYSTEM.

CIRCUMNUTATION. See NUTATION.

CIRRIPEDIA. Barnacles and their relatives. Subclass of CRUSTACEA. Typically marine, sessile and hermaphrodite. Unlike most of the Class in appearance, with a carapace comprising calcareous plates enclosing the trunk region. Usually a cypris larva, which becomes attached to the substratum by its 'head', remaining fixed throughout its adult life and filter-feeding using BIRAMOUS APPENDAGES on its thorax. Several parasitic forms occur.

CISTERNAE. Flattened sac-like vesicles of ENDOPLASMIC RETICULUM and GOLGI APPARATUS intimately involved in transport of materials via vesicles which either bud from or fuse with their membranous surfaces.

CIS-TRANS TEST (COMPLEMENTATION TEST). Genetic test to discover whether or not two mutations which have arisen on separate but usually homologous chromosomes are located within same CISTRON. See Fig 10. The two chromosomes, e.g. of phage or prokaryote origin, are artificially brought together in a single bacterial cell (e.g. by TRANSDUCTION) or in a diploid eukaryote by a sexual cross. If their co-presence in the cell rectifies their individual mutant expression, then COMPLEMENTATION is reckoned to have occurred

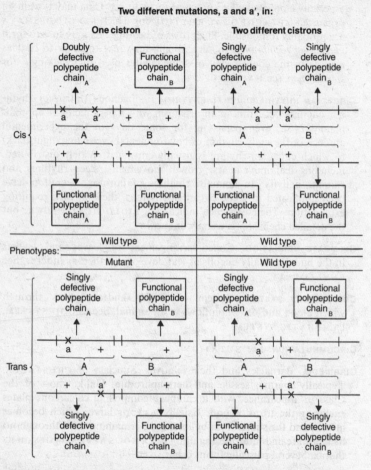

Fig. 10. *Theoretical basis of the cis-trans test. Where two mutations (—×—) occur within the same chromosome (*cis* configuration), complementation occurs through functional polypeptide production by the complementary cistron or cistrons of the other chromosome. Mutant phenotype only occurs where the two mutations occur within the same cistron but on different chromosomes. This effect enables precise mapping of the physical limits of cistrons within chromosomes.*

between the functional gene products of two distinct cistrons. However, if their resultant expression is still mutant, no such complementation has occurred and the two mutations are reckoned to lie within the same cistronic region. Two mutations lie in the *trans* condition if on separate chromosomes, but in the *cis* condition when on the same. *Trans*-complementation only occurs when two mutations lie in differ-

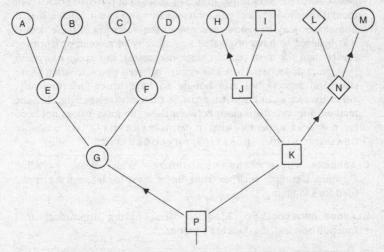

Fig. 11. *A cladogram illustrating terminology employed. Character states are represented by shapes of figures, and a change in these by an arrow. Character state in circles [○] is symplesiomorphous for taxa A, B, C and D but is synapomorphous for taxa E, F and G. It is homoplasous between taxa A–G (being an example of parallel development) and homoplasous between taxa A–H and M, but here it is also an example of convergent development.*

ent cistrons, and by careful mapping of mutations the boundary between two cistrons can be located from the results of the cis-trans test.

CISTRON. A region of DNA within which mutations affect the same functions by the criterion of the CIS-TRANS TEST. In molecular terms, the length of DNA (or RNA in some viruses) encoding a specific and functional product, usually a protein, in which case the cistron is 'read' via messenger RNA; but both ribosomal RNA and transfer RNA molecules have their own encoding cistrons. In modern terminology, 'cistron' is equivalent to 'gene', except that not all putative genes have been fully validated by complementation analysis.

CITRIC ACID CYCLE. See KREBS CYCLE.

CLADE. Phylogenetic lineage of related taxa originating from a common ancestral taxon. See CLADISTICS, GRADE.

CLADIST. Proponent of, or worker in, CLADISTICS.

CLADISTICS. Hennigian classification (after W. Hennig). Method of classification that attempts to infer phylogenetic relationships among organisms solely from PHENETIC characters, basing the classification solely upon resulting genealogies (the resulting diagram being a CLADOGRAM). See Fig. 11. A line of descent is characterized by the occurrence of one or more evolutionary novelties (*apomorphies*). Any character found in two or more taxa is HOMOLOGOUS in them if their most recent common ancestor also had it. Such a shared homologue may be *symplesiomorphous* (or *-morphic*) in these taxa if it is believed to have originated as a novelty in a common ancestor earlier than the most recent common ancestor, but *synapomorphous* (*-morphic*) if not. HOMOPLASY occurs between characters that share structural aspects but are thought to have arisen independently, either by PARALLEL EVOLUTION or CONVERGENCE: the cladistic method does not distinguish between these, because it does not need to; nor does it permit what it terms PARAPHYLETIC taxa. See CLASSIFICATION, PHENETICS, PHYLOGENETICS.

CLADOCERA. Order of BRANCHIOPODA. 'Water fleas', including *Daphnia*. Carapace encloses trunk limbs, used for feeding. Antennae used for swimming.

CLADODE (PHYLLOCLADE). Modified stem, having appearance and function of a leaf, e.g. butcher's broom.

CLADOGENESIS. Branching SPECIATION, in which an evolutionary lineage splits to yield two or more lineages. See CLADISTICS.

CLADOGRAM. See CLADISTICS.

CLADOPHYLL. Branch that resembles a foliage leaf.

CLAMP CONNECTION. Lateral connection between adjacent cells of a dikaryotic hypha, found in some of the BASIDIOMYCOTINA. Ensures that each cell of the hypha contains two genetically dissimilar nuclei. Compare CROZIER FORMATION.

CLASS. A taxonomic category in CLASSIFICATION. Of higher rank (more inclusive) than ORDER but of lower rank (less inclusive) than PHYLUM (or DIVISION). Thus there may be one or more classes in a phylum or division and one or more orders in a class. In employing any taxonomic category one aim is to ensure that all its members share a common ancestor which is also a member of the taxon, although opinions differ whether the resulting group should include all *descendants* of the common ancestor. Thus evolutionary taxonomists recognize the Class Reptilia, whereas cladists do not.

CLASSIFICATION. Any method organizing and systematizing the diversity of organisms, living and extinct, according to a set of rules. Belief in the existence of a pre-arranged (divine?) natural order, which it was the role of the scientist to discover, was common until the early 19th century, but has dwindled since the publication of Darwin's *The Origin of Species* in 1859. However, there is still considerable dialogue between ESSENTIALIST and NOMINALIST accounts of biological classification as to whether the groups which different classifications recognize are 'natural' (real) or merely human constructs (artificial). Characters are not randomly distributed among organisms but tend to cluster together with high predictability, suggesting that all the taxonomist has to do is discover the various nested sets of characters to attain a 'natural' classification. However, even the category which seems to have most to recommend its objective reality in nature (the species) lacks some of the features of a NATURAL KIND.

The charge of arbitrariness (artificiality) over the rules of classification has led to the search for an objective methodology. The solution of *numerical taxonomists* has been to select not just one or a few characters which are given added weight (often apparently arbitrarily) in comparisons between organisms, but to give all phenetic characters equal weight, in the expectation that natural (as opposed to artificial) groups will automatically emerge as clusters through overall phenetic similarity. The taxonomist then arranges these clusters into a rule-governed hierarchy of groups.

But this approach also has arbitrary elements, although favoured by some mathematically minded taxonomists. Critics argue that it fails to achieve a genealogically based classification: they deny that overall similarity alone is a sure guide to recency of common ancestry. Thus, crocodiles and lizards share more common features than either does with birds; yet crocodiles and birds share a more recent common ancestor than either does with lizards. If there is anything like an objectifying principle available to taxonomists it must surely be genealogy. Two principal schools of taxonomy which endeavour to objectify their methods by acknowledging the process of evolution in this way are *cladism* and *evolutionary taxonomy*. The principal difference between them (see CLADISTICS) is that cladists include *all* descendant species along with the ancestral species within taxonomic groups; evolutionary taxonomists hold that different rates of adaptation within the descendant groups of a single ancestor should be reflected in the classification. This may require that those that have diverged more from the ancestral stock are given special (often higher) taxonomic ranking compared with those that have diverged less, and that consequently *not all* descendant species will be included in the taxon of the ancestral species. Much depends on the accepted definition of MONOPHYLETIC. All biological classifications are hierarchical: the higher the taxonomic category, the more inclusive. In descending order of inclusiveness, and omitting intermediate (sub- and super-) taxa, the sequence is: kingdom, phylum (division in botany), class, order, family, genus, species. There are however enormous difficulties in establishing accurate genealogical classifications, not least with taxa which are extinct. See IDENTIFICATION KEYS.

CLATHRIN. One of the main proteins covering COATED VESICLES, interconnecting molecules (triskeletons) forming varying numbers of lattice-like pentagonal/hexagonal facets on the cytosolic surfaces of coated pits, causing invagination to form the vesicle. The clathrin coat is shed after the vesicle is formed and internalized.

CLAVICLE. MEMBRANE BONE of ventral side of PECTORAL (shoulder) GIRDLE of many vertebrates. Collar-bone of man.

CLEARING. Process used to prepare many histological slides in light microscopy. The object is to remove any alcohol used in the dehydration of the material; the preparation is soaked in two or three changes of *clearing agent* (e.g. benzene, xylene, or oil of cloves). Clearing makes the material transparent and permits embedding in paraffin wax (insoluble in alcohol) prior to sectioning.

CLEAVAGE (SEGMENTATION). Repeated subdivision of egg or zygote cytoplasm associated with, but not always accompanied by, mitoses. In animals it often produces a mass (the blastula) of small cells (blastomeres) which subsequently enlarge. Bilateral (radial) cleavage, in which ANIMAL POLE blastomeres tend to lie directly on top of vegetal blastomeres, occurs in echinoderms and chordates. Spiral cleavage, in which the first four animal blastomeres lie over the junctions of the first four vegetal blastomeres, is characteristic of other animal phyla. Cleavage may be *deterministic* or *indeterministic*, depending respectively upon whether the fates of blastomeres are already fixed or are plastic. Cleavage is complete (*holoblastic*) in eggs with little yolk; partial (*meroblastic*) in yolky eggs where only the non-yolky portion engages in cell division; *superficial* in centrolecithal eggs, where nuclear division produces many nuclei towards the centre of the cell and which then migrate to the cytoplasmic periphery to become partitioned by cell membranes. See MOSAIC DEVELOPMENT.

CLEIDOIC EGG. Egg of terrestrial animal (e.g. bird or insect) enclosed within protective shell, largely isolating it from its surroundings and permitting gaseous exchange and minor water loss or gain. Contrasts with most aquatic eggs, in which exchange of water, salts, ammonia, etc., occurs fairly freely. See URICOTELIC.

CLEISTOCARP (CLEISTOTHECIUM). Completely closed spherical fruit body (ascocarp) of some of the ASCOMYCOTINA, e.g. powdery mildews, from which spores are eventually liberated through decay or rupture of its wall.

CLEISTOGAMY. Fertilization within an unopened flower; e.g. in violet.

CLIMAX COMMUNITY. Community of organisms, composition more or less stable, in equilibrium with existing natural environmental conditions; e.g. oak forest in lowland Britain. Compare PLAGIOCLIMAX.

CLINE. Continuous gradation of phenotype or genotype in a species population, usually correlated with a gradually changing ecological variable. See INFRASPECIFIC VARIATION, ECOTYPE.

CLISERE. Succession of CLIMAX COMMUNITIES in an area as a result of climatic changes.

CLITELLUM. Saddle-like region of some annelid worms (Oligochaeta, Hirudinea), prominent in sexually mature animals. Contains mucus glands secreting a sheath around copulating worms binding them together; the resultant cocoon houses the fertilized eggs during their development.

CLITORIS. Small erectile organ of female amniotes, homologous to the male's penis; anterior to vagina and urethra.

CLOACA. Terminal region of the gut of most vertebrates into which kidney and reproductive ducts open. There is only one posterior opening to the body, the cloacal aperture, instead of separate anus and urogenital openings (e.g. placental mammals). Also terminal part of intestine of some invertebrates, e.g. sea cucumbers.

CLONAL SELECTION THEORY. The theory, originated by N. Jerne and M. Burnet, that during their development both T-CELLS and B-CELLS acquire specific antigen receptors, being activated (selected) to proliferate into clones of appropriate effector cells only on binding this antigen.

CLONE. (1) A group of organisms of identical genotype, produced by some kind of ASEXUAL reproduction and some sexual processes, such as haploid selfing, or inbreeding a completely homozygous line. Nuclear transplantation techniques introducing genetically identical nuclei into enucleated eggs can also produce clones in some animals, even *in utero*. (2) A group of cells descended from the same single parent cell. Often used of sub-populations of multicellular organisms (e.g. see CLONAL SELECTION THEORY) rather than the entire organism, which may in any case be a MOSAIC. (3) Nucleic acid sequences are said to be cloned when they are inserted into vectors such as PLASMIDS and copied within host cells during GENE MANIPULATION.

CLOTTING. See BLOOD CLOTTING.

CLUB MOSS. See LYCOPHYTA.

CNIDARIA. Subphylum of the COELENTERATA containing hydroids, jellyfish, sea anemones and corals. Gut incomplete (one opening); ectoderm containing CNIDOBLASTS. Two structural forms: (i) attached, sessile *polyp*, (ii) free-swimming *medusa*. Former is a cylindrical sac with mouth and tentacles at opposite end to the attachment; latter is umbrella-shaped, with flattened enteron, and mouth in middle of concave under-surface. Sometimes the phases alternate in a single life-cycle; sometimes only one phase occurs. Compare CTENOPHORA.

CNIDOBLAST (THREAD CELL). Specialized stinging cell found only in

CNIDARIA and a few of their predators which incorporate them. Several kinds exist with different functions (e.g. adhesion, penetration, injection). This cell produces an inert NEMATOCYST, and is commonly regarded as an *independent effector* (see EFFECTOR). Compare LASSO CELL.

CNS. See CENTRAL NERVOUS SYSTEM.

CoA. See COENZYME A.

COACERVATE. Inorganic colloidal particle (e.g. clay) on to which have been adsorbed organic molecules, maybe acting as an important concentrating mechanism in prebiotic evolution. See ORIGIN OF LIFE for discussion of illites and kaolinites.

COATED PIT. Region of many CELL MEMBRANES to which clathrin and associated proteins bind to initiate formation of a COATED VESICLE.

COATED VESICLE. Membranous vesicle (about 50 nm in diameter) budded off endocytotically from the plasma membrane, Golgi apparatus and endoplasmic reticulum; some probably become primary LYSOSOMES. Characterized by lattice-like coat of CLATHRIN molecules associated with other proteins. The clathrin is usually jettisoned soon after vesicle formation, prior to fusion with its target membrane, when its cargo of glycoprotein, neurotransmitter, etc., is released. See CELL MEMBRANES, CELL LOCOMOTION.

COBALAMIN. See CYANOCOBALAMIN.

COCCIDIA. Sporozoan protozoa parasitic in guts of vertebrates and invertebrates. Probably ancestral to haemosporidians (e.g. malarial parasites). Give rise to diseases termed *coccidioses*.

COCCOLITH. A scale composed of calcium carbonate ($CaCO_3$) on an organic base, occurring in the PRYMNESIOPHYTA.

COCCOLITHOPHORIDACEAE. See PRYMNESIOPHYTA.

COCCYX. Fused tail vertebrae. In man comprises two to three bones.

COCHLEA. Diverticulum of the sacculus of inner ears of crocodilians, birds and mammals, usually forming a coiled spiral. Contains the organ of Corti, involved in sound detection and pitch analysis, a longitudinal mound of HAIR CELLS running the length of the cochlea supported on the *basilar membrane* with a membranous flap (*tectorial membrane*) overlaying the hair cells. Vibrations in the round window caused by vibrations in the EAR OSSICLES are transmitted to the perilymph on one side of the basilar membrane, vibrating it and stimulating hair cells of the organ of Corti. The apex of the cochlea (spiral top) is most sensitive to lower frequency vibrations, the base to higher frequencies. See VESTIBULAR APPARATUS.

COCOON. Protective covering of eggs, larvae, etc. Eggs of some annelids are fertilized and develop in a cocoon. Larvae of many endopterygotan insects spin cocoons in which pupae develop (cocoon of silkworm moth is source of silk). Spiders may also spin cocoons for their eggs.

CODE, GENETIC. See GENETIC CODE.

CODOMINANT. See DOMINANCE.

CODON. Coding unit of MESSENGER RNA, comprising a triplet of nucleotides which base-pairs with a corresponding triplet (*anticodon*) of an appropriate TRANSFER RNA molecule. Some codons signify termination of the amino acid chain (the *nonsense codons* UAG, UAA and UGA, respectively *amber, ochre* and *opal*). For role of AUG codon, see PROTEIN SYNTHESIS. See GENETIC CODE.

COEFFICIENT OF SELECTION, s. Proportionate reduction in contribution to the gene pool (at a specified time t) made by gametes of a particular genotype, compared with the contribution made by the standard genotype, which is usually taken to be the most favoured. If $s = 0.1$, then for every 100 zygotes produced by the favoured genotype only 90 are produced by the genotype selected against, and the genetic contribution of the unfavoured genotype is $1 - s$. See FITNESS.

COELACANTHINI. Large suborder of CROSSOPTERYGII, mostly fossil (Devonian onwards). Freshwater, but with living marine representatives (*Latimeria*) in the Indian Ocean. Thought to have been extinct since the Cretaceous; but since 1938 several have been found. CHOANAE absent.

COELENTERATA. Phylum of diploblastic and radially symmetrical aquatic animals, comprising the subphyla CNIDARIA and CTENOPHORA. Ectoderm and endoderm separated by *mesogloea* (jelly-like matrix of variable thickness) and enclosing the gut cavity (*coelenteron*), with a single opening to the exterior. Peculiar cell types include MUSCULO-EPITHELIAL CELLS, and either CNIDOBLASTS (cnidarians) or LASSO CELLS (ctenophores).

COELENTERON. See ENTERON.

COELOM. Main (secondary) body cavity of many triploblastic animals, in which the gut is suspended. Lined entirely by mesoderm. Principal modes of origin are either by separation of mesoderm from endoderm as a series of pouches which round off enclosing part of the archenteron (ENTEROCOELY), or *de novo* by cavitation of the embryonic mesoderm (SCHIZOCOELY). Contains fluid (coelomic fluid), often receiving excretory wastes and/or gametes, which reach the exterior via ciliated funnels and ducts (*coelomoducts*). May be subdivided by septa into *pericardial, pleural* and *peritoneal* coeloms enclosing respectively the heart, lungs and gut. Reduced in arthro-

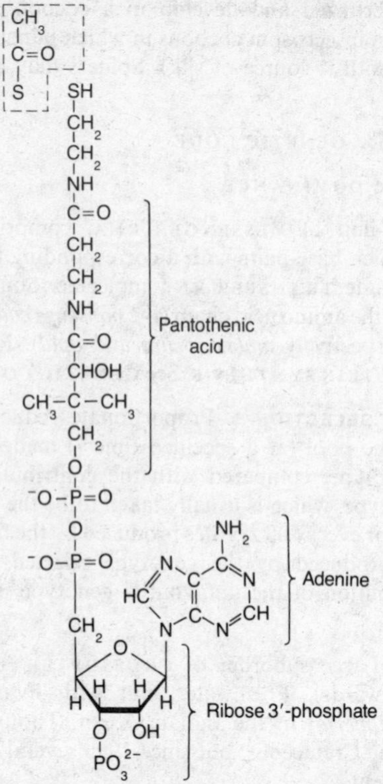

Fig. 12a. *Structure of the coenzyme A molecule. The modification to form acetyl-coenzyme A is indicated at top left.*

pods, restricted to cavities of gonads and excretory organs, the main body cavity being the HAEMOCOELE, as in Mollusca.

COELOMODUCT. Mesodermal ciliated duct (its lumen never intracellular) originating in the COELOM and growing outwards from the gonad or wall of the coelomic cavity to fuse with the body wall. Sometimes conveys gametes to exterior (see MULLERIAN DUCT, WOLFFIAN DUCT); sometimes excretory, e.g. kidneys of molluscs. Compare NEPHRIDIUM. See KIDNEY.

COENOBIUM. Type of algal colony where the number of cells is determined at its formation. Individual cells are incapable of cell division, are arranged in a specific manner, are coordinated and behave as a unit; e.g. *Volvox, Pandorina, Pediastrum, Hydrodictyon*.

COENOCYTE. (Bot.) Multinucleate mass of protoplasm formed by division of nucleus, but not cytoplasm, of an original cell with single nucleus; e.g. many fungi and some green algae. Compare SYNCYTIUM. See ACELLULAR.

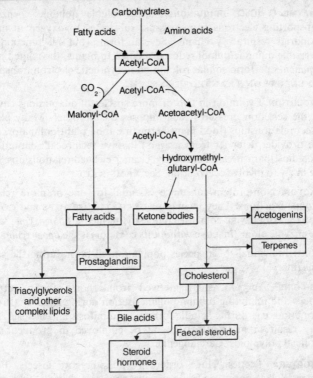

Fig. 12b. Some metabolic pathways in which acetyl-coenzyme A is involved.

COENOSPECIES. Group of related species with the potential, directly or indirectly, of forming fertile hybrids with one another.

COENZYME. Organic molecule (often a derivative of a mononucleotide or dinucleotide) serving as COFACTOR in an enzyme reaction, but, unlike a PROSTHETIC GROUP, binding only temporarily to the enzyme molecule. Often a recycled vehicle for a chemical group needed in or produced by the enzymic process, reverting to its original form when the group is removed – often by another enzyme in a pathway. Removal by coenzymes of reaction products from the enzyme environment may be essential to prevent end-product inhibition of the enzyme. In heterotrophs they are frequently derivatives of water-soluble VITAMINS. See COENZYME A, COENZYME Q, FAD, FMN, NAD, NADP.

COENZYME A (CoA, CoA-SH). Mononucleotide phosphate ester of pantothenic acid (a vitamin for vertebrates). Carrier of acyl groups in fatty acid oxidation and synthesis, pyruvate oxidation (see KREBS CYCLE) and various acetylations. When carrying an acyl group, referred to as acetyl coenzyme A (acetyl-CoA). See Figs. 12a and 12b.

COENZYME Q (CoQ, UBIQUINONE). Lipid-soluble quinone coenzyme transporting electrons from organic substrates to oxygen in mitochondrial respiratory chains. Several forms; but all function by reversible quinone/quinol redox reactions. In plants, the related *plastoquinones* perform similar roles in photosynthetic electron transport. See ELECTRON TRANSPORT SYSTEM.

COEVOLUTION. Evolution in two or more species of adaptations caused by the selection pressures each imposes on the other. Many plant/insect relationships (food plant/herbivore, food plant/pollinator, nest site provider/defender from grazers) involve reciprocal adaptations. Most host/parasite, predator/prey, cleaner/cleaned relationships, etc., are likely to involve coevolution. See ARMS RACE.

COFACTOR. Non-protein substance essential for one or more related enzyme reactions. They include PROSTHETIC GROUPS and COENZYMES. An enzyme-cofactor complex is termed a *holoenzyme*, while the enzyme alone (inactive without its cofactor) is the *apoenzyme*.

COHORT. Individuals of a species population all born during the same time interval.

COLCHICINE. An ALKALOID derived from roots of the autumn crocus and inhibiting tubulin polymerization during MICROTUBULE formation, e.g. during spindle formation in mitosis and meiosis. This may result in POLYPLOIDY, and be employed in production of artificial polyploids. Carcinogenic.

COLEOPTERA. Beetles. Huge order of endopterygote insects. Forewings (elytra) horny, covering membranous and delicate hind-wings (which may be small or absent) and trunk segments. Biting mouthparts. Larvae may be active predators (*campodeiform*), caterpillar-like (*eruciform*), or grub-like (*apodous*). Many larvae and adults are serious pests of crops, stored produce and timber. Some borers of live wood may transmit fungal disease (e.g. *Scolytus* and Dutch elm disease). Size range is probably greater than that of any other insect order.

COLEOPTILE. Protective sheath surrounding the apical meristem and leaf primordia (plumule) of the grass embryo; often interpreted as the first leaf.

COLEORHIZA. Protective sheath surrounding RADICLE in grass seedlings.

COLICINS. See PLASMID.

COLLAGEN. Major fibrous (structural) protein of CONNECTIVE TISSUE, occurring as *white fibres* produced by fibroblasts. Forms up to one third of total body protein of higher vertebrates. Provides high tensile strength (e.g. in tendon), without much elasticity (unlike ELASTIN). Collagen fibres are composed of masses of *tropocollagen* molecules, each a triple helix of collagen monomers. Yields *gelatin* on boiling.

COLLAR CELL. See CHOANOCYTE.

COLLATERAL BUNDLE. See VASCULAR BUNDLE.

COLLEMBOLA. Springtails. Order of small primitively wingless insects (see APTERYGOTA). Two caudal furcula fold under the abdomen and engage the hamula (another pair of abdominal appendages) prior to explosive release resulting in the spring. First abdominal segment carries ventral tube for adhesion. Compound eyes absent. No Malpighian tubules, and usually no tracheal system. Immensely abundant in soil, under bark, on pond surfaces, etc., forming vital link in detritus food chain.

COLLENCHYMA. Tissue providing mechanical support in many young growing plant structures (stems, petioles, leaves), but uncommon in roots. Consists of tapered living cells with walls strengthened by cellulose thickening, usually in the corners; still capable of extension. Commonly found in cortex of herbaceous stems and along veins of leaves. Compare SCLERENCHYMA.

COLLOID. A substance having particles of about 100–10 000 nm diameter, which remain dispersed in solution. Such *colloidal solutions* are intermediate in many of their properties between true solutions and suspensions. Brownian motion prevents colloid particles from sedimenting under gravity, and in lyophilic colloids (solvent-loving) such as aqueous protein solutions, each particle attracts around it a 'shell' of solvent forming a hydration layer, preventing them from flocculating. Competition for this solvent by addition of strong salt solution will precipitate the colloid.

COLON. Large intestine of vertebrates, excluding narrower terminal rectum. In amniotes and some amphibians, but not fish, is clearly marked off from small intestine by a valve. In mammals at least, has essential water-absorbing role in preparation of faeces. Bacteria housed within it produce vitamins (esp. VITAMIN K).

COLONY. (1) Several plant and animal organizations where various more or less (and often completely) distinct individuals live together and interact in mutually advantageous ways. Sometimes, as in the alga *Volvox* and certain ciliates, the organization approaches multicellularity and may even have been a transitional stage in its attainment; but there is usually insufficient communication or division of labour between cells for full multicellular status (see COENOBIUM). In colonial CNIDARIA and ECTOPROCTA there may be considerable POLYMORPHISM between the individuals (termed ZOOIDS) with some associated division of labour. In all these there are good asexual budding abilities, and it is likely that the colonial habit originated by failure of buds to separate. Not so in colonial insects (e.g. HYMENOPTERA, ISOPTERA), although here too division of labour is the rule (see CASTE). Among vertebrates, several bird and mammal species live and/or breed in colonies, and many behavioural adaptations reflect this. (2) A group of microorganisms (bacteria, yeasts, etc.) arising from a single cell and lying on surface of food source – as in culture on agar.

COLONY STIMULATING FACTOR (CSF). Glycoprotein active in stimulating blood granulocyte and macrophage formation. Lung tissue is a major source. Production is a component of INNATE IMMUNE RESPONSE. Some can force leukaemic cells to differentiate and so stop dividing.

COLOSTRUM. Cloudy fluid secretion of mammary glands during the first few days after birth of young and before full-scale milk production. Important source of antibodies (passive immunity) too large to cross the placenta. Rich in proteins, but low in fat and sugar.

COLUMELLA. (Bot.) (1) Dome-shaped structure present in sporangia of many Zygomycotina (fungi) of the order Mucorales (pin moulds); produced by formation of convex septum cutting off sporangium from hypha bearing it. (2) Sterile central tissue of moss capsule. (Zool.) The *columella auris*, or stapes. The EAR OSSICLE of land vertebrates (often complex in reptiles and birds) homologous with the HYOMANDIBULAR bone of crossopterygians, and transmitting air vibrations from the ear drum (to which it is primitively attached) directly to the oval window of the inner ear (amphibia, primitive reptiles). In mammals it no longer attaches to the ear drum, articulating with the incus at its outer edge.

COLUMELLA AURIS. See COLUMELLA.

COMMENSALISM. See SYMBIOSIS.

COMMISSURE. Nerve tissue tract joining two bilaterally symmetrical parts of the central nervous system. In arthropods and annelids, they connect ganglia of the paired ventral nerve cords, and the supraoesophageal ganglia (brain) with the suboesophageal. Commissures unite the vertebrate cerebral hemispheres (see CORPUS CALLOSUM).

COMMON. (Of vascular bundles) passing through stem and leaf. Compare CAULINE.

COMMUNITY. Term describing an assemblage of populations living in a prescribed area or physical habitat, inhabiting some common environment. An organized unit in possessing characteristics additional to its individual and population components, functioning as a unit in terms of flow of energy and matter. The biotic community is the living part of the ecosystem. It remains a broad term, describing natural assemblages of variable size, from those living upon submerged lake sediments to those of a vast rain forest. See ASSOCIATION, CONSOCIATION, SOCIETY.

COMPANION CELLS. Small cells characterized by dense cytoplasm and prominent nuclei, lying side-by-side with sieve tube cells in the phloem of flowering plants and arising with them by unequal longitudinal division of a common parent cell. One of their functions is to transport soluble food molecules into and out of sieve tube elements.

COMPARTMENT. Anatomical region in some (maybe many) animals, its boundaries well defined in development by cell–cell recognition. In *Drosophila*, a compartment comprises cells forming more than one CLONE (or POLYCLONE), whose growth respects a compartment boundary even when one of the component clones is a rapid-growing mutant. Fate of compartment polyclones is associated with the pattern of expression of HOMOEOTIC genes.

COMPENSATION POINT. Light intensity at which rate of respiration by a photosynthetic cell or organ equals its rate of photosynthesis. At this intensity there is no net gain or loss of oxygen or carbon dioxide from the structure. Compensation points for most plants occur around dawn and dusk, but vary with the species.

COMPENSATORY HYPERTROPHY. See REGENERATION.

COMPETENT. Describing embryonic cells while still able to differentiate into wide variety of cell types. Pluripotent, and undetermined. See PRESUMPTIVE, DETERMINED.

COMPETITION. The effect (result) of a common demand by two or more organisms upon a limited supply of resource, e.g. food, water, minerals, light, mates, nesting sites, etc. When *intraspecific*, it is a major factor in limiting population size (or density); when *interspecific*, it may result in local extinction of one or more competing species. An integral factor in Darwinian theory. See COMPETITIVE EXCLUSION PRINCIPLE, DENSITY-DEPENDENCE, NATURAL SELECTION.

COMPETITIVE EXCLUSION PRINCIPLE. The empirical generalization, often regarded as axiomatic, probably first enunciated by J. Grinnell (1913) but generally attributed to G. F. Gause (1934), that as a result of interspecific competition, no two ecological niches can be precisely the same (see NICHE); i.e. niches are mutually exclusive, and mutual coexistence of two species will require that their niches be sufficiently different. See CHARACTER DISPLACEMENT.

COMPLEMENT. Nine interacting serum proteins (beta globulins, C1–C9), mostly enzymes, activated in a coordinated way and participating immunologically in bacterial lysis and macrophage chemotaxis. Genetic loci responsible map in the S region of the H-2 COMPLEX in mice, and the HLA-B region of the MHC region in man. The principal event in the system is cleavage of the plasma protein C3, with subsequent attachment of the larger cleavage product (C3b) to receptors present on neutrophils, eosinophils, monocytes, macrophages and B-CELLS. C3b combines with immune complexes and causes them to adhere to these white cells, promoting ingestion. C3 cleavage (*fixation*) terminates a CASCADE, itself initiated by immune complexes.

COMPLEMENTARY DNA. See cDNA.

COMPLEMENTARY MALES. Males, often degenerate and reduced, living attached to females; e.g. in barnacles (Cirripedia), ceratioid angler fish.

COMPLEMENTATION. Production of a phenotype resembling wild-type (normal) when two mutations are brought together in the same cell in the *trans* configuration (on separate chromosomes); it occurs infrequently if the same functional unit (CISTRON or GENE) is defective on both chromosomes. The CIS-TRANS TEST is the rigorous method for defining the limits of cistrons, but often only the *trans* complementation test is made, the *cis* combination of individually recessive mutations almost always giving a wild phenotype.

COMPLEMENT FIXATION. See COMPLEMENT.

COMPOST AND COMPOSTING. See DECOMPOSER.

COMPOUND LEAF. Leaf whose blade is divided into several distinct leaflets.

CONCENTRIC BUNDLE. See VASCULAR BUNDLE.

CONCEPTACLE. Cavity containing sex organs, occurring in groups on terminal parts of branches of thallus in some brown algae, e.g. bladderwrack, *Fucus vesiculosus*.

CONDENSATION REACTION. Reaction in which two molecules are joined together (often by a covalent bond) with elimination of elements of water (i.e. one H_2O molecule). Occurs in all cells, during polymerization and many other processes. Contrast HYDROLYSIS.

CONDITIONED REFLEX (CONDITIONAL REFLEX). A REFLEX whose original unconditioned stimulus (US) is replaced by a novel conditioned stimulus (CS), the response remaining unaltered. Classically, the CS is presented to an animal either just prior to or in conjuction with the US; they are *associated* (presumably by formation of new neural pathways), and will each elicit the response when presented separately. Although Ivan Pavlov (1849–1936) is best remembered for conditioning dogs to salivate in response to the sound of a bell (CS) instead of the sight or smell of food (US), his dogs also became excited at the CS and moved to where it or the eventual food were delivered. A simple form of LEARNING. See CONDITIONING.

CONDITIONING. Associative learning. There are two broad categories. (1) *Classical conditioning*, in which an animal detects correlations between external events, one of which is either *reinforcing* or *aversive*, modifying its behaviour in such a way that appropriate *consummatory responses* (usually) are elicited. (2) *Instrumental* or *operant* conditioning, in which aspects of consummatory responses are modified as an

animal correlates and learns how variations in those responses affect
its success in attaining a reinforcing, or avoiding an aversive, stimulus.
Both are adaptive, since opportunities for successful instrumental
conditioning may depend upon appropriate prior classical conditioning. See CONSUMMATORY ACT, LEARNING, REINFORCEMENT.

CONDYLARTHRA. Extinct order of Paleocene and Eocene ungulate
(hooved) mammals, possibly ancestral to most other ungulates and
even to carnivores.

CONDYLE. Ellipsoid knob of bone, fitting into corresponding socket
of another bone. Condyle and socket form a joint, allowing movement in one or two planes, but no rotation; e.g. condyle at each side
of lower jaw, articulating with skull; *occipital condyles* on tetrapod
skull, fitting into atlas vertebra.

CONE. (Bot.) *Strobilus.* Reproductive structure comprising a number
of SPOROPHYLLS more or less compactly grouped on a central axis;
e.g. cone of pine tree. (Zool.) (1) Light-sensitive RECEPTOR of most
vertebrate retinas, though not usually of animals living in dim light.
The cone-shaped outer segment appears to be a modified CILIUM,
with the 9 + 2 microtubule arrangement where the outer segment
joins the rest of the cell. Pigment in the cone (*retinene*) requires bright
light before it bleaches; three different classes of cone each contain a
combination of retinene with a different genetically determined protein (opsin) bleaching at different wavelengths (red, green, blue).
Cones are thus responsible for colour vision. There are about 6
million cones per human RETINA, mostly concentrated in the FOVEA.
(2) Any of the cusps of mammalian molar teeth.

CONGENITAL. Of a property present at birth. In one sense a synonym
of HERITABLE (e.g. disorders arising from mutation such as DOWN'S
SYNDROME, PHENYLKETONURIA). Other infectious diseases (e.g.
German measles) may be transmitted across the placenta from mother
to foetus, as can some venereal diseases (e.g. congenital syphilis)
which are otherwise transmitted by sexual intercourse. In some cases,
however, the newborn may acquire a venereal disease by infection
from the mother at birth; this would be congenital infection.

CONIDIOPHORE. See CONIDIUM.

CONIDIUM. Asexual spore of certain fungi, cut off externally at apex
of specialized hypha (conidiophore).

CONIFER. Informal term for CONE-bearing tree. See CONIFEROPHYTA.

CONIFEROPHYTA. Conifers. Seed-plants with active cambial growth
and simple leaves; ovules and seeds exposed; sperm not flagellated.

CONJUGATED PROTEIN. Protein to which a non-protein portion (PROS-
THETIC GROUP) is attached.

CONJUGATION. (1) Union in which two individuals or filaments fuse together to exchange or donate genetic material. The process can involve a *conjugation tube* as in members of the Zygnematales (e.g. *Spirogyra*), or a *copulation tube* as seen in some Bacillariophyta, or no tube at all as in the Desmidiales. Process also occurs in fungi. In bacterial conjugation only a portion of the genetic material is transferred from the donor cell. Sex pili form (see PILUS), followed by transmission of one or more plasmids from one cell to the other; less commonly the chromosome itself is transferred (see F. FACTOR for *Hfr* strain), when conjugation may be employed in CHROMOSOME MAPPING. Conjugation occurs in some animals; in ciliates (e.g. *Paramecium*) interesting varieties of the process occur. In the simplest case, after partial cell fusion, macronuclei disintegrate and each of the two micronuclei undergoes meiosis, after which three of the four nuclei from each cell abort. The remaining nucleus in each cell divides mitotically, one of the two nuclei from each cell passing into the other in reciprocal fertilization. The cells separate, their nuclei each divide mitotically twice, two nuclei reforming the macronucleus while the other two regenerate the micronucleus. The significance of this is obscure, but cultures of just one mating type seem to 'age' and die out sooner than those that can conjugate; CYTOPLASMIC INHERITANCE is probably involved.

CONJUNCTIVA. Layer of transparent epidermis and underlying connective tissue covering the anterior surface of the vertebrate eyeball to the periphery of the cornea, and lining the inner aspect of the eyelids.

CONNECTIVE TISSUE. A variety of vertebrate tissues derived from MESENCHYME and the ground substance these cells secrete. A characteristic cell type is the FIBROBLAST, producing fibres of the proteins *collagen* and *elastin*, providing tensile strength and elasticity respectively. Another protein, *reticulin*, is associated with polysaccharides in the BASEMENT MEMBRANES underlying epithelia and surrounding fat cells in ADIPOSE TISSUE. Loose connective tissue (*areolar tissue*) binds many other tissues together (e.g. in capsules of glands, meninges of the CNS, bone periosteum, muscle perimysium and nerve perineurium). Collagen fibres align along the direction of tension, as in TENDONS and LIGAMENTS. The viscosity of many connective tissues is due to the 'space-filling' *hyaluronic acid*. The PERITONEUM, PLEURA and PERICARDIUM (serous membranes) are modified connective tissues, as are the skeletal tissues BONE and CARTILAGE. Besides supportive roles, connective tissue is defensive, due largely to presence of HISTIOCYTES (macrophages), which may be as numerous as fibroblasts. These and MAST CELLS constitute part of the RETICULO-ENDOTHELIAL SYSTEM. Connective tissues are frequently well vascularized, and permeated by tissue fluid.

CONSENSUS SEQUENCE. Conserved base sequences common in regulatory regions of DNA (e.g. promoter regions), where each base occurs in a particular position in a large proportion of genomes of that species. TATA BOX and PRIBNOW BOX sequences are examples.

CONSOCIATION. (Of plants) CLIMAX COMMUNITY of natural vegetation, with an ASSOCIATION dominated by *one* particular species; e.g. beech wood, dominated by the common beech tree.

CONSPECIFIC. Of individuals that are members of the same species.

CONSTITUTIVE ENZYME. An enzyme synthesized continuously, regardless of substrate availability. Compare INDUCIBLE ENZYME.

CONSTITUTIVE MUTANT, CONSTITUTIVE PHENOTYPE. Of mutants, initially studied in the context of the LAC OPERON in *E. coli*, whose continuous and abnormal production of a cell product is caused by a mutation in the cistron coding for the REPRESSOR molecule of the appropriate PROMOTER region, or from a mutation in the promoter region itself.

CONSUMMATORY ACT. An act, often stereotyped, terminating a behavioural sequence and leading to a period of quiescence. Compare APPETITIVE BEHAVIOUR. Factors, sometimes very specific, which lead to termination of the sequence are sometimes referred to as *consummatory stimuli*.

CONTACT INHIBITION. Phenomenon in which cells (e.g. fibroblasts) grown in culture in a monolayer normally cease movement at point of contact with another cell. When all available space is filled by cells, they also cease dividing – a phenomenon known as *contact inhibition of cell division*. Probably a normal self-regulatory device in regulating tissue and organ size. It tends to be lost in transformed CANCER CELLS.

CONTACT INSECTICIDE. Insecticide whose mode of entry to the body is via the cuticle rather than the gut. DDT and dieldrin are notorious examples, their fat-solubility (often needed to penetrate waxy epicuticle) resulting in accumulation in fat reserves of animals in higher TROPHIC LEVELS.

CONTINENTAL DRIFT. Theory, widely accepted since about 1953 but better described nowadays as *plate tectonics*, that crustal plates bounded by zones of tectonic activity, with the continents upon them, move slowly but cumulatively relative to one another. Helps explain distribution of many fossil and present-day forms formerly interpreted by invoking supposed land-bridges, and elevations and depressions of the sea bed. See GONDWANALAND, LAURASIA, PANGAEA.

CONTINUITY OF THE GERM PLASM. See GERM PLASM.

CONTOUR FEATHER. See FEATHER.

CONTRACEPTION. Deliberate prevention of fertilization and/or pregnancy, usually without hindering otherwise normal sexual activity. Includes: preventing sperm entry by use of a *condom*, a protective sheath over the penis; preventing access of sperm to the cervix by means of a *diaphragm* placed over it manually before copulation (later removed); preventing IMPLANTATION by means of an *intrauterine device* (*IUD*), a plastic or copper coil inserted under medical supervision into the uterus; and the CONTRACEPTIVE PILL. Withdrawal of the penis prior to ejaculation (*coitus interruptus*) is *not* an effective method. *Abstention* during the phase in the MENSTRUAL CYCLE when fertilization is likely is a further method. Use of condoms is fairly effective, with the added advantage of reducing the risk of infection by microorganisms during intercourse. IUDs can cause extra bleeding during menstruation and may not be tolerated by some women.

CONTRACEPTIVE PILL. An oral pill for women, usually containing a combination of OESTROGEN and PROGESTERONE. Taken each day for the first 21 days after completion of menstruation, but not for the next seven. Resultant fall in blood oestrogen and progesterone levels allows menstruation to occur. Oral contraceptive pills may have unpleasant side effects, and in some women contribute to thrombosis. Ovulation is inhibited through preventing normal gonadotropic effects of the HYPOTHALAMUS. Alternative methods of contraception are often advisable. Postcoital pills and oral contraceptives for men are among other hormonal contraceptives under study.

CONTRACTILE RING. Bundle of filaments, largely ACTIN, which assembles just beneath the plasma membrane at anaphase during animal cell divisions and in association with MYOSIN generates the force pulling opposed membrane surfaces together prior to 'pinching-off' and completion of cytokinesis.

CONTRACTILE ROOT. Root undergoing contraction at some stage, causing a change in position of the shoot relative to the ground. Some corm-bearing plants produce large, thick, fleshy roots possessing few root hairs. These roots store large amounts of carbohydrate in the cortex, which may be rapidly absorbed by the plant. The cortex then collapses and the root contracts downwards, pulling the corm deeper into the soil.

CONTRACTILE VACUOLE. Membrane-bound organelle of many protozoans (esp. ciliates), sponge cells and algal flagellates, In flagellated algae there are usually two anterior, contractile vacuoles. A vacuole will fill with an aqueous solution (diastole) and then expel the solution outside the cell (systole); this procedure is rhythmically repeated, and if two are present (e.g. in Volvocales) they usually fill and empty alternately. The process of filling is ATP-dependent, and therefore active. An alternative theory is that these vacuoles remove waste products from cells. Dinoflagellates have a similar but more complex structure, called a *pusule*, which may have a similar function.

CONTROL, CONTROLLED EXPERIMENT. See EXPERIMENT.

CONVERGENCE (CONVERGENT EVOLUTION). The process whereby complete organisms (or their parts) come to resemble each other more closely than their ancestors (or ancestral structures) did. The result is *phenetic* similarity, with little or no convergence at the genetic level implied. Structures which come to resemble each other in this way are said to be ANALOGOUS. Convergence poses problems for any purely phenetic CLASSIFICATION. Compare PARALLEL EVOLUTION. See CLADISTICS.

COPEPODA. Large subclass of CRUSTACEA. No compound eyes or carapace, and most only a few mm long. Usually six pairs of swimming legs on thorax; abdominal appendages absent. Filter-feeders, using appendages on head. Some marine forms, e.g. *Calanus*, occur in immense numbers in plankton and are vital in grazing food-webs. About 4500 species; some (e.g. fish louse) parasitic.

COPIA. TRANSPOSABLE ELEMENTS in *Drosophila* which resemble integrated retroviral proviruses. Extractable from eggs and cultured cells as double-stranded extrachromosomal circular DNA.

CORACOID. A cartilage bone of vertebrate shoulder girdle. Meets scapula at glenoid cavity, but reduced to a small process in non-monotreme mammals. See PECTORAL GIRDLE.

CORAL. See ACTINOZOA.

CORALLINE. A term referring to some members of the red algae (Rhodophyta) which become encrusted with lime (e.g. Corallinaceae).

CORDAITALES. Order of extinct palaeozoic gymnophytes that flourished particularly during the Carboniferous. Tall, slender trees, with dense crown of branches bearing many large, simple, elongated leaves. Sporophylls distinct from vegetative leaves, much reduced in size and arranged compactly in small, distinct, male and female cones. Microsporophylls stamen-like, interspersed among sterile scales; megasporophylls similarly borne, each consisting of a stalk bearing a terminal ovule.

CORIUM. See DERMIS.

CORK (PHELLEM). Protective tissue of dead, impermeable cells formed by cork cambium (phellogen) which, with increase in diameter of young stems and roots, replaces the epidermis. Developed abundantly on the trunk of certain trees, e.g. cork, oak, from which it is periodically stripped for commercial use. See BARK.

CORM. Organ of vegetative reproduction; swollen stem-base containing food material and bearing buds in the axils of scale-like remains of leaves of previous season's growth; food reserves not stored in leaves (compare BULB). Examples include crocus and gladiolus.

CORMOPHYTES. Refers to plants that possess a stem, leaf and roots (e.g. ferns, seed plants). Contrast BRYOPHYTA.

CORNEA. Transparent exposed part of the sclerotic layer of vertebrate and cephalopod eyes. Flanked in former by conjunctiva and responsible for most refraction of incident light (a 'coarse focus'), the lens producing the final image on the retina. Composed of orderly layers of COLLAGEN, lacking a blood supply, its nutrients derived via aqueous humour from CILIARY BODY.

CORNIFICATION (KERATINIZATION). Process whereby cells accumulate the fibrous protein *keratin* which eventually fills the cell, killing it. Occurs in the vertebrate epidermis, in nails, feathers and hair. See CYTOSKELETON.

COROLLA. Usually conspicuous, often coloured, part of a flower within calyx, consisting of a group of petals. See FLOWER.

COROLLA TUBE. Tube-like structure resulting from fusion of petals along their edges.

CORONARY VESSELS. Arteries and veins of vertebrates carrying blood to and from the heart.

CORPORA ALLATA. Small ectodermal endocrine glands in the insect head, connected by nerves to the CORPORA CARDIACA (to which they may fuse) and producing juvenile hormone (neotenin) which is responsible for maintenance of the larval condition during moulting. Decreasing concentration of their product is associated with progressive sequence of larval stages. Their relative inactivity in final larval stage of ENDOPTERYGOTA brings about pupation, and their complete inactivity in the pupa is responsible for differentiation into the final adult stage. Removal is termed *allatectomy*.

CORPORA CARDIACA (OESOPHAGEAL GANGLIA). Transformed nerve ganglia derived from the insect foregut, usually closely associated with the heart. Connected by nerves to, or fusing with, the CORPORA ALLATA and producing their own hormones; but mainly storing and releasing brain neurosecretory hormones, particularly *thoracotrophic hormone* which stimulates thoracic (prothoracic) glands to secrete ECDYSONE, initiating moulting.

CORPUS CALLOSUM. Broad tract of nerve fibres (commissures) connecting the two CEREBRAL HEMISPHERES in mammals in the neopallial region.

CORPUS LUTEUM. Temporary endocrine gland of ovaries of elasmobranchs, birds and mammals. In mammals develops from a ruptured GRAAFIAN FOLLICLE, and produces PROGESTERONE (as in elasmobranchs). Responsible in mammals for maintenance of uterine endometrium until menstruation, also during pregnancy. Its normal life (14 days in humans) is prolonged in pregnancy by *chorionic gonadotrophin* (HCG in humans). Its initial growth is due to LUTEINIZING HORMONE from the anterior pituitary. See OESTROUS CYCLE.

CORTEX. (Bot.) In some brown and red algae, tissue internal to epidermis but not central in position; in lichens, compact surface layer(s) of the thallus; in vascular plants, parenchymatous tissue located between vascular tissue and epidermis. (Zool.) (1) Outer layers of some animal organs, notably vertebrate ADRENALS, KIDNEYS and CEREBRAL HEMISPHERES. (2) Outer cytoplasm of cells (*ectoplasm*) where this is semi-solid (see CYTOSKELETON).

CORTICOSTEROIDS (CORTICOIDS). Steroids synthesized in the ADRENAL cortex from CHOLESTEROL. Some are potent hormones. Divisible into *glucocorticoids* (e.g. CORTISOL, cortisone, corticosterone), and *mineralocorticoids* (e.g. ALDOSTERONE). Some synthetic drugs related to cortisone (e.g. prednisone) reduce inflammation (e.g. in chronic bronchitis, relieving airway obstruction).

CORTICOTROPIN. See ACTH.

CORTISOL (HYDROCORTISONE). Principal glucocorticoid hormone of many mammals, humans included. (Corticosterone is more abundant in some small mammals.) Promotes GLUCONEOGENESIS and raises blood pressure. Low plasma cortisol level promotes release of *corticotropin releasing factor* (*CRF*) from the HYPOTHALAMUS, causing release in turn of ACTH from the anterior PITUITARY.

CORYMB. INFLORESCENCE, more or less flat-topped and indeterminate.

COSMOID SCALE. Non-placoid scale, having a thinner and harder outer layer composed of enamel-like material (*ganoine*) with hard non-cellular cosmine layer beneath. Growth is from the edge on the underside, since no living cells cover the surface. Characteristic of living and extinct lobefin fish (CROSSOPTERYGII), such as *Latimeria*. See GANOID SCALE, PLACOID SCALE.

COSTA. (1) In some members of the Bacillariophyta (diatoms), a ridge in the silica cell wall formed by well-defined siliceous ribs. (2) The midrib, or multilayered area, of a bryophyte leaf.

COSTAL. Relating to ribs.

COST OF MEIOSIS. The disadvantage which most (amphimictic) sexual individuals seem to incur in contributing copies of only half their genomes to any of their offspring (through meiosis) whereas greater genetic fitness would seem to come from producing parthenogenetic offspring. See SEX.

COTYLEDON (SEED LEAF). Leaf, forming part of seed EMBRYO; attached to embryo axis by hypocotyl. Structurally simpler than later formed leaves and usually lacking chlorophyll. Monocotyledons have one, dicotyledons two, per seed; the number varies in gymnophytes. Play important role in early stages of seedling development. In nonendospermic seeds, e.g. peas, beans, they are storage organs from which the seedling draws nutrients; in other seeds, e.g. grasses, compounds stored in another part of the seed, the ENDOSPERM, are absorbed by transfer cells on the outer epidermis of the cotyledons and passed to the embryo. Cotyledons of many plants (epigeal) appear above the soil, develop chlorophyll, and photosynthesize. See SCUTELLUM.

COTYLOSAURIA (MESOSAURIA). The 'stem reptiles' of the late Palaeozoic and Triassic. Limbs splayed sideways from the body; superficially rather amphibian. Probably a heterogeneous (polyphyletic) order.

COUNTERCURRENT SYSTEM. System where two fluids flow in opposite directions, one or both along vessels so apposed to one another that exchange of contents, heat, etc., occurs resulting in the level dropping progressively in one fluid while it rises progressively in the other. It may involve active secretion, as in the *countercurrent multiplier* system of the loop of Henle in the vertebrate KIDNEY, or be passive, as in the *countercurrent exchange* of respiratory gases in the teleost GILL. See RETE MIRABILIS and Fig. 23.

COUNTERSTAINING. See STAINING.

COV. See CROSS-OVER VALUE.

COXA. Basal segment of insect leg, linking trochanter and thorax.

COXAL BONES. See PELVIC GIRDLE.

COXAL GLANDS. Paired arthropod COELOMODUCTS. In Arachnida, opening on one or two pairs of legs; in Crustacea, a pair of coelomoducts (antennal glands) on the third (antennal) somite, or on the somite of the maxillae; sometimes both. In Onychophora, a pair in most segments. Excretory.

cpDNA. Chloroplast DNA. Larger than its mitochondrial counterpart mtDNA, but like it circular. Encodes the chloroplast's ribosomal RNAs and transfer RNAs, and part of ribulose bisphosphate (RuBP) carboxylase. Nuclear DNA encodes much of the rest of chloroplast structure. Many cpDNA mutations affecting the chloroplast are transmitted maternally, e.g. some forms of VARIEGATION. See MATERNAL INHERITANCE.

C₃, C₄ PLANTS. See PHOTOSYNTHESIS.

CRANIAL NERVES. Peripheral nerves emerging from brains of vertebrates (i.e. within the skull); distinct from SPINAL NERVES, which emerge from the spinal cord. Dorsal and ventral roots of nerves from several segments are involved, but (unlike those of spinal nerves) these remain separate. Each root is numbered and named as a separate nerve; numbering bears little relation to segmentation, but does to the relative posteriority of emergence. There are 10 pairs of cranial nerves in anamniotes; 11 or 12 pairs in amniotes. Nerves I and II (*olfactory* and *optic* nerves) are largely sensory; III (*oculomotor*) innervates four of the six eye muscles; IV (*trochlear*) innervates the superior oblique eye muscle; V (*trigeminal*) is sensory from the head, but motor to the jaw muscles; VI (*abducens*) innervates the posterior rectus eye muscle; VII (*facial*) is partly sensory, but mainly motor to facial muscles in mammals; VIII (*vestibulocochlear*) is sensory from the inner ear; IX (*glossopharyngeal*) is mainly sensory from tongue and pharynx; X (*vagus*) is large, including sensory and motor fibres to and from viscera; XI (*accessory*) is a motor nerve accessory to the vagus; XII (*hypoglossal*) is motor, serving the tongue.

CRANIATA. See VERTEBRATA.

CRANIUM. The vertebrate skull. See NEUROCRANIUM, SPLANCHNO-CRANIUM.

CRASSULACEAN ACID METABOLISM (CAM). Variant of the C_4 pathway of PHOTOSYNTHESIS, occurring particularly in succulents (e.g. cacti). Phosphoenolpyruvate carboxlyase fixes CO_2 into C_4 compounds at night, and then the fixed CO_2 is transferred to ribulose bisphosphate of the CALVIN CYCLE within the same cell during the day.

CREATINE. Nitrogenous compound ($NH_2.C[NH].N[CH_3].CH_2.COOH$), derivative of arginine, glycine and methionine; reversibly phosphorylated to *phosphocreatine*, which transfers its phosphate to ADP via the enzyme *creatine kinase*. Found in muscle. Its anhydride breakdown product, *creatinine*, is excreted in mammalian urine. See PHOSPHAGEN.

CREATININE. See CREATINE.

CREODONTS. Order of extinct mammals. Very varied, lasting into the Miocene, and ancestral to Fissipedia (dogs, cats, etc.). See CARNIVORA.

CRETACEOUS. GEOLOGICAL PERIOD, lasting roughly from 135–70 Myr BP. Much chalk deposited; anthophytes abundant; large DINOSAURS radiated, but along with ammonites and aquatic reptiles became extinct by the close. See EXTINCTION.

CRINOIDEA. Feather stars; sea lilies. Primitive class of ECHINODERMATA. Have long, branched, feathery arms; well-developed skeleton; tube feet without suckers; usually sedentary and stalked, with mouth upwards; microphagous; most modern forms free as adults. Long and important fossil history from Ordovician onwards (providing crinoid marble).

CRISTA (pl. CRISTAE). See (1) VESTIBULAR APPARATUS, (2) MITO-CHONDRION.

CRITICAL GROUP. Group of evidently closely related organisms, not easily categorized taxonomically. Used in the context of those apomicts which also reproduce by normal amphimictic means.

CROCODILIA. Sole surviving order of ARCHOSAURS. Alligators and crocodiles, appearing in the Triassic. Ancestors probably bipedal. Internal nares (CHOANAE) open far back in the mouth owing to presence of long bony FALSE PALATE. Some exhibit considerable parental care. Close affinities with birds.

CRO-MAGNON MAN. Earliest 'anatomically modern' humans. See HOMO.

CROP. In vertebrates, distensible expanded part of oesophagus in which food is stored (esp. birds). In invertebrates, an expanded part of the gut near the head, in which food may be stored or digested.

CROP MILK. Secretion comprising sloughed crop epithelium of both sexes in pigeons, on which nestlings are fed. Production influenced by PROLACTIN, like mammalian milk.

CROP ROTATION. The practice of growing different crops in regular succession to assist control of insect pests and diseases, increase soil fertility (especially when one season's growth includes nitrogen-fixing leguminous plants), and decrease erosion.

CROSS. The process or product of cross-fertilization. Contrast SELFING.

CROSSING-OVER. Mutual exchange of sections of homologous chromatids in first meiotic prophase. Under the influence of enzymes (see GENE CONVERSION, SPLICING). Responsible for *chiasmata* visible in meiotic chromosomes, and for non-random RECOMBINATION, its frequency and distribution are under genetic control (see SUPPRESSOR MUTATION). Non-homologous cross-overs and cross-overs between sister chromatids in MITOSIS (see PARASEXUAL cycle) also occur. Responsible for the CHIASMATA observed in first meiotic prophase. See CROSS-OVER VALUE, DNA REPAIR MECHANISMS, SYNAPTONEMAL COMPLEXES, TRANSPOSABLE ELEMENTS.

CROSSOPTERYGII. Order (sometimes superorder, or subclass) of OSTEICHTHYES, in the heterogeneous subclass CHOANICHTHYES. One known living form (*Latimeria*), the COELACANTH; fossil forms included ancestors of land vertebrates. Bony skeleton; paired fins, with central skeletal axes; COSMOID SCALES, or derivatives. First appeared in Devonian around 400 Myr BP, the fossil record disappearing around 70 Myr BP. Differ from DIPNOI in having normal conical teeth.

CROSS-OVER VALUE (COV, RECOMBINATION FREQUENCY). The percentage of meiotic products that are recombinant in an organism heterozygous at each of two linked loci. In diploid organisms, most easily measured by crossing the double heterozygote to the double recessive. This value gives the MAP DISTANCE between the two loci, used in CHROMOSOME MAPPING. The percentage can never exceed 50%, which value would indicate absence of LINKAGE between the loci. The CHI-SQUARED TEST can give the probability that this is so.

CROSS-POLLINATION. Transfer of pollen from stamens to stigma of a flower of a different plant of the same species. Compare SELF-POLLINATION.

CROWN GALL. See AGROBACTERIUM.

CROZIER FORMATION (HOOK FORMATION). (1) Similar to formation of CLAMP CONNECTION in Basidiomycotina, but occurring in dikaryotic cells of certain Ascomycotina; a hook develops in the ascogenous hypha where conjugate nuclear division takes place and is followed by cytokinesis; crozier formation may or may not immediately precede formation of an ascus. (2) In ferns and allies, the coiled juvenile leaf or stem.

CRUSTACEA. Class of ARTHROPODA, including shrimps, crabs, water fleas, etc. Mostly aquatic, with gills for gaseous exchange (and often nitrogenous excretion). Many segments with characteristic BIRA-MOUS APPENDAGES. Head bears single pairs of ANTENNULES, ANTENNAE (both primarily sensory), mandibles and maxillae (both for feeding). A pair of compound eyes common. Trunk composed of thorax and abdomen, often poorly distinct. Chitinous cuticle (COPE-PODS produce several million tons of CHITIN per year) often impregnated with calcium carbonate and excretory wastes (calcium reabsorbed prior to moulting). Small coelom, partly represented by 'kidneys' (antennal glands, maxillary glands) located in the head. Sexes usually separate; development usually via a NAUPLIUS larva. Includes hugely important microphagous filter-feeders in freshwater and marine food webs. Major subclasses are BRANCHIOPODA, OS-TRACODA, COPEPODA, CIRRIPEDIA, MALACOSTRACA.

CRYOBIOLOGY. Study of effects of very low temperatures on living systems. Some organisms or their parts (e.g. corneas, sperm) can be preserved under these conditions.

CRYOPHYTES. Plants growing on ice and snow; micro-plants, mostly algae but including some mosses, fungi and bacteria. Algal forms may be so abundant as to colour substratum, as in 'red snow', due to species of *Chlamydomonas*.

CRYPSIS. A relational term, indicating that an individual organism in a particular environment setting tends to be overlooked by one or more potential predators, through its having some combination of size, shape, colour, pattern and behaviour. Such characters may be polymorphic, as in the banding and shell colour polymorphism of the snail genus *Cepaea*. To the extent that such a combination can be shown to improve the bearer's fitness by reducing attention from predators, it is regarded as cryptic. There are similarities between crypsis and MIMICRY. See INDUSTRIAL MELANISM.

CRYPTIC SPECIES. See SPECIES.

CRYPTOGAM. Archaic term of early systematic botanists for all plants except gymnophytes and flowering plants (*phanerogams*), because their organs of reproduction are not as prominent as in the latter groups.

CRYPTOPHYTA. Division of Algae. Small group of unicellular algae, mostly motile, with two slightly unequal flagella emerging from a sub-apical pit (gullet). Cells dorsally-ventrally compressed, the majority naked with a firm cell membrane; others with cellulose wall. Reproduction by cell division. Variously pigmented, with chlorophylls *a* and *c*, α- and ε-carotene, xanthophylls. In some genera, phycobilin pigments are distinct from those of the Cyanophyta and Rhodophyta. Food reserve starch or starch-like. A few heterotrophs occur, either saprotrophs or ingesting food particles. Occur in marine and fresh water.

CRYPTORCHID. With testes not descended from abdominal cavity into scrotum.

CSF. See CEREBROSPINAL FLUID, COLONY STIMULATING FACTOR.

CTENIDIA. Gills of MOLLUSCA, situated in the mantle cavity. Involved in gaseous exchange, excretion, and/or filter-feeding.

CTENOPHORA. Comb-jellies, sea-gooseberries. Subphylum of COELENTERATA; with LASSO CELLS, but no cnidoblasts. Body neither polyp nor medusa; movement by cilia fused in rows (*combs* or *ctenes*); no asexual or sedentary phase. See CNIDARIA.

CULTIVAR. Variety of plant found only under cultivation.

CUSP. Pointed projection on biting surface of mammalian molar tooth. Each cusp is termed a *cone*. Cusping pattern may have taxonomic value.

CUTICLE. Superficial non-cellular layer, covering and secreted by the epidermis of many plants and invertebrates (esp. terrestrial species). In plants, an external waxy layer covering outer walls of epidermal cells; in bryophytes and vascular plants comprises waxy compound, cutin, almost impermeable to water. In algae may contain other compounds. In higher plants cuticle is only interrupted by stomata and lenticels. Its function is to protect against excessive water loss as well as protecting against mechanical injury.

The arthropod cuticle contains alpha-CHITIN, proteins, lipids, and polyphenol oxidases involved in its tanning (see SCLEROTIZATION). Normally subdivisible into an outer non-chitinous *epicuticle* and an inner chitinous *procuticle* (itself comprising an outer *exocuticle* and an inner *endocuticle*). The epicuticle (1 μm thick in insects) is composed of cemented and polymerized lipoproteins, and affords good waterproofing and resistance to desiccation (less so in Onychophora), having been of utmost importance in terrestrialization by insects. The endocuticles of decapod crustaceans are highly calcified, and often thick. In many arthropod larvae and some adults the cuticle remains soft and flexible (due largely to the protein *arthropodin*) but at some hinges (e.g. insect wing bases) another protein, *resilin*, enables greater flexibility still. Tanning hardens much of the arthropod cuticle, producing SCLERITES. Hardening is also brought about by water loss as the water-soluble arthropodin is converted to the insoluble protein sclerotin.

CUTICULARIZATION. Process of CUTICLE formation.

CUTIN. Fatty substance deposited in many plant cell walls, especially on outer surface of epidermal cells, where it forms a layer called CUTICLE.

CUTINIZATION. Impregnation of plant cell wall with CUTIN.

CUTIS. Skin; in vertebrates comprises dermis and epidermis.

CUTTING. Artificially detached plant part used in vegetative propagation.

CUVIER, GEORGES (1769–1832). Professor of vertebrate zoology at the *Musée d'Histoire Naturelle* in Paris (see LAMARCK). Polymath, specializing in geology and palaeontology. Formed the premise that any animal is so adapted to its environment (or conditions or existence) that it can *function* successfully in that environment. All parts of the animal therefore had to interrelate to form a viable whole; but certain parts are relatively invariant between organisms, and may therefore have value in CLASSIFICATION. His was not, however, an evolutionary system of classification, unlike Lamarck's; it is fairly clear that the two did not enjoy a cordial relationship. Cuvier had a genius for 'reconstructing' a whole vertebrate skeleton from a single bone.

CUVIERIAN DUCT. Paired major (common cardinal) vein of fish and tetrapod embryos returning blood to heart from CARDINAL VEINS (under gut) in a fold of coelomic lining forming posterior wall of pericardial cavity. Becomes part of the superior vena cava of adult tetrapods.

C VALUE. Total amount of DNA in a viral, bacterial, or haploid eukaryotic genome. Usually expressed in picograms (pg) per cell.

CYANELLE. Endosymbiotic blue-green alga, usually in association with protozoa.

CYANOBACTERIA (CYANOPHYTA, MYXOPHYTA). Blue-green algae. Division of the Monera, sharing general PROKARYOTIC properties and including unicellular, colonial and filamentous forms. Fossil blue-greens make up layered formations called STROMATOLITES. Probably first oxygen-producing organisms to evolve, responsible for early accumulation of atmospheric oxygen. Related to BACTERIA, their photosynthetic apparatus includes both chlorophyll *a*, carotenoids and phycobiliprotein pigments contained in flattened membranous sacs (thylakoids) scattered throughout the cell. Contain glycogen as stored carbohydrate. Gelatinous sheath usually envelops the cells, distinctly red or blue depending on whether the cell is growing in respectively highly acid or alkaline conditions. Cell wall construction and chemistry resemble those of Gram-negative bacteria.

Many planktonic blue-greens possess gas vesicles for cell buoyancy, and many benthic species move when in contact with a solid surface, without evident organs of locomotion but probably either through secretion of mucilage from pores in the cell wall, or through contractile waves. Some species reproduce by cell division; in some filamentous forms specialized fragments of the filament (hormogonia) serve in reproduction. Several unicellular reproductive agents are also formed by filamentous types: AKINETES, endospores, exospores and HETEROCYSTS, the latter being the site of production of the nitrogen-fixing enzyme *nitrogenase*. Sexual reproduction is not typical, but genetic recombination resembling that in bacterial conjugation may take place.

Widely distributed, occurring in waters of varying salinity, nutrient status and temperature. Species found in plankton, benthos and aerial habitats. Generally more abundant in neutral or slightly alkaline water; but also in alkaline hot springs. In nutrient-rich (eutrophic) lakes, several species often form late-summer BLOOMS, e.g. *Anabaena flos-aquae, Microcystis aeruginosa*. Some strains of these species poison animals which drink the water. Can cause noxious odours in stagnant waters. Many species form symbiotic relationships with other organisms, e.g. with fungi (forming lichens), bryophytes, ferns, seed plants and animals. Other than some bacteria, blue-greens are the only group of organisms capable of fixing atmospheric nitrogen (see NITROGEN CYCLE).

CYANOCOBALAMIN (VITAMIN B$_{12}$). Cobalt- and nucleotide-containing vitamin, synthesized only by some microorganisms. In vertebrates, carried across the gut wall by a glycoprotein (*intrinsic factor*) of gastric juice. Essential for erythrocyte maturation and nucleotide synthesis (through derivative, *coenzyme B$_{12}$*). Absence in diet or lack of intrinsic factor cause pernicious anaemia. See VITAMIN B COMPLEX.

CYANOPHYCEAN STARCH. A storage polysaccharide found in cells of the CYANOBACTERIA; considered to be amylopectin.

CYANOPHYCIN GRANULE. A protein storage body found in the cells of the CYANOBACTERIA.

CYCADOFILICALES (PTERIDOSPERMAE). Order of extinct, palaeozoic gymnophytes that flourished mainly during the Carboniferous. Of great phylogenetic interest. Reproduced by seeds, but with fern-like leaves; internal anatomy combined fern-like vascular system with development of secondary wood. Micro- and mega-sporophylls little different from ordinary vegetative fronds; not arranged in cones. Compare CYCADOPHYTA.

CYCADOPHYTA. The cycads; gymnophytes indigenous to tropical and sub-tropical regions, living up to a great age (up to 1000 years). With sluggish cambial growth and pinnately compound, palm-like or fern-like leaves; ovules and seeds exposed. Sperm flagellated and motile, but carried to vicinity of the ovule in a pollen tube. Stem unbranched, tuberous or columnar and up to 20 m in height, bearing crown of fern-like leaves. Dioecious, microsporophylls distinct from vegetative leaves and arranged in a compact cone. Megasporophylls leaf-like, loosely grouped, or highly modified, grouped in a compact cone. Compare CYCADOFILICALES.

CYCLIC AMP (cAMP). See AMP, SECOND MESSENGER.

CYCLOHEXIMIDE. Antibiotic, inhibiting TRANSLATION of nuclear mRNA on cytoplasmic ribosomes. Prokaryotic translation unaffected, so used to distinguish mitochondrial/chloroplast from nuclear encoding of cell protein.

CYCLOSIS. Circulation of protoplasm (cytoplasmic streaming) in many eukaryotic cells, especially large plant cells. Sometimes restricted, as in small plant cells and animal cells, to jerky movement of organelles and granules in the cytoplasm. Cytoplasmic streaming commonly involves ATP-dependent propulsion of organelles along actin microfilaments under the influence of MYOSIN, although individual organelle movement usually involves them in a DYNEIN- or KINESIN-dependent sliding along MICROTUBULES.

CYCLOSTOMES. (1) AGNATHA whose sole living representatives are lampreys and hagfishes. Eel-like but jawless, with a sucking mouth and one nostril (lampreys), or two (hagfishes); without bone, scales or paired fins. Sometimes placed in a single order (CYCLOSTOMATA), but more usually regarded as two long-separated orders (*Petromyzontiformes*, lampreys, *Myxiniformes*, hagfishes) in subclasses Monorhina and Diplorhina respectively. Lampreys are ectoparasitic on vertebrates and ANADROMOUS; hagfishes are colonial burrowers, feeding largely on polychaete worms or corpses. Lamprey larva is the AMMOCOETE. (2) A suborder of ECTOPROCTA.

CYME. Branched, flat-topped or convex INFLORESCENCE where the terminal flower on each axis blooms first.

CYPSELA. Characteristic fruit of Compositae (sunflower, daisy, etc.). Like an achene (and usually so described), but formed from an inferior ovary and thus sheathed with other floral tissues outside ovary wall. Strictly, a pseudo-nut, being formed from two carpels.

CYSTICERCUS. Bladderworm larva of some tapeworms. Some (e.g. *Echinococcus*) are asexual in an encapsulated cysticercus known as a *hydatid cyst*. Here, brood capsules form on the inner wall, each budding off scolices producing new hydatid cysts. Human liver may become infected with these cysts. See POLYEMBRYONY.

CYSTOCARP. Structure developed after fertilization in red algae; consisting of filaments bearing terminal CARPOSPORES produced from the fertilized carpogonium, the whole enveloped in some genera by filament arising from neighbouring cells.

CYSTOLITH. Stalked body, consisting of ingrowth of cell wall, bearing deposit of calcium carbonate; found in epidermal cells of certain plants, e.g. stinging-nettle. See STATOLITH.

CYTOCHALASINS. Anti-cytoskeletal drugs binding reversibly to ACTIN monomers.

CYTOCHROMES. System of electron-transferring proteins, often regarded as enzymes, with iron-porphyrin or (in *cyt c*) copper-porphyrin as prosthetic groups; unlike in haemoglobin, the metal atom in the porphyrin ring must change its valency for the molecule to function. Located in inner mitochondrial membranes, thylakoids of chloroplasts and endoplasmic reticulum. See CHEMIOSMOTIC THEORY, ELECTRON TRANSPORT SYSTEM.

CYTOGENETICS. Study of linking cell structure, particularly number, structure and behaviour of chromosomes (e.g. their rate of replication) to data from breeding work. Often provides evidence for phylogeny. See TAXONOMY.

CYTOKINESIS. Division of a cell's cytoplasm, as opposed to its nucleus. Distinguished, therefore, from MITOSIS and MEIOSIS. In eukaryotic cells, onset is usually in late anaphase, when the plasma membrane in middle of cell is drawn in to form a *cleavage furrow*, formed by a contractile ring of ACTIN microfilaments; this enlarges and finally breaks through the remains of the spindle fibres to leave two complete cells.

CYTOKININS (PHYTOKININS). Group of plant GROWTH SUBSTANCES recognized and named because of their stimulatory effect (requiring AUXIN) on plant cell division; but of diverse origin. Chemically identified as purines; first discovered (kinetin) was isolated from yeast, animal tissues and sweet corn kernels. Substances with similar physiological action occur in fruitlets, coconut milk and other liquid endosperms; also in microorganisms causing plant tumours, witches' brooms and infected tissues. Other growth-promoting influences include cell enlargement, seed germination, stimulation of bud formation, delay of senescence, and overcoming apical dominance. Effects are thought to involve increased nucleic acid metabolism and protein synthesis. Their mode of transport in plants is unresolved. See GIBBERELLINS.

CYTOLOGY. Study of cells, particularly through microscopy.

CYTOLYSIS. Dissolution of cells, particularly by destruction of plasma membranes.

CYTOPATHIC. Damaging to cells.

CYTOPLASM. All cell contents, including the plasma membrane, but excluding any nuclei. Comprises cytoplasmic matrix, or CYTOSOL, in which ORGANELLES are suspended, some membrane-bound and some not, plus crystalline or otherwise insoluble granules of various kinds. In amoeboid cells, there is a distinction between a semi-solid outer *plasmagel* (ectoplasm) and a less viscous inner *plasmasol*. A highly organized aqueous fluid, where enzyme localization is of paramount importance. See CELL MEMBRANES, CYTOSKELETON.

CYTOPLASMIC GROUND SUBSTANCE. Least differentiated part of the cytoplasm, as seen with the electron microscope; the portion surrounding the nucleus and various organelles; also called the *hyaloplasm*. Equivalent to the CYTOSOL.

CYTOPLASMIC INHERITANCE. (1) Eukaryotic genetics involving DNA lying outside the nucleus, often in organelles (see cpDNA, mtDNA, PLASMID) or endosymbionts (see KAPPA PARTICLES). Patterns of inheritance from such a source characteristically fail to observe Mendelian ratios. (2) Inheritance of a cytoplasmic pattern, apparently independently of both nuclear and organelle DNA. Compare MATERNAL INHERITANCE.

CYTOPLASMIC MALE STERILITY (CMS). Trait of higher plants (e.g. maize) determined by either a mitochondrial PLASMID gene or a mtDNA gene, and an example of MATERNAL INHERITANCE. Pollen production aborts in development and plants are therefore self-sterile. Important agronomically, since hybrid plant lines can be produced combining desirable characters from different inbred parents.

CYTOPLASMIC STREAMING. See CYCLOSIS, MICROTUBULE.

CYTOSINE. A PYRIMIDINE base found in the nucleic acids DNA and RNA, as well as in appropriate nucleotides and their derivatives.

CYTOSKELETON. Network of ACTIN microfilaments, tubulin MICRO-

TUBULES and INTERMEDIATE FILAMENTS, much of it just beneath the plasma membrane, conferring upon a eukaryotic cell (especially an animal cell) its shape and generating the spatial organization within it, providing its attachment capabilities and enabling it to move materials both within it and out from it. It provides several structures involved in locomotion (e.g. in muscle contraction, amoeboid and ciliary locomotion). Actin is the major constituent in cell surface MICROVILLI, MICROSPIKES and stereocilia, while tubulin is the main constituent in cilia and flagella. Beneath the cell surface, belt DESMOSOMES, STRESS FIBRES and CONTRACTILE RINGS all involve actin, sometimes in association with myosin or other protein filaments. Animal epithelial cells often have keratin-attachment sites (spot desmosomes) helping to link cells together. When such cells die, cross-linked keratinized cytoskeletons may form a protective surface layer, as in HAIR and NAILS. See CELL LOCOMOTION, CYCLOSIS, INTERCELLULAR JUNCTION.

CYTOSOL. The fluid and semi-fluid matrix of the cytoplasm, including the CYTOSKELETON, in which are suspended the organelles.

CYTOTAXONOMY. See TAXONOMY.

CYTOTOXIC. Poisonous, or lethal, to cells. See T-CELL.

D

DARWIN, DARWINISM. Charles Robert Darwin (1809–1882) began studying medicine at Edinburgh in 1825, but left after two years to study for the clergy at Cambridge. In 1831 the botanist J. S. Henslow suggested to him that he might join HMS *Beagle* on its survey of the South American coast, and in the same year he set sail on a momentous voyage lasting almost five years. In 1842 the Darwin family moved to Down House in Kent, where his most famous work *The Origin of Species by Means of Natural Selection* (published in 1859), and other influential books, were written. In 1857 Darwin received a letter from Alfred Russel WALLACE indicating that the two men were thinking along similar lines as to the mechanism of evolution – natural selection. After many years of ill health, Darwin died at Down House on April 19th, 1882.

In essence, *Darwinism* is the thesis that species are not fixed, either in form or number; that new species continue to arise while others become extinct; that observed harmonies between an organism's structure and way of life are neither coincidental nor necessarily proof of the existence of a benevolent deity, but that any apparent design is inevitable given: (a) the tendency of all organisms to overproduce, despite the limited nature of resources (e.g. food) available to them; (b) that few individuals of a species are precisely alike in any measurable variable; (c) that some at least of this variance is heritable; and (d) that some of the differences between individuals must result in a largely unobserved selective mortality, or 'struggle for existence', to which (a) must lead. Moreover, he argued, the 'struggle' is likely to be most intense between individuals of the same species, their needs being most similar. Darwin's metaphor 'struggle' caused much confusion, but his more abstract phrase for the mechanism of evolution, NATURAL SELECTION, has survived the test of time.

Much of the theoretical input for the theory of evolution by natural selection came from Lyell's *Principles of Geology* (1830), which Darwin took on HMS *Beagle*, and Malthus's *Essay on the Principle of Population* (1798), which he read 'for amusement' in 1838. The chief empirical influences included observations on the fauna and flora of the Galapagos archipelago, so similar to – yet distinct from – those on the mainland, the fossil armadillos and ground sloths of Patagonia, and the effects of ARTIFICIAL SELECTION on domesticated plants and animals. Darwinism argues for common genealogical links between living and fossil organisms. Darwin gave an unconvincing account of the origin and nature of phenotypic variation, so essential to his theory, and had rather little to say on the strict title of his major work; but Darwinism remains the single most powerful and unifying theory in biology. See BIOGEOGRAPHY, EVOLUTION, NEO-DARWINISM.

DAUGHTER CELLS and NUCLEI. Cells and nuclei resulting from the division of a single cell.

DAY-NEUTRAL PLANTS. Plants that flower without regard to daylength. See SHORT-DAY PLANTS, LONG-DAY PLANTS.

DEAMINATION. Removal of an amino (-NH₂) group, frequently from an AMINO ACID, by *transaminase* ENZYMES. In mammals, occurs chiefly in the liver, where the amino group is used in production of UREA. The process is important in GLUCONEOGENESIS, where resulting carbon skeleton yields free glucose.

DECAPODA. (1) Order of MALACOSTRACA, including prawns and lobsters (long abdomens), and crabs (reduced abdomens). Three anterior pairs of thoracic appendages are used for feeding; remaining five are for walking (hence name) or swimming, although first and second of these may bear pincers (chelae). Fused CEPHALOTHORAX covered by a CARAPACE, which may be heavily calcified. (2) Suborder of MOLLUSCA, of the class Cephalopoda, having ten arms; squids, cuttlefish. Compare OCTOPODA.

DECEREBRATE RIGIDITY. See RETICULAR FORMATION.

DECIDUA. Thickened and highly vascularized mucus membrane (endometrium) lining the uterus in many mammals (not in *ungulates*) during pregnancy. Some or all of the decidua comes away with the PLACENTA at birth.

DECIDUOUS. (Of plants) shedding leaves at a certain season, e.g. autumn. Compare EVERGREEN.

DECIDUOUS TEETH (MILK TEETH, PRIMARY TEETH). First of the two sets of teeth which most mammals have; similar to second (permanent) set which replaces it, but having grinding teeth corresponding only to the PREMOLARS, not to the MOLARS, of the permanent set.

DECOMPOSER. Any HETEROTROPH breaking down dead organic matter to simpler organic or inorganic material. In some ECOSYSTEMS the *decomposer food chain* is energetically more important than the *grazing food chain*. All release a proportion of their organic carbon intake as CO₂ and the heat release, as evidenced by compost heaps, can be considerable; a useful aspect is that most pathogenic bacteria and cysts, eggs, and immature forms of plant and animal parasites that may have been present, as in sewage, are killed.

An important factor in composting is the carbon to nitrogen ratio. A ratio of about 30 : 1 (by weight) is optimal: any higher and microbial growth slows. Because composting greatly reduces the bulk of plant wastes, it can be very useful in waste disposal.

DECUSSATION. Crossing of nerve tracts (e.g. some COMMISSURES) from one side of the brain to the other. Some fibres of the optic nerves and corticospinal tracts of vertebrates cross over in this way, so that parts of the visual field from each retina are transmitted to contralateral optic tecta; each side of the body is served by its contralateral cerebral hemisphere. See OPTIC CHIASMA.

DEFICIENCY. Chromosomal mutation in which a detectable length of chromosome has been lost. See ANEUPLOIDY, DELETION. Contrast DUPLICATION.

DEFICIENCY DISEASE. Disease due to lack of some essential nutrient, in particular a VITAMIN, trace element or essential AMINO ACID.

DEGENERATION. Reduction or loss of whole or part of an organ during the course of evolution, with the result that it becomes VESTIGIAL. In the context of cells (e.g. nerve fibres) it usually implies their disorganization and death. Compare ATROPHY.

DEHISCENT. (Of fruits) opening to liberate the seeds; e.g. pea, violet, poppy.

DEHYDRATION. Elimination of water. (a) Prior to STAINING, usually achieved by soaking for up to 12 hours in successively stronger ethanol (ethyl alcohol), with at least two changes of 100%, in the preparation of tissues for microscopical examination. Failure to dehydrate properly leads to shrinkage and brittleness on embedding in paraffin, and deterioration of histological structure. See CLEARING. (b) Dry mass of soil and biological material is found after dehydration by heating to constant weight in an oven at 90–95°C.

DEHYDROGENASE. Enzyme catalysing a REDOX REACTION involving removal of hydrogen from one substrate and its transfer to another, often a COENZYME (e.g. NAD, FAD). Many respiratory enzymes are dehydrogenases.

DELETION. Type of chromosomal MUTATION in which a section of chromosome is lost, usually during MITOSIS or MEIOSIS. Unlike some mutations, deletions are not usually reversible or correctable by a SUPPRESSOR MUTATION. See CHROMOSOME MAPPING.

DEME, -DEME. Denotes a group of individuals of a specified taxon – the specificity given by the prefix used. The term deme on its own is not generally advocated, but where found usually refers to a group of individuals below, at or about the SPECIES level. Thus, groups of individuals of a specified taxon in a particular area (*topodeme*); in a particular habitat (*ecodeme*); with a particular chromosome condition (*cytodeme*); within which free gene exchange in a local area is possible (*gamodeme*), and of those in a gamodeme which are believed to interbreed more or less freely under specified conditions (*hologamodeme*). In zoology, the term deme without prefix tends to be used in the sense of gamodeme. See INFRASPECIFIC VARIATION.

DEMOGRAPHY. Numerical and mathematical analysis of populations and their distributions.

DENATURATION. Changes occurring to molecules of globular proteins and nucleic acid in solution in response to extremes of pH or temperature, or to urea, alcohols or detergents. Most visible effect with globular proteins is decrease in solubility (precipitation from solution), non-covalent bonds giving the molecule its physiological secondary and tertiary structures being broken but the covalent bonds providing primary structure remaining intact. Solutions of double-stranded DNA become less viscous as denaturation by any of the above factors results in strand separation by rupture of hydrogen bonds. This *melting* of DNA occurs with just very small increases in temperature, and may be reversed by the same temperature drop, the *re-annealing* being used in DNA HYBRIDIZATION techniques for assessing the degree of genetic similarity between individuals from different taxa.

DENDRITE. One of many cytoplasmic processes branching from the CELL BODY of a nerve cell and synapsing with other neurones. Several hundred BOUTONS may form synaptic connections with a single cell body and its dendrites.

DENDROCHRONOLOGY. Use of isotopes and annual rings of trees to assess age of tree; in fossils, used to date the stratum and/or make inferences concerning palaeoclimate. See GROWTH RING.

DENDROGRAM. Branching tree-like diagram indicating degrees of phenetic resemblance between organisms (a *phenogram*), or their phylogenetic relationships. In the latter, the vertical axis represents time, or relative level of advancement. A CLADOGRAM is a dendrogram representing phylogenetic relationships as interpreted by CLADISTICS. See TAXONOMY.

DENITRIFICATION. Process carried out by various facultative and anaerobic soil bacteria, in which nitrate ions act as alternative electron acceptors to oxygen during respiration, resulting in release of gaseous nitrogen. This nitrogen loss accounts in part for the lack of fertility of constantly wet soils that support nitrate-reducing anaerobes and for lowered soil fertility generally, products of denitrification not being assimilable by higher plants or most microorganisms. Bacteria such as *Pseudomonas*, *Achromobacter* and *Bacillus* are particularly important. See NITROGEN CYCLE.

DENITRIFYING BACTERIA. See DENITRIFICATION.

DENSITY-DEPENDENCE. Widely observed and important way in which populations of cells or organisms are naturally regulated. One or more factors act as (a) increasing brakes on population increase with increased population density, and/or (b) decreasing brakes on population increase with decreased population density. There must be a *proportional* increase or decrease in the effect of the factor on population density as density rises or falls respectively. For example, the proportion of caterpillars parasitized by a fly must increase with increase in caterpillar density if the fly is to act as a density-dependent control factor. Since the caterpillar and fly may be regarded as a kind of NEGATIVE FEEDBACK system, some have suggested that ECOSYSTEMS might self-regulate by this sort of process. See BALANCE OF NATURE. CONTACT INHIBITION by cells is a form of density-dependent inhibition of tissue growth.

DENSITY GRADIENT CENTRIFUGATION. Procedure whereby cell components (nuclei, organelles) and macromolecules can be separated by ultracentrifugation in caesium chloride or sucrose solutions whose densities increase progressively along (down) the centrifuge tube. Can be used to separate and hence distinguish different DNA molecules on the basis of whether or not they have incorporated heavy nitrogen (^{15}N) atoms. Cell components or macromolecules cease sedimenting when they reach solution densities that match their own buoyant densities.

DENTAL FORMULA. Formula indicating for a mammal species the number of each kind of tooth it has. The number in upper jaw of one side only is written above that in lower jaw of the same side. Categories of teeth are given in the order: incisors, canines, premolars, molars. The formulae for the following mammals are:

	i	c	pm	m
hedgehog	3 . 1 . 3 . 3			
	2 . 1 . 2 . 3			
grey squirrel	1 . 0 . 2 . 3			
	1 . 0 . 1 . 3			
ruminants	0 . 0-1 . 3 . 3			
	3 . 1 . 3 . 3			
cats	3 . 1 . 2 . 0			
	3 . 1 . 2 . 1			
man	2 . 1 . 3 . 3			
	2 . 1 . 3 . 3			

DENTARY. One of the tooth-bearing MEMBRANE BONES of the vertebrate lower jaw, and the only such bone in lower jaws of mammals, one on each side.

DENTICLE. See PLACOID SCALE.

DENTINE. Main constituent of teeth, lying between enamel and pulp cavity. Secreted by ODONTOBLASTS (hence mesodermal in origin), and similar in composition to BONE, but containing up to 70% inorganic material. Ivory is dentine. See TOOTH.

DENTITION. Number, type and arrangement of an animal's TEETH. Where teeth are all very similar in structure and size (most non-mammalian and primitive insectivore mammals) the arrangement is termed *homodont*, where there is variety of type and size of teeth, the arrangement is termed *heterodont*. See DECIDUOUS TEETH, DENTAL FORMULA, PERMANENT TEETH.

DEOXYRIBONUCLEASE. See DNase.

DEOXYRIBONUCLEIC ACID. See DNA.

DEPHOSPHORYLATION. See PHOSPHORYLATION.

DERMAL BONE (MEMBRANE BONE). Vertebrate bone developing directly from mesenchyme rather than from pre-existing cartilage (cartilage bone). Largely restricted in tetrapods to bones of CRANIUM, JAWS and PECTORAL GIRDLE. See OSSIFICATION.

DERMAPTERA. Small order of orthopterous, exopterygote insects, including earwigs. Fan-like hind wings folding under short, stiff, forewings (resembling elytra); but wingless forms common; biting mouthparts; forceps-like cerci at end of abdomen.

DERMATOGEN. See APICAL MERISTEM.

DERMATOPHYTE. Fungus causing disease of the skin or hair of humans and other animals. Two of the most common diseases are athlete's foot and ringworm.

DERMIS (CORIUM). Innermost of the two layers of vertebrate skin, much thicker than the EPIDERMIS, and comprising CONNECTIVE TISSUE with abundant collagen fibres (mainly parallel to the surface); scattered cells including CHROMATOPHORES; blood and lymph vessels and sensory nerves. Sweat glands and hair follicles project down from the epidermis into the dermis, but are not strictly of dermal origin. Responsible for tensile strength of skin. May contain SCALES or BONE. See DERMAL BONE.

DERMOPTERA. Small order of placental mammals (one genus, *Cynocephalus*); so-called flying lemurs, although they have no close affinities with other mammals, but are probably an offshoot of a primitive insectivore stock. Also called colugos. The lower incisors are each divided and comb-like. They glide by means of the *patagium*, a hairy membranous skin fold stretching from neck to webbed finger tips and thence to webbed toes and tip of longish tail.

DESERT. A major BIOME, characterized by little rainfall and consequently little or no plant cover. Included are the *cold deserts* of polar regions, such as tundra and areas covered by permanent snow and ice. *Hot deserts* have very high temperatures, often exceeding 36°C in the summer months. Rainfall may be less than 100 mm per year. Deserts can be extensive: the African Sahara is the world's largest; Australian deserts cover some 44% of the continent. Annual plants (desert ephemerals) are most important in these conditions, both numerically and in kind. They have a rapid growth cycle which can be completed quickly when water is available, seeds surviving in desert soils during periods of drought. Perennials that do occur are mostly bulbous and dormant for most of the time. Taller perennials are either succulents (e.g. cacti) or possess tiny leaves that are leathery or shed during periods of drought. Many succulents exhibit CRASSULACEAN ACID METABOLISM, absorbing carbon dioxide at night. See OSMOREGULATION.

DESMID. Informal term referring to two orders of freshwater green algae (Chlorophyta) whose cell walls are porate and in two sections united by a narrow constriction or isthmus. Cells may be solitary, joined end to end in filamentous colonies, or united in amorphous colonies. Their taxonomy is complicated by polymorphism, species sometimes having more than one appearance. Cells possess a single nucleus and two chloroplasts. Mucilage is secreted through pores in the cell wall and is responsible for movement and, when copiously secreted, for adherence to the substrate (commonly as epiphytes). Barium sulphate crystals occur within a vacuole at the tip of cells in some genera. Sexual reproduction involves CONJUGATION and production of amoeboid gametes. Spines develop upon the zygote walls and germinate meiotically, completing the life cycle. Desmids are usually indicators of relatively unpolluted water, with low calcium and magnesium and slightly acidic pH. Compare diatoms (BACILLAR-IOPHYTA).

DESMIN. An INTERMEDIATE FILAMENT protein characteristic of smooth, striated and cardiac muscle cells and fibres. In sarcomeric muscle it may help to link the Z-discs together; in smooth muscle it probably serves to anchor cells together. Also a prominent component of fibres on the cytoplasmic side of DESMOSOMES.

DESMOSOMES. One kind of INTERCELLULAR JUNCTION (see Fig. 37) found typically where animal cells need firm attachment to one another against severe stress which would tear or shear tissues. A *belt desmosome* (zonula adhaerens) comprises a band of contractile ACTIN filaments near the apical end of each epithelial cell, just under the cell membrane. A sheet of cells so united may roll up to form a tube by contraction of these filaments – as in neural tube formation. *Spot desmosomes* are sites of keratin filament attachment on the inner cell membrane surface, the filaments forming a network within the cell and connected to those in other cells, spot desmosomes being paired in adjacent cells. *Hemidesmosomes* link keratin attachment sites of epithelial cells to the underlying BASAL LAMINA; like spot desmosomes, they are rivet-like, and probably transfer stress from the epithelium to the underlying connective tissue via the basal lamina. See CYTOSKELETON.

DESMOTUBULE. The tubule that traverses a plasmodesmatal canal, uniting the endoplasmic reticulum of two adjacent plant cells.

DETERMINANT. For antigenic determinant, see ANTIGEN.

DETERMINATE GROWTH. (Bot.) Growth of limited duration; characteristically seen in floral meristems and leaves of plants.

DETERMINED. Term applied to an embryonic cell after its fate has been irreversibly fixed. See COMPETENCE, ORGANIZER, POSITIONAL INFORMATION, PRESUMPTIVE.

DETRITUS. Organic debris from decomposing organisms and their products. The source of nutrient and energy input for the *detritus food chain*. See DECOMPOSER.

DEUTEROMYCOTINA. Formal subdivision of the fungi (EUMYCOTA), commonly known as fungi imperfecti. They all lack a sexual stage, thereby lying outside the remainder of fungal classification, heavily based as it is upon sexual reproduction. Some species may be secondarily non-sexual, others possibly never possessed a sexual stage. Generally believed to be non-sexual stages (anamorphs) of fungi belonging to the ASCOMYCOTINA and BASIDIOMYCOTINA, the largest number in the former category. Frequently, a sexual stage is discovered later, both stages (anamorph and telomorph) then being transferred to the group to which the telomorph belongs. Deuteromycetes occur in virtually every habitat and on every type of substratum. Many are saprotrophs in soil; many are plant and animal parasites (e.g. causing such diseases as ringworm and athlete's foot); some occur in flowing water. Many moulds are included, some of commercial importance (e.g. *Penicillium roquefortii* and *P. camembertii* in cheese production; *Aspergillus oryzae* in production of soy paste, or miso). Deuteromycete moulds grow prolifically on artificial media and are widely used in genetic, biochemical and nutritional research. Strains of the genus *Penicillium* produce antibiotics, while large species of *Aspergillus* produce citric acid when grown under very acidic conditions.

DEUTEROSTOMIA. That assemblage of coelomate animals (some call it an infragrade) in which the embryonic BLASTOPORE becomes the anus of the adult, a separate opening emerging for the mouth (see STOMODAEUM). It thus includes the POGONOPHORA, ECHINODER-

MATA, HEMICHORDATA, UROCHORDATA and CHORDATA. Compare PROTOSTOMIA.

DEUTOPLASM. Nutritional substances, comprising YOLK, within an egg.

DEVELOPMENT. Complex processes and events whereby a multicellular organism reaches its full size and form. Involves both genetic and environmental influences; but the distinction is somewhat arbitrary due to EPISTASIS. See EPIGENESIS.

DEVELOPMENTAL COMPARTMENTALIZATION. See COMPARTMENT.

DEVELOPMENTAL HOMEOSTASIS. See HOMEOSTASIS.

DEVONIAN. GEOLOGICAL PERIOD lasting from approx. 400–350 Myr BP. Noted for Old Red Sandstone deposits, for the variety of fossil fish, including ACTINOPTERYGII and CHOANICHTHYES, primitive AMPHIBIA (i.e. ichthyostegids), and a variety of pteridophytes.

DEXTRANS. Storage polysaccharides of yeasts and bacteria in which D-glucose monomers are linked by a variety of bond types, producing branched molecules.

DEXTRIN. Polysaccharide formed as intermediate product in the hydrolysis of STARCH (e.g. to maltose) by AMYLASES.

DEXTROSE. Alternative name for GLUCOSE.

DIABETES. (1) *Diabetes insipidus.* An uncommon disorder in which a copious urine arises usually owing to a person's inability to secrete ANTIDIURETIC HORMONE. (2) *Diabetes mellitus* (or simply, *diabetes*). The insulin-dependent form of this disease generally presents fairly early in life, due either to inability of the pancreas to secrete INSULIN, or to insensitivity of the appropriate target tissues to it. A classic symptom is presence of glucose in the urine (glycosuria). Twin studies indicate that the HERITABILITY of this form is not simple, and that it may be triggered by various factors. Twin studies also indicate that the form of the disorder which presents later in life, which is not insulin-dependent and may be treated by dietary pattern, has high heritability.

DIADELPHOUS. (Of stamens) united by their filaments to form two groups, or having one solitary and the others united; e.g. pea. Compare MONADELPHOUS, POLYADELPHOUS.

DIAGEOTROPISM. Orientation of plant part by growth curvature in response to stimulus of gravity, so that its axis is at right angles to direction of gravitational force, i.e. horizontal; exhibited by rhizomes of many plants. See GEOTROPISM, PLAGIOTROPISM.

DIAKINESIS. Final stage in the first prophase of MEIOSIS.

DIALYSIS. Method of separating small molecules (e.g. salts, urea) from large (e.g. proteins, polysaccharides) when in mixed solution, by placing the mixture in or repeatedly passing it through a semi-permeable bag or *dialysis tube*, e.g. made of cellophane, surrounded by distilled water (which itself may be removed and replaced). Small molecules will diffuse out of the mixture into the surrounding water, whereas large molecules are prevented by size from doing so. The principle underlies the design of artificial kidneys, which work by *renal dialysis*.

DIAPAUSE. Term used to indicate period of suspended development in insects (and occasionally in other invertebrates). In insects, usually a true DORMANCY, implying a *condition* rather than a *stage* in morphogenesis. Insect diapause can occur at any stage in development, perhaps most commonly in eggs or pupae, but usually only once in any life cycle.

DIAPHRAGM. Sheet of tissue, part muscle and part tendon, covered by a serous membrane and separating thoracic and abdominal cavities in mammals only. It is arched up at rest, its flattening during inspiration reducing the pressure within the thorax, helping to draw air into the lungs. See VENTILATION.

DIAPHYSIS. Shaft of a long limb bone, or central portion of a vertebra, in mammals. Contains an extended OSSIFICATION centre. See EPIPHYSIS.

DIAPSID. Vertebrate skull type in which two openings, one in the roof and one in the cheek, appear on each side. The feature is found in many reptiles (e.g. ARCHOSAURS) and all birds.

DIASTASE. See AMYLASES.

DIASTOLE. Brief period in the vertebrate HEART CYCLE when both atria and ventricles are relaxed, and the heart refills with blood from the veins. The term may also be used of relaxation of atria and ventricles separately; in which case the terms *atrial diastole* and *ventricular diastole* are used, and are not to be confused with true diastole. Compare SYSTOLE.

DIATOM. The common name for the algae of the Division BACILLARI-OPHYTA.

DICHASIUM. A CYME possessing two axes running in opposite directions; the type of cyme produced in those plants having opposite branching in the inflorescence.

DICHLAMYDEOUS (DIPLOCHLAMYDEOUS). (Of flowers) having perianth segments in two whorls.

DICHOGAMY. Condition in which male and female parts of a flower mature at different times, ensuring that self-pollination does not occur. See OUTBREEDING, PROTANDRY, PROTOGYNY. Compare HOMOGAMY.

DICHOTOMOUS VENATION. Branching of leaf veins into two more or less equal parts, without any fusion after they have branched.

DICHOTOMY. Branching, or bifurcation, into two equal portions. *Dichotomous keys* are employed in those identification manuals where one passes along a path dictated by consecutive decisions, each choice being binary (there being just two alternatives), one route involving the strict negation of the other. See IDENTIFICATION KEYS.

DICOTYLEDONAE. Larger of the two classes of ANTHOPHYTA (flowering plants); distinguished from Monocotyledonae by presence of two leaves (COTYLEDONS) in the embryo, by usually net-like leaf venation, by stem vascular tissue in the form of a ring of open bundles, and by flower parts in multiples of four or five. Pollen is usually tricolpate (having three furrows or pores), and commonly there is true secondary growth with a vascular cambium present. There are about 170 000 species, including many types of forest tree, potatoes, beans, cabbages, and such ornamentals as roses, clematis and snapdragon.

DICTYOPTERA. Insect order containing the cockroaches and mantids, a group sometimes included in the ORTHOPTERA. Largely terrestrial; wings often reduced or absent, and in general poor fliers; fore wings modified to form rather thick leathery tegmina (similar to elytra). Specialized stridulatory and auditory apparatus absent.

DICTYOSOME. See GOLGI APPARATUS.

DICTYOSTELE. An amphiphloic SIPHONOSTELE, comprising independent vascular bundles occurring as one or more rings. Present in stems of certain ferns. Individual bundles here are termed *meristeles*. See STELE, PROTOSTELE.

DICTYOSTELIUM. See MYXOMYCOPHYTA.

DIENCEPHALON. Posterior part of vertebrate forebrain. See BRAIN.

DIFFERENTIATION. The process whereby cells or cell clones assume specialized functional biochemistries and morphologies previously absent. Such *determined* cells usually lose the ability to divide. Usually associated with the selective expression of parts of the genome previously unexpressed, brought on e.g. by cell contact, cell density, the extracellular matrix and molecules diffusing in it, etc. Division of labour thus achieved is one evolutionary by-product of MULTICELLULARITY.

DIFFUSE POROUS WOOD. Wood in which pores, or vessels, are fairly uniformly distributed throughout the growth layers, or in which the size of pores changes only slightly from early to late wood; e.g. tulip tree (*Liriodendron tulipifera*).

DIFFUSION. Tendency for particles (esp. atoms, molecules) of gases, liquids, and solutes to disperse randomly and occupy available space. Process is accelerated by rise of temperature, the source of movement being thermal agitation. Cells and organisms are dependent on the process at many of their surfaces and interfaces; on its own it is often inadequate for their needs. See ACTIVE TRANSPORT, FACILITATED DIFFUSION, OSMOSIS, WATER POTENTIAL.

DIGENEA. Order of the TREMATODA. Includes those flukes which are usually vertebrate endoparasites as adults and mollusc endoparasites as sporocysts and rediae. Suckers simple. E.g. *Fasciola,* SCHISTO-SOMA.

DIGESTION. Breakdown by organisms, ultimately to small organic compounds, of complex nutrients that are either acted upon outside the organism (e.g. by saprotrophs), or have entered some organelle (e.g. food vacuole) or organ (enteron, gut) specialized for the purpose. Often includes the physical events of chewing and emulsification besides chemical breakage of covalent bonds by mineral acids and enzymes. Food molecules are often too large simply to diffuse across CELL MEMBRANES and their digestion is first required. A gut forms a tube to confine ingested material while extracellular enzymes hydrolyse it, given appropriate conditions (e.g. pH, temperature). Later stages of digestion may occur through enzymes located in the brush borders of intestinal epithelia (as with nucleotidases and disaccharidases; see MICROVILLUS). After digestion, molecules are incorporated (assimilated) into cells of the body. Although plants lack guts, their cells can digest contained material (e.g. polysaccharides, lipids; see LYSOSOME). Although both digestion and RESPIRATION are catabolic, and digestion like some respiration is anaerobic, digestion does not release significant amounts of energy.

DIGIT. Finger or toe of vertebrate PENTADACTYL LIMB. Contains phalanges. May bear nails, claws or hooves.

DIGITIGRADE. Walking on toes, rather than on whole foot (PLANTI-GRADE). Only the ventral surfaces of digits used. E.g. cat, dog.

DIHYBRID CROSS. A cross between two organisms or stocks heterozygous for the same allergies at the same two loci under study. A classic example was MENDEL's crossing of F1 pea plants obtained as progeny from plants *homozygous* for different alleles at two such loci. This gave his famous 9:3:3:1 ratio of phenotypes (often called the *dihybrid* ratio); but by no means all dihybrid crosses give this ratio. The phrase *dihybrid segregation* is often used to describe the production of these ratios. *Dihybrid selfing* may be possible in some hermaphrodites (e.g. peas) and monoecious plants. Contrast MONOHYBRID CROSS. See LINKAGE.

DIKARYON (DICARYON). Fungal hypha or mycelium in which cells occur containing two haploid nuclei which undergo simultaneous division during formation of each new cell. This forms a third, dikaryotic phase (*dikaryophase*) interposed between haploid and diploid phases in the life cycle. Occurs in Ascomycotina, where it is usually brief, and in the Basidiomycotina, where it is of relatively long duration. The paired nuclei may be genetically identical, or non-identical. See MONOKARYON.

DIMORPHISM. Of members of a species, structures, etc., existing in two clearly separable forms. E.g. *sexual dimorphism*, often very pronounced, between the two sexes; *heterophylly* in some plants (e.g. water crowfoot), in which leaves in two different environments have different morphologies. See DOSAGE COMPENSATION, POLYMORPHISM.

DINOFLAGELLATE. See PYRROPHYTA.

DINOPHYCEAE. See PYRROPHYTA.

DINOSAUR. See ORNITHISCIA, SAURISCHIA.

DIOECIOUS. Unisexual, male and female reproductive organs being borne on different individuals. Compare MONOECIOUS, HERMAPHRODITE. See OUTBREEDING.

DIPLANETISM. (Of fungi) succession of two morphologically different zoospore stages separated by a resting stage (e.g. some Oomycetes).

DIPLOBLASTIC. Level of animal organization in which the body is composed of two cell layers (germ layers), the outer ECTODERM and inner ENDODERM. Found only in the COELENTERATA, in which a jelly-like mesogloea separates the layers.

DIPLOID. Nuclei (and their cells) in which the chromosomes occur as homologous pairs (though rarely *paired up*), so that twice the HAPLOID number is present. Also applicable to appropriate tissues, organs, organisms and phases in a life cycle (see SPOROPHYTE). Most SOMATIC CELLS of animals are diploid (see MALE HAPLOIDY), but some of the cells of the GERMINAL EPITHELIUM engage in MEIOSIS, giving haploid products. See ALTERNATION OF GENERATIONS, DIKARYON.

DIPLOID APOGAMY. See APOMIXIS, PARTHENOGENESIS.

DIPLONT, DIPLOPHASE. The DIPLOID stage of a LIFE CYCLE. See ALTERNATION OF GENERATIONS, HAPLONT.

DIPLOPODA. Class (or subclass) of ARTHROPODA, containing millipedes. Abdominal trunk segments fused in pairs to form *diplosegments*, each with two pairs of legs; exoskeleton calcareous; ocelli and one pair of club-like antennae present; young usually hatch with three pairs of legs, suggesting possible relations with the Insecta (see NEOTENY). Development gradual. See CHILOPODA, MYRIAPODA.

DIPLOSPORY. (Bot.) Form of *apomixis* in which a (diploid) megaspore mother cell gives rise directly to the embryo. See PARTHENOGENESIS.

DIPLOTENE. Stage in first prophase of MEIOSIS.

DIPNOI (SARCOPTERYGII). The order of CHOANICHTHYES including lungfishes.

DIPTERA. Two-winged (or true) flies. Large order of the INSECTA, with enormous specialization and diversity among its members. Endopterygote. The hind pair of wings is reduced to form balancing HALTERES. Head very mobile; compound eyes and ocelli present; mouthparts suctorial, usually forming a *proboscis* and sometimes adapted for piercing; larvae legless and eruciform.

DISACCHARIDE. A carbohydrate comprising two monosaccharide groups joined covalently by a *glycosidic bond*. The group includes LACTOSE, MALTOSE and SUCROSE.

DISC FLOWER. Actinomorphic tubular flowers (florets) composing the central portion of the flower head (capitulum) of most Asteraceae (Compositae); contrasted with the flattened, zygomorphic ray-shaped florets on the margins of the head.

DISINFECTANT. Substance used particularly on inanimate surfaces to kill microorganisms, thus sterilizing them. Hypochlorites, phenolics, iodophores (complexes of iodine less staining, toxic and irritant than iodine solutions) and detergents all have disinfectant ability. The phenolic *hexachlorophene* is widely used in the food industry and hospital wards to reduce pathogenic staphylococci. See ANTISEPTIC.

DISPLACEMENT ACTIVITY. Act expressive of internal *ambivalence* and seemingly irrelevant or inappropriate to the context in which it occurs. Tends to occur when an animal is subject to opposing motivations or when some activity is thwarted.

DISRUPTIVE COLOURATION. Colouration in animals tending to break up their outlines, thus avoiding visual predation.

DISRUPTIVE SELECTION. See NATURAL SELECTION, POLYMORPHISM.

DISTAL. Situated away from; e.g. from place of attachment; from the head, along an antero-posterior axis; from the source of a gradient. Thus, the insect abdomen is *distal* to the head. Contrast PROXIMAL.

DIURESIS. Increased output of urine by kidney, as occurs after drinking much water or taking *diuretic* drugs. See ANTIDIURETIC HORMONE.

DIURNAL RHYTHM. See CIRCADIAN RHYTHM.

DIVERSITY (SPECIES DIVERSITY). The number of species in a community (its *richness*) is a poor indicator of community structure. Few species out of the total present are usually abundant (i.e. have a large population size, a large biomass, productivity, or some other measure of importance).The large number of rare species mainly determines the *species diversity*, and ratios between the number of species and their 'importance values' are termed *species diversity indices*. The particular index employed varies with the type of community studied.

DIVERTICULUM. Blind-ending tubular or sac-like outpushing from a cavity, often from the gut.

DIVISION. Major group in the Linnean hierarchy used in classifying plants. Includes closely related classes and is the taxonomic category between kingdom and class; equivalent to *phylum* in animal classification.

DIZYGOTIC TWINS. See FRATERNAL TWINS.

DNA (DEOXYRIBONUCLEIC ACID). The nucleic acid forming the genetic material of all cells, some organelles, and many viruses; a major component of CHROMOSOMES and the sole component of PLASMIDS. A polymer (polynucleotide), which is formed in cells by enzymatic dephosphorylation and CONDENSATION of many (deoxyribo)NUCLEOSIDE TRIPHOSPHATES (esp. dATP, dGTP, dCTP, dTTP). The product is a long chain of (deoxyribo)-NUCLEOTIDES, bonded covalently by *phosphodiester bonds*. *Duplex DNA* comprises two such *antiparallel* strands (running in opposite directions; see Fig. 13) but complementary in base composition, and held together in a double helix by hydrogen bonds between the complementary bases, by electronic interactions between bases, and by hydrophobic interactions. Each strand comprises a *sugar-phosphate backbone* from which the bases project inwardly. (Each end of the double helix has one 5'-ending strand paired to a 3'-ending strand, where the 5' and 3' indicate which carbon atoms of the two terminal deoxyriboses are bonded to their terminal phosphate group.) It is as duplex DNA that nuclear DNA is normally found, and this form is the ultimate store of *molecular information* for all cells (but single-stranded DNA BACTERIOPHAGES occur). Unlike RNA, DNA is not hydrolysed by dilute alkali. Two of the bases abundant in the nucleotides of DNA are purines (adenine and guanine), that form hydrogen bonds with two common pyrimidines (thymine and cytosine respectively, see BASE PAIRING). This ability of one strand of duplex DNA to act as template for the other enables DNA to be replicated by DNA POLYMERASE in a *semiconservative* way (see DNA REPLICATION). A combination of X-ray crystallographic and chemical data (see BASE RATIO) led J. Watson and F. Crick in 1953 to propose the three-dimensional model of duplex DNA held today. See DENATURATION, GENE, GENE MANIPULATION, GENETIC CODE, PROTEIN SYNTHESIS, REVERSE TRANSCRIPTASE, RNA.

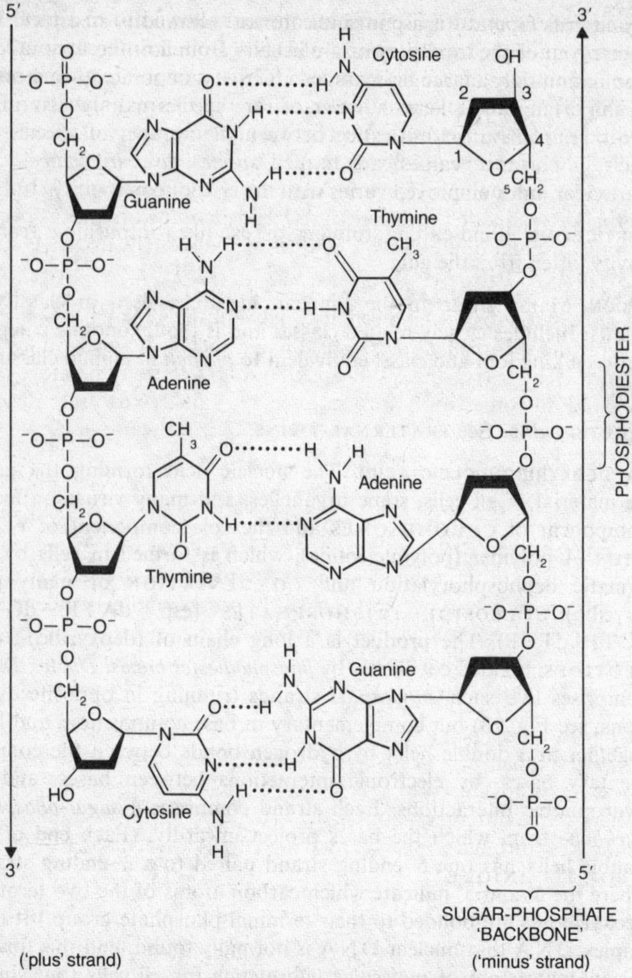

Fig. 13. The two antiparallel strands of a short section of a DNA duplex. The strands actually twist round each other in a double helix. The dots linking the central base-pairs represent hydrogen bonds. Primed numbers (5′, 3′) indicate the carbon atom numbers on the deoxyribose moiety involved in the phosphodiester bonding of each chain. These carbon atoms are numbered in the deoxyribose moiety attached to cytosine. The left-hand chain runs from 5′ to 3′ top to bottom, the right-hand chain runs from 5′ to 3′ bottom to top.

DNA CLONING. See GENE MANIPULATION.

DNA HYBRIDIZATION. A technique often used in experimental taxonomy, in which a source of duplex DNA is 'melted' (see DENATURATION) by slight temperature rise in solution, and allowed to re-anneal with either a similarly treated sample of DNA from a different source, or else commonly a single-stranded RNA sample (e.g. messenger RNA). The time taken to re-anneal, or the thermal stability of the hybrid duplex, indicates the degree of complementarity of the original strands. This can be used to indicate whether a piece of duplex DNA codes for the polypeptide translated from the messenger RNA, or perhaps to estimate the degree of relatedness of the organisms providing the original duplex DNA sources. See also DNA LIGASE, GENE MANIPULATION.

DNA LIGASE. Enzyme which repairs 'nicks' in the DNA *backbone*; i.e. where the *phosphodiester bond* linking adjacent nucleotides has yielded 3'-hydroxyl and 5'-phosphate groups. Its role therefore overlaps that of some DNA POLYMERASES. Valuable for hybridization (insertion) of DNA fragments with appropriate overlapping or 'sticky' ends. See GENE MANIPULATION.

DNA METHYLATION. See DNA REPAIR MECHANISMS.

DNA POLYMERASE (DNA pol). MULTIENZYME COMPLEX which incorporates appropriate nucleoside triphosphates into a DNA chain. Bacterial and eukaryotic cells generally possess more than one such complex. Are not sufficient for DNA REPLICATION, but carry out *chain elongation*; other proteins are required for *chain initiation*. See DNA REPAIR MECHANISMS.

DNA PROBE. A defined and fairly short DNA sequence, isotopically or otherwise labelled, which can be propagated by GENE MANIPULATION and introduced to DNA from the same or another individual (often from a different taxon) in order to detect complementary DNA sequences through DNA HYBRIDIZATION. See also SOUTHERN BLOT TECHNIQUE.

DNA REPAIR MECHANISMS. Some bacterial and eukaryotic DNA POLYMERASES can replace a nucleotide they insert incorrectly. DNA LIGASE then seals the phosphodiester bond. To avoid removing the nucleotide from the wrong (i.e. error-free) strand, cells methylate DNA which has been formed some while; repair enzymes thus distinguish old from new DNA, and repair only the new strand error. Mutants lacking such repair mechanisms are likely to be more susceptible to irradiating sources and to express (somatic) mutations so induced. In *photoreactivation*, cells of many organisms (but apparently not of placental mammals) repair radiation-damaged DNA (as from UV light) using an enzyme that functions when exposed to strong visible light. The main damage products are *pyrimidine dimers* formed by linking adjacent pyrimidines in the same DNA strand. The photoreactivating enzyme monomerizes these dimers again. See CROSSING-OVER.

DNA REPLICATION. Almost universal biological processes, in which DNA duplexes are catalytically and *semiconservatively* replicated by a DNA POLYMERASE (see MULTIENZYME COMPLEX) at rates of between 50–500 nucleotides per second. The duplex is first 'unzipped' by breaking the hydrogen bonds holding base-pairs in the duplex together. The resulting Y-shaped molecule is termed a *replication fork*. DNA polymerases then move down the two single-stranded arms in a 5′-to-3′ direction (see DNA), incorporating nucleotides in accordance with BASE PAIRING rules. Energy is supplied by hydrolysis of substrate nucleoside triphosphates, also catalysed by the polymerase. There are usually several simultaneous replication forks on one replicating chromosome, and newly-synthesized sections are joined up by the DNA *ligase* component of the polymerase (see DNA REPAIR MECHANISMS). Failure to replicate exactly results in a MUTATION.

DNase (DNAase, DEOXYRIBONUCLEASE). An enzyme (of which there are many forms) breaking down DNA by hydrolysis of the phosphodiester bonds of its sugar-phosphate backbone. Depending on the enzyme, it does this at either the 3′- or the 5′- end of the bond. As with peptidases, there are *endonucleases* and *exonucleases*, cleaving respectively terminal and non-terminal nucleotides from either a single strand or from both strands of the duplex, depending on the type of DNase. Pancreatic juice contains DNases. Valuable in GENE MANIPULATION. See RESTRICTION ENDONUCLEASE.

DNA SEQUENCING. Determination of the sequence of nucleotides making up a length of DNA. RESTRICTION ENDONUCLEASES digest the strand; the fragments are isolated by gel ELECTROPHORESIS, and then the sequence can be determined by rendering the DNA single-stranded and using it as a template for DNA POLYMERASE to resynthesize the complementary strand with labelled nucleoside triphosphates, or by chemical analysis of the fragments. Some subtle and elaborate methods are available.

DOLLO'S LAW. The generalization that evolution does not proceed back along its own path, or repeat routes.

DOMINANCE. (1) In genetics, one character is said to be completely dominant to another when it is expressed equally in the *homozygous* and *heterozygous* conditions; the other character is said to be completely RECESSIVE to it, and is only expressed in the homozygous condition. Normally, the two characters would form what Mendel termed a 'pair of contrasting characters'; they would, in other words, be determined by alternative alleles at the same locus. Genetic dominance is not synonymous with 'commonest character type in the population': that will depend upon SELECTION, amongst other factors. The term is often used of *genes* (or alleles); but since a 'gene' can have more than one effect (see PLEIOTROPY) accuracy requires use of the term to be restricted either to the context of characters, or to one particular aspect of phenotype affected by the gene. Dominance is usually a property of a normally functional ('wild type') allele; defective (mutant) alleles are usually, but not always, recessive. In some cases the degree of dominance is altered by selection when it is an evolving property of characters (see DOMINANCE MODIFICATION, MODIFIER). Two characters are said to be *codominant* when the respective homozygotes are distinguishable both from each other and from the heterozygote, and where the effects of both alleles can be detected in the phenotype; two characters are said to be *incompletely dominant* to one another when the heterozygote is distinguishable from both homozygotes, but distinct effects of the two alleles in the phenotype are not recognizable.

There can be no dominance in the HAPLOID state, or in the HEMIZYGOUS condition generally.

(2) In animal behaviour, a relational property indicating one individual's priority over another in contexts where some resource (e.g. food, mate, shelter) either is, or has in the past been, contested.

(3) In ecology, out of hundreds of organisms present in a community only a relatively few species or species groups generally exert the major controlling influence by virtue of their numbers (abundance), size, production, etc.; species or species groups which largely control the energy flow as well as affecting the environment of all other species are known as the dominant species, e.g. beech trees in a beech wood. When more than one dominant species or species group occurs in a particular plant community, they are called *codominants*.

DOMINANCE MODIFICATION. Phenomenon whereby different populations of a species evolve different genetic backgrounds (see MODIFIER) by which phenotypic effects of the same genetic mutation are expressed as either DOMINANT or RECESSIVE. Crossing between individuals from such populations may result in breakdown of dominance, producing an unclassifiable range of phenotypes.

DONOR. Source of material being grafted onto, or somehow inserted into, some other individual.

DOPA. See L-DOPA.

DOPAMINE. Intermediate in the biosynthesis of NORADRENALINE and ADRENALINE. In vertebrate brain, a NEUROTRANSMITTER whose low concentration produces symptoms of Parkinson's disease. See L-DOPA.

DORMANCY. (Bot.) See DORMANT. (Zool.) Term sometimes used of insect and other animal DIAPAUSE.

DORMANT. In a resting condition. Alive, but with relatively inactive metabolism and cessation of growth. Dormancy may involve the whole organism (higher plants and animals) or be confined to reproductive bodies (e.g. resting spores such as statoblasts, fungal sclerotia, bacterial spores). May be due to unfavourable conditions, and end as these ameliorate. Many seeds (e.g. pea, wheat), though capable of germinating after harvesting, do not do so unless kept moist. On the other hand, a dormant period is part of an annual rhythm for most plants. Often has survival value (e.g. winter dormancy in deciduous trees). After vegetative growth and flowering in spring, many bulbs (e.g. snowdrop, daffodil) have a dormant period coinciding with conditions favourable to growth of other plants. This is common in plants of moist, tropical climates. Dormancy of seeds in conditions otherwise favourable to germination is common (e.g. hawthorn, the weed wild oats) and is associated with incomplete development of the embryo, impermeable seed coats, limiting entry of water and/or oxygen, inhibitors and absence of growth stimulators. Dormancy in some seeds and deciduous trees is regulated by photoperiod (see PHYTOCHROME). See AESTIVATION, DIAPAUSE, HIBERNATION.

DORMIN. See ABSCISIC ACID.

DORSAL. (Zool.) Designating the surface of an animal normally directed away from the substrate; in chordates, the surface (posterior) in which the NEURAL TUBE forms, lying closest to the eventual nerve cord. In flatfish, the apparent adult dorsal surface is in fact lateral. (Bot.) Also used of leaves; synonymous with ABAXIAL.

DORSAL AORTA. See AORTA.

DORSAL LIP. See BLASTOPORE, ORGANIZER.

DORSAL PLACENTATION. Attachment of ovules to midrib of carpels in apocarpous gynoecia.

DORSIVENTRAL (DORSOVENTRAL). Term generally used to indicate some gradient or morphological feature associated with the AXIS linking the upper and lower parts of an organism or its parts. As with leaves, it often indicates some difference in structure along the axis. Compare ISOBILATERAL.

DOSAGE COMPENSATION. Mechanism existing in organisms with SEX CHROMOSOME imbalance (e.g. XY/XX, XO/XX) tending to equalize the effects of sex-linked loci in the two sexes. MODIFIER loci on the X-chromosome (*dosage compensators*) act either to enhance biosynthetic activity of sex-linked loci in the heterogametic sex, or to repress such activity in the homogametic sex. See BARR BODY, SEX DETERMINATION.

DOUBLE CIRCULATION. See HEART.

DOUBLE HELIX. See DNA.

DOUBLE RECESSIVE. Individual or stock in which each of two loci involved in breeding work is homozygous for alleles bringing about expression of RECESSIVE characters. See BACKCROSS.

DOUBLING RATE. Time required for a population of a given size to double in number.

DOWN FEATHER. See FEATHER.

DOWN'S SYNDROME (MONGOLISM). CONGENITAL disorder of people caused by TRISOMY of chromosome 21 (often by non-disjunction). Characterized by mental retardation, *mongoloid* facial features, simian palm and reduced life expectancy. Has a frequency of about one per 700 live births.

DPN. Former acronym for NAD.

DRIVE. Specific causal explanations are now sought for most animal activities, so *general drive theories* of motivation have been surpassed by investigation of the control of behaviour rather than its powering. Those specific causal influences promoting an action may be regarded as a part of that activity's *specific drive mechanism*.

DROSOPHILA. Genus of fruit flies (Diptera). Probably the best described animal genetically, and of enormous significance to studies of LINKAGE, CYTOGENETICS, SPECIATION and, most recently, developmental biology (e.g. see COMPARTMENT, HOMOEOTIC GENE).

DRUPE. Succulent FRUIT in which the wall (pericarp) comprises an outer skin (epicarp), a thick fleshy mesocarp, and a hard stony endocarp enclosing a single seed. Commonly called a stone-fruit; e.g. plum, cherry. Compare BERRY. In some drupes the mesocarp is fibrous; e.g. in the coconut the pericarp has tough, leathery epicarp, thick fibrous mesocarp and hard endocarp enclosing the seed and forming with it the nut we buy. Compare NUT.

DRYOPITHECINE. Term given to several Miocene and early Pliocene (down to 8–9 Myr BP) fossil ape (pongid) remains from Europe, India and East Africa. Includes the genus *Dryopithecus*. Generalized anthropoids, with cranial bones similar to those of tarsiers and hominids; lack of brachiating limb adaptations; lack of supraorbital torus; primitive monkey-like nasal aperture. Simian shelf lacking; incisors small; canine large. Probably ancestral to present great apes; and the early group (19 Myr BP) known as *Proconsul* (*Dryopithecus africanus*) may have been ancestral to hominids. See RAMAPITHECUS.

DUCTLESS GLAND. See ENDOCRINE GLAND.

DUCTUS ARTERIOSUS (DUCT OF BOTALLO). Vascular connection between pulmonary trunk (AORTIC ARCH VI) and AORTA (AORTIC ARCH IV) in amniote embryos, serving as a bypass for most blood from the right ventricle past the lungs while they are deflated and functionless. When the pulmonary circuit opens at birth the ductus closes and atrophies.

DUCTUS CUVIERI. See CUVIERIAN DUCT.

DUODENUM. Most anterior region of small intestine of mammals, its origin guarded by the pyloric sphincter. Receives the bile duct and pancreatic duct. Characterized by alkaline-mucus-secreting *Brunner's glands* in the submucosa. So-called because it is about 12 finger-breadths, about 25 cm, long in man; site of active digestion and absorption; like the rest of the small intestine, its luminal surface has numerous villi.

DUPLEX. Of a molecule composed of two chains or strands, usually held together by hydrogen bonds; e.g. double-stranded (duplex) DNA.

DUPLICATION. Chromosomal MUTATION in which a piece of chromosome is copied next to an identical section, increasing chromosome length. Can result from non-homologous CROSSING OVER in which two homologous chromosomes pair up imprecisely and a cross-over transfers an abnormally large piece of one chromosome to its homologue, resulting in a DELETION on one chromosome and a duplication on the other. See GENE DUPLICATION.

DURA MATER. See MENINGES.

DYNEIN. Accessory protein of the axoneme microtubules of eukaryotic CILIUM and flagellum. Also believed to be associated with microtubules during anaphase movement of chromatids/chromosomes. Has ATPase activity, and may be responsible for force-generating steps in both systems. See CELL LOCOMOTION.

DYSGENESIS. See HYBRID DYSGENESIS.

DYSTROPHIC. Term applied to certain lakes receiving large amounts of organic matter from elsewhere, having heavily stained brown water as a result. There is a high humic organic content.

E

EAR, INNER. Membranous labyrinth. Vertebrate organ which detects position with respect to gravity, acceleration, and sound. Lies in skull wall (auditory capsule); impulses transmitted to brain via AUDITORY NERVE. Comprises the VESTIBULAR APPARATUS and the COCHLEA. See LATERAL LINE SYSTEM.

EAR, MIDDLE. Tympanic cavity. Cavity between eardrum and auditory capsule of tetrapod vertebrates (but not urodeles, anurans or snakes). Derived from a gill pouch (spiracle). Communicates with pharynx via eustachian tube, and is filled with air, ensuring atmospheric pressure is maintained on both sides of the eardrum. In it lie the EAR OSSICLES.

EAR OSSICLES. Bones in middle ear connecting eardrum to INNER EAR in tetrapod vertebrates. Instead of just the single auditory bone (stapes, see COLUMELLA) of amphibia and primitive reptiles, mammals have in addition the incus and malleus. The first retains its original attachment to the oval window of the inner ear, but here articulates via the incus with the malleus which attaches to the eardrum (tympanum). These last two bones have evolved respectively from the quadrate and articular bones of mammal-like reptiles, in which they were involved in jaw suspension. By this articulation the pressure of the stapes on the oval window is amplified 22 times compared with that of the pressure waves on the tympanum: vibrations are damped, but produce larger forces.

EAR, OUTER (or EXTERNAL). That part of the tetrapod ear, absent from amphibians and some reptiles, external to the eardrum. Comprises a bony tube (*external auditory meatus*). In addition in mammals there is a flap of skin and cartilage (the pinna) at the outer opening which amplifies and focuses pressure waves upon the eardrum. Well developed in nocturnal mammals (e.g. bats).

EARDRUM (TYMPANUM, TYMPANIC MEMBRANE). Thin membrane stretching across the aperture between skull bones at the surface of the head (most anurans and turtles) or within an external meatus (most reptiles, birds and mammals). Vibrates, often aperiodically, transmitting external air pressure changes to EAR OSSICLES of middle ear cavity.

ECAD. (Bot.) A habitat form, showing characteristics imposed by habitat conditions and non-genetic. Compare ECOTYPE.

ECDYSIS. Moulting in arthropods. Periodic shedding of the CUTICLE in the course of growth. In insects this includes much of the lining of the tracheal system. The number of larval moults varies (up to 14 in APTERYGOTA); in endopterygotes there is one pupal moult (producing adult), but among insects only apterygotans moult as adults. In most crustaceans it proceeds throughout adult life. Insect ecdysis is under the control of ECDYSONE.

ECDYSONE (MOULTING HORMONE, GROWTH-AND-DIFFERENTIATION HORMONE). Hormone produced by insect *thoracic* (*prothoracic*) *glands*, and possibly also by the crustacean *Y-organ*. In insects its release is under the control of *thoracotropic hormone* produced by neurosecretory cells in the brain and released from the CORPORA CARDIACA. In crustaceans the brain neurosecretion is produced in the X-organs and transported to the *sinus glands* of the eye-stalk. Its release inhibits release of moulting hormone by the Y-organs. In insects at least ecdysone induces 'puffing' of selected chromosome regions, the sequence being tissue-specific. This is associated with selective gene transcription, notably by the epidermis; but one of its major effects is to make appropriate cells sensitive to *juvenile hormone* from the CORPORA ALLATA, with which ecdysone works to bring about moulting to the appropriate developmental stage. See DIAPAUSE.

ECESIS. Germination and successful establishment of colonizing plants; the first stage in succession.

ECHIDNA. Spiny anteater. See MONOTREMATA.

ECHINODERMATA. Phylum of marine and invertebrate deuterostomes; typically with pentaradiate symmetry as adults; an internal skeleton of calcareous plates in the dermis; TUBE FEET; nervous system typically one circular and five longitudinal nerve cords, lacking brain and ganglia; surface epithelium often ciliated, and sensory; coelom well developed, including peculiar WATER VASCULAR SYSTEM; no excretory organs; larvae typically pelagic, roughly bilaterally symmetrical, with tripartite coelom (*oligomerous*) and an often dramatic metamorphosis. Affiliations with HEMICHORDATA. Includes classes Stelleroidea (including subclasses Asteroidea, the starfish, and Ophiuroidea, the brittlestars); Echinoidea (sea urchins, etc.); Holothuroidea (sea cucumbers); Crinoidea (crinoids); and the new class Concentricycloidea (sea daisies).

ECHINOIDEA. Sea urchins, heart urchins, etc. Class of ECHINODERMATA; lacking separate arms; more or less globular in shape; mouth downwards; with rigid calcareous *test* of plates in dermis bearing spines and defensive PEDICELLARIAE; browsers and scavengers, often in enormous numbers, on sea bed.

ECHOLOCATION. Method used by several nocturnal, cave-dwelling or aquatic animals for determining positions of objects by reflection of high-pitched sounds. Many bats and dolphins use it, as do oil birds and the platypus.

ECODEME. See DEME.

ECOLOGICAL NICHE. See NICHE.

ECOLOGY. Term deriving from the Greek οικος (house, or place to live); the study of relationships of organisms or groups of organisms to their environments, both animate and inanimate. Increasingly quantitative, employing modelling and computer simulations. See ECOSYSTEM, TROPHIC LEVEL, COMPETITION, and cross-references included there.

ECOSPECIES. Group within a species comprising one or more ECOTYPES, whose members can reproduce amongst themselves without loss of fertility among offspring. Approximates to a *hologamodeme* (see -DEME), or to an ideal 'biological' SPECIES, and as a term is used more in botanical than in zoological contexts. See INFRA-SPECIFIC VARIATION.

ECOSYSTEM. COMMUNITY of organisms, interacting with one another, plus the environment in which they live and with which they also interact; e.g. a lake, a forest, a grassland, tundra. Such a system includes all abiotic components such as mineral ions, organic compounds, and the climatic regime (temperature, rainfall and other physical factors). The biotic components generally include representatives from several TROPHIC LEVELS; primary producers (autotrophs, mainly green plants), macroconsumers (heterotrophs, mainly animals) which ingest other organisms or particulate organic matter, microconsumers (saprotrophs, again heterotrophic, mainly bacteria and fungi) which break down complex organic compounds upon death of the above organisms, releasing nutrients to the environment for use again by the primary producers. See BALANCE OF NATURE, FOOD CHAIN, PYRAMID OF BIOMASS.

ECOTONE. The transition between two or more diverse communities, as between forest and grassland. Zone which may have considerable length, yet be far narrower than adjoining communities.

ECOTYPE. Term generally employed in botanical contexts, referring to a species population exhibiting genetic adaptation to the local environment, the phenotypic expression of which withstands transplantation of the plant, or of its offspring, to a new environment. See INFRASPECIFIC VARIATION, ECOSPECIES. Compare ECAD.

ECTEXINE. See EXINE.

ECTODERM. Outermost GERM LAYER of metazoan embryos, developing mainly into epidermal and nervous tissue and, when present, NEPHRIDIA.

ECTOPARASITE. See PARASITE.

ECTOPHLOIC SIPHONOSTELE. A siphonostele with phloem external to the xylem.

ECTOPLASM (ECTOPLAST). See cell CORTEX.

ECTOPROCTA (POLYZOA, formerly BRYOZOA). Phylum of colonial and often polymorphic coelomates, retaining continuity by coelomic tubes (cyclostomes) or merely by a tissue strand (ctenostomes, cheilostomes). Feeding (polyp) individuals do so by microphagy using a LOPHOPHORE of tentacles, and secreting a calcareous zooecium (together termed a ZOOID). See STATOBLAST.

ECTOTHERMY. See POIKILOTHERMY.

ECTOTROPHIC. (Of mycorrhizas) with the mycelium of the fungus forming an external covering to the root and penetrating only between the outer cortical cells; e.g. in pine trees. See MYCORRHIZA; compare ENDOTROPHIC.

EDAPHIC FACTORS. Environmental conditions determined by physical, chemical and biological characteristics of the soil.

EDENTATA (XENARTHRA). Aberrant order of eutherian Mammalia, mainly of South American history and distribution. Includes tree sloths, anteaters, armadillos and extinct glyptodonts. Only anteaters are truly toothless (hence ordinal name), the others having molars at least.

EFFECTOR. Cell or organ by which an animal responds to internal or external stimuli, often via the nervous system. Include muscles, glands, chromatophores, cilia. Cnidoblasts are often regarded as *independent effectors* in that they do not seem to require stimulation from other cells (e.g. of the nervous system) for their activity.

EFFERENT. Leading away from; e.g. from the central nervous system (motor nerves), from the gills (blood vessels) or from a glomerulus (arteriole).

EGESTION. Removal of undigested material and associated microorganisms of the gut flora (up to 50% dry weight in man) from the anus. This material has never been inside body cells. A quite different process from EXCRETION, with which it may be confused. The voided material is termed *egesta*.

EGG CELL. See OVUM.

EGG MEMBRANES. Few animal eggs, if any, have just a plasma membrane separating the cytoplasm from the external environment. Additional membranes are: (1) *Primary membranes*: the vitelline membrane, or thicker chorion. (2) *Secondary membranes*: consisting of or formed by the follicle cells around the egg. (3) *Tertiary membranes*: secreted by accessory glands, oviduct, etc., including albumen, shell membranes, egg 'jelly', etc. Protective against mechanical damage, desiccation.

ELAIOPLAST. Colourless plastid (leucoplast) in which oil is stored; common in liverworts and monocotyledons.

ELASMOBRANCHII. Subclass of CHONDRICHTHYES, appearing in the middle Devonian. Includes sharks (SELACHII), skates and rays (Rajiformes) and angel sharks (Squatiniformes). Cartilaginous skeleton; dermal denticles probably the remnants of ancestral bony placoderm armour; upper jaws independent of braincase (hyostylic jaw suspension) or in some sharks with anterior attachment to braincase (amphistylic jaw suspension). Gills border gillslits (usually five); spiracle present. Internal fertilization, male having *claspers*, modified pelvic fins, acting as intromittant organs. Tail heterocercal; teeth in rows, replacing in turn those lost. The HOLOCEPHALI (Chimaeras) form a second chondrichthyan subclass.

ELASTIN. Principal fibrous protein of the yellow fibres of animal CONNECTIVE TISSUE. Numerous in lungs, walls of large arteries and in ligaments. Highly extensible and elastic. Compare COLLAGEN.

ELATER. (1) Elongated cell with wall reinforced internally by one or more spiral bands of thickening, occurring in numbers among spores in capsules of liverworts. Assist in discharge of spores by movements in response to humidity changes. (2) Appendage of spores of horsetails; formed from outermost wall layer, coiling and uncoiling as the air is dry or moist; possibly assisting in spore dispersal.

ELECTRIC ORGANS. Organs of certain fishes which produce electric currents by means of modified muscle cells (*electrocytes*) which no longer contract but generate ion current flow on nervous stimulation. Two basic kinds: those producing strong stunning current (e.g. electric eel, electric ray, electric catfish), and those producing currents of low voltage (e.g. in Mormyridae, Gymnotidae excepting electric eel) as a continuous series of pulses for locating prey and obstacles in muddy water, and for mate location.

ELECTROCARDIOGRAM (ECG). Record of electrical changes associated with the HEART CYCLE, usually by means of electrodes placed on the patient's skin. Can also monitor foetal heart in the uterus.

ELECTROENCEPHALOGRAM (EEG). Record of changes in electrical potential ('brain waves') produced by the cerebral cortex; detected through the skull and picked up by electrodes placed on the scalp. The waves are then amplified. Four main types: *alpha* (produced when awake, but disappearing when asleep); *beta* (appear when nervous system is active – as in mental activity); *theta* (produced in children, and in adults in emotional stress situations); *delta* (occur in sleeping adults; in awake adults they indicate brain damage).

ELECTRON MICROSCOPE. See MICROSCOPE.

ELECTRON TRANSPORT SYSTEM (ETS, ELECTRON TRANSPORT CHAIN).
Chain of specifically and adjacently arranged enzymes (mostly conjugated proteins) and associated COENZYMES embedded in an ion-
impermeable membrane, along which electrons pass by REDOX reactions from one to the next. During this passage, electrons fall from
higher to lower *redox potentials* (energy states). Because of the positioning of these proteins in the membrane, energy so released is used
to pump protons across the membrane, resulting in a proton gradient
which itself acts as a store of potential energy. As the protons return
down their electrochemical gradient through specific channels in the
membrane, they provide the *protonmotive force* needed for ATP synthesis by ATPase, itself associated with the proton channels. This
process is common to the inner membranes of MITOCHONDRIA and
the thylakoid system of CHLOROPLASTS. In chloroplasts the electrons
are first boosted to a high energy level by photons; in mitochondria
they are derived from hydrogen atoms (also the proton source)
covalently bonded in electron-rich respiratory substrates. Associated
coenzymes, not all intrinsic to the membrane, may include NAD,
NADP, FAD, flavoproteins, plastoquinone and ubiquinone. Similar
ETSs occur in bacterial membranes (e.g. see BACTERIORHODOPSIN).

ELECTROPHORESIS. Technique for separating charged molecules in
buffer solution, particularly proteins, nucleic acids and their degradation products, based on their different mobilities (caused by their
different net charges at a given pH) in an electric field generated by
direct current through the buffer. Substances for separation are
usually allowed to move through a porous medium such as a gel (e.g.
starch, agar, polyacrylamide) or paper (e.g. filter, cellulose acetate).
Separated substances occur in bands on the medium and may be
stained or identified by some labelling device, by fluorescence, by
comparisons with knowns, or by removal and subsequent analysis. In
immunoelectrophoresis antigens are placed in wells cut in agar gel.
After separation of antigens by electrophoresis a trough is cut between the wells, filled with antibody, and diffusion allowed to take
place. Where antigen meets appropriate antibody, arcs of precipitin
form, allowing complex antigen mixtures to be compared.

EMASCULATION. Removal of stamens from hermaphrodite flowers
before they have liberated their pollen, usually as a preliminary to
artificial hybridization.

EMBEDDING. Method employed in the preparation of permanent
microscope slides of thin tissue sections. After DEHYDRATION and
CLEARING, the material is put into molten paraffin wax (usually for
1–3 hours, with one or two changes of wax) which impregnates the
tissue. After setting, the wax block is sectioned using a MICROTOME.
The wax is removed by xylene, itself removed by absolute alcohol,

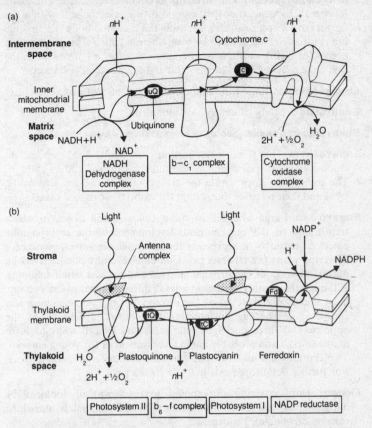

Fig. 14 *Hypothesized arrangements of electron-transporting molecules within (a) mitochondrial and (b) chloroplast membranes. Protons are extruded from mitochondria during activity but taken into thylakoids during the light reactions of photosynthesis.*

and gradual rehydration of the section is achieved by passing for a few minutes through progressively more dilute alcohols. Staining can then proceed. In electron microscopy, Araldite ® is frequently used for the embedding.

EMBRYO. (Bot.) Young plant developed from an ovum after sexual (including parthenogenetic) reproduction. In seed plants, it is contained within the seed and comprises an axis bearing at its apex either the apical meristem of the future shoot or, in some species, a young bud (the plumule), while at the other end is the root (the radicle). From the centre of the axis grow one or more seed leaves (cotyledons). (Zool.) The structure produced from an egg (usually fertilized), by generations of mitotic divisions while still within the EGG MEMBRANES, or otherwise inside the maternal body. Embryonic life is usually considered to be over when hatching from membranes occurs (or birth); in humans an embryo becomes a FOETUS when the first bone cells appear in cartilage (at about 7 weeks of gestation).

EMBRYOGENESIS. Formation and development of an EMBRYO.

EMBRYOLOGY. Study of embryo development.

EMBRYONIC MEMBRANE. See EXTRAEMBRYONIC MEMBRANES.

EMBRYOPHYTA. In some classifications, a plant Subkingdom including all plants possessing multicellular sex organs and an embryo. The embryo develops within the archegonium in mosses, liverworts, ferns and their relatives, but within the embryo sac in seed plants.

EMBRYO SAC. Large oval cell in the nucellus of the ovule, in which fertilization of the egg cell and development of the embryo take place. At maturity, it represents the entire female gametophyte of a flowering plant (ANTHOPHYTA). Contains several nuclei derived by mitotic division of the original MEGASPORE nucleus (itself haploid). Although the number of nuclei varies in different types of embryo sac, most commonly there is, at the micropylar end, an *egg-apparatus* consisting of the egg nucleus and two others, *synergids*. At the opposite end three nuclei become separated by cell walls to form *antipodal cells* and probably aid in nourishment of the young embryo. Two central *polar nuclei* fuse to form the *primary endosperm nucleus*. For further details, see DOUBLE FERTILIZATION.

ENAMEL. Hard covering of exposed part (crown) of tooth; 97% inorganic material (two thirds calcium phosphate crystals, one third calcium carbonate), 3% organic.

ENATION. Outgrowth produced by local hyperplasia on a leaf as a result of viral infection.

ENDARCH. Type of primary xylem maturation, characteristic of most stems, where the oldest xylem elements (protoxylem) are closer to centre of axis than those formed later. Compare EXARCH.

ENDEMISM. Occurrence of organisms or taxa (termed *endemic*) whose distributions are restricted to a geographical region or locality, such as an island or continent. (2) Continual occurrence in a region of a particular (endemic) disease, as opposed to sporadic outbreaks of it (epidemics).

ENDERGONIC. (Of a chemical reaction) requiring energy; as in synthesis by green plants of organic compounds from water and carbon dioxide by means of solar energy. Compare EXERGONIC. See THERMODYNAMICS for more detail.

ENDEXINE. Inner layer of EXINE of bryophyte spores and vascular plant pollen grains. See INTINE.

ENDOCARP. Innermost layer of the carpel wall, or pericarp of fruit, in flowering plants. Frequently used to denote the 'stone' of drupes.

ENDOCRINE GLAND (DUCTLESS GLAND). Gland whose product, one or more HORMONES, is secreted directly into the blood and not via ducts (compare EXOCRINE GLAND). The gland may be a discrete organ, or comprise more scattered and diffuse tissue. Examples include: ADRENAL, OVARY, PANCREAS, PITUITARY, TESTIS, THYROID. See ENDOCRINE SYSTEM.

ENDOCRINE SYSTEM. Physiologically interconnected system of ENDOCRINE GLANDS occurring within an animal body. Compared to neurotransmitters, the more diffuse hormonal outputs can take more time to reach effective concentrations, and therefore require a longer physiological half life (i.e. persistence in the body). Hormones generally exert effects over longer timescales, appropriate in growth, timing of breeding and control of blood and tissue fluid composition. Hormonal effects depend as much on distributions of RECEPTOR SITES on target cells as on the molecules secreted. See NERVOUS SYSTEM for a further comparison of roles, and NEUROENDOCRINE COORDINATION. See NEUROHAEMAL ORGAN, NEUROSECRETION.

ENDOCRINOLOGY. Study of the structure and function of the ENDOCRINE SYSTEM.

ENDOCYTOSIS. Collective term for PHAGOCYTOSIS and PINOCYTOSIS. An essential process in much eukaryotic CELL LOCOMOTION. See also EXOCYTOSIS, CELL MEMBRANES, PHAGOCYTE, COATED VESICLE, and Fig. 44.

ENDODERM (ENTODERM). Innermost GERM LAYER of an animal embryo. Composed like mesoderm (when present) of cells which have moved from the embryo surface to its interior during GASTRULATION. Develops into greater part of gut lining and associated glands, e.g. where applicable, liver and pancreas, thyroid, thymus and much of the branchial system. Not to be confused with ENDODERMIS.

ENDODERMIS. Single layer of cells forming sheath around the vascular region (stele), most clearly seen in roots; in some stems identifiable by its content of starch grains (the *starch sheath*). Usually regarded as innermost layer of cortex. In roots, most characteristic feature of very young endodermis is band of impervious wall material, the CASPARIAN STRIP, in radial and transverse walls of cells. With age, especially in monocotyledons, endodermis cells (except PASSAGE CELLS) may become further modified by deposition of layers of suberin over entire wall surface followed, particularly on the inner tangential wall, by a layer of cellulose, sometimes lignified. Endodermis is important physiologically in control of transfer of water and solutes between cortex and vascular cylinder, since these must pass through protoplasts of endodermis cells.

ENDOGAMY. See INBREEDING.

ENDOLYMPH. Viscous fluid occurring within the vertebrate COCHLEA and VESTIBULAR APPARATUS. These are separated from the skull wall by PERILYMPH.

ENDOMETRIUM. Glandular MUCOUS MEMBRANE lining the uterus of mammals. Undergoes cyclical growth and regression or destruction during the period of sexual maturity. Receives embryo at IMPLANTATION. See OESTROUS CYCLE, MENSTRUAL CYCLE, PLACENTA.

ENDOMITOSIS (ENDOREDUPLICATION). Process whereby all the chromosomes of an INTERPHASE nucleus replicate and separate within an intact nuclear membrane (which does not divide). No spindle or other mitotic apparatus found. Resulting nuclei are ENDOPOLYPLOID, the degree of ploidy sometimes exceeding 2000. Compare POLYTENY, in which chromosomes do not separate after duplication.

ENDONUCLEASE. See DNase.

ENDOPARASITE. See PARASITE.

ENDOPELON. Community of algae living and moving within muddy sediments. See BENTHOS.

ENDOPEPTIDASE. Proteolytic enzyme hydrolysing certain peptide bonds in protein molecule, e.g. pepsin. Compare EXOPEPTIDASE.

ENDOPHYTON. Community of algae growing between cells of other plants, or in cavities within plants. Well known associations occur in some liverworts. See BENTHOS.

ENDOPLASM. That part of a cell's cytoplasm distinguished from the ECTOPLASM (if any) by greater fluidity; may be termed *plasmasol*.

ENDOPLASMIC RETICULUM (ER). Eukaryotic cytoplasmic organelle comprising a complex system of membranous stacks (cisternae) and not unlike chloroplast thylakoids in appearance, but often being continuous with the outer of the two nuclear membranes and, like this membrane, bearing attached ribosomes on the cytosol side (when termed *rough ER*). A ribosome-free system of tubules (*smooth ER*), continuous with the cisternae, projects into the cytosol and pinches off transport vesicles. ER is not physically continuous with the GOLGI APPARATUS, but is functionally integrated with it. A large rough ER is indicative of a metabolically active (e.g secretory) cell.

ER seems to be the sole site of membrane production in a eukaryotic cell, membrane proteins and phospholipids being incorporated from precursors in the cytosol. Enzymes in the lipid bilayer pick up fatty acids, glycerol phosphate and choline and create LECITHIN, while protein components are fed into the ER lumen as they are produced at ribosomes bound to attachment sites on the cisternae. GLYCOSYLATION of newly synthesized proteins occurs within the cisternae through activity of *glycosyl transferase* located in the ER membrane. Only rough ER is involved in PROTEIN SYNTHESIS. Smooth (*transcisternal*) ER is generally a small component but from it TRANSPORT VESICLES (some of them COATED VESICLES) are budded off to carry protein and lipid to other parts of the cell. Some proteins are processed in the Golgi apparatus after the vesicles have fused there. See Figs. 3 and 29.

ENDOPODITE. See BIRAMOUS APPENDAGE.

ENDOPOLYPLOIDY. The result of ENDOMITOSIS.

ENDOPTERYGOTA. Insects with complete metamorphosis (pupal stage in life cycle) and with wings developing within the larva (see IMAGINAL DISC), although first visible externally in the pupa. Sometimes regarded as a subclass. Includes orders Neuroptera (lacewings); Coleoptera (beetles); Strepsiptera (stylopids); Mecoptera (scorpion flies); Siphonaptera (fleas); Diptera (true flies); Lepidoptera (butterflies, moths); Trichoptera (caddis flies); Hymenoptera (bees, ants, wasps). Often used synonymously with Holometabola, but see THYSANOPTERA. Compare EXOPTERYGOTA.

END ORGAN. Structure at peripheral end of a nerve fibre; usually either a RECEPTOR or a motor end-plate (see NEUROMUSCULAR JUNCTION).

ENDORPHINS. Peptide NEUROTRANSMITTERS, isolated from the PITUITARY GLAND, having morphine-like pain-suppressing effects. Also implicated in memory, learning, sexual activity, depression and schizophrenia. See ENKEPHALINS.

ENDOSKELTON Skeleton lying within the body. Vertebrate cartilage and bone provide support, protection and a system of levers enabling manipulation of the external environment; arthropods have internal projections of their cuticle (*apodemes*) for muscle attachment; echinoderms and annelids, among other invertebrates, use a *hydrostatic skeleton* to greater or lesser extent, and these too are endoskeletons.

ENDOSPERM. Nutritive tissue surrounding and nourishing the embryo in seed plants. (1) In flowering plants (ANTHOPHYTA), formed in embryo sac by division of usually triploid endosperm nucleus after fertilization. In some seed plants (non-endospermic, exalbuminous), it is entirely absorbed by the embryo by the time seed is fully developed (e.g. pea, bean seeds); in other seeds (endospermic, albuminous), part of the endosperm remains and is not absorbed until seed germinates (e.g. wheat, castor oil). (2) Also applied to tissue of female gametophyte in conifers and related plants which is formed by cell division within the embryo sac before fertilization, outer layers persisting in the seed. Compare PERISPERM.

ENDOSPORE. Spore formed within a parent cell; in bacteria, a thick-walled resistant spore; in blue-green algae, a thin-walled spore. Term also used for inner layer of spore wall.

ENDOSTYLE. Ciliated and mucus-secreting groove or pocket in ventral wall of pharynxes of urochordates, hemichordates, cephalochordates and ammocoete larvae of lampreys. The vertebrate thyroid is probably homologous with it.

ENDOSYMBIOSIS. Symbiotic association between cells of two or more different species, one inhabiting the other, the larger being host for the smaller. In *serial symbiosis*, one after another such symbiotic associations may occur telescoped within the largest cell. It is believed to account for the occurrence of eukaryotic chloroplasts (ancestor a cyanobacterium?), mitochondria (ancestor a purple photosynthetic bacterium?) and, some believe, cilia. See GLAUCOPHYTA, KAPPA PARTICLES.

ENDOTHELIUM. Single layer of flattened, polygonal cells lining verte-brate heart, blood and lymph vessels. Mesodermal in origin.

ENDOTHERMIC. See HOMOIOTHERMIC.

ENDOTOXIN. Glycolipids attached to cell walls of certain Gram-negative bacteria, giving them pathogenicity (e.g. *Salmonella typhi*, causing typhoid fever). Often complexed with protein. Released during autoly-sis. Not bound by antibody. Compare EXOTOXIN.

ENDOTROPHIC. (Of mycorrhizas) with mycelium of the fungus within cells of root cortex; e.g. orchids (where it may be the sole means of nutrient support, host cells digesting the hyphae). See MYCORRHIZA; compare ECTOTROPHIC.

END-PRODUCT INHIBITION (RETROINHIBITION, FEEDBACK INHIBITION). The inhibition of an ENZYME, often the first in a metabolic pathway, by the product of the last enzyme in the pathway. Ensures against overproduction of the final product. See ALLOSTERIC, REGULATORY ENZYME.

ENERGY FLOW. The passage of energy through an ECOSYSTEM from source (generally the sun), through the various TROPHIC LEVELS (within organic compounds), and ultimately out to the atmosphere as respiratory heat loss. There is about 90% loss of energy between one trophic level and the next in the grazing food chain. See PYRAMID OF BIOMASS.

ENHANCER. Site on eukaryotic DNA at which a protein may bind and turn on transcription of a particular gene (cistron), which may be either close-by or relatively distant (e.g. some tens of kilobases away) on the same chromosome.

ENKEPHALINS. Peptide NEUROTRANSMITTERS isolated from the thalamus and parts of the spinal cord and concerned with pain-related pathways. Morphine-like pain reducers. See ENDORPHINS.

ENRICHMENT CULTURE. Microbiological technique allowing selection and isolation from a natural, mixed population of microorganisms of those having growth characteristics desired by the investigator. Involves culturing on a medium whose composition is adjusted for selective growth of desired organisms, by altering nutrients, pH, temperature, aeration, light intensity, etc. Employed in bacteriology, mycology and phycology.

ENTEROCOELY. Method of COELOM formation within pouches of mesoderm budded off from embryonic gut wall. Develops this way in echinoderms, hemichordates, brachiopods and some other animals.

ENTEROKINASE (ENTEROPEPTIDASE). Enzyme (peptidase) secreted by vertebrate small intestine, converting inactive trypsinogen to active trypsin. Removes a small peptide group. Component of succus entericus. See KINASE.

ENTERON (COELENTERON). The gut (gastrovascular) cavity within the body wall of coelenterates, having a single opening serving as both mouth and anus. May be subdivided by mesenteries (as in sea anemones); sometimes receives the gametes (as in jellyfish). May serve as hydrostatic skeleton. See ARCHENTERON.

ENTHALPY. See THERMODYNAMICS.

ENTOMOGENOUS. (Of fungi) parasitic of insects.

ENTOMOLOGY. Study of insects.

ENTOMOPHAGOUS. Insect-eating.

ENTOMOPHILY. Pollination by insects.

ENTOPROCTA. Phylum of pseudocoelomate and mostly marine invertebrates, of uncertain relationships. Trochophore larva attaches by its oral surface; stolon grows out from the new aboral surface and produces a colony of adult individuals. These feed by ciliated tentacles which are simply folded away inside their protective cover, not withdrawn into a body cavity as in ENTOPROCTA. Excretion by protonephridia. Anus opens within tentacular ring.

ENTRAINMENT. Synchronization of an endogenous rhythm with an external cycle such as that of light and dark. See CIRCADIAN RHYTHM.

ENTROPY. See THERMODYNAMICS.

ENVIRONMENT. Collective term for the conditions in which an or-

ganism lives, both biotic and abiotic. Compare INTERNAL ENVIRON-
MENT.

ENZYME. A protein catalyst produced by a cell and responsible for the
high rate and specificity of one or more intracellular or extracellular
biochemical reactions. Enzyme reactions are always reversible.
Almost all enzymes are globular proteins consisting either of a single
polypeptide or of two or more polypeptides held together (in quaternary structure) by non-covalent bonds. By virtue of their three-dimensional configurations in solution, enzymes act upon other molecules
(substrates), and thus catalyse one type of (but not necessarily just
one) chemical reaction.

Their shapes provide them with one or more *active sites* (domains)
which bind temporarily and usually non-covalently with compatible
substrate molecules to form one or more *enzyme-substrate* (*ES*) *complexes*, catalysis occurring only during the brief existence of the
complex. One or more *products* are then released as the active site is
freed again to bind fresh substrate. Active sites have conformations
and charge distributions which are substrate-specific and their component amino acids commonly alter their relative three-dimensional
positions (termed an *induced fit*) as the substrate binds, enabling
several sub-reactions involved in catalysis to proceed.

Enzymes do nothing but speed up the rates at which the *equilibrium
positions* of reversible reactions are attained. In some poorly understood way, ultimately explicable in terms of THERMODYNAMICS,
enzymes reduce the *activation energies* of reactions, enabling them to
occur much more readily at low temperatures – essential for biological
systems.

It is now known that RNA molecules can act as catalysts of
reactions, sometimes involving themselves as substrates (see SPLIC-
ING). When they involve non-self RNA molecules as substrates, as
some do, they can be regarded as enzymes in the full sense (see
RIBOZYMES).

In general, cells can do only what their enzymes enable them to do.
During both evolution and multicellular development, cells come to
look and function differently from each other because they come to
have different biochemical capabilities. An enzyme's presence in a cell
is dictated by the expression of one or more cistrons encoding it; thus
DIFFERENTIATION is understood through molecular biology (see
GENE EXPRESSION).

Because enzyme molecules are generally globular proteins, their
shapes and functions may be affected by pH changes in their aqueous
environments (see DENATURATION). Denaturation by extremes of
pH is usually reversible; not so denaturation by heat. Temperature
increase will raise the rate of collision of enzyme and substrate
molecules, thus increasing the rate of ES complex formation and
raising the reaction rate. This is opposed by increased enzyme de-

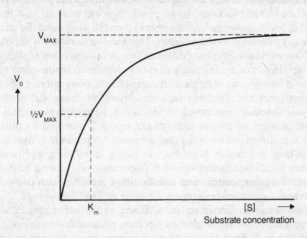

Fig. 15. *Effect of increasing substrate concentration on velocity of enzyme–substrate reaction.*

naturation as the *optimum temperature* for the reaction is exceeded. Eventually the reaction ceases, sometimes only at temperatures well in excess of 100°C (see ARCHAEBACTERIA).

At any one instant, the proportion of enzyme molecules bound to substrate will depend upon the substrate concentration. As this is increased, the initial velocity of the reaction (V_o) on addition of enzyme increases up to a maximum value, V_{max} (see Fig. 15), at which substrate level the enzyme is said to be *saturated* (all active sites maximally occupied), and no further addition of substrate will increase V_o. The value of substrate concentration at which $V_o = \frac{1}{2}V_{max}$ is known as the MICHAELIS CONSTANT (K_m) for the enzyme–substrate reaction. Low K_m indicates high affinity of the enzyme for the substrate.

Some enzymes (e.g. aspartase) bind just one very specific substrate molecule; others bind a variety of the same kind (e.g. all terminal peptide bonds in the case of exopeptidases). The difference arises from the degree of *stereospecificity* of the enzyme. Many need an attached PROSTHETIC GROUP or a diffusible COENZYME for activity. In such enzymes the protein component is termed the *apoenzyme* and the whole functional enzyme-cofactor complex is termed the *holoenzyme*. Enzymes requiring metal ions are sometimes termed *metalloenzymes*, the commonest ions involved being Zn^{2+}, Mg^{2+}, Mn^{2+}, Fe^{2+} or Fe^{3+}, Cu^{2+}, K^+ and Na^+. These ions commonly provide a needed charge within an active site.

Some enzymes occur as part of a MULTIENZYME COMPLEX. In nearly all cases, the shape of the enzyme alters as the ES complex forms, and this brings appropriate groups into such proximity that they are obliged to react. In so doing their electrostatic and hydrophobic bondings to the enzyme break, they fall away, and the enzyme returns to its original shape again. This *induced fit theory* is supported by X-ray crystallographic evidence. The suffix -*ase* often replaces the last few letters of a substrate's name to give the common name of the enzyme using it as substrate: thus sucrase digests sucrose. But an international code for enzymes recognizes six major categories of enzyme function, numbered as follows: 1. *oxido-reductases* (e.g. dehydrogenases), catalysing REDOX REACTIONS; 2. *transferases*, transferring a group of atoms from one substrate to another; 3. *hydrolases*, catalysing hydrolysis reactions; 4. *lyases*, catalysing additions to double bonds (saturating them); 5, *isomerases*, performing isomerizations; 6. *ligases*, performing condensation reactions involving ATP cleavage.

Allosteric enzymes have, in addition to an active site, another stereo-specific site to which an *effector*, or *modulator* molecule can bind. When it does, the shape of the active site is altered so that it can or cannot bind substrate (allosteric stimulation or inhibition respectively). In this way the enzyme can be part of a fine control circuit, requiring the presence or absence of a substance – in addition to substrate presence – before enzyme activity proceeds. Some allosteric enzymes respond to two or more such modulators, permitting still finer control over timing of enzyme activity (see REGULATORY ENZYME).

Feedback (or *retro*-) *inhibition* of a biochemical pathway is often achieved by *allosteric inhibition* of the first enzyme in the sequence by the final product. The product binds non-covalently to the modulator site on the enzyme, closing the active site allosterically.

Enzyme inhibition of a simpler kind is achieved in competitive inhibition, where an inhibitor substance competes with the substrate for the enzyme's active site. The binding is reversible so that the percentage inhibition for fixed inhibitor level decreases on addition of substrate. An extremely important example of this involves probably the most abundant enzyme, *ribulose bisphosphate carboxylase*, the CO_2-fixing enzyme in C_3 PHOTOSYNTHESIS, in which O_2 molecules compete with CO_2 molecules for the active site (see PHOTORESPIRATION). In *uncompetitive inhibition* the inhibitor combines with the ES complex (one piece of evidence for the latter's existence), which cannot therefore yield normal product. In *noncompetitive inhibition* (a form of allosteric inhibition) the inhibitor binds at a non-active site on the enzyme and ES complex so as to deform the active site and prevent ES breakdown, a process unaffected by increasing substrate concentration, being either reversible or irreversible.

Some enzymes are *constitutive*, being synthesized independently of substrate availability; others are *inducible* (e.g. many liver enzymes), being synthesized only when substrate becomes available. The molecular biology of this is to some extent explained in GENE EXPRESSION.

Some enzymes are located randomly in the cytosols of cells; others have very restricted distributions and may be attached to particular membranes or within the matrices of particular organelles. One effect of the latter restriction is that initial velocities of reactions (V_o) can be quite high for a substrate level that would be too low if the molecules were randomized over the whole cell. Another advantage is that incompatible reactions can be kept physically separated.

ENZYME INHIBITION. See ENZYME.

ENZYME KINETICS. Study of the effects of substrate, inhibitor and modulator concentrations on the rate of an enzyme reaction, particularly on initial velocities (V_o). The interrelationships are normally expressed graphically, giving enzyme–substrate–inhibitor curve characteristics, one example being included in the entry for ENZYME, Fig. 15.

EOCENE. GEOLOGICAL EPOCH of the Tertiary period lasting approximately from 54–38 Myr BP.

EOSINOPHIL. One type of white blood cell. MYELOID cells with, in man, a bilobed nucleus, and cytoplasmic vesicles (granules) capable of fusing with the plasma membrane on appropriate stimulation and releasing a toxic protein against large targets, especially parasitic worms. Also release antihistamine, damping inflammatory responses. Migrate towards regions containing T-CELL products. Capable of limited phagocytosis.

EPHEMERAL. Plant with a short life cycle (seed germination to seed production), having several generations in one year; e.g. groundsel. *Desert ephemerals* pass the dry season as dormant seeds. Compare ANNUAL, BIENNIAL, PERENNIAL.

EPHEMEROPTERA. Mayflies. Order of exopterygote insects; with long-lived aquatic nymphs which may moult up to 23 times, adults living from a few minutes to a day since they have rudimentary mouthparts and neither eat nor drink. Final nymphal moult produces a unique *subimago*, which moults to produce the adult. Two pairs of membranous wings, held vertically at rest. One pair of CERCI, with or without additional third caudal prolongation.

EPHYRA. Pelagic larval stage in life cycle of Scyphozoa (jellyfish); develops into adult medusa. Budded asexually from sessile scyphistoma.

EPIBLAST. (Zool.) Most superficial layer of vertebrate BLASTODERM to form part of the embryo. Overlies hypoblast.

EPIBOLY. Process, observed in amphibian and other vertebrate embryos, during which the region occupied by cells of the animal half of the blastula expands over the vegetal half. In amphibians the cells migrate and roll under through the BLASTOPORE, the vegetal cells remaining as just a plug filling the blastopore.

EPICOTYL. Upper portion of the axis of an embryo or seedling, above the cotyledons and below the next leaf or leaves. Compare HYPOCOTYL.

EPIDEMIC. Large-scale temporary increase in prevalence of a disease due to a parasite or some health-related event. Compare ENDEMIC.

EPIDERMIS. Outermost layer of cells of a multicellular organism. (Bot.) Primary tissue, one cell thick, forming protective cell layer on surface of plant body, covered in aerial parts by a non-cellular protective CUTICLE. (Zool.) In invertebrates, often one cell thick, secreting a protective non-cellular CUTICLE. In vertebrates there is no non-cellular cuticle, and the epidermis is composed of several layers of cells; the outermost ones often undergo CORNIFICATION and die.

EPIDIDYMIS. Long (6 m in man) convoluted tube, one attached to each testis in amniotes. Receives sperm from seminiferous tubules and houses them during their maturation, reabsorbing them if they are not ejaculated (in four weeks in man). Peristaltic contractions of the epididymides propel sperm into the sperm duct during ejaculation. Derived embryologically from the mesonephric (Wolffian) duct.

EPIGAMIC. (Of animal characters) attractive to the opposite sex and therefore subject to SEXUAL SELECTION. Often concerned with courtship and mating.

EPIGEAL. (1) Seed germination in which the seed leaves (cotyledons) appear above the ground; e.g. lettuce, tomato. Compare HYPOGEAL. (2) Of animals, inhabiting exposed surface of land, as distinct from underground.

EPIGENESIS. Theory of reproduction and development deriving from Aristotle and espoused by William Harvey (1651) that the parts of an embryo are not all present and *preformed* at the start of development but arise anew one after the other during it. See EPISTASIS, PREFORMATION.

EPIGENETICS. Study of causal interactions between genes and their products which bring the phenotype into being.

EPIGLOTTIS. Cartilaginous flap on ventral wall of mammalian pharynx. The glottis pushes against it during swallowing, preventing food, etc., from entering the trachea.

EPIGYNOUS. See RECEPTACLE.

EPINASTY. (Bot.) More rapid growth of upper side of an organ. In a leaf, would result in a downward curling leaf blade. The growth substance ETHENE has been implicated. Compare HYPONASTY.

EPINEPHRINE. American term for ADRENALINE.

EPIPELON. Extremely widespread community of algae occurring in all waters where sediments accumulate onto which light penetrates. The species are almost all microscopic, living on and in the surface millimetres of the sediment, being unable to withstand long periods of darkness and anaerobic conditions. Motile species exhibit endogenous vertical migration rhythm. An important algal community, particularly in shallow ponds and lakes, as well as in highly transparent oligotrophic and montane lakes. See BENTHOS.

EPIPETALOUS. (Of stamens) borne on the petals, with stalks (filaments) more or less fused with the petals and appearing to originate from them.

EPIPHYSIS. (1) Separately ossified end of growing bone, forming part of joint; peculiar to mammalian limb bones and vertebrae. Separated from rest of bone (DIAPHYSIS) by cartilaginous plate (epiphysial cartilage). Epiphysis and diaphysis fuse when growth is complete. (2) Synonym for PINEAL GLAND.

EPIPHYTE. Plant attached to another plant, not growing parasitically upon it but merely using it for support; e.g. various lichens, mosses, algae, ivy, and orchids, all commonly epiphytes of trees.

EPIPHYTON. Community of organisms living attached to other plants, sometimes in very large populations; well developed in aquatic habitats where algae attach to other plants.

EPISOME. A genetic element (DNA) that may become established in a cell either autonomously of the host genome, replicating and being transferred independently, or else as an integrated part of the host genome, participating with it in recombination and being transferred with it. Term first applied to temperate BACTERIOPHAGE, but includes PLASMIDS. See F FACTOR, TRANSPOSON.

EPISTASIS. Interaction between non-allelic genetic elements or their products, sometimes restricted to cases in which one element suppresses expression of another (*epistatic dominance*). Analogous to genetic DOMINANCE. Segregation of epistatic genes in a cross can modify expected phenotypic ratios among offspring for characters they affect. See HYPOSTASIS, MODIFIER, SUPPRESSOR, MUTATION, POLYGENES, GENETIC VARIATION.

EPITHELIUM. (Zool.) Sheet or tube of firmly coherent cells (see DESMOSOME) with minimal material between them, of ectodermal or endodermal origin, lining cavities and tubes and covering exposed surfaces of body; one surface of epithelium is therefore free, the other usually resting on a BASEMENT MEMBRANE over connective tissue. Its cells are frequently secretory, secretory parts of most glands being epithelial. Classified according to: height relative to breadth (e.g. *columnar*, *cuboidal*, or *squamous*, in order of diminishing relative height); whether the sheet is one cell thick (*simple*) or many (*stratified* or *pseudostratified*); and presence of cilia (*ciliated*). When morphologically identical tissue is derived from mesoderm, it is either ENDOTHELIUM or MESOTHELIUM. (Bot.) Layer of cells lining schizogenously formed secretory canals and cavities, e.g. in resin canals of pine.

EPITOPE. Antigenic determinant. See ANTIGEN.

EPITREPTIC BEHAVIOUR. Behaviour by one individual tending to cause the approach of a member of the same species (a conspecific).

EPIZOITE. Non-parasitic sedentary animal living attached to another animal. Compare EPIPHYTE.

EQUATORIAL PLATE. Plane in which the chromosomes of a cell lie during metaphase of mitosis and meiosis; the equator of the spindle.

EQUILIBRIUM POTENTIAL. Potential (voltage gradient) at which a particular ion type passes equally easily in either direction across a cell membrane. Different ions have different equilibrium potentials. See MEMBRANE POTENTIAL.

EQUISETALES. Horsetails. See SPHENOPHYTA.

ERGASTOPLASM. Defunct term for ENDOPLASMIC RETICULUM.

ERGOT. (1) Disease of cereal and wild grass inflorescences caused by the fungus *Claviceps purpurea* (Ascomycotina). (2) Dark spurshaped sclerotium developing in place of a healthy grain in a diseased inflorescence. Ergots contain substances poisonous to humans and domestic animals. Some (e.g. ergotamine) are used medicinally.

ERYTHROBLAST. Nucleated bone marrow cell which undergoes successive mitoses, develops increasing amounts of haemoglobin, and gives rise to a *reticulocyte*, and finally the fully differentiated RED BLOOD CELL.

ERYTHROCYTE. See RED BLOOD CELL.

ERYTHROPOIESIS. Red blood cell formation. See HAEMOPOIESIS.

ESCAPE. Cultivated plant found growing as though wild.

ESCHERICHIA COLI (*E. coli*). Motile, Gram-negative, rod-shaped bacterium (Enterobacteriaceae) used most extensively in bacterial genetics and molecular biology. Normal inhabitant of the human colon; is usually harmless although some strains can cause disease. See BACTERIA, CHROMOSOME, JACOB-MONOD THEORY, GRAM'S STAIN.

ESSENTIAL AMINO ACID. See AMINO ACID.

ESSENTIAL FATTY ACID. Fatty acids required in the diet for normal growth. In mammals, include linoleic and gamma-linolenic acids, obtained from plant sources, without which poor growth, scaly skin, hair loss and eventually death occur. Precursors of arachidonic acid and PROSTAGLANDINS.

ESSENTIALISM. The view, associated in particular with Aristotle, that for any individual there is a definitive set of properties, individually necessary and collectively sufficient, rendering it the kind of individual that it is. This approach has at times been adopted in the context of the taxa used in classification, sometimes rhetorically and in opposition to evolutionary theories. See NATURAL KIND, NOMINALISM.

ETAERIO. (Of fruits) an aggregation; e.g. of achenes, in buttercup; of drupes, in blackberry.

ETHENE (ETHYLENE). Simple gaseous hydrocarbon (C_2H_4) produced in small amounts by many plants and acting as a plant hormone, or GROWTH SUBSTANCE. Production often stimulated by AUXINS, but its release commonly inhibits auxin synthesis (negative feedback) and transport. Normally inhibits longitudinal growth, but promotes radial enlargement of tissues. Its effect on fruit ripening has agricultural importance, and is used to promote ripening of tomatoes picked green and stored in ethene until marketed. Also used to promote ripening of grapes. Ethene also promotes abscission of leaves, flowers and fruits in a variety of plant species, and is used commercially to promote fruit loosening in cherries, grapes and blueberries. Also appears to play roles in sex determination of flowers in MONOECIOUS plants and in EPINASTY.

ETHIOPIAN. Designating a zoogeographical region comprising Africa south of the Sahara. Sometimes Madagascar is treated as a separate region (the Malagasy Region).

ETHOLOGY. Study of animal behaviour in which the overriding aim is to interpret behavioural acts and their causes in terms of evolutionary theory. The animal's reponses are interpreted within the context of its actual environmental situation.

ETHYLENE. See ETHENE.

ETIOLATION. Phenomenon exhibited by green plants when grown in darkness. Such plants are pale yellow because of absence of chlorophyll, their stems are exceptionally long owing to abnormal lengthening of internodes, and their leaves are reduced in size.

EUBACTERIALES. Eubacteria; a large and diverse order of BACTERIA, lacking photosynthetic pigments. Simple, undifferentiated cells with rigid cell walls, either spherical or straight rods. If motile, move by peritrichous flagella. Thirteen recognized families. Includes the important genera *Azotobacter* and *Rhizobium* (both nitrogen-fixers), *Escherichia*, etc.

EUCARPIC. (Of fungi) with a mature thallus differentiated into distinct vegetative and reproductive portions. Compare HOLOCARPIC.

EUCARYOTE. See EUKARYOTE.

EUCHROMATIN. Eukaryotic chromosomal material (chromatin) staining maximally during metaphase and less so in the interphase nucleus, when it is less condensed. See CHROMOSOME, HETEROCHROMATIN.

EUGENICS. Study of the possibility of improving the human GENE POOL. Historically associated with some extreme political tendencies and with encouragement of breeding by those presumed to have favourable genes and discouragement of breeding by those presumed to have unfavourable genes; nowadays the more humanitarian GENETIC COUNSELLING has largely replaced talk of eugenics.

EUGLENOPHYTA. Division of ALGAE. Characterized by possession of chlorophylls *a* and *b*, paramylon as the cytoplasmic storage product, one membrane of *chloroplast endoplasmic reticulum*, a MESOKARYOTIC nucleus, flagella with fibrillar hairs in a row, and lacking sexual reproduction. Colourless (non-photosynthetic) forms occur, and can arise irreversibly from pigmented forms. Occur in marine and brackish water and most freshwater habitats and are often abundant in water polluted by organic waste.

EUKARYOTE (EUCARYOTE). Organism in whose cell or cells chromosomal genetic material is (or was) contained within one or more nuclei and so separated from the cytoplasm by two nuclear membranes. Some eukaryotic cells (e.g. mammalian erythrocytes, phloem sieve tubes) lose their nuclei during development; but all are distinguished from prokaryotic cells by generally much larger size, by presence of the proteins ACTIN, MYOSIN, TUBULIN and HISTONES, somewhat denser (80S) RIBOSOMES, and by a greater variety of membrane-bound organelles. Cell division is by MITOSIS and/or MEIOSIS. Where CELL LOCOMOTION occurs, some of the above proteins are involved. Compare PROKARYOTE. See CELL, NUCLEUS, MESOKARYOTE.

EUMYCOTA. Fungal division containing the true fungi (MASTIGOMYCOTINA, ZYGOMYCOTINA, DEUTEROMYCOTINA, ASCOMYCOTINA, BASIDIOMYCOTINA). Are basically mycelial in organization and are often contrasted with the plasmodial slime fungi (MYXOMYCOTA).

EUPHOTIC ZONE (PHOTIC ZONE). Uppermost zone of lakes, seas and rivers, with sufficient light for active photosynthesis. In clear water, may extend to 120 metres.

EUPLOID. Term describing cells whose nuclei have an exact multiple of the HAPLOID set of chromosomes, there being no extra or fewer than that multiple. Thus, DIPLOID, TRIPLOID, TETRAPLOID, etc., cells are all euploid. Compare ANEUPLOID.

EURYHALINE. Able to tolerate a wide variation of osmotic pressure of environment. Compare STENOHALINE, OSMOREGULATION.

EURYPTERIDA. Fossil subclass of the MEROSTOMATA, appearing in the Ordovician. Free-swimming, marine, brackish and freshwater forms; prosoma with six pairs of ventral appendages, the first being CHELICERAE, the others modified for grasping, walking and swimming. Larva resembled trilobite larva of king crab. Active predators, about two metres in length. See ARACHNIDA.

EURYTHERMOUS. Able to tolerate wide variations of environmental temperature. Compare STENOTHERMOUS.

EUSOCIAL. Term applied generally to certain colonial insects which exhibit cooperative brood care, overlap between generations, and reproductive CASTES. Includes termites, bees, ants and wasps.

EUSPORANGIATE. (Of sporangia in vascular plants) arising from a group of parent cells and possessing a wall of two or more layers of cells. Spore production greater than in the LEPTOSPORANGIATE type.

EUSTACHIAN TUBE. Tube connecting middle ear to pharynx in tetrapod vertebrates. Allows equalization of air pressure on either side of eardrum. See EAR, MIDDLE; SPIRACLE.

EUSTELE. Stele in which primary vascular tissues are arranged in discrete strands around a pit.

EUSTIGMATOPHYTA. Division of the ALGAE containing basically unicellular forms, found in freshwater or soil. Most species form zoospores possessing a single emergent flagellum, although a second basal body is present. Named after the large orange-red EYESPOT at anterior end of the zoospore, independent of the chloroplast (main difference from the Xanthophyta). Chlorophyll a and β-carotene are present, with xanthophylls.

EUTHERIA (PLACENTALIA). Placental mammals. Infraclass of the MAMMALIA, and the dominant mammals today. Most of the 3800 species occur within about six orders: Insectivora (e.g. shrews, hedgehogs), Chiroptera (bats), Rodentia (e.g. mice, rats), Artiodactyla (e.g. deer, pigs), Carnivora (e.g. cats, dogs, weasels) and Primates (e.g. lemurs, monkeys, apes, humans). Appear in Upper Cretaceous, at time of dinosaur extinction. Connection between embryo and uterus intimate and complex; amnion and chorion present; umbilicus links embryo to chorio-allantoic PLACENTA; scrotum posterior to penis. Gestation period of varying length; newborn young more advanced developmentally than in other mammals. Great ADAPTIVE RADIATION in early Cenozoic. See PROTOTHERIA, METATHERIA.

EUTROPHIC. (Of lakes) rich in nutrients; highly productive in terms of organic matter produced. Compare OLIGOTROPHIC.

EUTROPHICATION. Usually rapid increase in the nutrient status of a body of water, both natural and occurring as a by-product of human activity. May be caused by run-off of artificial fertilizers from agricultural land, or by input of sewage or animal waste. May occur when large flocks of migrating birds collect around watering holes. Leads to reduction in species diversity as well as change in species composition, often accompanied by massive growth of dominant species. Excessive production stimulates respiration, increasing dissolved oxygen demand and leading to anaerobic conditions, commonly with accumulation of obnoxious decay and animal death. Artificial eutrophication can be slowed or even reversed by removal of nutrients at source, but may require costly sewage treatment plants.

EVERGREEN. (Of plants) bearing leaves all year round (e.g. pine, spruce). Contrasted with DECIDUOUS.

EVOCATION. Ability of an inducer to bring forth a particular mode of differentiation in a tissue which is *competent*. It has been suggested that the inducer brings about release of a substance (the *evocator*) which initiates the differentiation. See INDUCTION, ORGANIZER.

EVOLUTION. (1) *Microevolution*: changes in appearance of populations and species over generations. (2) *Macroevolution* or *phyletic evolution*: origins and EXTINCTIONS of species and grades (see SPECIATION).

Microevolution includes changes in mean and modal phenotype, morph ratios, etc. such as occur within populations from one generation to the next. When statistically significant changes in such variables (or the genes responsible for them) occur with time, a population may be said to evolve. Macroevolution includes large-scale phyletic change over geological time (e.g. successive origins of crossopterygian fish, amphibians, reptiles, birds and mammals), as well as extinctions of taxa within such groups. It is usually accepted that causes of evolutionary change include NATURAL SELECTION and GENETIC DRIFT, and that macroevolutionary change can be explained by the same factors that bring about microevolution.

Debate has recently centred upon the rate of evolutionary change. Some biologists accept that evolution largely occurs by gradual ANAGENESIS; others stress the role of CLADOGENESIS and take the view that species persist unchanged for considerable periods of time, and that relatively rapid speciation events punctuate the fossil record (*punctuated equilibrium*). Darwin considered both to be possibilities. At the molecular level, controversy centres on the respective influences in evolution of random alterations in genetic material (the *neutralist* view) and of selective changes (the *selectionist* view). See MOLECULAR CLOCK. Opposed to evolutionary explanations of the composition of the Earth's fauna and flora is the group of views termed 'special creationism', which holds that there are no bonds of genetic relationship between species, past or present. See ORIGIN OF LIFE.

Although Anaximander (6th. cen. BC), Empedocles (5th. cen. BC) and Aristotle (4th. cen. BC) all held evolutionary views of some kind, they depended more on *a priorism* than on observation and testable theory. LAMARCK is often considered the most influential evolutionary thinker prior to Charles DARWIN and Alfred WALLACE but his theory was very different from theirs. They themselves drew apart on the question of human origins and the role of sexual selection.

Evidence for the fact of macroevolution comes principally from comparative morphology (especially anatomy and embryology), from geographical distributions of organisms, and from fossil records. The modern theory of evolution (NEO-DARWINISM) derives largely from the kind of genetical knowledge which Darwin lacked, principally the occurrence of Mendelian segregation, which helps explain how variations can be maintained in populations. Evidence for recency of common ancestry of taxa now comes especially from DNA sequencing and hybridization. Evidence for microevolution and Darwinian natural selection (amounting to his 'special theory of evolution') stems largely from population genetics (e.g. see INDUSTRIAL MELANISM), although Darwin himself drew heavily on the analogy of ARTIFICIAL SELECTION. See NATURAL SELECTION.

EVOLUTIONARILY STABLE STRATEGY (ESS). A heritable strategy (commonly but by no means always behavioural) which, if adopted by (expressed in) most members of a population, cannot be supplanted in evolution by an alternative (mutant) strategy. The strategy may be complex and involve a variety of different sub-responses in accordance with environmental changes, not least other organisms' behaviours. See GAME THEORY.

EVOLUTIONARY TAXONOMY. A school of biological CLASSIFICATION which makes use of both phenetic and phylogenetic data in classifying organisms. Because there is no theoretical guide as to when one approach should be used and when the other, this very influential school has been criticized by adherents of CLADISTICS. See PARALLEL EVOLUTION.

EVOLUTIONARY TRANSFORMATION SERIES. A pair of HOMOLOGOUS characters, one derived directly from the other. See PLESIOMORPHOUS, APOMORPHOUS, CLADISTICS.

EXARCH. Type of maturation of primary xylem in roots, in which the oldest xylem elements are located closest to the outside of the axis. Compare ENDARCH.

EXCRETION. (1) Any process by which an organism gets rid of waste metabolic products. Differs from EGESTION in that wastes removed are products of the organism's cells rather than simply undigested wastes; and from SECRETION since substances produced would generally be harmful if allowed to accumulate, and as a rule have no intrinsic value to the organism. The simplest excretory method is passive diffusion, either through the normal body surface or across organs with enlarged surface areas (gills, lungs). These may be supplemented or replaced by internal excretory organs, particularly where the body surface cannot be used. Excretory organs typically remove metabolic products from interstitial fluids (e.g. lymph, blood plasma). The gut occasionally serves as a route for excretory products, but is not an excretory organ. Nitrogenous excretion is usually in the form of ammonia (aquatic environments), urea (terrestrial environments) or uric acid (environments where water is at a premium). Common invertebrate excretory organs include FLAME CELLS, NEPHRIDIA, and MALPIGHIAN TUBULES, but in some cases (e.g. large crustaceans) excretion may be deposited in the exoskeleton, commonly to be lost during moulting. Vertebrate KIDNEYS work by filtration and selective reabsorption, and, like some invertebrate excretory organs, also have roles in OSMOREGULATION.

Excretion in plants includes GUTTATION and removal by diffusion of excess oxygen produced by photosynthesis, since oxygen may inhibit that process. Leaf fall also removes a number of metabolic wastes. (2) A substance, or mixture of substances, excreted: *excreta*.

EXERGONIC. (Of a chemical reaction) yielding energy. See THERMODYNAMICS.

EXINE. Outer layer of spores and pollen grains; usually divided into two main layers: an outer ectexine and an inner endexine. Often composed of SPOROPOLLENIN.

EXOCRINE GLAND. Any animal gland of epithelial origin which secretes, either directly or most commonly via a duct, onto an epithelial surface. See GLAND, ENDOCRINE GLAND.

EXOCYTOSIS. Process whereby a vesicle (e.g. secretory vesicle), often budded from the ENDOPLASMIC RETICULUM or GOLGI APPARATUS, fuses with the plasma membrane of the cell, with release of vesicle contents to exterior. Common process in SECRETION. When restricted to anterior region of cell it is an important stage in much eukaryotic CELL LOCOMOTION. Compare ENDOCYTOSIS. See SYNAPTIC VESICLES.

EXODERMIS. Layer of closely fitting cortical cells with suberized walls, replacing the withered piliferous layer in older parts of roots.

EXOENZYME. Enzyme secreted or produced externally by the protoplast and functioning outside the cell (or hypha).

EXOGAMY. See OUTBREEDING.

EXONS. Coding segments of DNA alternating with non-coding intervening sequences, or INTRONS.

EXONUCLEASE. Enzyme which removes nucleotides one by one from the end of a polynucleotide chain. See DNase.

EXOPEPTIDASE. Proteolytic enzyme which removes amino acids one by one from the end of a protein molecule. Compare ENDOPEPTIDASE.

EXOPODITE. See BIRAMOUS APPENDAGE.

EXOPTERYGOTA (HETEROMETABOLA). Winged insects with incomplete metamorphosis; sometimes regarded as a subclass of the INSECTA. No pupal stage. Wings develop outside the body; successive larvae (nymphs) become progressively adult-like. Includes palaeopteran orders Ephemeroptera and Odonata; orthopteroid orders Plecoptera, Grylloblattoidea, Orthoptera, Phasmida, Dermaptera, Embioptera, Dictyoptera, Isoptera and Zoraptera; and hemipteroid orders Psocoptera, Mallophaga, Siphunculata, Hemiptera and Thysanoptera. See ENDOPTERYGOTA.

EXOSKELETON. Skeleton covering the outside of the body, or located in the skin. In arthropods (see CUTICLE), secreted by the epidermis; in many vertebrates, e.g. tortoises, armadillos, the exoskeleton consists of bony plates beneath the epidermis. Many primitive jawless vertebrates (ostracoderms) and primitive jawed vertebrates (placoderms) had body armour comprising bony skin plates and scales. The scales and denticles of modern fish are remnants of this.

EXOTOXIN. Toxin released by a microorganism into surrounding growth medium or tissue during *growth phase* of infection. Generally inactivated by heat and easily neutralized by specific antibody. Produced mainly by Gram-positive bacteria, such as the agents of botulism, diphtheria, *Shigella* dysentery and tetanus. The alga *Prymnesium parvum* forms a potent exotoxin that causes extensive fish mortalities in brackish water conditions in many countries in Europe and in Israel. Compare ENDOTOXIN.

EXPERIMENT. The intentional manipulation of material conditions so as to elicit an answer to a question, often posed in the form: what is the effect of x on y? The aim of the experimenter is to isolate x as the only free variable, keeping constant all other variables which might affect the value of y. Values of x can then be paired off with values of y, when changes in x are said to be the cause of any changes in y. A similar approach compares the results of two experimental situations differing in just one initial condition, which often has zero value in one of the experimental situations (called the *control*) but is allowed free range over its values in the other experimental situation (called the *experiment*). The effects of this free-ranging variable are then compared with the effect of its absence (zero value), and since it is the only independent variable, any differences in effect can be said to have been caused by changes in its value. Controlled experiments must have this comparative element. The rationale is to eliminate all possible alternative causes of effects save the one under investigation. Without such controlled experiments, the material causes of phenomena could never be ascertained. It is often assumed, not always with justification, that methods used to study biological material do not themselves affect the properties being studied.

EXPLANATION. A phenomenon may be said to have been fully explained when all its component parts can be formally deduced as consequences of sets of actual initial conditions (the minor premises) satisfying the terms of whichever general law (the major premise) represents our most inclusive summary of the relevant experimental data to date. Attempts to explain biological phenomena solely in terms of the language employed in physics and chemistry exemplify what is termed *reductionism*. Most people believe this can only be achieved if terms peculiar to biology can be 'paired off' by identity or equivalence relations to terms in the physical sciences. It is highly contentious whether this can be achieved, even in principle.

EXPLANTATION. See TISSUE CULTURE.

EXPONENTIAL GROWTH. Growth of cells, populations, etc., in which rate of increase is dependent only upon the number of individuals and their potential net reproductive rate. In other words, no competition occurs between individuals for resources nor is there any other detrimental effect of individuals upon one other. Such a situation is characteristic of the initial growth phase of microorganisms in cultures, or of organisms introduced into regions where food is not limiting and where natural controls (e.g. predators, parasites) are absent. The exponential growth curve can be defined by the equation:

$$N_t = N_o e^{(b-d)t}$$

where t is a very short time interval
     N_t is the number of individuals after time t
     N_o is the number of individuals at the beginning of the time interval
     b is the 'birth' rate during time t
     d is the 'death' rate during time t
     e   is a constant, taken for convenience to be the base of Napierian logarithms, 2.718 (the exponential constant).

EXPRESSIVITY. Level to which the effect of a gene is realized in the phenotype. See PENETRANCE.

EXTENSOR. Muscle or tendon straightening a joint, antagonistic to FLEXOR.

EXTEROCEPTOR. A RECEPTOR detecting stimuli emanating from outside an animal. Compare INTEROCEPTOR.

EXTINCTION. Termination of a genealogical lineage. Used most frequently in the context of species, but applicable also to populations and to taxa higher than species. Agents of 'background rate' extinction include competition, predation and disease, alteration of habitat and random fluctuations in population size.

There have been four periods of so-called *mass extinction*, during which the Earth's fauna has suffered extinction rates far higher than the normal background rate. These occurred in the Ordovician, the late Devonian, the late Permian (225 Myr BP), the late Triassic (190 Myr BP) and the late Cretaceous/Tertiary (K/T, 65 Myr BP). Possible causes of greater than normal extinction rates include evolutionary competition, geological (e.g. volcanic) and climatic change and cometary or other impact. Victims of the K/T extinction included the dinosaurs and 60–75% of all marine species, and evidence (high iridium levels and soot in clays at the K/T boundary) suggests that cometary impact could have resulted in large-scale fire. This could have released huge volumes of oxides of nitrogen into the atmosphere, causing severe acid rain and reducing surface temperatures. One likely genetic factor in extinction as population size decreases is *inbreeding depression*. See PUNCTUATED EQUILIBRIUM.

EXTRACELLULAR. In general, occurring outside the plasma membrane; but where a CELL WALL is present, often refers to the region surrounding this. See GLYCOCALYX.

EXTRACHROMOSOMAL INHERITANCE. Inheritance of genetic factors not forming part of a chromosome. Examples include PLASMID, mitochondrial and chloroplast inheritance. Inheritance of a variety of intracellular symbionts may also be regarded as extrachromosomal. See CYTOPLASMIC INHERITANCE, EPISOME.

EXTRAEMBRYONIC COELOM. In amniote development, the space lying between the mesoderm layers lining inner surface of the chorion and outer surface of the amnion.

EXTRAEMBRYONIC MEMBRANES. The YOLK SAC, CHORION, AMNION and ALLANTOIS of amniote vertebrates; membranes derived from the zygote but lying outside the epidermis of the embryo proper. Have played a major part in evolution of vertebrate terrestrialization. See Fig. 16.

EYE. Sense organ responding to light. In invertebrates, either a simple scattering of light-sensitive pigment spots in the general epithelium but more often comprising an optic cup of receptor cells with screen-

(a)

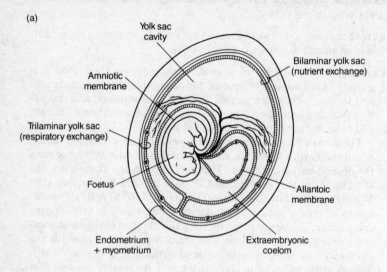

(b)

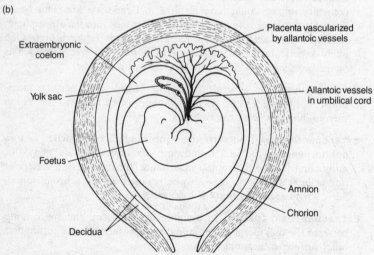

Fig. 16. Extraembryonic membranes during development of (a) wallaby (marsupial) and (b) human. The uterus wall is outermost in both.

ing pigment cells (each functional unit an *ocellus*), lacking a refractive surface so that no image can be formed. Although a lens may be present, most ocelli can only differentiate between light and dark. Nonetheless, this enables orientation with respect to light direction and intensity.

(a)

(b)

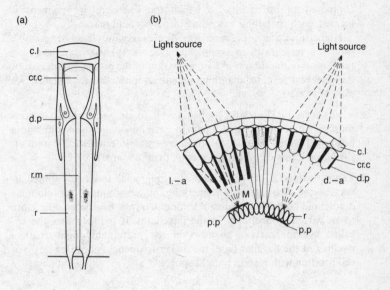

(c)

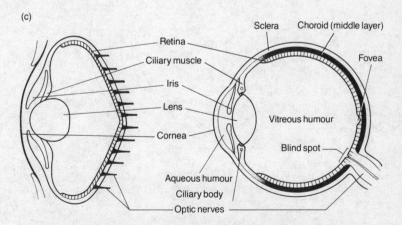

Fig. 17. (a) *An ommatidium from an insect compound apposition eye.* **(b)** *A superposition compound eye, light-adapted (l.–a) and dark-adapted (d.–a). The central three ommatidia show the proposed laminated structure of the cones; c. l = corneal lens, cr. c = crystalline cone, d. p = distal pigment cell, p. p = proximal pigment cell, r = retinula cell (photoreceptor), r. m = rhabdome.* **(c)** *Comparison of cephalopod eye (left) with human eye (right).*

The basic unit of the arthropod compound eye is the *ommatidium*, comprising a cornea lens, crystalline cone, a group of usually 7–8 sense (retinula) cells radially arranged around a central rhabdome formed from their innermost fibrillar surfaces (rhabdomeres composed of microvilli), in which the light-sensitive pigment is located, each rhabdome extending into a nerve fibre distally.

Higher molluscan (i.e. cephalopod) eyes (e.g. of *Octopus*) resemble those of vertebrates in complexity (see CONVERGENCE); however, there is an ommatidium-like organization in the retina. For details of the vertebrate eye, see diagram and entries for structures labelled. See also TAPETUM.

EYE MUSCLES. (a) *Extrinsic* (outside eyeball). In vertebrates six such muscles rotate the eyeball: a pair of anterior oblique and four, more posterior, rectus muscles. Supplied by cranial nerves III, IV and VI. (b) *Intrinsic* (inside eyeball); see IRIS, CILIARY BODY.

EYESPOT (STIGMA). (1) Light-sensitive pigment spots of some invertebrates. See EYE. (2) Something of a misnomer for globular carotenoid-containing region of some eukaryotic flagellated cells, close to or within chloroplast. In the flagellates, it is likely that the so-called eyespot casts a shadow on the presumably light-sensitive swelling at the flagellar base, the flicker frequency indicating angle of cell rotation with respect to light source.

F

F, (FIRST FILIAL GENERATION). Offspring obtained in breeding work after crossing the parental generation (P_1) or by selfing one or more of its members.

F₂ (SECOND FILIAL GENERATION). Offspring obtained after crossing members of the F_1 generation or by selfing one or more of its members.

FACIAL NERVE. See CRANIAL NERVES.

FACILITATED DIFFUSION. Carrier-mediated transport across CELL MEMBRANES, the transported molecule never moving against a concentration gradient. Only speeds up rate of equilibrium attainment across membrane. Examples include transport of glucose across plasma membranes of fat cells, skeletal muscle fibres, the microvilli of ileum mucosa and across proximal convoluted tubule cells of vertebrate kidneys. A T P hydrolysis is not involved. Compare ACTIVE TRANSPORT. See TRANSPORT PROTEINS.

FACILITATION. (1) Increase in responsiveness of a postsynaptic membrane to successive stimuli, each one leaving the membrane more responsive to the next. Compare temporal SUMMATION. (2) Social facilitation. The increased probability that other members of a species will behave similarly once one member has acted in a certain way. See NERVOUS INTEGRATION.

FACULTATIVE. Indicating the ability to live under altered conditions, or to behave adaptively under markedly changed circumstances. Thus a *facultative parasite* may survive in the free-living or parasitic mode (see MIXOTROPH); a *facultative anaerobe* may survive aerobically or anaerobically; a *facultative apomict* may reproduce either by APOMIXIS or by more conventional sexual means.

FAD (FLAVIN ADENINE DINUCLEOTIDE). PROSTHETIC GROUP of several enzymes (generally flavoproteins). Derived from the vitamin riboflavin and involved in several REDOX REACTIONS, e.g. as catalysed by various dehydrogenases (e.g. NADH dehydrogenase, succinate dehydrogenase) and oxidases (e.g. xanthine oxidase, amino acid oxidase). Reduced flavin dehydrogenase (FD, see ELECTRON TRANSPORT SYSTEM) can reduce methylene blue:

$$FD–FADH_2 + \text{methylene blue} = FD–FAD + \text{methylene blue}_{red}$$
$$\text{(blue)} \qquad\qquad\qquad\qquad \text{(colourless)}$$

FAECES. See EGESTION.

FAIRY RING. Circle of mushrooms or other sporocarps, mainly members of the Basidiomycotina, which is formed by radial growth of an underground mycelium from its initial starting point.

FALLOPIAN TUBE. In female mammals, the bilaterally paired tube with funnel-shaped opening just behind ovary, leading from perivisceral cavity (coelom) to uterus. By muscular and ciliary action it conducts eggs from ovary to uterus. Is frequently the site of fertilization. Represents part of MÜLLERIAN DUCT of other vertebrates.

FALSE ANNULUS. Discrete grouping of thick-walled cells on the jacket of some fern sporangia, not directly influencing dehiscence.

FAMILY. A taxonomic category, below ORDER and above GENUS. Typically comprises more than one genus. Familial suffixes normally end -*aceae* in botany, and -*idae* in zoology.

FASCIA. Sheet of connective tissue, as enclosing muscles.

FASCIATION. Coalescing of stems, branches, etc., to form abnormally thick growths.

FASCICLE. (1) Bundle of pine leaves or other needle-like leaves of gymnophytes. (2) Now obsolete term, formerly applied to vascular bundle.

FASCICULAR CAMBIUM. Cambium that develops within a vascular bundle.

FAT (NEUTRAL FAT). Major form of LIPID store in higher animals and some plants. Commonly used synonymously with TRIGLYCERIDE, which not only stores more energy per gram than any other cell constituent ($2\frac{1}{2}$ times the ATP yield of glycogen) but, being hydrophobic, requires less water of hydration than polysaccharide and is therefore far less bulky per gram to store. ADIPOSE TISSUE is composed of cells with little besides fat in them. Hydrolysed by lipases to yield fatty acids and glycerol. See CHYLOMICRON.

FAT BODY. (1) Organ in abdomen of many amphibia and lizards containing ADIPOSE TISSUE, used during hibernation. (2) In insects, diffuse tissue between organs, storing fat, protein, occasionally glycogen and uric acid.

FATE MAP. Diagram showing future development of each region of the egg or embryo. A series of such maps indicates the trajectories of each part from egg to adult. Construction may involve vital staining, cytological and genetic markers. Most easily constructed in cases of highly MOSAIC DEVELOPMENT.

FATTY ACID. Organic aliphatic and usually unbranched carboxylic acid, often of considerable length. Condensation with glycerol results in ester formation to form mono–, di–, and triglycerides (fat).

Commonly a component of other LIPIDS. Free fatty acids are transported in blood plasma largely by albumin. Saturated fatty acids include *palmitic acid*, $CH_3(CH_2)_{14}COOH$ and *stearic acid*, $CH_3(CH_2)_{16}COOH$; unsaturated fatty acids include *oleic acid*, $CH_3(CH_2)_7CH:CH(CH_2)_7COOH$. See FATTY ACID OXIDATION, LIPASE.

FATTY ACID OXIDATION (BETA-OXIDATION). Prior to oxidation, fatty acids undergo a complex activation in the cytosol followed by transport across the mitochondrial membranes, whereupon the acyl group binds to COENZYME A to form a fatty acyl-CoA thioester. The terminal two carbon atoms are removed enzymatically (forming acetyl CoA, for entry into the KREBS CYCLE) while another CoA molecule is bound to the remaining fatty acid chain. Each sequential 2-carbon removal is accompanied by dehydrogenation and production of reduced NAD for entry into the ELECTRON TRANSPORT SYSTEM of mitochondria and ATP production. See VITAMIN E.

FEATHER. Elaborate and specialized epidermal production characteristic of birds as a class, providing thermal insulation, colouration and, generally, lift and thrust of wings during flight. Develops from *feather germ*, a minute projection from the skin, within which longitudinal ridges of epidermal cells (*barb ridges*) form early on. On each ridge further cells of appropriate shape and position form the *barbules* after keratinization, some cell processes becoming *barbicels*, or hooks. A deep pit in the epidermis, the *feather follicle*, surrounds the bases of feathers. The first feathers are *down feathers* in which the quill is very short, a ring of barbs with minute and non-interlocking barbules sticking up from its top edge giving a soft and fluffy texture. Some follicles produce down feathers throughout life, but most are pushed out by new feathers during moulting. The adult (*contour*) feathers grow in definite tracts on the skin, with (except in penguins and RATITES) bare patches between.

Other feather types include: *intermediate feathers*, showing a combination of features of contour and down feathers; *filoplumes* (*plumulae*), which are hair-like, usually lacking vanes; *vibrissae*, stiff and bristle-like, often around the nares, and *powder down*, which is soft downy material giving off dusty particles used in feather cleaning.

The amount of keratin required to make a new set of contour feathers may cause the timing of moulting to be under strong selection pressure. Feathers may be pigmented or have a barbule arrangement which produces interference colours. See Fig. 18.

FECUNDITY. Reproductive output, usually of an individual. Number of offspring produced. See FITNESS.

FEEDBACK INHIBITION. See END-PRODUCT INHIBITION.

FEMUR (pl. FEMORA). (1) Thigh-bone of tetrapod vertebrates. (2) The third segment from the base of an insect leg.

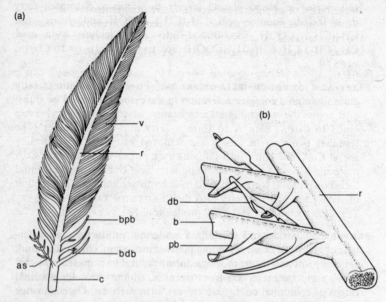

Fig. 18. (a) *Generalized contour feather; as = aftershaft, bdb = barb with distal barbule, bpb = barb with proximal barbule, c = calamous or quill, r = rachis, v = one side of vein.* **(b)** *Enlarged view of part of rachis of contour feather seen from dorsal side. Three proximal barbules (pb) have been cut short. These and the distal barbules (db) are less widely spaced than shown here.*

FERAL. Of domesticated animals, living in a wild state. See ESCAPE.

FERMENTATION. Enzymatic and anaerobic breakdown of organic substances (typically sugars, fats) by microorganisms to yield simpler organic products. Pasteur showed in about 1860 that microorganisms were responsible (contrary to view of Liebig). Kuhne called the 'active principle' an ENZYME in 1878, and Buchner first isolated a fermentative cell-free yeast extract in 1897.

The term is often used synonymously with anaerobic RESPIRA-TION. Classic examples include alcohol production by yeasts, and the conversion of alcohol to vinegar (acetic acid) by the bacterium *Acetobacter aceti*, a process commonly called *acetification*. Lactic acid production by animal cells is another example. In all cases the final hydrogen acceptor in the pathway is an organic compound. See BIOTECHNOLOGY.

FERNS. See PTEROPHYTA.

FERREDOXINS (IRON-SULPHUR PROTEINS). Proteins containing iron and acid-labile sulphur in roughly equal amounts; extractable from wide range of organisms, where they are components of the ELEC-TRON TRANSPORT SYSTEMS of mitochondria (involved in aerobic respiration) and of chloroplasts (involved in PHOTOSYNTHESIS), where they undergo reversible Fe(II)–Fe(III) transitions.

FERRITIN. Iron-storing protein (esp. in spleen, liver and bone marrow). The iron (Fe^{3+}) is made available when required for haemoglobin synthesis, being transferred by TRANSFERRIN.

FERTILIZATION (SYNGAMY). Fusion of two GAMETES (which may be nucleated cells or simply nuclei) to form a single cell (*zygote*) or fusion nucleus. Commonly involves cytoplasmic coalescence (*plasmogamy*) and pooling of nuclear material (*karyogamy*). With MEIOSIS it forms a fundamental feature of most eukaryotic sexual cycles, and in general the gametes that fuse are HAPLOID. When both are motile, as primitively in plants, fertilization is *isogamous*; when they differ in size but are similar in form it is *anisogamous*; when one is non-motile (and usually larger) it is termed *oogamous*. This is the typical mode in most plants, animals and many fungi. In many gymnophytes and all anthophytes neither gamete is flagellated, and a POLLEN TUBE is involved in the fertilization process. In animals, *external fertilization* occurs (typically in aquatic forms) where gametes are shed outside the body prior to fertilization; *internal fertilization* occurs (typically as an adaptation to terrestrial life) where sperm are introduced into the female's reproductive tract, where fertilization then occurs. After fertilization the egg forms a *fertilization membrane* to preclude further sperm entry. Sometimes the sperm is required merely to activate the egg (see ACTIVATION, PARTHENOGENESIS, PSEUDOGAMY). See also ACROSOME, DOUBLE FERTILIZATION.

FEULGEN METHOD. Staining method applied to histological sections, giving purple colour where DNA occurs.

F FACTOR (F PLASMID, F PARTICLE, F ELEMENT, SEX ELEMENT, SEX FACTOR). One kind of PLASMID found in cells of the bacterium *E.coli*, and playing a key role in its sexuality (i.e. inter-cell gene transfer). It encodes an efficient mechanism for getting itself transferred from cell to cell, like many drug-resistance plasmids. Rarely, the F plasmid integrates into the host chromosome (forming an *Hfr* cell), when the same transmission mechanism results in a segment of chromosome adjacent to the integrated F plasmid being transferred from donor cell to recipient. In this condition it behaves very like a λ PROPHAGE, replicating only when the host chromosome does. When this cell conjugates with a cell lacking an F particle, a copy of the *Hfr* chromosome passes along the conjugation canal, the F particle entering last (see CHROMOSOME MAPPING, TRANSDUCTION), if at all. The resulting diploid or partial diploid cells do not remain so for long since recombination (hence *high frequency recombinant, Hfr,* strain) between the DNA duplexes occurs and the emerging clones are haploid.

FIBRE. (Bot.) An elongated, tapering, thick-walled sclerenchyma cell of vascular plants, whose walls may or may not be lignified; may or may not possess a living protoplast. (Zool.) Term applied to thin, elongated cell (e.g. nerve fibre, muscle fibre), or the characteristic structure adopted by molecules of collagen, elastin and reticulin. See also FILAMENT.

FIBRIL. (1) Submicroscopic thread comprising cellulose molecules, in which form cellulose occurs in the plant cell wall; (2) thread-like thickening on the inner faces of large hyaline cells in the leaf or stem cortex of the moss *Sphagnum*.

FIBRIN. Insoluble protein meshwork formed on conversion of fibrinogen by thrombin. See BLOOD CLOTTING.

FIBRINOGEN. Plasma protein produced by vertebrate liver. See BLOOD CLOTTING.

FIBRINOLYSIS. One of the homeostatic processes involved in HAEMOSTASIS. As in BLOOD CLOTTING the major inactive participant is a plasma protein, here *plasminogen*, which is converted to the SERINE PROTEASE enzyme *plasmin* by a variety of activators. Plasmin dissolves blood clots and removes fibrin which may otherwise build up on endothelial walls.

FIBROBLAST. Characteristic cell type of vertebrate connective tissue, responsible for synthesis and secretion of extracellular matrix materials such as tropocollagen, which polymerizes externally to form COLLAGEN. Migrate during development to give rise to mesenchymal derivatives. See FILOPODIUM.

FIBROUS ROOT. Root system comprising a tuft of adventitious roots of more or less equal diameter arising from the stem base or hypocotyl and bearing small lateral roots; e.g. wheat, strawberry. Compare TAP ROOT.

FIBULA. The posterior of the two bones (other is TIBIA) in lower part of hind-limb of tetrapods. Lateral bone in lower leg of human.

FILAMENT. (1) Stalk of the STAMEN, supporting the anther in flowering plants. (2) Term used to describe thread-like thalli of certain algae and fungi. (3) Term used of many long thread-like structures or molecules. See ACTIN, GILL, MYOSIN.

FILARIAL WORMS. Small parasitic nematode worms of humans and their domestic animals, typically in tropical and semitropical regions. *Filariasis* is caused by blockage of lymph channels by *Wuchereria bancrofti* (up to 10 cm long), the young (microfilariae, 200 μm long) accumulating in blood vessels near the skin. Transmitted by various mosquitoes. Cause gross swellings of legs: *elephantiasis*. Another filarian, *Onchocerus volvulus*, transmitted by blackflies (*Simulium* spp.), causes *onchoceriasis* (river blindness). The flies need water to breed, and inject the worms when they bite humans. These cause fibrous nodules under the skin and inflammation of the eye, leading to blindness. See SUPERSPECIES.

FILICALES. Order of PTEROPHYTA (ferns), including the great majority of existing ferns and a few extinct forms. Perennial plants with a creeping or erect rhizome, or with an erect aerial stem several metres in height (e.g. tropical tree ferns). Leaves are characteristically large and conspicuous. Sporophylls either resemble ordinary vegetative leaves, bearing sporangia on the under surface, often in groups (sori), or else are much modified and superficially unlike leaves (e.g. royal fern). Generally homosporous, prothalli bearing both antheridia and archegonia; but includes a small group of aquatic heterosporous ferns. See LIFE CYCLE.

FILOPLUME (PLUMULE). See FEATHER.

FILOPODIUM. Dynamic extension of the cell membrane up to 50 μm long and about 0.1 μm wide, protruding from the surfaces of migrating cells, e.g. FIBROBLASTS, or growing nerve axons. Grow and retract rapidly, probably as a result of rapid polymerization and depolymerization of internal actin filaments, which have a paracrystalline arrangement. Possibly sensory, testing adhesiveness of surrounding cells. Smaller filopodia, up to 10 μm long, are termed *microspikes*. See CELL LOCOMOTION.

FILTER FEEDING. Feeding on minute particles suspended in water (MICROPHAGY) which are often strained through mucus or a meshwork of plates or lamellae. Water may be drawn towards the animal by cilia, or enter as a result of the animal's locomotion. Very common among invertebrates, and found among the largest fish (basking and whale sharks) and mammals (baleen whales).

FINGER DOMAIN. An amino acid sequence within a protein which binds a metal atom, producing a characteristic 'finger-like' conformation within the protein. Such domains tend to bind nucleic acid and may be found repeated tandemly as in *multifinger loops* (see TRANSCRIPTION FACTORS). Compare HOMOEOBOX. See NUCLEAR RECEPTORS, UBIQUITIN.

FIN RAY. See FINS.

FINS. (1) Locomotory and stabilizing projections from the body surface of fish and their allies. Include unpaired medial AGNATHAN *fin-folds* with little or no skeletal support; but the term generally refers to the medial and paired ray fins of the CHONDRICHTHYES and OSTEICHTHYES, in which increasingly extensive skeletal elements (*fin rays*) articulate with the vertebrae, and PECTORAL and PELVIC GIRDLES. The pectoral and pelvic fins are paired and are used for steering and braking. The dorsal, anal and caudal fins are unpaired and medial, opposing yaw and roll. The caudal fin (tail) is generally also propulsive (see HETEROCERCAL, HOMOCERCAL). Fins are segmented structures, seen clearly in the muscle attachments of the ray fins of ACTINOPTERYGII. (2) Paired membranous and non-muscular stabilizers along the sides of arrow worms (Chaetognatha). (3) Horizontal and muscular fringe around the mantle of cephalopods such as the cuttlefish (*Loligo*) by means of which its gentler swimming is achieved.

FISH. General term, covering AGNATHA (jawless fish), CHONDRICHTHYES (cartilaginous fish) and OSTEICHTHYES (bony fish).

FISSION. Form of ASEXUAL reproduction, involving the splitting of a cell into two (*binary fission*) or more than two (*multiple fission*) separate daughter cells. In prokaryotes, without a nucleus, the circular chromosome attaches to the cell membrane, and a membrane furrow separates the replicated strands into the two daughter cells. In eukaryotes, binary fission involves one mitotic division followed by cytokinesis, and is common among Protozoa. Multiple fission (as in many parasitic protozoans) usually involves several rounds of mitosis followed by cytokinesis to form spores (*sporulation*). See CELL DIVISION.

FISSIPEDIA. Suborder of CARNIVORA, including all land carnivorous mammals. Canines large and pointed; jaw joint a transverse hinge (preventing grinding); carnassial teeth often present. Includes cats (Felidae), foxes, wolves and dogs (Canidae), weasels, badgers, otters (Mustelidae), civets, genets and mongooses (Viverridae), hyaenas (Hyaenidae), racoons and pandas (Procyonidae), and bears (Ursidae).

FITNESS (SELECTIVE VALUE). Factor describing the difference in reproductive success of an individual or genotype relative to another. Usually symbolized by w. Often regarded as compound of survival (longevity) and annual fecundity.

(1) Of individuals. Lifetime reproductive success; either 'lifetime reproductive output' (the lifetime fecundity), or the number of offspring reaching reproductive age. Both omit information on the reproductive output of these offspring, and hence on the number of grandchildren reaching reproductive age, or the number of great grandchildren doing so, and so on. Fecundity alone is therefore only one component of fitness: an individual may leave more descendants in the long term by producing fewer total offspring but by ensuring a greater probability of their survival to reproductive age (e.g. by provisioning fewer seeds with more food reserves). Likewise, natural selection will favour any heritable factor that improves the chances of a gene's representa-

tion in subsequent generations. Thus an individual may promote future representation of its own genes, even if it leaves no offspring itself, by contributing to the fitness of close relatives. Any actions which do so contribute improve an agent's *inclusive fitness*, calculated from that individual's reproductive success plus its effects upon the reproductive success of its relatives, each effect weighted by the relative's coefficient of RELATEDNESS to the agent. Likewise, an individual's fitness may be improved by the effects of its relatives. See HELPER, UNIT OF SELECTION.

(2) Of genotypes. Usually applied to a single locus, where the fitness value, w, of a genotype such as Aa is defined as $1-s$, where s is the *selection coefficient* against the genotype. The existence of fitness differences between genotypes creates selection for the evolution of the genetic system itself. See COEFFICIENT OF SELECTION.

FIXATION. (1) In microscopy, the first step in making permanent preparations of organisms, tissues, etc., for study. Aims at killing the material with the least distortion. Solutions of formaldehyde and osmium tetroxide often used. Some artifacts of structure usually produced.

(2) Of genes. The spread of an allele of a gene through a population until it comes to occupy 100% of available sites (i.e. until it is the only allele found at that locus). It is then *fixed* in the population.

(3) Of elements, e.g. carbon, nitrogen. Conversion of an inorganic source of the element to an organic source. C-fixation occurs in photosynthesis; N-fixation occurs in soils, ponds, etc., through the action of prokaryotes (e.g. bacteria, blue-green algae).

FLAGELLATA. See MASTIGOPHORA.

FLAGELLIN. See FLAGELLUM.

FLAGELLUM. (1) Extension of the cell membranes of certain eukaryotic cells, with internal axoneme, basal body, etc., identical to those of a CILIUM (see Fig. 9), but the whole more variable in length, and generally longer. Flagella beat in wave-like undulations, unlike cilia, whose down-beat power-stroke is followed by an up-stroke offering less resistance. In some algae and fungi they have a locomotory role, propelling the organisms through water. In plants such as mosses, liverworts, ferns and a few gymnophytes (e.g. *Ginkgo*) they are found only in the gametes; they are absent from flowering plants. Outer surface may be smooth (*whiplash flagellum*), or may bear one or more rows of minute scales (*tinsel flagellum*). (2) In some prokaryotes, a hollow, membrane-less filament, 3–12 μm long and 10–20 nm in diameter, composed of helically-arranged subunits of the protein *flagellin*. The attachment of the flagellum is by 'hook', 'bearing' and 'rotor'. The flagellum is in the form of a fixed helix, several often rotating in unison. Powered by *protonmotive force* (see BACTERIORHODOPSIN). Involved in chemotactic responses by the cell. See CELL LOCOMOTION.

FLAME CELL (SOLENOCYTE). Cell bearing a bunch of flickering flagella (hence name) and interdigitating with a *tubule cell* (which forms a hollow tube by wrapping itself around the extracellular space). Combined, they form the excretory units (*protonephridia*) of the PLATYHELMINTHES, nemertine worms and the ENTOPROCTA.

FLATWORMS. See PLATYHELMINTHES.

FLAVIN. Term denoting either of the nucleotide COENZYMES (FAD, FMN) derived from RIBOFLAVIN (vitamin B_2) by the enzymes *riboflavin kinase* and FMN *adenylyltransferase*. ATP hydrolysis accompanies the reactions. See FLAVOPROTEINS.

FLAVOPROTEINS. A group of conjugated proteins in which one of the *flavins* FAD or FMN is bound as prosthetic group. Occur as dehydrogenases in ELECTRON TRANSPORT SYSTEMS.

FLEXOR. Muscle or tendon involved in bending a joint; antagonizes extensors.

FLORA. (1) Plant population of a particular area or epoch. (2) List of plant species (with descriptions) of a particular area, arranged in families and genera, together with an IDENTIFICATION KEY.

FLORAL APEX. Apical meristem that will develop into a flower or inflorescence.

FLORAL DIAGRAM. Diagram illustrating relative positions and number of parts in each of the sets of organs comprising a flower. See FLORAL FORMULA, Fig. 19.

FLORAL FORMULA. Summary of the information in a FLORAL DIAGRAM. The floral formula of buttercup (*Ranunculaceae*), $K_5C_5A\infty G\infty$, indicates a flower with a calyx (K) of five sepals, corolla (C) of five petals, androecium (A) of an indefinite number of stamens and a gynoecium (G) of an indefinite number of free carpels. The line below the number of carpels indicates that the gynoecium is superior. The floral formula of the campanula (Campanulaceae), $K_5C_{(5)}A_5G_{\overline{(5)}}$, shows that the flower has five free sepals, five petals united () to form a gamopetalous corolla, five stamens, and five carpels united () to form a syncarpous gynoecium. The line above the carpel number indicates that the gynoecium is inferior (see RECEPTACLE).

FLORAL TUBE. Cup or tube formed by fusion of basal parts of sepals, petals and stamens, often in flowers possessing inferior ovaries. (See RECEPTACLE.)

FLORET. One of the small flowers making up the composite inflorescence (Compositae), or the spike of grasses. In the former, *ray florets* are often female while *disc* florets are often hermaphrodite. See GYNOMONOECIOUS.

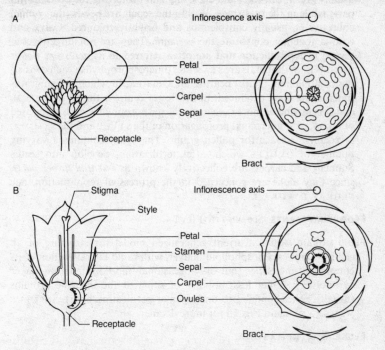

Fig. 19. *Diagrams illustrating flower structure; half-flower (median vertical section) and floral diagram (right). **A** Buttercup, **B** Campanula.*

FLORIDEAN STARCH. Polysaccharide storage product occurring in red algae (RHODOPHYTA); somewhat similar to amylopectin.

FLORIGEN. Hypothetical plant 'hormone' (see GROWTH SUBSTANCE), invoked to explain transmission of flowering stimulus from leaf, where it is perceived, to growing point.

FLORISTICS. Study of composition of vegetation in terms of species (FLORA) present in a particular region, which may be a political entity such as a country, province, or large natural region such as the boreal forest. Floristics aims to account for all plants of the region, with keys, descriptions, ranges, habitats and phenology, and to offer analytical explanations of the flora's origin and geohistorical development.

FLOWER. Specialized, determinate, reproductive shoot of flowering plants (ANTHOPHYTA), consisting of an axis (RECEPTACLE) on which are inserted four different sorts of organs, all evolutionarily modified leaves. Outermost are SEPALS (the *calyx*, collectively), usually green, leaf-like, and enclosing and protecting the other flower parts while in the bud stage. Within the sepals are petals (the *corolla*, collectively), usually conspicuous and brightly coloured. Calyx and corolla together constitute the *perianth*. They are not directly concerned in reproduction and are often referred to as *accessory flower parts*. Within the petals are STAMENS (microsporophylls), each consisting of a filament (stalk) bearing an ANTHER, in which pollen grains (microspores) are produced. In the flower centre is the GYNOECIUM, comprising one or more CARPELS (megasporophylls), each composed of an OVARY, a terminal prolongation of the STYLE and the STIGMA, a receptive surface for pollen grains. The ovary contains a varying number of OVULES which, after fertilization, develop into seeds. Stamens and carpels are collectively known as *essential flower parts*, since they alone are concerned in the process of reproduction. See FLORAL DIAGRAM.

FLOWERING PLANTS. See ANTHOPHYTA.

FLUID MOSAIC MODEL. Current generalized model for structure of all cell membranes. Phospholipid bilayer, with rapid lateral motion of its component molecules (but only occasional 'flip-flop' from one layer to another); rather less mobility of some of the intrinsic proteins (usually either confined to one layer or permeating both). See CELL MEMBRANES and Fig. 5d for more detail.

FLUKE. See TREMATODA.

FLUORESCENT ANTIBODY TECHNIQUE. Cells or tissues are treated with an antibody (specific to an antigen) which has been labelled by combining it with a substance that fluoresces in UV light and can thereby indicate the presence and location of the antigen with which it combines.

FMN (FLAVIN MONONUCLEOTIDE). A FLAVIN; derivative of riboflavin. Prosthetic group of some FLAVOPROTEINS.

FOETAL MEMBRANES. See EXTRAEMBRYONIC MEMBRANES.

FOETUS. In mammals, the stage in intrauterine development subsequent to the appearance of bone cells (osteoblasts) in the cartilage, indicating the onset of OSSIFICATION. In humans, this occurs after seven weeks of gestation. See EMBRYO.

FOLIC ACID (PTEROYLGLUTAMIC ACID). Vitamin of the B-complex (water-soluble) whose coenzyme form (tetrahydrofolic acid, FH_4) is a carrier of single carbon groups (e.g. $-CH_2OH$, $-CH_3$, $-CHO$) in many enzyme reactions. Involved in biosynthesis of purines and the pyrimidine thymine. Very little in polished rice, but widely distributed in animal and vegetable foods. Often given to pregnant women because deficiency of folic acid causes megaloblastic anaemia.

FOLLICLE. (Bot.) Dry fruit derived from a single carpel which splits along a single line of dehiscence to liberate its seeds; e.g. of larkspur, columbine. (Zool.) See GRAAFIAN FOLLICLE, HAIR FOLLICLE, OVARIAN FOLLICLE.

FOLLICLE-STIMULATING HORMONE (FSH). Gonadotrophic glycoprotein hormone secreted by vertebrate anterior PITUITARY gland. Stimulates growth of follicular cells of GRAAFIAN FOLLICLES in the ovary and formation of spermatozoa in testis. See MATURATION OF GERM CELLS, MENSTRUAL CYCLE.

FOLLICULAR PHASE. Phase in mammalian OESTROUS and MENSTRUAL CYCLES, in which Graafian follicles grow and the uterine lining proliferates due to increasing oestrogen secretion.

FONTANELLE. Gap in the skeletal covering of the brain, either in the chondrocranium or between the dermal bones, covered only by skin and fascia. Present in new-born babies between frontal and parietal bones of the skull; closes at about 18 months.

FOOD CHAIN. A metaphorical chain of organisms, existing in any natural community, through which energy and matter are transferred. Each link in the chain feeds on, and hence obtains energy from, the one preceding it and is in turn eaten by and provides energy for the one succeeding it. Number of links in the chain is commonly three or four, and seldom exceeds six. At the beginning of the chain are green plants (autotrophs). Those organisms whose food is obtained from green plants through the same number of links are described as belonging to the same TROPHIC LEVEL. Thus green plants occupy one level (T^1), the PRODUCER level. All other levels are CONSUMER levels: T^2 (herbivores, or primary consumers); T^3 and T^4 (secondary consumers; the smaller and larger carnivores respectively). At each trophic level, much of the energy (and carbon atoms) are lost by respiration and so less biomass can be supported at the next level. Bacteria, fungi and some protozoa are consumers that function in decomposition of all levels (see DECOMPOSER). All the food chains in a community of organisms make up the FOOD WEB. See PYRAMID OF BIOMASS.

FOOD VACUOLE. Vacuole, usually enclosing potential food objects, produced during ENDOCYTOSIS by phagocytic cells. Temporary, normally becoming a heterophagosome (see LYSOSOME).

FOOD WEB. The totality of interacting FOOD CHAINS within a community of organisms. See ECOSYSTEM.

FORAMEN. Natural opening (e.g. FORAMEN MAGNUM of skull, FORAMEN OVALE of foetal heart). Foramina in bones permit nerves and blood vessels to enter and leave.

FORAMEN MAGNUM. Opening at back of vertebrate skull, at articulation with vertebral column, through which spinal cord passes.

FORAMEN OVALE. Opening between left and right atria of hearts of foetal mammals, normally closing at birth (failure to do so resulting in 'hole' in the heart). While open, it permits much of the oxygenated blood returning to the foetal heart from the placenta to pass across to the left atrium, thus bypassing the pulmonary circuit (which in the absence of functional lungs is largely occluded). From there blood passes via the left ventricle to the foetal body.

FORAMINIFERA. Order of mainly marine protozoans whose shells form an important component of chalk and of many deep sea oozes (e.g. *Globigerina ooze*). Shells may be calcareous, siliceous, or composed of foreign particles. Thread-like pseudopodia protrude through pores in the shell and may or may not exhibit cytoplasmic streaming. See HELIOZOA, RADIOLARIA.

FOREBRAIN (PROSENCEPHALON). Most anterior of the three expansions of the embryonic vertebrate brain. Gives rise to *diencephalon* (thalamus and hypothalamus) and *telencephalon* (cerebral hemispheres). Also the origin of the eyestalks. Associated originally with olfaction.

FORM. (Bot.) Smallest of the groups used in classifying plants. Category within species, generally applied to members showing trivial variations from type, e.g. in colour of the corolla. See INFRASPECIFIC VARIATION. (Zool.) Used more or less synonymously with *morph*, to indicate one of the forms within a dimorphic or polymorphic species population.

FORMATION. See BIOME.

FOSSIL. Remains of an organism, or direct evidence of its presence, preserved in rock, ice, amber, tar, peat or volcanic ash. Animal hard parts (hard skeletons) commonly undergo *mineralization*, a process which also turns sediment into hard rock (both regarded as diagenesis). The aragonite (a form of $CaCO_3$) of molluscs and gastropods may recrystallize as the common alternative form, calcite; or it may dissolve to leave a void. This mould may then be filled later by *replacement*, involving precipitation of another mineral (possibly calcite or silica). Partial replacement and impregnation of the original hard parts in both plants and animals by mineral salts (*permineralization*) may occur, especially if the material is porous – as are wood and bone. Fossils may occur *in situ*, or else (derived fossils) be released by erosion of the rock and subsequent reposition in new sediments. Sometimes, the fossil imprints of different locomotory styles (gaits) of the same individual animal are given different taxonomic names. Fossils may be dated by various methods, and provide direct evidence for EVOLUTION, as well as telling us about past conditions on Earth. See GEOLOGICAL PERIODS. For fossils of mankind's ancestors, see HOMINID.

FOSSORIAL. Of animals adapted to digging, burrowing.

FOUNDER EFFECT. Effects on a population's subsequent evolution attributable to the fact that founder individuals of the colonizing population have only a small and probably non-representative sample of the parent population's GENE POOL. Subsequent evolution may take a different course from that in the parent population as a result of this limited genetic variation. Likely to occur where colonization is a rare event, as on oceanic islands, and where the colonizer is not noted for mobility. May be combined with effects of GENETIC DRIFT.

FOVEA. (Bot.) Pit in the wall of palynomorphs, such as spores, pollen, or dinoflagellate cysts. (Zool.) Depression in retina of some vertebrates, containing no rod cells but very numerous cone cells. May lie in a circular region termed the MACULA. Blood vessels absent, and no thick layer of nerve fibres between cones and incoming light as in rest of inverted retina. It is a region specialized for acute diurnal vision. Found in diurnal birds, lizards and primates, including man.

FRACTIONATION. See CELL FRACTIONATION.

FRATERNAL TWINS. Dizygotic twins who develop as a result of simultaneous fertilization of two separate ova. Such twins are no more alike genetically than other siblings. See MONOZYGOTIC TWINS.

FREE ENERGY. See THERMODYNAMICS.

FREEMARTIN. Female member of unlike-sexed twins in cattle and occasionally other ungulates. Sterile, and partially converted towards hermaphrodite condition by hormonal (or possibly H–Y ANTIGEN) influence of its twin brother reaching it through anastomosis of their placentae.

FREE NUCLEAR DIVISION. Stage in development in which unwalled nuclei result from repeated division of primary nucleus.

FREE NUCLEAR ENDOSPERM. Endosperm in which there are many nuclear divisions without cell division (cytokinesis) before cell walls start to form.

FREEZE DRYING. Method of preserving unstable substances by drying when deeply frozen.

FREEZE-ETCHING. Technique used in electron MICROSCOPY for examining the outer surfaces of membranes.

FREEZE-FRACTURE. Technique used in electron MICROSCOPY for examining the inner surfaces of membranes.

FREQUENCY-DEPENDENT SELECTION. Form of SELECTION occurring when the advantage accruing to a character trait in a species population is inversely proportional to the trait's frequency in the popula-

β-form (pyranose)

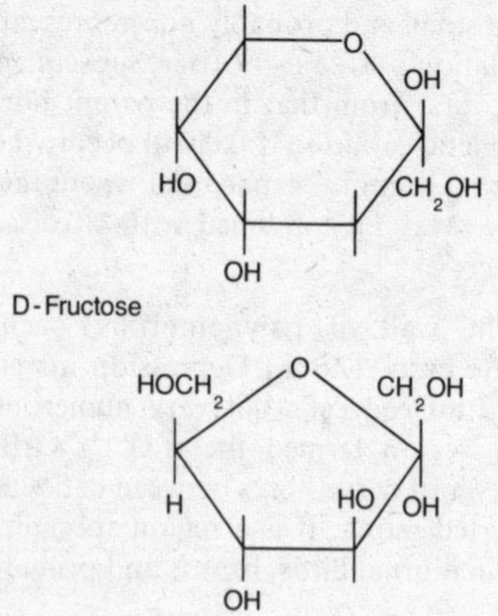

D-Fructose

α-form (furanose)

tion. When rare, it will be favoured by selection; when common, it will be at a disadvantage compared with alternative traits. Two or more traits determined by the same genetic mechanism (locus, or loci) may thus coexist in the population in a condition of POLYMOR-PHISM.

FROND. Term applied to leaf of a fern as well as divided leaves of other plants (e.g. palm).

FRONTAL BONE. A MEMBRANE BONE, a pair of which covers the front part of the vertebrate brain (forehead region in man). Air spaces (*frontal sinuses*) extend from nasal cavity into frontal bones of mammals.

FRONTAL LOBE. Major part of the CEREBRAL CORTEX of the primate brain, including human's. Behind frontal bone. Has numerous connections with many parts of the brain.

FRUCTOSE. A ketohexose reducing sugar, $C_6H_{12}O_6$. In combination with glucose, forms sucrose (non-reducing). The sweetest of sugars.

FRUIT. Ripened ovary of the flower, enclosing seeds.

FRUITIFICATION. Reproductive organ or fruiting structure, often used in the context of fungi, myxomycetes and bacteria.

FRUSTULE. Silica elements of the diatom cell wall.

FRUTICOSE. Lichen growth form where the thallus is shrub-like and branched.

FSH. See FOLLICLE-STIMULATING HORMONE.

FUCOIDIN. Commercially marketed phycocolloid in cell walls and intercellular spaces of brown algae (PHAEOPHYTA).

FUCOSAN VESICLES. Refractive vesicles, usually around the nucleus, containing a tannin-like compound in the brown algae (PHAEOPHYTA). Also called *physodes*.

FUCOSERRATEN. A sexual attractant (gamone) produced by macrogamete (egg cell) in the brown algal genus *Fucus*.

FUCOXANTHIN. Carotenoid pigment present with chlorophyll in various algal groups; e.g. PHAEOPHYTA, CHRYSOPHYTA, PRYMNESIOPHYTA, BACILLARIOPHYTA.

FUNCTION. In one sense, the function of a component in an organism is the contribution it makes to that organism's FITNESS. Therefore it may also be the ultimate reason for that component's existence in the organism, having been selected for in previous generations. This does not exclude the possibility that a component may arise by mutation in an individual and have immediate selective value (and hence function) in that individual; but its function would not then be the reason for its existence in that individual. See TELEOLOGY.

FUNGI. Kingdom containing those eukaryotes lacking chlorophyll (being either saprotrophs, parasites or symbionts), with either an acellular (commonly coenocytic) or a relatively simple tissue-like organization; organizational unit is the *hypha* (matted to form a *mycelium*), and the most complex structures are reproductive bodies involved in spore production. Considered to deserve kingdom status on account of: chitinous wall material; a basically ACELLULAR organization usually involving vacuolated hyphae rather than distinct cells; storage of oil and glycogen (but not starch); some very complicated parasitic life cycles involving several spore types, and some peculiar genetic mechanisms (see CLAMP CONNECTION, CROZIER FORMATION, DIKARYON). Considered to have evolved from a filamentous algal stock. See EUMYCOTA, MYXOMYCOTA.

FUNGICIDE. A compound destructive to fungi.

FUNGI IMPERFECTI. See DEUTEROMYCOTINA.

FUNICULUS. (Bot.) Stalk attaching ovule to the placenta in an ovary.

FURCULA. See WISHBONE.

FUSIFORM INITIALS. Vertically elongated cells in vascular cambium that give rise to cells of the axial system in secondary xylem and phloem.

G

G₀, G₁, G₂ PHASES. See CELL CYCLE.

GABA (GAMMA-AMINOBUTYRIC ACID). Amino acid, related to glycine and taurine, but restricted to the central nervous system. Depresses neurone activity in spinal cord and brain by hyperpolarizing nerves. Released in cerebral cortex in amounts related to level of cortical activity.

G-ACTIN. See ACTIN.

GALACTOSE. An aldohexose sugar; constituent of LACTOSE, and commonly of plant polysaccharides (many gums, mucilages and pectins) and animal GLYCOLIPIDS and GLYCOPROTEINS.

GALL BLADDER. Muscular bladder arising from BILE DUCT in many vertebrates, storing bile between meals. Bile is expelled under influence of the intestinal hormone CHOLECYSTOKININ.

GAMETANGIAL CONTACT. Form of CONJUGATION in which, following growth and contact of the gametangia, nuclei are transferred from the antheridium through a fertilization or copulation tube; e.g. in oomycete fungi.

GAMETANGIAL COPULATION. Fusion of entire gametangial protoplasts, as occurs during CONJUGATION in zygomycete fungi and some algae (Zygnemaphyceae).

GAMETANGIUM. (Bot.) Gamete-producing cell; most commonly in the contexts of algae and fungi. However more complex antheridia, oogonia and archegonia are sometimes cited as examples too. Compare SPORANGIUM.

GAMETE (GERM CELL). Haploid cell (sometimes nucleus) specialized for FERTILIZATION. Gametes which so fuse may be identical in form and size (*isogamous*) or may differ in one or both properties (*anisogamous*). The terms 'male' and 'female' are often applied to gametes, but serve only to indicate the sex of origin, for gametes do not have sexes. Where they differ in size it is customary to refer to the larger gamete as the *macrogamete*, and to the smaller as the *microgamete*. Sometimes plasmogamy is absent in fertilization, in which case the nuclei which fuse may be regarded as gametes. See AUTOGAMY, AUTOMIXIS, MATURATION OF GERM CELLS, OVUM, PARTHENOGENESIS, SPERM.

GAME THEORY. In biology, denotes all approaches to the study of 'decision making' by living systems (usually lacking conscious overtones) in which an organism's responses to its conditions are viewed as *strategies* whose (evolutionary) goal is maximization of the organism's FITNESS. Often convenient to regard each organism as having at any time a decision procedure for responding to future circumstances in such a way as to maximize any possible pay-off to itself while minimizing the pay-offs to others (but see INCLUSIVE FITNESS). Decision procedures which cannot be superseded by rival procedures will be EVOLUTIONARILY STABLE STRATEGIES. See ARMS RACE, OPTIMIZATION THEORY.

GAMETOCYTE. Cell (e.g. oocyte, spermatocyte) undergoing meiosis in the production of gametes. Primary gametocytes undergo the first meiotic division; secondary gametocytes undergo the second meiotic division. See MATURATION OF GERM CELLS.

GAMETOGENESIS. Gamete production. Frequently, but by no means always, involves MEIOSIS. In eukaryotes there are often haploid organisms and stages in the life cycle where gametes can only be produced by MITOSIS. See SPERMATOGENESIS, OOGENESIS.

GAMETOPHORE. In bryophytes, a fertile stalk bearing gametangia.

GAMETOPHYTE. In plants showing ALTERNATION OF GENERATIONS, the haploid (n) phase; during it, gametes are produced by mitosis. Arises from a haploid spore, produced by meiosis from a diploid SPOROPHYTE. See LIFE CYCLE.

GAMMA GLOBULINS (IMMUNE SERUM GLOBULINS). Class of globular serum proteins. Includes those with ANTIBODY activity, and some without.

GAMONE. Compound involved in bringing about fusion of gametes in some brown algae (Phaeophyta); e.g. ectocarpan (in *Ectocarpus*), multifidin and aucanten (in *Cutleria*), fucoserraten (in *Fucus*).

GAMOPETALOUS (SYMPETALOUS). (Of a flower) with united petals; e.g. primrose. Compare POLYPETALOUS.

GAMOSPALEOUS. (Of a flower) with united sepals; e.g. primrose. Compare POLYSEPALOUS.

GANGLION. Small mass of nervous tissue containing numerous CELL BODIES with synapses for integration. CENTRAL NERVOUS SYSTEMS of many invertebrates contain many such ganglia, connected by nerve cords. In vertebrates the CNS has a different overall structure, but ganglia occur in the peripheral and AUTONOMIC NERVOUS SYSTEMS, where they may be encapsulated in connective tissue. Some of the so-called nuclei of the vertebrate brain are ganglia.

GANGLIOSIDE. Type of glycolipid common in nerve cell membranes.

GANOID SCALE. Scale characteristic of primitive ACTINOPTERYGII.

Outer layer is hard inorganic enamel-like ganoine, thicker than in otherwise similar COSMOID SCALE. Grows in thickness by addition of material both above (ganoine) and below (laminated bone). Found today in e.g. *Polypterus*, *Lepisosteus* and sturgeons.

GAP GENES. A class of *Drosophila* segmentation genes, mutants of which delete several adjacent segments and create gaps in the antero-posterior pattern. They are the first zygotic genes to be expressed in *Drosophila* development. Some genes in the class, such as *Krüppel* (*Kr*), *hunchback* (*hb*) and *knirps* (*kni*), encode proteins with DNA-binding FINGER DOMAINS, whose regulation is exerted at the transcriptional level (see BICOID GENE, OSKAR GENE). The patterns of their expression determine those of PAIR-RULE GENES, which in turn determine those of SEGMENT-POLARITY GENES although mutual interactions complicate this picture. The *Drosophila* gap gene *knirps* encodes a hormone receptor-like protein of the steroid-thyroid superfamily essential for abdominal segmentation, similar to ligand-dependent DNA-binding proteins of vertebrates (see RECEPTOR PROTEINS). The *Krüppel* product (K_r) and *hb* product (Hb) bind to different DNA sequences upstream of the two *hb* promotors.

GAP JUNCTION. See INTERCELLULAR JUNCTION.

GAS BLADDER (SWIM BLADDER, AIR BLADDER). Elongated sac growing dorsally from anteror part of gut in most of the ACTINOPTERYGII. In fullest development (in ACANTHOPTERYGII) acts as hydrostatic organ; but may also act as an accessory organ of gaseous exchange, as a sound producer, or as a resonator in sound reception. Opinions differ as to whether the gas bladder or the vertebrate lung is the ancestral structure; they are certainly homologous.

GASTRIC. Of the stomach. *Gastric juice* is a product of vertebrate *gastric glands*, and contains hydrochloric acid, proteolytic enzymes and mucus.

GASTRIN. Hormone secreted by mammalian stomach and duodenal mucosae in response to proteins and alcohol. Stimulates gastric glands of stomach to secrete large amounts of gastric juice. Relaxes pyloric sphincter and closes cardiac sphincter. Oversecretion may result in gastric ulcers. See SECRETIN, CHOLECYSTOKININ.

GASTROPODA. Large class of the MOLLUSCA. Marine, freshwater and terrestrial. Head distinct, with eyes and tentacles; well-developed, rasping tongue (radula). Foot large and muscular, used in locomotion. Visceral hump coiled, and rotated on the rest of body (torsion) so that the anus in the mantle cavity points forward; some forms undergo a secondary detorsion. Visceral hump commonly covered by a single (univalve) shell. Often a trochosphere larva. Includes subclasses Prosobranchia (e.g. limpets), Opisthobranchia (e.g. sea hares) and Pulmonata (e.g. snails, slugs).

GASTROTRICHA. Class of the ASCHELMINTHES (or a phylum in its own right), probably closely related to nematode worms. Composed of a small number of cells, these minute aquatic invertebrates have an elastic cuticle but unlike nematodes have a ciliated but acellular hypodermis. Hermaphrodite or parthenogenetic. No larval stage. See ROTIFERA.

GASTRULA. Stage of embryonic development in animals, succeeding BLASTULA, when the primary GERM LAYERS are laid down as a result of the morphogenetic processes of GASTRULATION.

GASTRULATION. Phase of embryonic development in animals during which the primary GERM LAYERS are laid down; its onset is characterized by the morphogenetic movements of cells, typically through the BLASTOPORE, forming the ARCHENTERON. Movements may result in EPIBOLY, but frequently also *emboly* in which cells invaginate, involute and ingress.

GAS VACUOLE. Structure comprising gas vesicles, or hollow cylindrical tubes with conical ends, found in the cytoplasm of all orders of CYANOBACTERIA (blue-green algae) except Chamaesiophonales. A gas vesicle comprises protein ribs or spirals arranged like hoops of a barrel. Gas vacuoles may function in light shielding and/or buoyancy.

GATED CHANNELS. TRANSPORT PROTEINS of membranes, not constitutively (permanently) open to the passage of molecules, but capable of closure. *Ligand-gated channels*, such as those responding to NEUROTRANSMITTERS, open only in response to an extracellular ligand; *voltage-gated channels* (e.g. the SODIUM PUMP of nerve and muscle fibres) are dependent for opening and closure upon an appropriate membrane potential. Others may only open when concentrations of certain ions in the cell are appropriate. See IMPULSE, MUSCLE CONTRACTION.

GEL. Mixture of compounds, some commonly polymeric, having a semisolid or solid constitution. See CHROMATOGRAPHY, ELECTROPHORESIS.

GEL ELECTROPHORESIS. See ELECTROPHORESIS.

GEL FILTRATION. See CHROMATOGRAPHY.

GEMMA. Organ of vegetative reproduction in mosses, liverworts and some fungi. Consists of a small group of cells of varying size and shape that becomes detached from the parent plant and develops into a new plant; often formed in groups, in receptacles known as gemmae-cups.

GEMMATION. Asexual reproduction involving formation of a group of cells (a GEMMA in plants) which develops into a new individual, or a new member of a colony of connected individuals. May develop before its complete or partial separation from the parent. Occurs in many bryophytes, coelenterates and ascidians. Referred to as *budding* in animals.

GEMMULE. (1) Of sponges, a bud formed internally as a group of cells, which may become free by decay of the parent and subsequently form a new individual. Freshwater sponges over-winter in this way. (2) See PANGENESIS.

GENE. Usually regarded as the smallest physical unit of heredity encoding a molecular cell product; commonly considered also to be a UNIT OF SELECTION. The term gene (coined by W. Johannsen in 1909) may be used in more than one sense. These include: a) ALLELE, b) LOCUS, and c) CISTRON. What MENDEL treated as algebraic units ('factors') or 'atoms of heredity' obeyed his laws of inheritance and were considered to be the physical determinants of discrete *phenotypic* characters. This may be called the classical gene concept (see GENETICS). In 1903, W. S. Sutton pointed out that the segregation and recombination of Mendelian factors studied in heredity found a parallel in the behaviour of chromosomes revealed by the microscope. Through the work of T. H. Morgan in the period 1910–20, chromosomes came to be regarded as groups of linked genes (or, more abstractly, of their loci), and the positions of loci and their representative alleles were first mapped on the chromosomes of *Drosophila* in this period. Morgan found that alternative genes (alleles) at a locus could mutate from one to another.

The importance of genes in enzyme production first emerged through work on the chemistry and inheritance of eye colour in *Drosophila*, and through work on AUXOTROPHIC mutants of the mould *Neurospora crassa* by G. W. Beadle and E. L. Tatum (1941). It heralded the modern phase of genetics and molecular biology.

Genes soon became accepted as the heritable determinants of enzymes (one gene:one enzyme). However, the correspondence between the nucleotide composition of a gene and the amino acid composition of its encoded product was first revealed in variants of haemoglobin (a non-enzymic protein) and its genes, and their precise sequential correspondence (*colinearity*) was first established in detailed studies of the bacterial enzyme *tryptophan synthetase* and its gene. Great precision was by now being achieved in fine genetic mapping of BACTERIOPHAGE chromosomes using the CIS-TRANS TEST, and genes were soon regarded as nucleic acid sequences, mappable geographically on a chromosome, each encoding a specific enzyme, or (as in the polypeptide subunits of haemoglobin) non-enzyme protein. Subsequent work on *tryptophan synthetase* of the bacterium *E. coli* showed that two genes were required to encode this enzyme, and that their different polypeptide products associated to give the *quaternary structure* of the functional enzyme (see PRO-TEIN). The *functional gene concept* thus denoted a nucleic acid sequence encoding a single polypeptide chain. Nowadays, the term 'gene' is used to indicate the length of nucleic acid encoding any molecular cell product, be it a polypeptide, transfer RNA or ribosomal RNA molecule, and can usually be equated with CISTRON (but see ALLELIC COMPLEMENTATION). Most such sequences, at least as they occur naturally in eukaryotic chromosomes, contain one or more INTRONS. See CHROMOSOME, PROTEIN SYNTHESIS and genetic references below.

GENE AMPLIFICATION. Process in which a small region of the GENOME of a cell is selectively copied many times while the rest remains unreplicated. Occurs in some specialized cell lines where large quantities of a particular cell product are needed rapidly. In rRNA cistrons, up to 1000 extra nucleoli may arise in amphibian oocytes in this way, with consequent large-scale ribosome production. Cistrons for rRNA are amplified in all cells with nucleoli. Gene amplification is associated with some kinds of dry resistance in cell cultures; amplication of cellular ONCOGENES is a fairly common feature of tumour cells. The whole phenomenon of selective DNA amplification is rich in theoretical interest. See NUCLEOLUS, GENE DUPLICATION, POLYTENY.

GENE BANK (GENE LIBRARY). Term given to the collection of DNA fragments resulting from digestion of a genome by a RESTRICTION ENDONUCLEASE. Each fragment is clonable by inserting it into an appropriate phage vector and introducing it into an appropriate host cell (e.g. *E. coli*) for copying.

GENECOLOGY. Study of population genetics with particular reference to ecologies of populations concerned.

GENE CONVERSION. Phenomenon, in eukaryotes occurring mainly at synapsis during meiosis, whereby a donor DNA sequence, a few hundred bases or perhaps a kilobase in length, is transferred from one gene to another having substantial sequence homology (usually between homologous loci, but sometimes between related sequences at non-homologous loci, notably those of dispersed MULTIGENE FAMILIES). The donor sequence is repaired back to its original form. It may be responsible for much of the diversity in some mammalian immunogobulin production (see ANTIBODY DIVERSITY). In one model this involves 'nicking' (cutting) of a single-strand invading DNA sequence and melting (unzipping) of the invaded duplex DNA so that *heteroduplex* base pairing between the two can occur. The ousted sequence is enzymatically degraded while the invading sequence is cut and then annealed into its new position. Its original complementary strand is then used as template for its resynthesis to form the original duplex again. In another model, increasingly favoured, the recipient DNA duplex is nicked and gapped in both strands, the gap being filled by copy-synthesis using both strands of the donor duplex as templates. This would generate heteroduplexes only in regions flanking the gap. In a heterozygote of the yeast *Saccharomyces*, where conversion occurs at a rate of several per cent per gene per meiosis, each allele can usually convert the other with about equal frequency, but examples of strongly biassed conversion are known which could, in principle, lead to fixation of the favoured allele.

Initiation of the cutting, and hence of the recombination, seems in one form of the process to occur within a gene promoter region. There is growing support for the view that the sites of heteroduplex formation (*Holliday junctions*) are responsible for much of eukaryotic crossing-over: there is about 30 – 50% association of gene conversion with crossing-over. Gene conversion in prokaryotes involves similar processes, although the initial alignment of homologous duplexes is less highly organized (see recA, recB and recC). See RECOMBINATION.

GENE DOSAGE. Effective number of copies of a gene in a cell or organism. See DOSAGE COMPENSATION, GENE AMPLIFICATION.

GENE DUPLICATION. Mechanisms resulting in tandem duplication of loci along a chromosome. One of the possible evolutionary consequences of diploidy as opposed to haploidy is that with two functional representatives of a locus per cell it may not matter if one mutates and loses its original function. This is very likely also the evolutionary significance of gene duplication: one copy is free to mutate and take on a new function, the other functioning as normal. The enzymes of the glycolytic pathway may have arisen this way from a common ancestral gene sequence, as most certainly do the various types of globin in haemoglobin. Non-homologous CROSSING-OVER is one mechanism for producing gene duplication. See MULTIGENE FAMILIES, GENE AMPLIFICATION, DUPLICATION.

GENE EXPRESSION/GENE REGULATION. The effect of those mechanisms which dictate whether or not a particular genetic element is transcribed (acts as a template for mRNA synthesis) at any particular time. In prokaryotes it may best be explained by some variant of the JACOB-MONOD THEORY; in eukaryotes too this theory may find application, although different CHROMOSOME structure here raises fresh problems. It is common in eukaryotes to recognize two classes of regulatory phenomena involving gene expression: short-term (reversible) regulation, and long-term (often irreversible) regulation. Short-term regulation often relates to a cell's production of inducible and repressible enzymes. Steroid hormones ('effectors', see

ECDYSONE) frequently bind to receptor proteins (see NUCLEAR RECEPTORS) in the cell prior to entry into the nucleus and activate transcription of selected genes (see TRANSCRIPTION FACTORS). Long-term eukaryote regulation includes those processes involved in: a) rendering a cell DETERMINED, prior to differentiation, b) MATERNAL EFFECTS, c) the origins of facultative and constitutive HETEROCHROMATIN. The precise roles of NUCLEOSOMES and of nucleosome-free regions of eukaryotic chromosomes in the control of gene expression have still to be clarified.

GENE FIXATION. See FIXATION.

GENE FLOW. The spread of genes through populations as affected by movements of individuals and their propagules (e.g. spores, seeds, etc.), by NATURAL SELECTION and GENETIC DRIFT. See VAGILITY, PANMIXIS.

GENE FREQUENCY. Frequency of a gene in a population. Affected by MUTATION, SELECTION, emigration, immigration and GENETIC DRIFT. See HARDY-WEINBERG EQUILIBRIUM.

GENE LIBRARY. See GENE BANK.

GENE MANIPULATION (GENETIC ENGINEERING). Set of procedures by which selected pieces (genes) of one genome (e.g. human) can be enzymatically cut out from it, spliced into a vector (e.g. a PLASMID) and inserted into an appropriate host microorganism (e.g. *E. coli*, yeast, etc.) in which it is replicated and passed on to all daughter cells of the microorganism forming a clone. Various enzymes used include RESTRICTION ENDONUCLEASES, DNA LIGASES, etc. If the aim is mass production of the substance encoded by the transposed gene, then insertion is so arranged that transcription of the gene occurs within the clone of cells, producing large amounts of the desired substance (e.g. insulin). Clones can be screened for their activities and selected appropriately. The recombinant DNA thus artificially produced might be hazardous unless properly contained, and strict precautions are applied in such work. Thus only non-pathogenic strains are used as hosts, or else strains that can only grow in laboratory conditions. See BIOTECHNOLOGY.

GENE POOL. Sum total of all genes in an interbreeding population (gamodeme) at a particular time. See DEME.

GENERATIVE CELL. (Bot.) In the pollen grain, the cell of the male GAMETOPHYTE which divides mitotically to produce two generative nuclei (gametes). See DOUBLE FERTILIZATION.

GENERATIVE NUCLEUS. See DOUBLE FERTILIZATION, GENERATIVE CELL.

GENERATOR POTENTIAL. Initial depolarization of the membrane of an excitable cell (receptor, nerve or muscle) by stimulus or transmitter, which triggers an ACTION POTENTIAL when threshold depolarization is reached.

GENE REGULATION. See GENE EXPRESSION.

GENERIC. (Adj.) Of GENUS.

GENET. The genetic individual. Particularly employed in the context of vegetative (clonal) reproduction and growth. Compare RAMET.

GENE TARGETING. Technique whereby organisms of virtually any desired genotype may be produced. In mice, where the procedure seems destined to define a wide range of gene functions, it commonly involves culturing embryonic stem (ES) cells as undifferentiated pluripotent cell lines. A vector is then used to introduce a modified (exogenous) DNA sequence into the targeted locus in vitro. The ES cells, which have the ability to generate all cell types including germ line cells, can then be microinjected into blastocysts to produce germ line chimaeras. Cells which have integrated the modified sequence into a homologous stretch of DNA by homologous recombination can then be detected and selected for by using appropriate genetic markers in the vector. It may be possible to obtain individuals homozygous for the altered DNA sequence by crossing siblings heterozygous for it. Gene targeting in yeast results in almost entirely homologous recombination of exogenous DNA. See GENE CONVERSION.

GENETIC. Concerned with genes, or their effects. Compare HEREDITARY.

GENETIC ASSIMILATION. Phenomenon involving conversion of an acquired character (resulting maybe from transference of individuals from one environment to another) into one with greater HERITABILITY than it had before, where the causal mechanism involved is selection acting on the genotypes of the transferred population. The significance of the process in evolutionary terms is debatable: there is no assurance that the initial acquired character will be adaptive in the conditions bringing it about. See MUTATION, PHENOCOPY.

GENETIC CODE. Table of correspondence between (a) all possible triplet sequences (codons) of messenger RNA and (b) the amino acid which each triplet causes to be incorporated into protein during PROTEIN SYNTHESIS. In addition, certain triplets cause termination of polypeptide chain synthesis and occur regularly at the 3′-ends of polypeptide-encoding sequences (open reading frames). They may also arise as motivating mutations within an encoding sequence (see CODON). DNA is sometimes spoken of as a 'code', but this is shorthand. Indeed, the genetic code itself is really a *cipher*, since the

1st position (5′ end) ↓	2nd position				3rd position (3′ end) ↓
	U	**C**	**A**	**G**	
U	Phe	Ser	Tyr	Cys	U
	Phe	Ser	Tyr	Cys	C
	Leu	Ser	STOP	STOP	A
	Leu	Ser	STOP	Trp	G
C	Leu	Pro	His	Arg	U
	Leu	Pro	His	Arg	C
	Leu	Pro	Gln	Arg	A
	Leu	Pro	Gln	Arg	G
A	Ile	Thr	Asn	Ser	U
	Ile	Thr	Asn	Ser	C
	Ile	Thr	Lys	Arg	A
	Met	Thr	Lys	Arg	G
G	Val	Ala	Asp	Gly	U
	Val	Ala	Asp	Gly	C
	Val	Ala	Glu	Gly	A
	Val	Ala	Glu	Gly	G

Fig. 20. Diagram of the amino acids encoded by RNA triplets (codons). The triplets run from the 5′-end to the 3′-end of the RNA. The three nucleotides for any triplet are found by taking one from the left column, one from the horizontal row and one from the right column. The amino acid indicated at their intersection is that encoded by that triplet. Thus CCC encodes proline. Three stop codons are also indicated.

amino acids in a polypeptide correspond to the letters of an alphabet rather than to words. Since more than one triplet may encode some amino acids, the code is said to be degenerate. The translation of mRNA into protein is only possible because of the specificity of transfer RNA molecules for particular amino acids – itself the result of specificity of the enzymes which activate amino acids and bind them to appropriate tRNAs.

The code is remarkably uniform from prokaryotes to eukaryotes; but in mitochondria there are fewer codons and slightly different reading rules; in *Mycoplasma capricolum* the codon UGA is read as tryptophan rather than as a STOP CODON. Until the 1970s it was thought that the reading frame for translating mRNA into polypeptide never overlapped: that there were unequivocal initiation sites for any mRNA molecule. However, overlapping reading frames (overlapping genes) have been discovered in some viruses (e.g. ψX 174). See WOBBLE HYPOTHESIS, START CODON.

GENETIC COUNSELLING. Service, generally provided by specialists in human genetic disorders. Seeks to explain to parents with children already affected by genetic disorders the nature of those disorders, and the probability of their (and their children) having further affected offspring, and helps families to reach decisions and take appropriate action in the light of this information. Many carriers of genetic disorders (those heterozygous for the condition, but who do not themselves exhibit it) can be diagnosed through screening procedures. The probability of someone being a carrier can often be ascertained from information about the occurrence of the disorder in close relatives and, to an increasing extent, by DNA analysis.

GENETIC DRIFT (SEWALL WRIGHT EFFECT). Statistically significant change in population gene frequencies resulting not from selection, emigration or immigration, but from causes operating randomly with respect to the fitnesses of the alleles concerned. Such random sampling error might for example occur if, in a population of beetles, a disproportionately large number of those killed by a wandering elephant happened to be heterozygous for a recessive eye colour. The frequency of this, and maybe other, alleles could now alter significantly even though no selection had taken place. Genetic drift is expected to be of significance only in small populations, where alleles may easily go to extinction or fixation by chance alone. In large populations, effects of sampling error are usually considered to be small in comparison with those of NATURAL SELECTION. See FOUNDER EFFECT.

GENETIC ENGINEERING. See GENE MANIPULATION.

GENETIC MAPPING. See CHROMOSOME MAPPING.

GENETICS. Study of heredity and variation in biological systems. The origin of the modern, particulate, theory of inheritance is marked by the work of G. MENDEL, with many other contributors to this classical phase of genetics in which phenotypic ratios in breeding tests were ultimately explained in terms of chromosome behaviour. Post-classical work, largely with microorganisms and phages, directed attention towards a biochemical understanding of genetics, isolation of DNA and determination of its structure, eventually enabling GENE MANIPULATION. Population (ecological) genetics attempts to quantify the roles of selection and genetic drift in shaping the GENETIC VARIATION within populations. See GENE.

GENETIC VARIATION. Occurrence of genetic differences between individuals, most commonly studied in species populations. Upon such

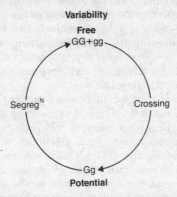

Fig. 21a. *Free and potential variability. Free variability (open to direct selection) is represented by differences in phenotype between GG and gg genotypes and is converted to potential variability of Gg by crossing and is released again by segregation.*

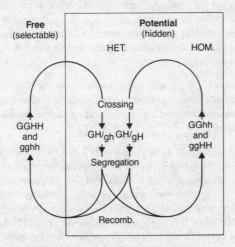

Fig. 21b. *The states of variability in a system where the G and H loci are additive in effect on phenotype, so that GgHh and GGhh are phenotypically identical. The extremes of phenotype (free variability) are expressed by the double dominant and double recessive.*

differences, when expressed, can natural selection act. MUTATION is the ultimate source of genetic variation; but in most sexual populations MEIOSIS then results in recombination both between and within parentally-derived chromosomes, generating enormous genetic diversity among gametes and, at least potentially, among offspring phenotypes. The BREEDING SYSTEM of the population is an important consideration, affecting the relative levels of free and potential variability. *Free variability* is genetic variation which is expressed phenotypically and approximates to the proportion of homozygotes in the population; *potential variability* is genetic variation which does not express itself phenotypically and approximates to the proportion of heterozygotes in the population.

However, two or more loci often affect the same character (polygenic inheritance), and in the simplest case (where both loci are additive with respect to effect on phenotype) may generate the same phenotypes in the double heterozygote, GgHh, as in both the double homozygotes, GGhh and ggHH, thus 'protecting' the latter from selection, even though their *homozygous potential variation* contributes to the total potential variation in the population. Alleles always exert their effects against a 'genetic background' which can modify their expressions.

Only through crossing, with its subsequent segregation, can a major part of the potential variability in a population be freed and become available for selection. Inbred populations, comprising all possible types of homozygotes, will harbour almost as much potential variability as outbred ones with the same genes in the same frequencies; but this variability is never freed in inbred populations and they tend to be evolutionarily static. See VARIATION, POLYMORPHISM, HERITABILITY.

GENITALIA. (Zool.) External reproductive organs. In many arthropods, especially insects, structures of male and female genitalia are often species-specific and serve as a prezygotic mechanism preventing hybridization.

GENOME. The total genetic material within a cell or individual, depending upon context. A bacterial genome comprises a circular chromosome containing an upper limit of about 0.01 pg (1 picogram = 10^{-12} g) of DNA; a haploid mammalian genome contains from about 3–6 pg of DNA, while some amphibia and psilopsid plants may contain well over 100 pg of DNA per haploid genome. The DNA content may or may not correlate with chromosome number, the fit being generally better in plants than in animals. Much eukaryote DNA is not part of structural genes coding for detectable cell products. The evolution of genome size is much debated.

GENOMIC IMPRINTING. A transiently heritable modification of chromatin, probably mediated by methylation, with the result that, in mammals, maternal and paternal genomes contribute differently to early development.

GENOTYPE. Genetic constitution of a cell or individual, as distinct from its PHENOTYPE.

Era	Period	Epoch	Millions of years (Myr) since start of period
Cenozoic	Quaternary	Holocene	(11 000yr)
		Pleistocene	2–3
	Tertiary	Pliocene	12
		Miocene	25
		Oligocene	40
		Eocene	60
		Palaeocene	70
Mesozoic	Cretaceous		135
	Jurassic		195
	Triassic		225
Palaeozoic	Permian		270
	Carboniferous		350
	Devonian		400
	Silurian		440
	Ordovician		500
	Cambrian		600

Table 4. Main fossil-bearing geological periods and approximate time-scales since the beginnings of the periods.

GENUS. Taxonomic category, between FAMILY and SPECIES, which may include one or more examples of species. See BINOMIAL NOMENCLATURE, CLASSIFICATION.

GEOLOGICAL PERIODS, EPOCHS. See Table 4. The fossil record commences in pre-Cambrian times with organisms resembling bacteria and blue-green algae (see STROMATOLITES) in deposits 3 billion years old. A few green algae have been identified from an upper pre-Cambrian formation about 1 billion years old. In Cambrian rocks, in addition to algae and bacteria, a diverse range of aquatic invertebrate animals is found (e.g. brachiopods, trilobites and the onychophoran *Aysheaia*), including many which apparently left no descendants. Fungi and spores with wall markings like those of various land plants were also present. In the Ordovician period, fish-like vertebrates (ostracoderms) appeared, and primitive land plants are known from the Silurian period. Colonization of land by plants was an event of enormous importance, making it possible for animals to live there permanently. By the late Devonian period, arthropods (including insects) were established on land and the first terrestrial vertebrates, amphibians, had appeared. Club mosses, horsetails and ferns, many of them tree-like, were now abundant on land, providing an enormously rich flora in the Carboniferous period. Gymnophytes, mosses and liverworts, algae and fungi were also present. Tremendous accumulations of remains of Carboniferous plants, partially decayed and subjected to intense pressure, formed the coal seams of today.

The Mesozoic era was the age of the great dinosaurian reptiles, while gymnophytes were the dominant land plants. Flowering plants appeared in the Jurassic period, becoming increasingly dominant towards the end of the Cretaceous period as the gymnophytes declined, and remained the dominant plant group to the present. The ascendancy of mammals also began with the end of the Mesozoic era. See CONTINENTAL DRIFT, and under individual periods, epochs and groups of organisms.

GEOPHYTES. Class of RAUNKIAER'S LIFE FORMS; plants possessing perennating buds below the soil surface on a corm, bulb, tuber or rhizome.

GEOTAXIS. TAXIS in which the stimulus is gravitational force.

GEOTROPISM (GRAVITROPISM). (Bot.) Orientation of plant parts under stimulus of gravity. Main stems, *negatively geotropic*, grow vertically upwards and, when laid horizontally, exhibit increased elongation of cells in growth region on lower side at tip of stem, which turns upwards and resumes its vertical position. Main roots, *positively geotropic*, grow vertically downwards and if laid horizontally exhibit increased elongation of cells on upper side of growth region, the root turning down again as a result. In shoots placed in a horizontal position, differences in GIBBERELLIN and AUXIN concentrations develop between upper and lower sides. Together, these cause lower side of the shoot to elongate more than upper side, giving the observed upward growth. When it eventually resumes vertical growth, lateral asymmetry in growth substance concentrations disappears, and growth continues vertically. In roots, growth substance asymmetries are less well understood. There is evidence that gradients of these substances within the root cap are brought about by relatively tiny movements of starch-containing plastids (amyloplasts), for when a plant is placed in a horizontal plane, these plastids move from the transverse walls of vertically growing roots and come to rest near what were previously vertically orientated walls. Then, after several hours, the root curves downwards and these plastids return to their original positions. It is not yet clear how these plastid movements translate into growth substance gradients. There is little evidence for the role of auxin (IAA) in roots; some have been unable to find it in the root cap at all. Instead, ABSCISIC ACID is found there, and has been shown to be redistributed from the cap to the root itself, and to act as an inhibitor on cells in the region of elongation on lower side of a horizontally positioned root. See DIAGEOTROPISM, PLAGIOTROPISM.

GERM CELL. Any of the cells forming a germinal epithelium, plus its cell products, the GAMETES. In vertebrates, *primordial germ cells* migrate from the early gut or yolk sac to sites of the eventual genital

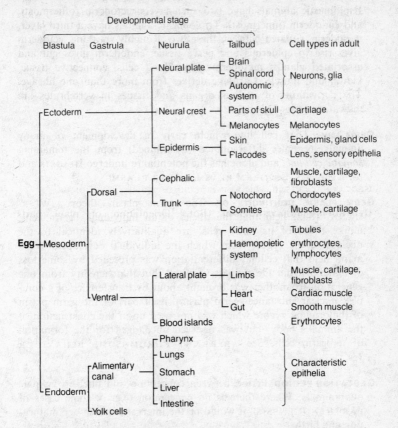

Fig. 22. *Times of production and germ layer origins of the basic tissues and organs of a vertebrate. The diagram represents a decision-making plan for differentiation although many more decision points are actually involved. Some cell types, e.g. cartilage, have more than one derivation.*

ridges in ovaries or testes where, after rounds of mitosis, some of their daughter cells will undergo meiosis and differentiate into gametes. See GERM PLASM, OVARY, TESTIS.

GERMINAL EPITHELIUM. See OVARY, TESTIS.

GERMINAL VESICLE. Enlarged oocyte nucleus formed during diplotene of first meiotic prophase. The mammalian oocyte nucleus may persist in this state for a considerable time (in humans from fifth month of prenatal life until eventual fertilization in the adult).

GERM LAYER. One of the main layers or groups of cells distinguishable in an animal embryo during and immediately after gastrulation.

Diploblastic animals have two such layers: ectoderm (outermost), and endoderm (innermost). Triploblastic animals have a third layer, mesoderm, situated between these two. Roughly speaking, ectoderm gives rise to epidermis and nerve tissue; endoderm gives gut and associated glands; mesoderm gives blood cells, connective tissue, kidney and muscle. Cartilage derives from more than one lineage. The derivations of various organs and tissues in vertebrates are indicated in Fig. 22.

GERM LINE. That cell line which, early in development of many animals, becomes differentiated (determined) from the remaining *somatic cell line*, and alone has the potential to undergo MEIOSIS and form gametes. See GERM PLASM, POLAR PLASM.

GERM PLASM, CONTINUITY OF. The germ plasm theory (WEISMANN, 1892) held that the nuclei of an individual's germ cells, unlike those of its body cells, are qualitatively identical to the nucleus of the zygote from which the individual developed. Weismann held that cellular differentiation was preceded by some loss of material from the nucleus of a cell, but that heredity from one generation to another was brought about by transference of a complex nuclear substance (germ plasm), itself part of the germ plasm of the original zygote which was not used up in the construction of the animal's body, but was reserved unchanged for the formation of its germ cells. See ABERRANT CHROMOSOME BEHAVIOUR (3).

GESTATION PERIOD. Period between conception and birth in viviparous animals. Where there is no conception (e.g. in some cases of PARTHENOGENESIS), it would be the interval between egg maturation and birth.

GIANT CHROMOSOME. See POLYTENY.

GIANT FIBRES (GIANT AXONS). Nerve axons of very large diameter (e.g. 1 mm in squids). Occur in many invertebrates (e.g. annelids, crustacea, and nudibranch and cephalopod molluscs) and some vertebrates where rapid conduction is required (e.g. for escape) and achieved through reduced electrical resistance of the axoplasm with larger axon diameter. May be either a single enormous cell or the result of fusion of many cells. There are fewer synaptic barriers in these nerves than is usual, increasing speed of conduction.

GIBBERELLINS. Class of plant GROWTH SUBSTANCES, originally isolated from the fungus *Gibberella fujkoroi* when it caused abnormal elongation of infected rice plants in the 1930s. The best-studied is *gibberellic acid*, GA$_3$, which has the following structure:

Gibberellins have spectacular effects upon stem elongation in certain plants; they cause dwarf beans to grow to the same height as tall varieties. Are involved in seed germination, and will substitute in many species (e.g. lettuce, tobacco, wild oats) for the dormancy-breaking cold or light requirement, promoting early growth of the embryo. Specifically, they enhance cell elongation, making it possible for the root to penetrate growth-restricting seed coat or fruit wall, which has practical application in ensuring uniformity of germinating barley in production of barley malt used in the brewing industry. Can be used to promote early seed production of biennial plants; stimulate pollen germination and growth of pollen tubes in some species; and can promote fruit development (e.g. almonds, peaches, grapes). In barley and other grass seeds, the embryo releases gibberellins causing the aleurone layer of the endosperm to produce enzymes (e.g. α-amylase), digesting the starch store to mobilize sugars for germination. The gibberellin causes expression of the gene encoding the amylase (i.e. de-repression), the mechanism resembling that by which ECDYSONE exerts its effects. Some commercial plant growth retardants achieve their results by blocking gibberellin synthesis.

GILL. (Bot.) See LAMELLA, GILL FUNGI. (Zool.) Any of several organs of gaseous exchange in aquatic animals, such as the vascularized projections of the external surface (*external gills*) of many annelids and arthropods. Parapodia of polychaetes and thin-walled trunk limbs of branchiopod crustaceans increase the surface area for diffusion of dissolved gases and probably function as gills; but in some *tracheal gills* of insect larvae and pupae, these leaf-like projections, despite their rich supply of tracheae (though lacking open spiracles), absorb less oxygen than the remaining body surface. *Spiracular gills* occur in some aquatic insect pupae where one or more pairs of spiracles is drawn out into long processes and generally supplied with a PLASTRON. Vertebrate gills are either *external* (ectodermal) or *internal* (endodermal). The former occur in larvae of a few bony fish (*Polypterus* and some lungfish) and of amphibians, and are almost always soon lost or replaced by internal gills. Fish gills may serve as organs of OSMOREGULATION. See VENTILATION and Fig. 23.

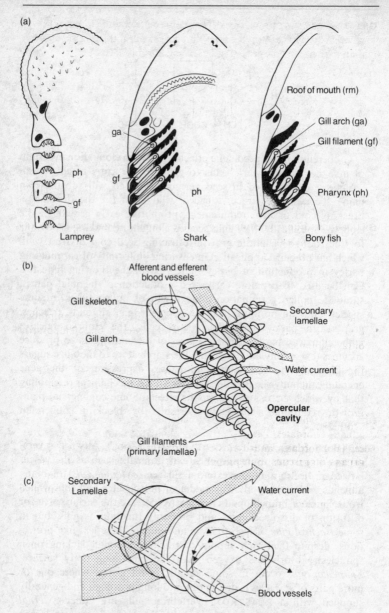

Fig. 23. (**a**) *The arrangement of gills in three types of fish.* (**b**) *The arrangement of secondary gill lamellae in a bony fish.* (**c**) *The countercurrent flow of blood through the lamellae.*

GILL ARCHES. VISCERAL ARCHES between successive gill slits in jawed fish (typically five pairs) and larval amphibians (never more than four pairs) comprising gill bars and attendant tissues.

GILL BAR. Skeletal support of gill slits in chordates, containing in addition blood vessels and nerves. In jawed fish typically comprises a dorsal element (*epibranchial*) and a ventral element (*ceratobranchial*), the two bent forward on each other. Five pairs are present in most jawed fish, in addition to the jaws and HYOID ARCH.

GILL BOOK. Stacks of segmentally arranged vascularized leaf-like lamellae attached to the posterior faces of oscillating plates. This movement and locomotion of the limbs on which they occur (e.g. the swimming paddles of the merostomatid *Limulus*) provide ventilation of the gill book under water. See LUNG BOOK.

GILL FUNGI. Fungi belonging to the family Agaricaceae (BAS-IDIOMYCOTINA); possessing characteristic fruiting body comprising a stalk (stipe) supporting a cap (pileus), on the undersurface of which are radially arranged gills (lamellae) bearing the hymenium; e.g. mushrooms, 'toadstools'.

GILL POUCH. Outpushing of side-wall of pharynx towards epidermis in all chordate embryos. Precursor of gill slit in fish and some amphibia; in terrestrial vertebrates breaks through to exterior only temporarily or (as in humans) not at all. See SPIRACLE.

GILL RAKERS. Skeletal projections of inner margins of gill bars of fish, particularly elongated in those which strain incurrent water for food particles.

GILL SLIT. One of a series of bilateral pharyngeal openings of aquatic chordates. Usually vertically elongated and, in urochordates, cephalochordates and cyclostome larvae, primarily concerned with FILTER-FEEDING, but probably with additional role in gaseous exchange. In fish and most larval amphibia they have the latter role; but presence of GILL RAKERS in filter-feeding fish may again give them a nutritional role. See GILL POUCH, from which they develop.

GINKGOPHYTA. Ginkgos. Gymnophytes possessing active cambial growth and fanshaped leaves with open, dichotomous venation. Ovules and seeds are exposed and seed coats fleshy. After pollination, sperm are transported to the vicinity of an ovule in a pollen tube, but are flagellated and motile. *Ginkgo bilboa* (maidenhair tree) is the sole surviving species of this once widespread and moderately abundant group, which flourished in mid-Mesozoic times but diminished during the later Mesozoic and Tertiary.

GIZZARD. Region of the gut in many animals, where food is ground prior to the main digestion. Walls very muscular, often with hard 'teeth' (e.g. crustaceans) or containing grit, stones, etc. (e.g. birds, reptiles).

GLAND. An organ (sometimes a single cell) specialized for secretion of a specific substance or substances. (Bot.) Superficial, discharging secretion externally, e.g. glandular hair (lavender), NECTARY, HYDATHODE, or embedded in tissue, occurring as isolated cells containing the secretion, or as layer of cells surrounding intercellular space (secretory cavity or canal) into which secretion is discharged; e.g. resin canal of pine. (Zool.) In animals, glands are either *exocrine* (secreting onto an epithelial surface, usually via a duct), or *endocrine* (secreting directly into the blood, not via ducts). Fig. 24 illustrates one classification of exocrine glands.

GLAUCOPHYTA. Division of the ALGAE. Eukaryotic cells, lacking chloroplasts but harbouring instead endosymbiotic and modified blue-green algae. Apparently on the main evolutionary route to algae possessing chloroplasts, and thought to represent an intermediary stage. See ENDOSYMBIOSIS.

GLENOID CAVITY. Cup-like hollow on each side of pectoral girdle (on the scapula, and the coracoid when present) into which head of the humerus fits, forming tetrapod shoulder-joint.

GLIAL CELLS (GLIA, NEUROGLIA). Non-conducting nerve cells, performing supportive and protective roles for neurones. Include *astrocytes* (attaching neurones to blood vessels), *oligodendrocytes* (forming myelin sheaths of axons of central nervous system), SCHWANN CELLS, *microglia* (phagocytic) and *ependyma* cells (lining ventricles of the brain and cerebrospinal canal).

GLOBIGERINA OOZE. Calcareous mud, covering huge areas (about one third) of ocean floor. Formed mainly from shells of FORAMINIFERA, *Globigerina* being an important genus.

GLOBULINS. Group of globular proteins, soluble in aqueous salt solutions and of wide occurrence in plants (e.g. in seeds) and animals (e.g. in vertebrate blood plasma). In humans, ELECTROPHORESIS can separate $alpha_1$-, $alpha_2$-, BETA- and GAMMA-GLOBULINS. Normal individuals belong to one of three genetic types, separable by electrophoretograms of their plasma proteins. Some globulins are involved in lipid and iron transport. See TRANSFERRIN.

GLOMERULUS. (1) Small knot of capillaries covered by basement membrane and surrounded by Bowman's capsule, forming part of a NEPHRON of vertebrate KIDNEY. Through it small solute molecules (i.e. not cells or plasma proteins) are filtered under pressure from the blood to form the *glomerular filtrate*. (2) *Caudal glomeruli*; small tissue masses containing RETIA MIRABILIA in mammalian tails, some involved in heat conservation.

GLOSSOPHARYNGEAL NERVE. See CRANIAL NERVES.

GLOTTIS. Slit-like opening of trachea into pharynx of vertebrates.

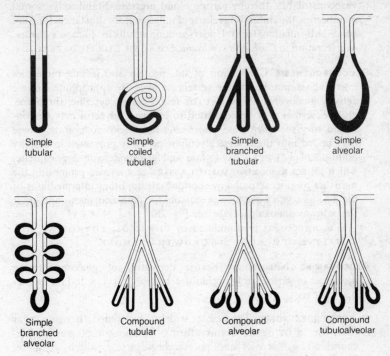

Simple
tubular

Simple
coiled
tubular

Simple
branched
tubular

Simple
alveolar

Simple
branched
alveolar

Compound
tubular

Compound
alveolar

Compound
tubuloalveolar

Fig. 24. Diagram illustrating the major types of animal exocrine gland.

Can usually be closed by muscles. In mammals, opens between vocal cords. See LARYNX, EPIGLOTTIS, TRACHEA.

GLUCAGON. Polypeptide hormone of vertebrates (29 amino acids), produced by alpha-cells of pancreas in response to drop in blood glucose level. Results in activation of ADENYLATE CYCLASE in target cells (e.g. liver, adipose tissue) with resultant rise in cyclic AMP in those cells and appropriate enzyme activation to ensure glycogenolysis (with release of glucose from glycogen), GLUCONEOGENE-SIS and lipolysis (with release of free fatty acids), restoring levels of these metabolites in the blood plasma. Also stimulates INSULIN release from beta-cells, the two hormones acting antagonistically in control of blood glucose and free fatty acid levels. Its effects are *hyperglycaemic*.

GLUCOCORTICOIDS. Steroid adrenal cortex hormones (principally COR-TISOL and cortisone) concerned with normal metabolism and resistance to stress conditions (e.g. long-term cold, starvation) by promoting deposition of glycogen in liver, GLUCONEOGENESIS and release of fatty acids from fat reserves. They render blood vessels more sensitive to vasoconstrictors, thereby raising blood pressure. Stabilize lysosomal membranes, thus inhibiting release of inflammatory substances (and are hence anti-inflammatory). Undersecretion results in *Addison's disease*, oversecretion in *Cushing's syndrome*. See ACTH, CORTICOSTEROIDS.

GLUCONEOGENESIS. Conversion of fat, protein and lactate molecules into glucose, notably by the vertebrate liver. By appropriate enzyme activity glucogenic AMINO ACIDS (e.g. alanine, cysteine, threonine, glycine, serine) may be converted to *pyruvate*; glycerol may be converted to *glyceraldehyde–3–phosphate*; fatty acids to *acetate*. These may be fed into the reverse glycolytic pathway promoted in cells by build-up of ACETYL–COA, citrate and glyceraldehyde–3–phosphate, which all act as positive MODULATORS of enzymes promoting the pyruvate-to-glucose pathway. Besides citrate, other intermediates of the KREBS CYCLE are also precursors for gluconeogenesis and enter the pathway via oxaloacetate (see Fig. 25).

Gluconeogenesis is stimulated by CORTISOL, THYROXINE, ADRENALINE, GLUCAGON and GROWTH HORMONE.

GLUCOSAMINE. Nitrogenous hexose derivative of glucose forming monomer of various polysaccharides, notably CHITIN and HYALURONIC ACID.

GLUCOSE (DEXTROSE). The most widely distributed hexose sugar (*dextrose* in its dextrorotatory form). Component of many disaccharides (e.g. sucrose) and polysaccharides (e.g. starch, cellulose, glycogen). An *aldohexose* reducing sugar, and (as glucose-1-phosphate) the initial substrate of GLYCOLYSIS for most, if not all, cells. When completely oxidized (in combined glycolysis and aerobic respiration) sufficient energy is released per glucose molecule under intracellular conditions for the generation of 36 molecules of ATP from ADP and P_i, in the overall equation:

$$C_6H_{12}O_6 + 6O_2 = 6CO_2 + 6H_2O$$

GLUME (SERILE GLUME). Chaffy bract, a pair of which occurs at base of grass spikelet, enclosing it.

GLUTEN. Protein occurring in wheat giving firmness to risen dough in bread-making. Those allergic to gluten suffer from *coeliac disease*, resulting in poor absorption of dietary components and in the consequent *malabsorption syndrome*. Gluten-free products are available for such people.

GLYCAN. Synonym of POLYSACCHARIDE.

GLYCERIDE. Fatty acid ester of glycerol. When all three –OH groups of glycerol are so esterified the result is a *triglyceride*. See FAT.

(a) GLYCEROL

(b) TRIGLYCERIDE (neutral fat)

GLYCEROL. A trihydric alcohol and component of many lipids, notably of glycerides.

GLYCOCALYX (CELL COAT). Carbohydrate-rich region at surfaces of most eukaryotic cells, deriving principally from oligosaccharide components of membrane-bound GLYCOPROTEINS and GLYCOLIPIDS, although it may also contain these substances secreted by the cell. Role of the cell coat is not properly understood yet. See CELL MEMBRANES.

GLYCOGEN. The chief polysaccharide store of animal cells and of many fungi; often called 'animal starch'. It resembles AMYLOPECTIN structurally in being an α-[1,4]-linked homopolymer of glucose units, although it is more highly branched. It can be isolated from tissues by digesting them with hot KOH solutions. As with amylopectin, it gives a red-violet colour with iodine/KI solutions. Its hydrolysis is termed *glycogenolysis*. Like starch it is osmotically inactive and therefore a suitable energy storage compound. See GLUCONEOGENESIS, GLYCOLYSIS, GLUCAGON, INSULIN.

GLYCOLIPID. Lipid with covalently attached mono- or oligosaccharides; found particularly in the outer half of phospholid bilayers of plasma membranes. Considerable variation in composition both between species and between tissues. All have a carbohydrate polar head end. Range in complexity from relatively simple galacto-cerebrosides of the Schwann cell MYELIN SHEATH to complex *gangliosides*. May be involved in cell-surface recognition. See GLYCOSYLATION, GLYCOPROTEIN, CELL MEMBRANES.

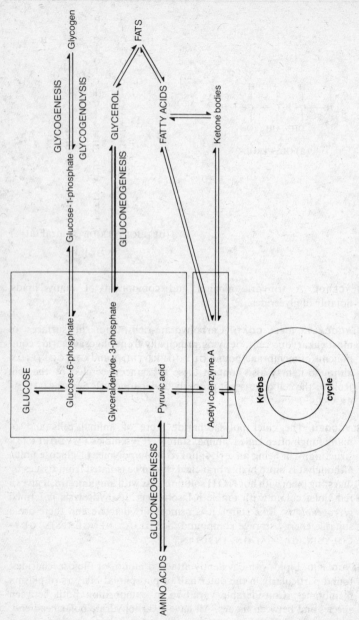

Fig. 25a. Diagram to illustrate where inputs to the glycolytic pathway occur for resynthesis of glucose.

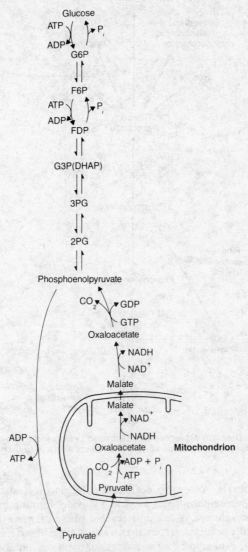

Fig. 25b. *Diagram indicating role of mitochondria in reversal of glycolytic pathway.*

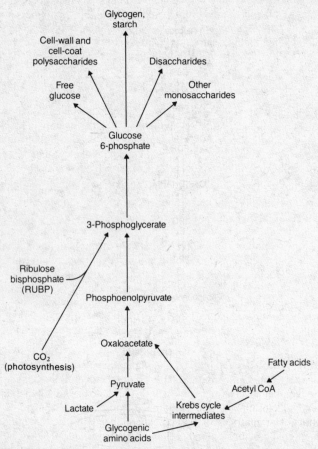

Fig. 26. *Diagram of biochemical pathways linking some non-carbohydrates to carbohydrates. Glucose-6-phosphate acts as a key branch-point.*

GLYCOLYSIS. Anaerobic degradation of glucose (usually in the form of glucose-phosphate) to yield pyruvate, forming initial process by which glucose is fed into aerobic phase of RESPIRATION, which usually occurs in MITOCHONDRIA. Cells without mitochondria (and those prokaryotes without a MESOSOME) rely on glycolysis for most of their ATP synthesis, as do facultatively anaerobic cells (e.g. striated muscle fibres) when there is a shortage of oxygen. The pathway is illustrated in Fig. 27. It generates a net gain of two molecules of ATP per molecule of glucose used, plus reducing power in the form of two $NADH_2$ molecules. The $NADH_2$ is available for reduction of pyru-

vate to lactate, or of acetaldehyde to alcohol, or for fatty acid and
steroid synthesis from acetyl coenzyme A, as occurs in liver cells.

Perhaps the most significant step in glycolysis is hydrolysis of each
fructose 1,6–bisphosphate molecule into two triose phosphate molecules, the remaining steps in the pathway thereby effectively occurring
twice for every initial glucose-phosphate molecule used. Conversion
of fructose–6–phosphate to fructose 1,6–bisphosphate is the main
rate-limiting step in glycolysis, and phosphofructokinase, the enzyme
involved, is a REGULATORY ENZYME, modulated by the ratio in the
levels of $(AMP + ADP):ATP$ in the cytosol so that high ATP
levels inhibit glycolysis. Enzymes involved in the glycolytic pathway
appear to have arisen by evolution from one ancestral enzyme by a
process involving GENE AMPLIFICATION. See PASTEUR EFFECT,
PENTOSEPHOSPHATEPATHWAY. For *aerobic glycolysis*, see CANCER
CELL.

GLYCOPROTEIN. Protein associated covalently at its N-terminal end
with a simple or complex sugar residue. In PROTEOGLYCANS the
carbohydrate forms the bulk of the molecule, with numerous long
and usually unbranched GLYCOSAMINOGLYCANS bound to a single
core protein. These important extracellular components contrast with
cell surface glycoproteins, which generally comprise short but often
complex non-repeating oligosaccharide sequences bound to an integral membrane protein. Proteins become glycosylated (i.e. have
their sugar residues added) in the ENDOPLASMIC RETICULUM and
GOLGI APPARATUS. See CELL MEMBRANES, GLYCOSYLATION,
GLYCOLIPIDS.

GLYCOSAMINOGLYCANS (GAGs). Long, unbranched polysaccharides
(formerly called mucopolysaccharides) of repeated disaccharide units,
one member always an amino sugar (e.g. N-acetylglucosamine, N-
acetylgalactosamine). They comprise varying proportions of the
extracellular matrices of tissues, where they are often numerously
bound to a core protein to become *proteoglycans*, e.g. HYALURONIC
ACID, CHONDROITIN, HEPARIN. See GLYCOCALYX, CELL MEM-
BRANES.

GLYCOSIDE. Substance formed by reaction of an *aldopyranose* sugar,
such as glucose, with another substance such that the aldehyde
moiety in the sugar is replaced by another group. Glycosidic bonds
form the links between monosaccharide units in the formation of
polysaccharides. Some plant glycosides, termed *cardiac glycosides*,
alter the excitability of heart muscle and may be defensive; examples
include *ouabain* and *digitalin*. See ANTHOCYANINS, TANNINS.

GLYCOSYLATION. Bonding of sugar residue to another organic compound. GLYCOPROTEINS are formed in the lumen of rough endoplasmic reticulum, but may be subsequently modified in the lumen
of the GOLGI APPARATUS, where other amino acids of the protein

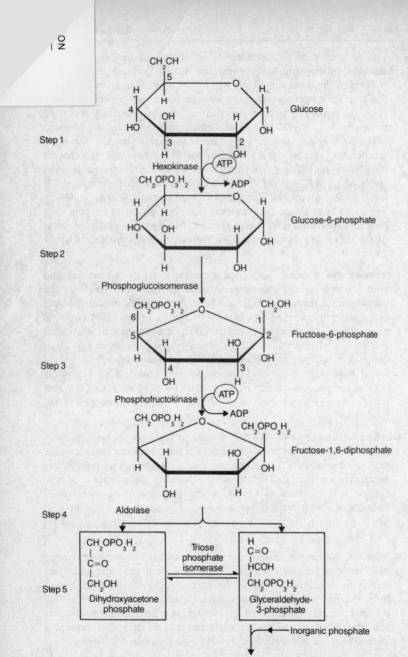

Fig. 27. The stages of vertebrate glycolysis and the enzymes involved. Because of the hydrolysis at Step 5 all later stages are represented by two molecules for every original glucose molecule.

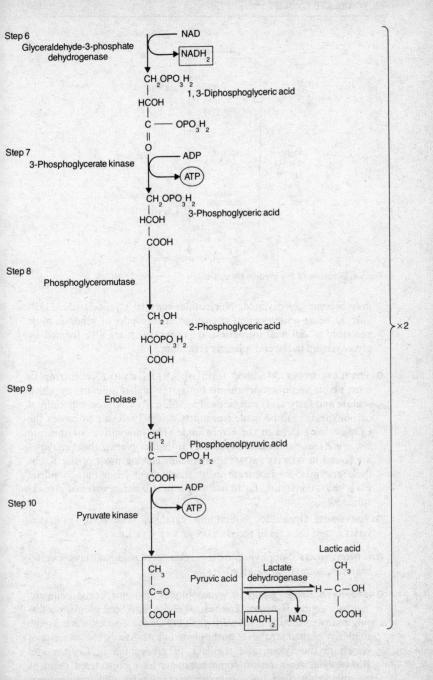

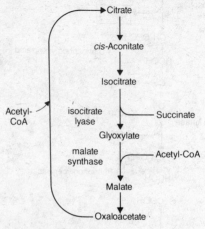

Fig. 28. *Diagram of the glyoxylate cycle.*

may become glycosylated. Nucleotides may be glycosylated, UDP-glucose being an important coenzyme in transport of glucose, most probably in cell wall formation. GLYCOLIPIDS are also formed by glycosylation in the endoplasmic reticulum.

GLYOXYLATE CYCLE. Modified form of KREBS CYCLE, occurring in most plants and microorganisms but not in higher animals, by which acetate and fatty acids can be used as sole carbon source, especially if carbohydrate is to be made from fatty acids. The cycle by-passes the CO_2-evolving steps in the Krebs cycle. The innovative enzymes are *isocitrate lyase* and *malate synthase*. In higher plants, these enzymes are found in GLYOXYSOMES, organelles lacking most of the Krebs cycle enzymes; so isocitrate must reach them from mitochondria. Plant seeds converting fat to carbohydrate are rich sources of glyoxysomes.

GLYOXYSOME. Organelles containing catalase, related to PEROXISOMES, and the sites of the GLYOXYLATE CYCLE.

GNATHOSTOMATA. Subphylum or superclass containing all jawed vertebrates. Contrast AGNATHA.

GNETOPHYTA. Small group of gymnophytes (gymnosperms) comprising three genera (*Gnetum*, *Ephedra*, *Welwitschia*). Seed plants, possessing many anthophyte (flowering plant) characteristics, such as the similarity of their strobili to anthophyte inflorescences, the presence of vessels in the xylem, and the lack of archegonia in *Gnetum* and *Welwitschia*. As a result, some botanists have considered them as possible connecting links between gymnophytes and anthophytes.

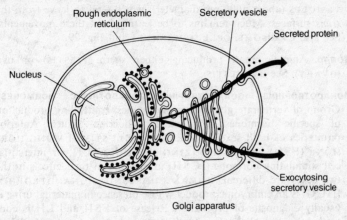

Fig. 29. *Illustration of the role of the Golgi apparatus in secretion of synthesized protein. Arrows show the route from site of production to release.*

Currently regarded as specialized, pointing off one line of gymnophyte evolution.

GOBLET CELL. Pear-shaped cell present in some epithelia (e.g. intestinal, bronchial), and specialized for production of MUCUS.

GOLGI APPARATUS (GOLGI BODY, GOLGI COMPLEX, DICTYOSOME). System of roughly parallel interconnecting flattened sacs (cisternae) situated close to the ENDOPLASMIC RETICULUM of eukaryote cells but physiologically separate from it. Numbers vary from one to hundreds per cell; tend to be interconnected in animal but not in plant cells. May be up to 30 cisternae per Golgi body, but normally about six. Each cisterna has a *cis* surface (towards nucleus) and a *trans* surface (away from nucleus). Transport vesicles from the endoplasmic reticulum seem to fuse on the *cis* surfaces, adding their membrane material to the cisternae and depositing GLYCOPROTEINS for processing within the cisternal lumina. Some of the oligosaccharide of the glycoprotein may be removed, while other sugar units are added, to yield mature glycoproteins of different sorts – possibly vesicle-specific. Many of these will be retained within the membrane during modification. In plant cells, scales, both organic and inorganic, are formed in the Golgi body before being transported to the periphery. In diatoms, the Golgi body gives rise to translucent vesicles which collect beneath the plasmalemma where they fuse to form a SILICALEMMA in which the siliceous cell wall forms.

Two types of vesicle are budded from the Golgi body: COATED

VESICLES (about 50 nm diameter) and the larger SECRETORY VESICLES (about 1000 nm diameter), which tend to leave from the *trans* surfaces. Much remains to be learnt about these movements. See also LYSOSOME, CELL MEMBRANES and Fig. 29.

GONAD. Animal organ producing either sperm (TESTIS) or ova (OVARY). See OVOTESTIS.

GONADOTROPHINS (GONADOTROPINS, GONADOTROPIC HORMONES). Group of vertebrate glycoprotein hormones, controlling production of specific hormones by gonadal endocrine tissues. Anterior pituitaries of both sexes produce FOLLICLE-STIMULATING HORMONE (FSH) and LUTEINIZING HORMONE (LH, or interstitial cell stimulating hormone (ICSH) in males); but their effects in the two sexes are different. HUMAN CHORIONIC GONADOTROPHIN (HCG) is an embryonic product whose presence in maternal urine is usually diagnostic of pregnancy. Release of FSH and LH is controlled by hypothalamic *gonadotrophic-releasing factors* (GnRFs). PROLACTIN is also gonadotrophic. See MENSTRUAL CYCLE.

GONDWANALAND. Southernmost of the two Mesozoic supercontinents (the other being LAURASIA) named after a characteristic geological formation, the *Gondwana*. Comprised future South America, Africa, India, Australia, Antarctica and New Zealand (the last breaking earliest from the supercontinent, with present-day examples of a relict Gondwana-like flora and fauna). Rifts between Gondwanaland and Laurasia were not effective barriers to movements of land animals until well into the Cretaceous. By the dawn of mammalian radiation Gondwanaland had largely split into its five major continental regions, each being the nucleus of radiation for its inhabitant fauna and flora. Flora was characterized by *Glossopteris*; podocarps and tree ferns still persist in New Zealand, as do the reptile *Sphenodon* (see RHYNCHOCEPHALIA), giant crickets and flightless birds (e.g. kiwi and, up to 5000 years ago, moas). See CONTINENTAL DRIFT, ZOOGEOGRAPHY.

GONOCHORISM. Condition of having sexes separate; individuals having functional gonads of only one type. Bisexual.

G₁, G₂ PHASE. Phases of the CELL CYCLE.

G-PROTEIN. Adaptor protein coupling activation of membrane receptor to activation of target protein (e.g. ADENYLATE CYCLASE) in the membrane. Different G-proteins may enable one hormone to initiate different responses in different cells. See SECOND MESSENGER, GTP, MEMBRANE RECEPTOR.

GRAAFIAN FOLLICLE. Fluid-filled spherical vesicle in mammalian ovary containing OOCYTE attached to its wall. Growth is under the control of FOLLICLE-STIMULATING HORMONE of anterior pituitary, its rupture (*ovulation*) also being a gonadotrophic effect (see

LUTEINIZING HORMONE). After ovulation the follicle collapses, but theca and granulosa cells grow and proliferate forming the CORPUS LUTEUM. Androgen precursors are made by the *theca cells* of the Graafian follicle, and aromatised to oestrogens by the *granulosa cells*; in primates *theca lutein cells* of the corpus luteum make oestrogen precursors. See MENSTRUAL CYCLE, OVARY.

GRADE. A given level of morphological organization sometimes achieved independently by different evolutionary lineages, e.g. the mammalian grade. See CLADE.

GRAFT. (1) To induce union between normally separate tissues. (2) A relatively small part of one organism transplanted either on to another part of the same organism (in animals often the whole organism) or on to a different organism, or a part of it. See AUTOGRAFT, ALLOGRAFT, ISOGRAFT, XENOGRAFT. See IMMUNOLOGICAL TOLERANCE, MHC.

GRAFT HYBRID. See CHIMAERA.

GRAFT-VERSUS-HOST RESPONSE. Reaction of immunocompetent donor cells to recipient tissues (e.g. skin, gut epithelia, liver), often destroying them. Particularly problematic in bone transplants.

GRAMICIDIN. See IONOPHORE.

GRAM'S STAIN. Stain devised by C. Gram in 1884 which differentiates between bacteria which may be otherwise similar morphologically. To a heat-fixed smear containing the bacteria is added crystal violet solution for 30 s which is then rinsed off with Gram's iodine solution; 95% ethanol is applied and renewed until most of the dye has been removed (20 s – 1 min). Those bacteria with the stain retained are *Gram-positive*, those without are *Gram-negative*. A counterstain (e.g. eosin red, saffranin, brilliant green) is then applied, colouring the Gram-negative bacteria but not the Gram-positive ones. Differentiation reflects differences in amount and ease of access of peptidoglycan in the bacterial envelope. Gram-negative bacteria have a second lipo-protein membrane outside the thin peptidoglycan layer covering the inner (cytoplasmic) membrane. Gram-positive bacteria lack this outer membrane, but have a thicker peptidoglycan coat. Among other differences, Gram-positive bacteria are more susceptible to penicillin, acids, iodine and basic dyes; Gram-negative bacteria are more susceptible to alkalis, ANTIBODIES and COMPLEMENT.

GRANA (sing. GRANUM). In CHLOROPLASTS, groups of disc-shaped, flattened vesicles (*thylakoids*) stacked like coins in a pile, whose membranes bear photosynthetic pigments. Most highly developed in chloroplasts of higher plants. The thylakoids are the sites of the photochemical reactions of PHOTOSYNTHESIS.

GRANULOCYTE (POLYMORPH). Granular LEUCOCYTE. Develops from MYELOID TISSUE and has granular cytoplasm. Include NEUTROPHILS, EOSINOPHILS and BASOPHILS.

GRANUM. See GRANA.

GRAPTOLITES. Extinct invertebrates of doubtful affinities (possibly with either COELENTERATA or HEMICHORDATA), whose name (literally, written on stone) indicates importance as fossils, notably of shales. Upper Cambrian–Lower Carboniferous; used to subdivide the Ordovician and Silurian.

GRAVITROPISM. See GEOTROPISM.

GREAT CHAIN OF BEING (SCALA NATURA). View proposed by Aristotle and incorporated by Leibniz in his metaphysics and by BUFFON (for whom it was a scale of degradation, from man at the top): that there is a linear and hierarchical progression of forms of existence, from simplest to most complex, lacking both gaps and marked transitions.

GREENHOUSE EFFECT. Effect in which short wavelength solar radiation entering the Earth's atmosphere is reradiated from the Earth's surface in the longer infrared wavelengths and is then reabsorbed by components of the atmosphere to become an important factor in heating the total atmosphere. Oxygen, ozone, carbon dioxide and water vapour all absorb in the infrared wavelengths, and increasing amounts of carbon dioxide from combustion of fossil fuels is a growing factor in raising the mean atmospheric temperature. Effect resembles heat reflection by greenhouse glass.

GREY MATTER. Tissue of vertebrate spinal cord and brain containing numerous cell bodies and dendrites of neurones, along with unmyelinated neurones synapsing with them, glial cells and blood vessels. Occurs as inner region of nerve cord, around central canal; in brain too it generally occupies the inner regions, but in some parts (e.g. CEREBRAL CORTEX of higher primates), some cell bodies of grey matter have migrated outwards to form a third layer on top of the white, axon zone.

GROOMING. Mammalian equivalent of preening in which licking, scratching and picking of fur occur, either of the animal's own, of its mate, or of another member of the social group. May occur, often solicited, as a DISPLACEMENT ACTIVITY, as an effective cleaning activity and as an indication of the relative social status of both groomer and groomed.

GROUND MERISTEM. Primary MERISTEM in which procambium is embedded and which is surrounded by protoderm; matures to form the GROUND TISSUES.

GROUND TISSUE. (Bot.) All tissues except the epidermis (or periderm) and the vascular tissues; e.g. those of the cortex and pith.

GROUP SELECTION. Postulated evolutionary mechanism whereby characters disadvantageous to individuals bearing them, but beneficial to the group of organisms they belong to, can spread through the

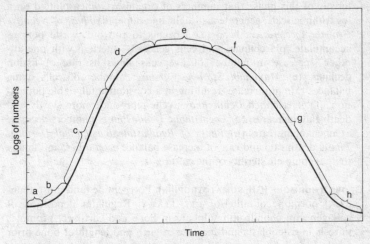

Fig. 30. *The growth curve of unicellular organisms under optimal growth conditions. Phases a to h are explained in the entry.*

population countering the effects of selection at the individual level. Invoked to explain apparent reproductive restraint by individuals when environmental resources are scarce, and other cases of apparent ALTRUISM which may have a simpler explanation in terms of NATURAL SELECTION. Some hold that group selection may have played a part in the evolution of SEX. See UNIT OF SELECTION.

GROWTH. Term with a variety of senses. At the individual level, usually involves increase in dry mass of an organism (or part of one), whether or not accompanied by size increase, involving differentiation and morphogenesis. Commonly involves cell division, but cell division without increase in cell size does not produce growth. Nor is uptake of water alone sufficient for growth. Usually regarded as irreversible, but ATROPHY of tissues and dedifferentiation of cells can occur. In most higher plants growth is restricted to MERISTEMS; in animals growth is more diffuse. See GROWTH CURVES, EXPONENTIAL GROWTH, ALLOMETRY.

GROWTH CURVES. Many general features of the growth of a population may be indicated by the growth curve of unicellular organisms under optimal conditions for growth. (See Fig. 30.) Stages (a) to (e) represent a logistic growth curve (see EXPONENTIAL GROWTH).

   (a) *Lag Phase*: latent phase, in which cells recover from new conditions, imbibe water, produce ribosomal RNA and subsequent proteins. Cells grow in size, but not in number. (b) *Phase of Accelerated Growth*: fission initially slow, cell size large. During this phase, rate of division increases and cell size diminishes. (c) *Exponential or Logarithmic Phase*: cells reach maximum rate of division. Characteristic of this phase that numbers of organisms, when plotted on a logarithmic scale, generate a straight-line slope. (d) *Phase of Negative Growth Acceleration*: food (e.g.) begins to run out, waste poisons accumulate, pH changes and cells generally interfere with one another. Increase in number of live cells slows as rate of fission declines. (e) *Maximum Stationary Phase*: number of cells dying balances rate of increase, resulting in a constant total viable population. (f) *Accelerated Death Phase*: cells reproduce more slowly and death rate increases. (g) *Logarithmic Death Phase*: numbers decrease at unchanging rate. (h) *Phase of Readjustment and Final Dormant Phase*: death rate and rate of increase balance each other, and finally there is complete sterility of the culture.

GROWTH HORMONE (GH, SOMATOTROPHIN). Polypeptide (and most abundant) hormone of anterior PITUITARY. Regulates deposition of collagen and chondroitin sulphate in bone and cartilage; promotes mitosis in osteoblasts and increase in girth and length of bone prior to closure of EPIPHYSES. Transported in plasma by a GLOBULIN protein. Its release is prevented by hypothalamic SOMATOSTATIN. Induces release from liver of the hormone SOMATOMEDIN, which mediates its effects at the cell level. Low levels in man result in *dwarfism*, high levels in *acromegaly*.

GROWTH RING. (Bot.) Growth layer in secondary xylem or secondary phloem, as seen in transverse section. The periodic and seasonally related activity of vascular cambium produces growth increments, or growth rings. In early and middle summer new xylem vessels are large and produce a pale wood, contrasting with the narrower and denser vessels of late summer and autumn, which produce a dark wood. Width of these rings varies from year to year depending upon environmental conditions, such as availabilities of light, water and nutrients and temperature. Age of a tree may be estimated by counting the number of growth rings, and in semi-arid regions trees are sensitive rain gauges (e.g. bristlecone pine *Pinus longaeva*). Each growth ring is different, and by comparing rings from dead and living trees information concerning the climatic history of the past 8200 years has been developed. See DENDROCHRONOLOGY.

GROWTH SUBSTANCE. (Bot.) Term used in preference to plant hormone, to include all natural (endogenous) and artificial substances with powerful and diverse physiological and/or morphogenetic effects in plants. Sites of natural production are often diffuse and rarely comprise specialized glandular organs – often just a patch of cells, commonly without physical contact. The minute quantities produced, and the notorious synergisms in their modes of action, pose profound research problems. See AUXINS, ABSCISIC ACID, CYTOKININS, ETHENE, GIBBERELLINS.

GTP (GUANOSINE 5'-TRIPHOSPHATE). Purine nucleoside triphosphate. Required for coupling some activated membrane receptors to ADENYLATE CYCLASE activity. A GTP-binding protein (G-PROTEIN) hydrolyses the GTP and keeps the enzyme's activity brief. Involved in tissue responses to various hormones (e.g. ADRENALINE), a precursor for NUCLEIC ACID synthesis and essential for chain elongation during PROTEIN SYNTHESIS.

GUANINE. Purine base of NUCLEIC ACIDS. See GTP, INOSINIC ACID.

GUANOSINE. Purine nucleoside. Comprises D-ribose linked to guanine by a beta-glycosidic bond. See GTP.

GUARD CELLS. (Bot.) Specialized, crescent-shaped, unevenly thickened epidermal cells in pairs surrounding a STOMA. Changes in shapes of guard cells, due to changes in their turgidities, control opening and closing of the stomata, and hence affect rate of loss of water vapour in TRANSPIRATION and amount of gaseous exchange.

GUILD. Group of organisms, often species within a higher taxon, having the same broad feeding habits; e.g. phytophages (plant-feeders, themselves divisible into chewers and suckers), parasitoids, scavengers, etc.

GUT. See ALIMENTARY CANAL.

GUTTATION. Excretion of water drops by plants through HYDATHODES, especially in high humidity, due to pressure built up within the xylem by osmotic absorption of water by roots. See ROOT PRESSURE.

GYMNOPHYTA. (Formerly Gymnospermae). Major class of seed-bearing vascular plants, with seeds not enclosed in an ovary (contrast ANTHOPHYTA), and with pollen deposited directly on the ovule. Each embryo may have several cotyledons. Conifers (CONIFEROPHYTA) the most familiar group, but including CYCADOPHYTA, GINKGOPHYTA and GNETOPHYTA. See SPERMATOPHYTA, TRACHEOPHYTA.

GYMNOSPERM. See GYMNOPHYTA.

GYNANDROMORPH. Animal, usually an insect, which is a genetic

MOSAIC in that some of its cells are genetically female while others are genetically male. Loss of an X-chromosome by a stem cell of an insect which developed from an XX zygote thus produces a clone of 'male' tissue. Sometimes expressed bilaterally, one half of the animal being phenotypically male, the other female. Also occurs in birds and mammals. See INTERSEX.

GYNANDROUS. (Bot.) (Of flowers) having stamens inserted on the gynoecium.

GYNOBASIC. (Bot.) (Of a style) arising from base of ovary (due to infolding of ovary wall in development); e.g. white dead nettle.

GYNODIOECIOUS. Having female and hermaphroditic flowers on separate plants; e.g. thymes (*Thymus*). Compare ANDRODIOECIOUS, GYNOMONOECIOUS.

GYNOECIUM (PISTIL). Collective term for the carpels of a flower; i.e. the female components of the flower. Compare ANDROECIUM.

GYNOGENESIS. Condition whereby a female animal must mate before she can produce parthenogenetic eggs. In some triploid thelytokous animals (the salamanders *Ambystoma*, the fish *Poeciliopsis*) the sperm penetrates the egg to initiate cleavage but contributes nothing genetically.

GYNOMONOECIOUS. (Bot.) With female and hermaphroditic flowers on the same plant; e.g. many Compositae. Compare ANDROMONOECIOUS, GYNODIOECIOUS.

GYROGONITES. Lime-encrusted, fossilized oogonia and encircling sheath cells (nucule) of the CHAROPHYTA.

H

HABITAT. Place or environment in which specified organisms live; e.g. sea shore. Compare NICHE.

HABITUATION. LEARNING, in which an animal's response to a stimulus declines with repetition of the stimulus at the same intensity. Needs to be distinguished experimentally from sensory ADAPTATION and muscular fatigue.

HAEM (HEME). Iron-containing PORPHYRIN, acting as PROSTHETIC GROUP of several pigments, including HAEMOGLOBIN, MYOGLOBIN and several CYTOCHROMES.

HAEMERYTHRIN. Reddish-violet iron-containing respiratory pigment of sipunculids, one polychaete, priapulids and the brachiopod *Lingula*. Prosthetic group is not a porphyrin, the iron attaching directly to the protein. Always intracellular.

HAEMOCOEL. Body cavity of arthropods and molluscs, containing blood. Continuous developmentally with the BLASTOCOELE. Unlike the COELOM, it never communicates with the exterior or contains gametes. Often functions as a hydrostatic skeleton. Contains haemolymph.

HAEMOCYANIN. Copper-containing protein (non-porphyrin) respiratory pigment occurring in solution in haemolymph of malacostracan and chelicerate arthropods and in many molluscs. Blue when oxygenated, colourless when deoxygenated.

HAEMOGLOBIN. Protein respiratory pigment with iron(Fe^{++})-containing porphyrin as prosthetic group. Tetrameric molecule, comprising two pairs of non-identical polypeptides associated in a quaternary structure, binding oxygen reversibly, forming *oxyhaemoglobin*. Occurs intracellularly in vertebrate ERYTHROCYTES, but when found in invertebrates (e.g. earthworms) is usually in simple solution in the blood. Also found in root nodules of leguminous plants (as *leghaemoglobin*), but only if *Rhizobium* is present. Scarlet when oxygenated, bluish-red when deoxygenated. The ability of the haemoglobin molecule to pick up and unload oxygen depends on its shape in solution, which varies allosterically with local pH. This in turn is a function of the partial pressure of CO_2 (see BOHR EFFECT). Haemoglobins are adapted for maximal loading and unloading of oxygen within the oxygen tension ranges occurring in their respective organisms. See Fig. 31.

Normal adult human haemoglobin (HbA) contains two α- and two β-globin chains; foetal haemoglobin (HbF) contains two α- and two

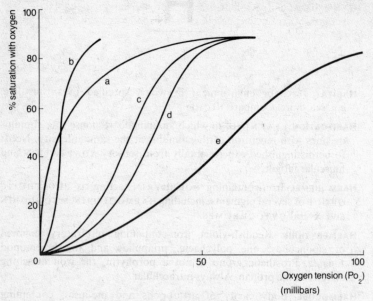

Fig. 31. *Oxygen equilibration curves for: (a) myoglobin, (b) Arenicola Hb, (c) human foetal Hb, (d) adult human Hb, (e) pigeon Hb.*

γ-globin chains. The transition from foetal to adult haemoglobin production begins late in foetal life and is completed in early infancy. Some carbon dioxide is carried by haemoglobin as *carbamino compounds*, and one consequence of inhaling cigarette smoke is that carbon monoxide produced binds irreversibly to haemoglobin, forming *carboxyhaemoglobin*. See MYOGLOBIN.

HAEMOLYSIS. Rupture of red blood cells (e.g. through osmotic shock) with release of haemoglobin.

HAEMOPHILIA. Hereditary disease, sex-linked and recessive in humans, in which blood fails to clot owing to absence of Factor VIII (see BLOOD CLOTTING).

HAEMOPOIESIS. Blood formation; in vertebrates includes both plasma and cells. In anoxia, the kidney produces a hormone, *erythropoietin*, stimulating red cell production in red bone marrow (see MYELOID TISSUE). In vertebrate embryos, erythropoiesis occurs commonly in yolk sac, liver, spleen, lymph nodes and bone marrow; in adults is restricted to red bone marrow (and yellow bone marrow in long bones under oxygen stress). Leucocytes also originate in myeloid tissue but from different stem cells from erythrocytes. Much of the plasma protein is formed in the liver, notably fibrinogen, albumen and α- and β-globulins.

HAEMOSTASIS. Several homeostatic mechanisms maintaining blood in a fluid state, and within blood vessels. Includes BLOOD CLOTTING and FIBRINOLYSIS.

HAIR. (Bot.) (1) Trichome. Single- or many-celled outgrowth from epidermal cell; usually a slender projection composed of cells arranged end to end whose functions are various, e.g. absorption (see ROOT HAIR), secretion (see GLAND), reduction in transpiration rate. (2) An appendage on a flagellum. (Zool.) Epidermal thread protruding from mammalian skin surface, composed of numerous cornified cells. Each develops from the base of a HAIR FOLLICLE (its '*root*') in the undersurface of which (*hair bulb*) cells are produced mitotically. Hair colour depends upon amount of MELANIN present; with age, increasing presence of air bubbles results in total internal reflection of light, making hair appear white. In most non-human mammals body hair is thick enough for hair erection (by *erector pili muscles*) to have a homeostatic effect on heat retention. Nerve endings provide hair with a sensory role (see SKIN). So-called hairs of arthropods are bristles.

HAIR CELL. Ectodermal cells with modified membranes found in vertebrate VESTIBULAR APPARATUS, acting as mechanoreceptors by responding to tension generated either by a gelatinous covering layer (in maculae, cristae), or by the tectorial membrane (in cochlea). They normally bear one long true cilium (the *kinocilium*) and a tuft of several large and specialized microvilli (*stereocilia*) of decreasing length. These produce a receptor potential when deformed.

HAIR FOLLICLE. Epidermal sheath enclosing the length of a hair in the skin. Surrounded by connective tissue serving for attachment of erector pili muscles. May house a SEBACEOUS GLAND. See SKIN.

HALOPHYTE. Plant growing in and tolerating very salty soil typical of shores of tidal river estuaries, saltmarshes, or alkali desert flats.

HALLUX. 'Big toe'; innermost digit of tetrapod hind foot. Often shorter than other digits. Turned to the rear in most birds, for perching. Compare POLLEX.

HALTERE. Modified hind wing of DIPTERA (two-winged flies) concerned with maintenance of stability in flight. Comprises basal lobe closest to thorax, a stalk and an end knob. Numerous *campaniform sensilla* forming plates on the basal lobe react to forces there, indicating vertical movements of the haltere (its only plane of movement) and torque produced by both turning movements of the fly and inertia of the haltere. Halteres vibrate with same frequency as fore wings, but out of phase. Halteres are like gyroscopes; their combined nervous input to the thoracic ganglion enables adjustment of wings to destabilizing forces.

HAMILTON'S RULE. Prediction that genetically determined behaviour which benefits another organism, but at some cost to the agent with the allele(s) responsible, will spread by SELECTION when the relation (rb − c) > 0 is satisfied; where r = degree of RELATEDNESS between agent and recipient, b = improvement of individual FITNESS of recipient caused by the behaviour and c = cost to agent's individual fitness as a result of the behaviour. See ALTRUISM, KIN SELECTION.

HAPLOCHLAMYDEOUS. See MONOCHLAMYDEOUS.

HAPLODIPLONTIC. (Of a LIFE CYCLE) in which both haploid and diploid mitoses occur.

HAPLOID. (Of a nucleus, cell, etc.) in which chromosomes are represented singly and unpaired. The haploid chromosome number, *n*, is thus half the DIPLOID chromosome number, 2*n*. Haploid cells are commonly the direct product of MEIOSIS, but haploid mitosis is relatively common too. No haploid cell can undergo meiosis. Diploid organisms generally produce haploid gametes. In humans, *n* = 23. See MALE HAPLOIDY, POLYPLOIDY, ALTERNATION OF GENERATIONS.

HAPLOID PARTHENOGENESIS. (Bot.) Development of an embryo into a haploid sporophyte from an unfertilized egg on a haploid gametophyte. See PARTHENOGENESIS.

HAPLONT. Organism representing the HAPLOID stage of a LIFE CYCLE. Compare DIPLONT.

HAPLONTIC. (Of a LIFE CYCLE) in which there is no diploid mitosis, but in which haploid mitosis does occur.

HAPLOSTELE. Solid cylindrical STELE in which a central strand of primary xylem is sheathed by a cylinder of phloem.

HAPLOTYPE. A haploid genotype. The gametes produced by a normal outbred diploid individual will be of a variety of haplotypes.

HAPTEN. Single isolated antigenic determinant. See ANTIGEN.

HAPTERON (HOLDFAST). Bottom part of some algae, attaching the plant to the substratum; may be discoid or root-like in structure.

HAPTONEMA. Appendage arising near a eukaryotic FLAGELLUM, but thinner and with different properties and structure. Found in the algal division PRYMNESIOPHYTA (HAPTOPHYTA). In transverse section, haptonemata comprise three concentric membranes surrounding a core containing seven microtubules. The function of this organelle is not fully understood, but it can serve in temporary attachment to a surface.

HAPTOTROPISM (THIGMOTROPISM). (Bot.) A TROPISM in which the stimulus is a localized contact, e.g. tendril in contact on one side with solid object such as a twig; response is curvature in that direction producing coiling around the object.

HARDY-WEINBERG THEOREM (H-W LAW, PRINCIPLE or EQUILIBRIUM). Theorem predicting for a normal amphimictic population the ratios of the three genotypes (e.g. AA:Aa:aa) at a locus with two segregating alleles, A and a, given the frequencies of these genotypes in the parent population. The theorem assumes: random (i.e. non-assortative) mating, no NATURAL SELECTION, GENETIC DRIFT or MUTATION, or immigration or emigration. Its utility is that once the parental population's genotypic ratios have been determined, their predicted ratio in the next generation can be checked against the observed values, and any departures from expectation tested for significance (e.g. by CHI-SQUARED TEST). If significant, and if all assumptions other than selection can be discounted, then this is *prima facie* evidence that selection caused the departures from expected values. If in the parental generation the frequencies of alleles A and a are p and q respectively, then the theorem states that the genotypic ratios in the next and all succeeding generations will be:
AA : Aa : aa
$p^2 : 2pq : q^2$

HATCH-SLACK PATHWAY. See PHOTOSYNTHESIS.

HAUSTORIUM. Specialized penetrative food-absorbing structure. Occurs (1) in certain plant-parasitic fungi, formed within a living host cell at end of a hyphal branch and (2) in some parasitic plants (e.g. dodder), withdrawing food material from host tissues.

HAVERSIAN SYSTEM (OSTEON). Anatomical unit of compact BONE. Comprises a central *Haversian canal*, which branches and anastomoses with those of other Haversian systems and contains blood vessels and nerves, surrounded by layers of bone deposited concentrically by osteocytes and forming cylinders. Blood is carried from vessels at the bone surface to the Haversian system by *Volkmann's canals*.

H-2 COMPLEX. Mouse major histocompatibility complex. See MHC.

HDL. High-density LIPOPROTEIN.

HEART. Muscular, rhythmically contracting pump forming part of the cardiovascular system and responsible for blood circulation. All hearts have valves to prevent back-flow of blood during contraction. Initiation of heart beat may be by extrinsic nerves (*neurogenic*), as in many adult arthropods, or by an internal pacemaker (*myogenic*), as in vertebrates and some embryonic arthropods.

Vertebrate heart muscle (CARDIAC MUSCLE) does not fatigue, and is under the regulation of nerves (see CARDIO-ACCELERATORY/ INHIBITORY CENTRES) and hormones (e.g. ADRENALINE). The

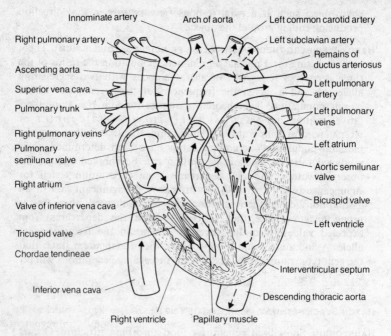

Fig. 32. *Mammalian heart showing direction of oxygenated* (--->) *and deoxygenated* (→) *blood.*

basic S-shaped heart of most fish comprises four chambers pumping blood unidirectionally forward to the gills. This *single circulation* has the route:

body → heart → gills → body.

Replacement of gills by lungs in tetrapods was associated with the need for a heart providing a *double circulation* in order to keep oxygenated blood (returning to the heart from the lungs) separate from deoxygenated blood (returning to the heart from the body). The blood route becomes:

body → heart → lungs → heart → body.

In amphibians, two atria return blood from these two sources to a single ventricle, and separation is limited; reptiles have a very complex ventricle with a septum assisting separation; birds and mammals have two atria and two ventricles, one side of the heart dealing with oxygenated and the other with deoxygenated blood (see HEART CYCLE). In annelids, the whole dorsal aorta may be contractile with, in addition, several vertical contractile vessels, or 'hearts'. The insect heart is a long peristaltic tube lying in the roof of the abdomen and perforated by paired segmental holes (*ostia*) through which blood enters from the haemocoele. There may be accessory hearts in the thorax. Blood is driven forwards into the aorta, which opens into the haemocoele. A similar arrangement occurs in other arthropods. The basic molluscan heart comprises a median ventricle and two atria. See PERICARDIUM.

HEART CYCLE (CARDIAC CYCLE). One complete sequence of contraction and relaxation of heart chambers, and opening and closing of valves, during which time the same volume of blood enters and leaves the heart. Chamber contraction (*systole*) is followed by its relaxation (*diastole*) when it fills again with blood. In mammals and birds ventricular diastole draws in most ot the blood from the atria; atrial systole adds only 30% to ventricular blood volume. The PACE-MAKER and its associated fibres ensure that the two atria contract simultaneously just prior to the two ventricles. Atrioventricular valves open when atrial pressure exceeds ventricular and close when ventricular pressure exceeds atrial. See CARDIO-ACCELERATORY CENTRE.

HEARTWOOD. Central mass of xylem tissue in tree trunks; contains no living cells and no longer functions in water conduction but serves only for mechanical support; its elements frequently blocked by TYLOSES, and frequently dark-coloured (e.g. ebony), impregnated with various substances (tannins, resins, etc.) that render it more resistant to decay than surrounding sapwood.

HeLa CELL. Cell from human cell line widely used in study of cancer. Original source was Helen Lane, a carcinoma patient, in 1952.

HELIOPHYTES. Class of RAUNKIAER'S LIFE FORMS.

HELIOTROPISM. See PHOTOTROPISM.

HELIOZOA. SARCODINA of generally freshwater environments, without shell or capsule, but sometimes with siliceous skeleton, and usually very vacuolated outer protoplasm. Locomotion by 'rolling', successive pseudopodia pulling the animal over in turn. Food is caught by cytoplasm flowing over axial supports of pseudopodia. Flagellated stage common. Some are autogamous; binary fission usual.

HELMINTH. Term usually applied to parasitic flatworms, but occasionally to nematodes also.

HELPER. An animal which helps rear the young of a conspecific to which it is not paired or mated. Commonly there are genetic bonds between the helper and its beneficiary 'family'. Many of the studies are on communal nesters in birds. Of considerable theoretical interest. See HAMILTON'S RULE, INCLUSIVE FITNESS, ALTRUISM.

HELPER CELL. See T-CELL and Fig. 41.

HEMICELLULOSES. Heterogeneous group of long-chain polysaccharides, distinct from cellulose. Basic units are arabinose, xylose, mannose or galactose. Integral component of plant CELL WALLS, especially in lignified tissue. More soluble and less ordered than cellulose; may function as food reserve in seeds (e.g. in endosperm of date seeds).

HEMICHORDATA. Group of marine organisms of disputed phylogenetic relationships. Either a distinct phylum, or a subphylum of the CHORDATA. Includes the pterobranchs, and the acorn worms (enteropneusts) such as *Balanoglossus*. Lack both ENDOSTYLE and any homologue of NOTOCHORD. Possess *proboscis pore* (detectable also in cephalochordates and craniates) and gill slits. Nerve cord(s) usually solid. Development indirect; the enteropneust larva is the *tornaria*.

HEMICRYPTOPHYTES. Class of RAUNKIAER'S LIFE FORMS.

HEMIDESMOSOME. See DESMOSOME.

HEMIMETABOLA. See EXOPTERYGOTA.

HEMIPTERA (RHYNCOTA). Large order of exopterygotan insects. Includes aphids, cicadas, bed bugs, leaf hoppers, scale insects. Of enormous economic importance. Usually two pairs of wings, the anterior pair either uniformly harder (*Homoptera*) or with tips more membranous than the rest of the wing (*Hemiptera*). Mouthparts for piercing and sucking. Many are vectors of pathogens.

HEMIZYGOUS. Term applied to cell or individual where at least one chromosomal locus is represented singly (i.e. its homologue is absent), in which case the locus is hemizygous. Sometimes a chromosome pair bears a non-homologous region (as in the HETEROGAMETIC SEX), or all chromosomes are present singly (as in HAPLOIDY).

HEPARIN. GLYCOSAMINOGLYCAN product of MAST CELLS; an anticoagulant, blocking conversion of prothrombin to thrombin. Reduces EOSINOPHIL degranulation. Stored with HISTAMINE in mast cell granules, and hence found in most connective tissues.

HEPATIC. (Adj.) Relating to the LIVER.

HEPATICAE. See HEPATICOPSIDA.

HEPATICOPSIDA (HEPATICAE). Liverworts. Class of BRYOPHYTA, whose sporophytes develop capsule maturation and undergo meiosis before the seta elongates. Consist of a thin prostrate, or creeping to erect body (thallus), a central stem with three rows of leaves, attached to the substratum by rhizoids. Sex organs antheridia and archegonia, variously grouped; microgametes flagellated and motile. Fertilization is followed by development of a capsule containing spores, which germinate on being shed to form most usually a short thalloid PROTONEMA from which new liverwort plants arise. Includes leafy and thallose species. Generally occur in moist soils, on rock, or epiphytically. Rarely aquatic. See LIFE CYCLE.

HEPATIC PORTAL SYSTEM. System of veins and capillaries conveying most products of digestion (not CHYLOMICRONS) in cephalo-chordates and vertebrates from the gut to the liver. Being a portal system, it begins and ends in capillaries.

HERB. Plant with no persistent parts above ground, as distinct from shrubs and trees.

HERBACEOUS. Having the characters of a herb.

HERBARIUM. Collection of preserved and diverse plant specimens, usually arranged according to a classificatory scheme. Used as a reference collection for checking identities of newly collected specimens, as an aid to teaching, as a historical collection, and as data for research.

HERBIVORE. Animal feeding largely or entirely on plant products. See FOOD WEB, CARNIVORE, OMNIVORE.

HEREDITARY. (Adj.) Of materials and/or information passed from individuals of one generation to those of a future generation, commonly their direct genetic descendants. Hereditary and genetic material are not identical: an egg cell, e.g., contains a great deal of cytoplasm that is non-genetic; material passed from mother to embryo across a placenta might also be termed hereditary but not genetic.

HERITABILITY. Roughly speaking, the degree to which a character is inherited rather than attributable to non-heritable factors; or, that component of the variance (in the value) of a character in a population which is attributable to genetic differences between individuals. Estimation of heritability is complex. May be regarded as the ratio of additive genetic variance to total phenotypic variance for the character in the population, where *additive genetic variance* is the variance of breeding values of individuals for that character, and where *breeding value* (which is measurable) is twice the mean deviation from the population mean, with respect to the character, of the progeny of an individual when that individual is mated to a number of individuals chosen randomly from the population.

HERMAPHRODITE (BISEXUAL). (Bot.) (Of a flowering plant or flower) having both stamens and carpels in the same flower. Compare UNISEXUAL, MONOECIOUS. (Zool.) (Of an individual animal) producing both sperm and ova, either simultaneously or sequentially. Does not imply self-fertilizing ability, but if self-compatible such individuals would probably avoid the COST OF MEIOSIS. Commonly, but not exclusively, found in animals where habit makes contact with other individuals unlikely (e.g. many parasites) or hazardous. Rare in vertebrates. See OVOTESTIS, PARTHENOGENESIS.

HETEROCERCAL. Denoting type of fish tail (caudal fin) characteristic of CHONDRICHTHYES, in which vertebral column extends into dorsal lobe of fin, which is larger than the ventral lobe. Compare HOMOCERCAL.

HETEROCHLAMYDEOUS. (Of flowers) having two kinds of perianth segments (sepals and petals) in distinct whorls. Compare HOMO-CHLAMYDEOUS.

HETEROCHROMATIN. Parts of, or entire, chromosomes which stain strongly basophilic in interphase. Such regions are transcriptionally inactive and highly condensed. *Facultative heterochromatin* (as in inactivated X-chromosomes of female mammals) occurs in only some somatic cells of an organism and appears not to comprise repeat DNA sequences. The resulting animal may thus be a MOSAIC of cloned groups of cells, each with different heterochromatic chromosome regions. *Constitutive heterochromatin* (e.g. around CEN-TROMERES of human chromosomes, and CHROMOCENTRES of e.g. *Drosophila*) comprises condensed chromosome regions that occur in all somatic cells of an organism and often consist of DNA repeat sequences.

HETEROCHRONY. Changes during ONTOGENY in the relative times of appearance and rates of development of characters which were already present in ancestors. Sometimes regarded as inclusive of two distinct processes: *progenesis*, in which development is cut short by precocious sexual maturity; and *neoteny*, in which somatic development is retarded for selected organs and parts. See ALLOMETRY.

HETEROCYST. Specialized cell found in some blue-green algae (CYA-NOBACTERIA). Larger than the vegetative cells and, when seen with the light microscope, appear empty. Are formed from vegetative cells by dissolution of storage products and formation of a multilayered envelope outside the cell wall, accompanied by breakdown of thylakoids and formation of new internal membranes. Are involved in NITROGEN FIXATION and house the enzyme nitrogenase. They increase in numbers as the nitrogen content of the medium decreases. In addition, they have been reported to be able to germinate to form filaments, but this is unusual.

HETERODONT. See DENTITION.

HETERODUPLEX. The double helix (duplex) formed by annealing of two single-stranded DNA molecules from different original duplexes so that mispaired bases occur within it. Heteroduplex regions are likely to occur as a result of most kinds of RECOMBINATION involving breakage and annealing of DNA, and may be short-lived, for when heteroduplex DNA is replicated any mispaired bases should base-pair normally in the newly synthesized strands. Furthermore, most organisms have DNA REPAIR MECHANISMS of greater or lesser efficiency for correcting base-pair mismatches by excision/replacement. See DNA HYBRIDIZATION, recA.

HETEROECIOUS. (Of rust fungi, Order Uredinales of Basidiomycotina) having certain spore forms of the life cycle on one host species and others on an unrelated host species. Compare AUTOECIOUS.

HETEROGAMETIC SEX. The sex producing gametes of two distinct classes (in approx. 1 : 1 ratio) as a result of its having SEX CHROMOSOMES that are either partially HEMIZYGOUS (as in XY individuals) or fully hemizygous (as in XO individuals). This sex is usually male, but is female in birds, reptiles, some amphibia and fish, Lepidoptera, and a few plants. Sometimes the XY notation is restricted to organisms having male heterogamety, female heterogamety being symbolized by ZW (males here being ZZ). See HOMOGAMETIC SEX, SEX DETERMINATION, SEX LINKAGE.

HETEROGRAFT. See XENOGRAFT.

HETEROKARYOSIS. Simultaneous existence within a cell (or hypha or mycelium of fungi) of two or more nuclei of at least two different genotypes to produce a *heterokaryon*. Usually these nuclei are from different sources, their association being the result of plasmogamy between different strains; they retain their separate identities prior to karyogamy. Heterokaryosis is extremely common in coenocytic filamentous fungi. Ascomycotina and Basidiomycotina have a dikaryotic phase in their life cycles (see DIKARYON). See PARASEXUALITY.

HETEROMEROUS. (Of lichens) where the thallus has algal cells restricted to a specific layer, creating a stratified appearance.

HETEROMETABOLA. See EXOPTERYGOTA.

HETEROMORPHISM. Occurrence of two or more distinct (heteromorphic) morphological types within a population, due to environmental and/or genetic causes. Examples include genetic POLYMORPHISM, HETEROPHYLLY, and the phenotypically distinct phases of those life cycles, particularly in the algae (e.g. the green alga *Derbesia*), with ALTERNATION OF GENERATIONS.

HETEROPHYLLY. Production of morphologically dissimilar leaves on the same plant. Many aquatic plants produce submerged leaves that are much dissected and delicate, while leaves floating upon the water surface are simple and entire. See HETEROMORPHISM, PHENOTYPIC PLASTICITY.

HETEROPTERA. Suborder of HEMIPTERA.

HETEROPYCNOSIS. The occurrence of HETEROCHROMATIN. Heterochromatic regions were formerly called heteropycnotic.

HETEROSIS. See HYBRID VIGOUR.

HETEROSPOROUS. (Bot.) Of individuals or species producing two kinds of spore, microspores and megaspores, that give rise respectively to distinct male and female gametophyte generations. Examples are found in some clubmosses and ferns and all seed plants. See ALTERNATION OF GENERATIONS, LIFE CYCLE.

HETEROSTYLY. Condition in which the length of style differs in flowers of different plants of the same species, e.g. pin-eyed (long style) and thrum-eyed (short style) primroses (*Primula*). Anthers in one kind of flower are at same level as stigmas of the other kind. A device for ensuring cross-pollination by visiting insects. Compare HOMOSTYLY. See POLYMORPHISM, SUPERGENE.

HETEROTHALLISM. (Of algae, fungi) the condition whereby sexual reproduction occurs only through participation of thalli of two different MATING TYPES, each self-sterile. In fungi, includes *morphological heterothallism*, where mating types are separable by appearance, and *physiological heterothallism*, where interacting thalli (often termed plus and minus strains) offer no easily recognizable differences by which to distinguish them. Includes forms in which both thalli bear male and female sex organs and others which have no sex organs, union dependent on hyphal fusions. Compare HOMOTHALLISM.

HETEROTRICHOUS. (Of algae) having a type of thallus comprising a prostrate creeping system from which project erect branched filaments. Common in filamentous forms.

HETEROTROPHIC (ORGANOTROPHIC). Designating those organisms dependent upon some external source of organic compounds as a means of obtaining energy and/or materials. All animals, fungi, and a few flowering plants are *chemoheterotrophic* (chemotrophic), depending upon an organic carbon source for energy. Some autotrophic bacteria (purple non-sulphur bacteria) are *photoheterotrophic*, using solar energy as their energy source but relying on certain organic compounds as nutrient materials. Both these groups contrast with those organisms (*photoautotrophs*) able to manufacture all their organic requirements from inorganic sources, and upon which all heterotrophs ultimately depend. Thus all herbivores, carnivores, omnivores, saprotrophs and parasites are heterotrophs. See DECOMPOSER.

HETEROZYGOUS. Designating a locus, or organisms, at which the two representatives (alleles) in any diploid cell are different. Organisms are sometimes described as *heterozygous for* a character determined by those alleles, or *heterozygous at* the locus concerned. Thus, where two alleles *A* and *a* occupy the *A-locus*, of the three genotypes possible (*AA*, *Aa*, *aa*), *Aa* is heterozygous while the other two are HOMOZYGOUS. See DOMINANCE.

HETEROZYGOUS ADVANTAGE. Selective advantage accruing to heterozygotes in populations and which may be responsible for some balanced POLYMORPHISMS. Mutations in diploids arise in the heterozygous condition and must therefore confer an advantage in that state if they are to spread. Theory has it that if a mutation is advantageous, selection will make its advantageous heterozygous effects dominant and its deleterious effects recessive, thus giving heterozygotes a higher FITNESS than homozygotes. There are remarkably few well-documented examples in which the evidence for heterozygous advantage is conclusive. See SICKLE-CELL ANAEMIA. Compare BALANCED LETHAL SYSTEM.

HEXOSE. Carbohydrate sugar (monosaccharide) with six carbon atoms in its molecules. Includes glucose, fructose and galactose. Combinations of hexoses make up most of the biologically important disaccharides and polysaccharides.

HIBERNATION. Far-reaching physiological adaptation of some homeothermic animals to prolonged cold. Marked drops in body temperature (to perhaps $1°C$ above ambient temperature) and basal metabolic rate (down to 1% of normal) are associated with the ability to elevate these again if conditions become dangerously cold (contrast DIAPAUSE in insects). Common in mammalian orders Insectivora, Chiroptera and especially Rodentia. See ADIPOSE TISSUE, AESTIVATION.

HIGH-ENERGY PHOSPHATES. Group of phosphorylated compounds transferring chemical energy required for cell work. Depends upon their tendency to donate their phosphate group to water (to be hydrolysed) as is indicated by their STANDARD FREE ENERGIES of hydrolysis (the more negative it is, the greater the tendency to be hydrolysed). *Phosphate-bond energy* (not *bond energy*, the energy required to *break* a bond) indicates the difference between free energies of reactants and products respectively before and after hydrolysis of a phosphorylated compound.

Fig. 33 indicates tendencies for phosphate groups (shown by arrows) to be transferred between commonly occurring phosphorylated compounds of cells. High-energy phosphate bonds arise because of *resonance hybridity* between single and double bonds of the phosphorus/oxygen atoms, which renders them more stable (i.e. they have less free energy) than expected from structure alone. All common phosphorylated compounds of cells, including ADP and AMP, are resonance hybrids at the phosphate bond; but the terminal phosphate of AMP has a standard free energy less than half as negative as those of ADP and ATP.

HILL REACTION. Light-induced transport of electrons from water to acceptors such as potassium ferricyanide (*Hill reagents*) which do not occur naturally, accompanied by the release of oxygen. During it,

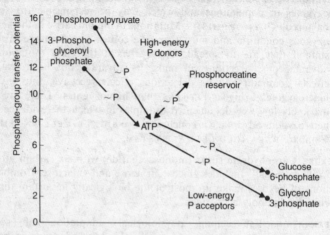

Fig. 33. *The flow of phosphate groups from high-energy donors to low-energy acceptors via ATP, assuming molar concentrations of reactants and products.*

electron acceptors become reduced. Named after R. Hill, who studied the process in 1937 in isolated chloroplasts. See PHOTOSYNTHESIS.

HILUM. Scar on seed coat, marking the point of former attachment of seed to funicle.

HINDBRAIN. Hindmost of the three expanded regions of the vertebrate brain as marked out during early embryogenesis, developing into the cerebellum and medulla oblongata. See BRAIN.

HIP GIRDLE. See PELVIC GIRDLE.

HIRUDINEA. Leeches. Class of ANNELIDA. Marine, freshwater and terrestrial predators and temporary ectoparasites with suckers formed from modified segments at anterior and posterior ends. Most remain attached to host only during feeding. Hermaphrodite; embryo develops directly within cocoon secreted by clitellum.

HISTAMINE. Potent vasodilator formed by decarboxylation of the amino acid *histidine* and released by MAST CELL degranulation in response to appropriate antigen. Degraded by *histaminase* released by EOSINOPHIL degranulation. Increases local blood vessel permeability in early and mild inflammation. Responsible for itching/sneezing during ALLERGY.

HISTIOCYTE. See MACROPHAGE.

HISTOCHEMISTRY. Study of the distributions of molecules occurring within tissues, within both cells and intercellular matrices. Besides direct chemical analysis, it involves sectioning, STAINING and AUTO-RADIOGRAPHY.

HISTOCOMPATIBILITY ANTIGEN. ANTIGEN (often cell-surface GLYCO-PROTEIN) initiating an immune response resulting in rejection of a HOMOGRAFT. See HL-A SYSTEM.

HISTOGENESIS. Interactive processes whereby undifferentiated cells from major GERM LAYERS differentiate into tissues.

HISTOLOGY. Study of tissue structure, largely by various methods of STAINING and MICROSCOPY.

HISTONES. Basic proteins of major importance in packaging of eukaryotic DNA. DNA and histones together comprise CHROMATIN, forming the bulk of the eukaryotic CHROMOSOME. Histones are of five major types: H1, H2A and H2B are lysine-rich; H3 and H4 are arginine-rich. H1 units link neighbouring NUCLEOSOMES while the others are elements of nucleosome structure. Prokaryotic cells lack histones.

HL-A SYSTEM. The most important human MAJOR HISTOCOMPATI-BILITY COMPLEX, located as a gene cluster on chromosome 6 and probably involving several hundred gene loci.

HnRNA. Heterogeneous nuclear RNA. See RNA PROCESSING.

HOLARCTIC. ZOOGEOGRAPHICAL REGION amalgamating the Palaearctic and Nearctic regions.

HOLOBLASTIC. Form of CLEAVAGE.

HOLOCARPIC. (Of fungi) having the mature thallus converted in its entirety to a reproductive structure. Compare EUCARPIC.

HOLOCENE (RECENT). The present, post-Pleistocene, epoch (system) of the QUATERNARY period.

HOLOCENTRIC. Of chromosomes with diffuse CENTROMERE activity, or a large number of centromeres. Common in some insect orders (Heteroptera, Lepidoptera) and a few plants (*Spirogyra, Luzula*).

HOLOCEPHALI. Subclass of the CHONDRICHTHYES, including the ratfish *Chimaera*. First found in Jurassic deposits. Palatoquadrate fused to cranium (*autostylic* jaw suspension). Grouped with elasmobranchs because of their common loss of bone.

HOLOCRINE GLAND. Gland in which entire cells are destroyed with discharge of contents (e.g. sebaceous gland). Compare APOCRINE GLAND, MEROCRINE GLAND.

HOLOENZYME. Enzyme/cofactor complex. See ENZYME, APO-ENZYME.

HOLOGAMODEME. See DEME.

HOLOMETABOLA. Those insects with a pupal stage in their life history. See ENDOPTERYGOTA. Compare THYSANOPTERA.

HOLOPHYTIC. Having plant-like nutrition; i.e. synthesizing organic compounds from inorganic precursors, using solar energy trapped by means of chlorophyll. Effectively a synonym of photoautotrophic. Compare HOLOZOIC; see AUTOTROPHIC.

HOLOSTEI. Grade of ACTINOPTERYGII which succeeded the chondrosteans as dominant Mesozoic fishes. Oceanic forms became extinct in the Cretaceous, but living freshwater forms include the gar pikes, *Lepisosteus*, and bowfins, *Amia*. Superseded in late Triassic and Jurassic by TELEOSTEI. Tendency to lose GANOINE covering to scales.

HOLOTHUROIDEA. Sea cucumbers. Class of ECHINODERMATA. Body cylindrical, with mouth at one end and anus at the other; soft, muscular body wall with skeleton of scattered, minute plates; no spines or pedicillariae; suckered TUBE FEET; bottom-dwellers, often burrowing; tentacles (modified tube feet) around mouth for feeding. Lie on their sides.

HOLOTYPE (TYPE SPECIMEN). Individual organism upon which naming and description of new species depends. See NEOTYPE, BINOMIAL NOMENCLATURE.

HOLOZOIC. Feeding in an animal-like manner. Generally involves ingestion of solid organic matter, its subsequent digestion within and assimilation from a food vacuole or gut, and egestion of undigested material via an anus or other pore. Compare HOLOPHYTIC.

HOMEOBOX. See HOMOEOBOX.

HOMEOSTASIS. Term given to those processes, commonly involving *negative feedback*, by which both positive and negative control are exerted over the values of a variable or set of variables, and without which control the system would fail to function.

(1) *Physiological*. Various processes which help regulate and maintain constancy of the internal environment of a cell or organism at appropriate levels. Each process generally involves: a) one or more sensory devices (misalignment detectors) monitoring the value of the variable whose constancy is required; b) an input from this detector to some effector when the value changes, which c) restores the value of the variable to normality, consequently shutting-off the original input (negative feedback) to the misalignment detector. In unicellular organisms homeostatic processes include osmoregulation by contractile vacuoles and movement away from unfavourable conditions of pH; in mammals (homeostatically sophisticated) the controls of blood glucose (see INSULIN, GLUCAGON), CO_2 and pH levels, and

its overall concentration and volume (see OSMOREGULATION), of ventilation, heart rate and body temperature provide a few examples.

(2) *Developmental*. Mechanisms which prevent the FITNESS of an organism from being reduced by disturbances in developmental conditions. The phrase *developmental canalization* has been used in this context.

(3) *Genetic*. Tendency of populations of outbreeding species to resist the effects of artificial SELECTION, attributable to the lower ability of homozygotes than heterozygotes in achieving *developmental homeostasis*.

(4) *Ecological*. Several ecological factors serving to regulate population density, species diversity, relative biomasses of trophic levels, etc., may be thought of as homeostatic. See DENSITY-DEPENDENCE, BALANCE OF NATURE, ARMS RACE.

HOMEOTHERMY. See HOMOIOTHERMY.

HOMEOTIC. See HOMOEOTIC.

HOME RANGE. That part of a habitat that an animal habitually patrols, commonly learning about it in detail; occasionally, identical to the animal's total range. Differs from a TERRITORY in that it is not defended, but may be geographically identical to it if it becomes a territory at some part of the year.

HOMINID. Member of the primate family Hominidae (superfamily Hominoidea), including mankind, whose most distinctive attribute is bipedalism – upright walking. Short face, small incisors and canines and tendency towards parabolic shape of dentary also characteristic. Probably originated in Upper Miocene/Pliocene. Includes the AUSTRALOPITHECINES, whose relationship with *Homo* requires more fossil material and analysis. Pre-molecular data suggest *Ramapithecus* as the earliest fossil ape with hominid affinities so far discovered; post-molecular data suggest a much more recent separation of the human lineage from the apes, at perhaps as little as 5 Myr BP. Some would even place African anthropoid apes (*Pan, Gorilla*) and humans in the same clade. See HOMO, PONGIDAE, RAMAPITHECUS.

HOMINOID. Member of the primate superfamily Hominoidea (see Fig. 34). Includes gibbons, the great apes and HOMINIDS. Distinguished from other CATARRHINE primates by widening of trunk relative to body length, elongated clavicles, broad iliac blades and broad, flat back. Normally no free tail after birth, when the spinal column undergoes curvature as an adaptation to partial or complete bipedalism. See ANTHROPOIDEA.

HOMO. Genus of CATARRHINE primates including mankind and its immediate relatives. The three species regularly (but not universally) recognized in the literature are *H. sapiens, H. erectus* and *H. habilis*, only the first of which, modern man, is extant. Usual to

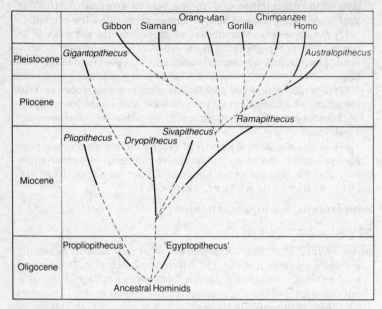

Fig. 34. Diagram of probable phylogeny of the Hominoidea. Dotted lines represent likely relationships although fossils have not been found to support them. Solid lines indicate fossil remains.

regard Neanderthal man as a subspecies of modern man, *H. sapiens neanderthalensis*. Characterized by a cranial capacity (brain volume) greater than 700 cm³, rising to in excess of 1600 cm³; post-orbital constriction never as marked as in *Australopithecus*; dental arcade evenly rounded with no diastema in most individuals; first lower premolar bicuspid; molar teeth smaller than in *Australopithecus*. Canines and incisors small. Pelvic girdle and hind-limb skeleton adapted for full bipedalism; hand capable of precision grip. The stock immediately ancestral to *Homo* is generally thought to have been AUSTRALOPITHECINE, possibly *A. africanus*.

Java Man (*H. modjokertensis*) was once regarded as a contemporary of *H. habilis*, but such an early date now seems unlikely. Various fossils are thought to form intermediates between *H. habilis* and *H. erectus* proper. *H. erectus* was probably widespread over Africa, S.E. Asia and, perhaps, Europe from about 1–0.35 Myr BP, but was probably restricted to Africa from about 1.6–1.0 Myr BP. Fossils from Peking (Peking man, 0.35 Myr BP), once regarded as forming a separate genus *Sinanthropus*, are now classified within the species *H. erectus*, which is widely, though not universally, accepted to have been ancestral to modern man, *H. sapiens*. The Neanderthals somehow intervened. There appear to have been two neanderthal stocks: a robust, extreme, form from the earlier part of the late Pleistocene of Europe and western Asia, and a more progressive form from Africa and eastern Asia. Some neanderthal skulls had a cranial capacity in excess of the average of modern man, albeit of somewhat different proportions. Their post-cranial skeletons overlap those of modern man in most morphological respects, but show an enhanced robustness and signs of adaptation to cold environments. The precise origins of anatomically modern man prior to 40 Kyr BP await clarification, but an African origin at about 100 Kyr BP has received recent fossil support. Modern humans were unknown in Europe or Australia prior to about 35 Kyr BP. See HOMINID.

HOMOCERCAL. Designating those outwardly symmetrical fish tails (caudal fins) in which the upper lobe is approximately of the same length as the lower. Typical of modern ACTINOPTERYGII; evolved from the HETEROCERCAL fin.

HOMOCHLAMYDEOUS. (Of flowers) having perianth segments of one kind (sepals) in two whorls. Compare HETEROCHLAMYDEOUS.

HOMODONT. See DENTITION.

HOMOEOBOX (HOMEOBOX). Conserved protein-encoding DNA sequence, first located in 1984 within several HOMOEOTIC gene sequences of *Drosophila* but since detected in the genomes of annelids, molluscs, echinoderms and vertebrates. In many cases, homoeobox products in very distantly related organisms share a high percentage of their amino acid sequences, and it tends to be these *homoeodomains* that have DNA-binding properties. These proteins tend therefore to be localized within nuclei. Expression of some homoeobox sequences may be instrumental in such developmental events as segmentation and cell determination so that their expression is often highly tissue-specific. There is evidence that some encode TRANSCRIPTION FACTORS. See COMPARTMENT.

HOMOEOTIC (HOMEOTIC). (1) Term describing a *control gene* which, by either being transcribed or remaining silent during development (according to decisions between alternative pathways of DEVELOP-MENT), can profoundly affect the developmental fate of a region of a plant or animal's body. As yet found only in insects, one nematode and a few plants A hierarchical sequence of binary decisions could provide clones of cells with 'genetic addresses' for differentiation. In the insect wing IMAGINAL DISC, such a decision sequence appears to be: anterior/posterior, dorsal/ventral, proximal/distal. A *homoeotic mutation* is a DNA sequence change comprehensively transforming its own and its descendant cells' morphologies into those of a different organ, in insects, normally one produced by a different imaginal disc. In *Drosophila*, examples include *engrailed*, *antennapedia* and *bithorax*.

Implications of homoeosis for HOMOLOGY are controversial. *Antennapedia* mutants have all or parts of their antennae converted into leg structures. Since antennae are regarded as the paired appendages of the second embryonic somite, their homology with and evolutionary development from paired ambulatory appendages receives some support from this source. The same applies to the segmentally-arranged mouthparts.

(2) Of organs whose positions are altered as a result of homoeotic mutation. See COMPARTMENT, HOMOEOBOX, IMAGINAL DISC.

HOMOGAMETIC SEX. The sex producing gametes which are uniform with respect to their SEX-CHROMOSOME complement. In mammals, this is the female sex (XX); in birds, reptiles and lepidopterans it is the male sex (ZZ). See HETEROGAMETIC SEX, SEX DETERMINATION.

HOMOGAMY. Condition in which male and female parts of a flower mature simultaneously. Compare DICHOGAMY.

HOMOGRAFT. See ALLOGRAFT.

HOMOIOTHERMY (HOMEOTHERMY). Maintenance of a constant body temperature higher than that of the environment. Involves physiological HOMEOSTASIS. Characteristic of mammals and birds. Some fish are able to keep some muscles considerably warmer than the surrounding water, and there is considerable evidence that many large extinct reptiles were homoiothermic. See POIKILOTHERMY.

HOMOKARYON (HOMOCARYON). Cell, fungal hypha or mycelium with more than one genetically identical nucleus in its cytoplasm; in fungi such nuclei commonly are haploid. See COENOCYTIC.

HOMOLOGOUS. (1) For homologous characters, see HOMOLOGY. (2) Homologous chromosomes: those capable, at least potentially, of pairing up to form BIVALENTS during first prophase of MEIOSIS, having approximately or exactly the same order of loci (but not normally of alleles). A normal diploid cell has pairs of such chromosomes. The chromosomes of a haploid gamete are normally homologous with those in the haploid gamete with which it fuses. The term may still be applied to chromosomes which are heterozygous for an INVERSION; but where translocation or reciprocal translocation have occurred the resulting chromosomes may be HEMIZYGOUS or only *partially homologous* with normal chromosomes. See ANEUPLOID, POLYPLOIDY.

HOMOLOGY. A controversial term. In evolutionary biology denotes common descent. Two or more structures, developmental processes, DNA sequences, behaviours, etc., usually occurring in different taxa, are said to be *homologous* if there is good evidence that they are derivations from (or identical to) some common ancestral structure, developmental process, etc. Very often a structure, etc., serving one function in one taxon has come, often with modification, to serve a different function in another. Two of the EAR OSSICLES of mammals are homologous with the articular and quadrate bones of ancestral reptiles; the vertebrate THYROID gland is almost certainly homologous with the ENDOSTYLE of urochordates, etc. The term may be applicable even when, as with the vertebrate PENTADACTYL LIMB, comparable structures do not always arise from the same embryonic segment. Where structures are repeated along the organism with little or no modification, as occurs in METAMERISM, they are termed *serially homologous* structures. Much evidence for homology is likely to come from work on HOMOEOTIC mutants. See ANALOGOUS.

In PHYLOGENETICS, or CLADISTICS, two characters, etc., are homologous if one (the *apomorphic character*) is derived directly from the other (the *plesiomorphic character*). The relationship is often termed special homology. Some workers in cladistics equate homology with SYNAPOMORPHY. In molecular biology, the term often indicates a significant degree of sequence similarity between DNA or protein sequences.

HOMONOMY. Characters of two or more taxa which have the same development, are found on different parts of the organism, and whose developmental pathways have a common evolutionary origin. E.g. each mammalian hair is *homonomous* with all other mammalian hair.

HOMOPLASY. In CLADISTICS, the term used to denote parallel or convergent evolution of characters.

HOMOPTERA. See HEMIPTERA.

HOMOSPOROUS. Having one kind of spore giving rise to gametophytes bearing both male and female reproductive organs, e.g. many ferns. Compare HETEROSPOROUS. See LIFE CYCLE.

HOMOSTYLY. Usual condition in which flowers of a species have styles of one length, as opposed to HETEROSTYLY.

HOMOTHALLISM. (Of algae, fungi) the condition whereby thalli are morphologically and physiologically identical, so that fusion can occur between gametes produced on the same thallus. Compare HETEROTHALLISM.

HOMOZYGOUS. Any LOCUS in a diploid cell, organism, etc., is said to be homozygous when the two ALLELES at that locus are identical. Organisms are said to be homozygous for a character when the locus determining that character is homozygous. Homozygous mutations are normally expressed phenotypically, unless the genetic background dictates otherwise (see PENETRANCE). Characters which are RECESSIVE are only expressed in the homozygous (or HEMIZYGOUS) condition. See HETEROZYGOUS.

HORMOGONIUM. Short piece of filament, characteristic of some filamentous blue-green algae, that becomes detached from the parent filament and moves away by gliding, eventually developing into a separate filament. Several may develop from one filament.

HORMONE. Term once applied in both botanical and zoological contexts, but now restricted to the latter (see GROWTH SUBSTANCE). Denotes any molecule, usually of small molecular mass, secreted directly into the blood by ductless glands and carried to specific target cells/organs by whose response they bring about a specific and adaptive physiological response. The term *chemical messenger* is still sometimes used in this context, SECOND MESSENGERS being molecular signals produced within, but not exported by, a cell. Neurotransmitters and neurosecretions customarily fall outside the compass of the term hormone, a distinction blurred by NEURO-HAEMAL ORGANS. Hormones tend to be either water-soluble peptides and proteins, or steroids. The latter have the longer physiological half-lives and are hydrophobic, being rendered soluble by binding to specific transport proteins (see NUCLEAR RECEPTORS, TRANSCRIPTION FACTORS). In this form they may enter nuclei to bring about selective expression and typically mediate long-term responses. Water-soluble hormones commonly bind to receptor sites on cell membranes (see ADENYLATE CYCLASE) and tend to mediate short-term responses. Examples of hormones include ADRENALINE, ECDYSONE, THYROXINE, INSULIN, TESTOSTERONE and OESTROGEN. See ENDOCRINE SYSTEM, PITUITARY GLAND, PROSTAGLANDINS, CASCADE.

HORN. Matted hair or otherwise keratinized epidermis of mammal, surrounding a knob-like core arising from a dermal bone of the skull. Neither the core nor the keratinized sheath is ever shed, nor do they ever branch, unlike ANTLERS.

HORSETAILS. See SPHENOPHYTA.

HOST. (1) Organism supporting a PARASITE in or on its body and to its own detriment. A *primary* (*definitive*) *host* is that in which an animal parasite reproduces sexually or becomes sexually mature; a *secondary* (*intermediate*) *host* is that in which an animal parasite neither reproduces sexually nor attains sexual maturity, but which generally houses one or more larval stages of the parasite. (2) Organism supporting (e.g. housing) a non-parasitic organism such as a commensal. See SYMBIOSIS.

HOST RESTRICTION. See PHAGE RESTRICTION.

HUMAN. See HOMO.

HUMAN CHORIONIC GONADOTROPHIN (HCG). Peptide hormone produced by developing human blastocyst and PLACENTA, prolonging the period of active secretion of oestrogens and progesterone by the CORPUS LUTEUM, which otherwise atrophies, inducing menstruation. Its presence in urine is usually diagnostic of early pregnancy. See MENSTRUAL CYCLE.

HUMAN PLACENTAL LACTOGEN (HPL). Hormone produced by human placenta after about five weeks of pregnancy. Most important effect is to switch maternal metabolism from carbohydrate to fat utilization. INSULIN antagonist.

HUMERUS. Bone of tetrapod fore-limb adjoining PECTORAL GIRDLE proximally, and both radius and ulna distally. See PENTADACTYL LIMB.

HUMORAL. Transported in soluble form, particularly in blood, tissue fluid, lymph, etc. Often refers to hormones, antibodies, etc. See HUMORAL IMMUNITY.

HUMORAL IMMUNITY. Immunity due to soluble factors (in plasma, lymph or tissue fluid). Production of ANTIBODY constitutes *humoral response* to an antigen. Contrasted with *cellular response* (see IMMUNITY).

HUMUS. Complex organic matter resulting from decomposition of dead organisms (plants, animals, decomposers) in the soil giving characteristic dark colour to its surface layer. Colloidal (negatively charged), improving cation absorption and exchange and preventing leaching of important ions, thus acting as a reservoir of minerals for plant uptake; water-retention of soil also improved. See SOIL PROFILE.

HYALURONIC ACID. Non-sulphated GLYCOSAMINOGLYCAN of D-glucuronic acid and N-acetylglucosamine; found in extracellular matrices of various connective tissues.

HYALURONIDASE. Enzyme hydrolysing hyaluronic acid, decreasing its viscosity. Of clinical importance in hastening absorption and diffusion of injected drugs through tissues. Some bacteria and leucocytes produce it. Reptile venoms and many sperm ACROSOMES contain it.

H-Y ANTIGEN (H-W ANTIGEN). Minor HISTOCOMPATIBILITY ANTIGEN encoded by locus on Y sex chromosome of most vertebrates (W sex chromosome in birds), and responsible for rejection of tissue grafted on to animal of opposite sex. Not now thought to be a product of the gene for testis-determining factor (TDF). See SEX DETERMINATION, SEX REVERSAL GENE.

HYBRID. In its widest sense, describes progeny resulting from a cross between two genetically non-identical individuals. Commonly used where the parents are from different taxa; but the term also has wider

applicability as with *inversion hybrids* (inversion heterozygotes) where
offspring are simply heterozygous for a chromosome INVERSION.
Where parents of a hybrid have little chromosome homology, particularly where they have different chromosome numbers, hybrid
offspring will be sterile (e.g. *mules*, resulting from horse × donkey
crosses) through failure of chromosomes to pair during MEIOSIS,
although one offspring sex may be partially or completely fertile.
Hybrid sterility is one factor maintaining species boundaries, and
selection against hybrids is a major factor in theories of SPECIATION.
Mitosis in hybrid zygotes is unlikely to be affected by lack of
parental chromosome homology but development may be thwarted
by imbalance of gene products. Sometimes hybridization (especially
between inbred lines of a species) may produce HYBRID VIGOUR.
See ALLOPOLYPLOIDY.

HYBRID DYSGENESIS (HD), or DYSGENESIS. Infertility and other
defects arising from crossing of certain genetic strains, notably in
Drosophila where high sterility and increased chromosome mutability
occur among offspring from crosses between laboratory female and
wild male stocks. The non-reciprocal nature of these results and the
discovery of chromosomal mutational 'hot-spots' housing extra
DNA implicated chromosomal insertions, and transposable elements
are now known to be responsible. These HD insertions are integrated
into wild stock genomes, encoding repressor molecules which inhibit
transposition and hence mutagenesis. Repressor concentration would
thus be high in wild eggs, preventing transposition of elements
donated by wild males. Absence of HD insertions from laboratory
strains means that incoming wild HD elements find a repressor-free
environment in the egg, transpose readily and cause chromosome
damage and consequent sterility in offspring.

HYBRIDIZATION. (1) Production of one or more HYBRID individuals. (2) Molecular hybridization (see DNA HYBRIDIZATION).
(3) See CELL FUSION.

HYBRIDOMA. Clone resulting from division of hybrid cell resulting
from artificial fusion of a normal antibody-producing B-CELL with a
B-cell tumour cell. Technique involved in production of *monoclonal
antibodies*. See ANTIBODY.

HYBRID SWARM. Continuum of forms resulting from hybridization of
two species followed by crossing and backcrossing of subsequent
generations. May occur when habitat is disturbed or newly colonized,
as with the oaks *Quercus ruber* and *Q. petraea* in Britain. See IN-
TROGRESSION.

HYBRID VIGOUR (HETEROSIS). Increased size, growth rate, productivity, etc., of the offspring resulting usually from a cross involving
parents from different inbred lines of a species, or occasionally from

two different (usually congeneric) species. Possibly results from HETEROZYGOUS ADVANTAGE or, probably more generally, from fixation of different deleterious recessives in the inbreds.

HYBRID ZONE. Area (zone) between two populations normally recognized as belonging to different species or subspecies and occupied by both parental populations and their phenotypically recognizable hybrids. Existence of a narrow hybrid zone may indicate that the parent populations are distinct evolutionary species; a wide zone may indicate that they are geographical variants of the same evolutionary species. Not to be confused with INTROGRESSION. See SEMI-SPECIES.

HYDATHODE. Water-excreting gland occurring on the edges or tips of leaves of many plants. See GUTTATION.

HYDATID CYST. Asexual multiplicative phase of some tapeworms (e.g. *Echinococcus*) within the secondary host (e.g. man, sheep, pig) in which a fluid-filled sac produces thousands of secondary cysts (brood capsules), each of which buds off a dozen or so retracted scolices. In humans the cysts may become malignant and send *metastases* around the body with sometimes fatal results. An example of POLYEMBRY-ONY.

HYDRANTH. See POLYP.

HYDROCORTISONE. See CORTISOL.

HYDROGEN BOND. Electrostatic attraction forming relatively weak non-covalent bond between an electronegative atom (e.g. O, N, F) and a hydrogen atom attached to some other electronegative atom. Responsible in large measure for secondary, tertiary and quaternary PROTEIN structures, for BASE PAIRING between complementary strands of nucleic acid, for the cohesiveness and high boiling point of water.

HYDROID. Member of the HYDROZOA, in its polyp form.

HYDROLASE. Enzyme catalysing addition or removal of a water molecule. See HYDROLYSIS.

HYDROLYSIS. Reaction in which a molecule is cleaved with addition of a water molecule. Some of the most characteristic biochemical processes (e.g. digestion, ATP breakdown and other dephosphorylations such as those in respiratory pathways) involve hydrolysis reactions. Chemically it is the opposite of a CONDENSATION REACTION.

HYDROPHILY. Pollination by means of water.

HYDROPHYTE. (1) Plant whose habitat is water, or very wet places; characteristically possessing AERENCHYMA. Compare MESOPHYTE, XEROPHYTE. (2) Class of RAUNKIAER'S LIFE FORMS.

HYDROPONICS. System of large-scale plant cultivation developed from water-culture methods of growing plants in the laboratory. Roots are allowed to dip into a solution of nutrient salts, or else plants are allowed to root in some relatively inert material (e.g. quartz sand, vermiculite) irrigated with nutrient solution. The external environment is commonly kept artificially constant.

HYDROSERE. SERE commencing in water or otherwise moist sites.

HYDROTROPISM. TROPISM in which the stimulus is water.

HYDROXYAPATITE. Crystalline calcium phosphate. Mineral component of BONE. Used in column chromatography for eluting proteins with phosphate buffers.

5-HYDROXYTRYPTAMINE. See SEROTONIN.

HYDROZOA. Class of CNIDARIA containing hydroids, corals, siphonophores, etc. Usually there is an ALTERNATION OF GENERATIONS in the life cycle between a sessile polyp (hydranth) phase and a pelagic medusoid phase; but one or other may be suppressed. Most polyp forms (but not *Hydra*) are COLONIAL, showing division of labour between feeding and reproductive individuals. Gonads ectodermal, unlike those of SCYPHOZOA.

HYGROSCOPIC. Readily absorbing and retaining moisture; applies to chemical substances as well as plant cells and other structures responding to changes in humidity. In many legumes (Leguminoseae), fruit dispersal involves a hygroscopic mechanism.

HYMENIUM. Layer of regularly arranged spore-producing structures found in the fruit bodies of many fungi (e.g. Ascomycotina, Basidiomycotina).

HYMENOPTERA. Large and diverse order of endopterygote insects, including sawflies (Symphyta), bees, ants, wasps and ichneumon flies (Apocrita). Fore wings coupled with hind wings by hooks. Mouthparts typically for biting, but sometimes for lapping or sucking (as in bumble bees). Ovipositor used, besides egg-laying, for sawing (sawflies), piercing or stinging. Abdomen often constricted to form a thin waist, its first segment fused with metathorax. Larvae generally legless. Bees, ants and wasps often EUSOCIAL.

HYOID ARCH. Vertebrate VISCERAL ARCH next behind jaws. Dorsal part forms HYOMANDIBULAR; ventral part in adults forms *hyoid bone*, usually supporting tongue. Contains facial nerve; gives rise to many face muscles.

HYOMANDIBULAR. Dorsal element (bone or cartilage) of hyoid arch, taking part in jaw attachment in most fish (see HYOSTYLIC). Becomes columella auris (stapes) in tetrapods (see EAR OSSICLES).

HYOSTYLIC. Method of jaw suspension of most modern fishes. Upper jaw has no direct connection with the braincase and the jaw is supported entirely by the hyomandibular. This widens the gape. See AUTOSTYLIC, AMPHISTYLIC.

HYPERPARASITE. Organism living parasitically upon another parasite. Provoked Jonathan Swift's doggerel expressing supposed infinite regress.

HYPERPLASIA. Abnormal increase in amount of tissue by cell division, e.g. in tumour growth.

HYPERTONIC. Relational term expressing the greater relative solute concentration of one solution compared with another. The latter is *hypotonic* to the former. A hypertonic solution has a lower WATER POTENTIAL than one hypotonic to it, and has a correspondingly greater OSMOTIC PRESSURE. See ISOTONIC, EURYHALINE.

HYPERTROPHY. Often interchangeable with HYPERPLASIA. Sometimes used of enlargement of individual components of tissues, organs, etc., without increase in cell division. See REGENERATION.

HYPHA. Filament or thread of a fungal thallus, often vacuolated. Tubular, increasing in length by growth at its tip (near which most enzymes are secreted) and giving rise to new hyphae by lateral branching. May be septate (with cross-walls) or non-septate. See COENOCYTIC, MYCELIUM.

HYPOCOTYL. Part of a seedling stem, below the cotyledon(s).

HYPODERMIS. Layer of cells immediately below the epidermis of leaves of certain plants, often mechanically strengthened (e.g. in pine), forming an extra protective layer, or forming water-storage tissue.

HYPOGEAL. (Of cotyledons) remaining underground when the seed germinates; e.g. broad bean, pea. Contrast EPIGEAL.

HYPOGLOSSAL NERVE. See CRANIAL NERVES.

HYPOGYNOUS. See RECEPTACLE.

HYPONASTY. (Bot.) More rapid growth of lower side of an organ than upper side; e.g. in a leaf, resulting in upward curling of leaf-blade. Compare EPINASTY.

HYPOPHYSECTOMY. Surgical removal of the PITUITARY GLAND.

HYPOPHYSIS. See PITUITARY GLAND.

HYPOSTASIS. Suppression of expression of a (*hypostatic*) gene by another non-allelic gene. Compare RECESSIVE. See SUPPRESSOR MUTATION, EPISTASIS.

HYPOTHALAMUS. Thickened floor and sides of the third ventricle of the vertebrate forebrain (diencephalon). Its nuclei control many activities, largely homeostatic. It integrates the AUTONOMIC NERVOUS SYSTEM, with centres for sympathetic and parasympathetic control; receives impulses from the viscera. Ideally situated to act as an integration centre for the endocrine and nervous systems, secreting various RELEASING FACTORS into the PITUITARY portal system and neurosecretions into the posterior pituitary. Releases substances inhibiting release of releasing factors (e.g. see SOMATOSTATIN). Contains control centres for feeding and satiety – the latter inhibiting the former after feeding. In higher vertebrates, is a centre for aggressive emotions and feelings and for psychosomatic effects. Contains a thirst centre responding to extracellular fluid volume; helps regulate sleeping and waking patterns; monitors blood pH and concentration and, in homeotherms, body temperature. See NEUROENDOCRINE COORDINATION.

HYPOTHALLUS. Thin, shiny, membranous adherent film at base of a slime mould (MYXOMYCOTA) fruitification.

HYPOTHESIS. A temporary working explanation or conjecture, commonly based upon accumulated data, which suggests some general principle or relationship of cause and effect; a postulated solution to a problem that may then be tested experimentally. See NULL HYPOTHESIS.

HYPOTONIC. Of a solution with a lower relative solute concentration (higher WATER POTENTIAL) than another. See HYPERTONIC, EURYHALINE.

I

IAA. Indole-3-acetic acid. Most common GROWTH SUBSTANCE in plants, produced in apical meristems of shoots and tips of coleoptiles. One of several AUXINS.

IAN. Indole-3-acetonitrile. Natural plant AUXIN.

I-BAND. See STRIATED MUSCLE.

ICHTHYOSAURIA. Extinct reptilian order (subclass Euryapsida), as fossils from Triassic to Cretaceous. They were not dinosaurs (see ARCHOSAURS), but were contemporaries. Vertebral column curved downwards to form reverse heterocercal tail; legs modified into paddles, with addition of extra digits. Fleshy dorsal fin and upper tail lobe lacked skeletal support. Jaws with homodont dentition. Became increasingly streamlined for aquatic locomotion; convergence with porpoises, etc.

ICSH. Interstitial-cell stimulating hormone. See LUTEINIZING HORMONE.

IDENTICAL TWINS. See MONOZYGOTIC TWINS.

IDENTIFICATION KEYS. Keys used in discovering the name of a specimen are commonly constructed so as to lead the investigator through a sequence of choices between mutually exclusive character descriptions, so chosen as to eliminate all but the specimen under observation. The format is commonly DICHOTOMOUS. A disadvantage arises when not all characters of the dichotomous key are observable, as through damage or incompleteness. A *polyclave* overcomes this, placing reliance only upon characters observable in the specimen to hand. It commonly comprises a set of punch-cards, each representing a different form or state that a character can take. Each species within the set dealt with by the cards is located on a master sheet and is given a unique number representing its set of character-states. When sufficient cards with character descriptions appropriate to the specimen are held up together only one punch-hole remains open, and the number corresponding to that hole is the number in the master sheet which identifies the species (or other taxonomic unit being identified).

IDIOBLAST. (Bot.) Cell clearly different in form, structure or contents from others in the same tissue; e.g. cystolith-containing parenchyma cells.

IDIOGRAM. See KARYOTYPE.

IDIOTYPE. Antigenic constitution of the variable (V) region of an immunoglobulin molecule. See ANTIBODY, ANTIBODY VARIATION.

IgA. Monomeric or polymeric immunoglobulin, often dimeric (composed of two polypeptides). Most abundant in seromucous secretions such as saliva, milk, and such as occur in urogenital regions. See ANTIBODY.

IgD. Low titre immunoglobulin found attached to B-CELL membranes. Of uncertain function.

IgE. Low titre immunoglobulin located on basophil and mast cell surfaces. Possibly involved in immunity to helminths; also involved in asthma and hay fever hypersensitivity.

IGF. Insulin-like growth factor. See GROWTH HORMONE.

IgF. Class of monomeric immunoglobulin proteins with four subclasses (IgGl–4), accounting for at least 70% of human immunoglobulin titre. Each molecule contains two heavy and two light chains. The sole antitoxin class, and major antibodies of secondary immune responses (see B-CELL). The only antibody class to cross the mammalian PLACENTA. See ANTIBODY.

IgM. Class of large pentameric immunoglobulin molecules (five linked subunits), largely confined to plasma. Produced early in response to infecting organisms, whose surface antigens are often complex.

ILEUM. Region of mammalian small intestine closest to colon and developing from region occupied by embryonic yolk sac; not anatomically distinct from JEJUNUM.

ILIUM. Paired bone forming dorsal part of tetrapod PELVIC GIRDLE (present in rudimentary form even in fishes) and articulating with one or more sacral vertebrae.

IMAGINAL DISC. Organ-specific PRIMORDIA of holometabolous insects, derived from blastoderm and distributed mainly in larval thorax. Composed of imaginal cells. There are 19 such discs in *Drosophila* larvae, which evaginate at metamorphosis and differentiate largely into adult epidermal structures: eyes and antennae from one pair; front legs from another pair; wings and halteres from two more pairs; genitals from a midline unpaired disc, and so on. Most discs are DETERMINED in the larval insect and remain so through artificial subculture; but *transdetermination* of cells may occur. Much remains to be learnt of the evolution of this mode of development. Cues for differentiation of discs into adult tissues are apparently hormonal; but the origin of the determined state appears to depend upon POSITIONAL INFORMATION, expressed as genetic addresses. See HOMOEOTIC GENE, COMPARTMENT.

IMAGO. Sexually mature adult insect.

IMBIBITION. Tendency of COLLOIDS, and substances forming colloidal gels, to adsorb water passively (*colloidal imbibition*); often responsible for swelling of organs, such as seeds during germination.

IMBRICATE. (Of leaves, petals, etc.) closely overlapping.

IMMUNE TOLERANCE (IMMUNOLOGICAL TOLERANCE). Acquired inability to react to particular self- or non-self-antigens. Both B-CELLS and T-CELLS display tolerance, generally to their specific antigen classes. The concentration of antigen required to induce tolerance in neonatal B-cells is 100-fold less than for adult B-cells. Responsible for suppression of transplant rejection. First noticed in non-identical twin cattle which shared foetal circulations (i.e. were synchorial). See THYMUS.

IMMUNITY. Ability of animal or plant to resist infection by parasites and effects of other harmful agents. Essential requirement for survival, since most of these organisms are perpetually menaced by viruses, bacteria and fungi, or parasitic animals.

In animals there are two functional divisions of the immune system: *innate* (*non-specific*) *immunity* and *adaptive* (*specifically acquired*) *immunity*. The former includes physical and chemical barriers to entry of pathogens (e.g. lysozyme, mucus, intact skin/cuticle, sebum, colony-stimulating factors, stomach acid, ciliary respiratory lining, commensal gut competitors, non-lymphocytic leucocytes of the RETICULO-ENDOTHELIAL SYSTEM and neutrophils. Adaptive immune responses, unlike innate immune responses, differ in quality and/or quantity of response on repeated exposure to antigen: the primary response to antigen takes longer to achieve significant antibody titre than does the secondary response. They include *active natural immunity*, in which the animal's MEMORY CELLS respond to a secondary natural contact with antigen by multiplication and specific antibody release; and *active induced immunity*, in which a VACCINE (see also INOCULATION) initially sensitizes memory cells. *Passive immunity* may be either natural, as by acquisition of antibodies via the placenta or colostrum in mammals, or induced, usually via specific antibodies injected intravenously. There is sometimes a distinction made between *cell-mediated* (lymphocytic and phagocytic) responses and *humoral* (antibody) responses in immunity, but it is never clear-cut: cells are involved in initiation of antibody responses and cell-mediated responses are unlikely in the complete absence of antibody. Cell-mediated immunity tends nowadays to refer to any immune response in which antibodies play a relatively minor role. For distinctions between *primary* and *secondary immune responses*, see B-CELL. See ANTIGEN-PRESENTING CELL, T- CELL, INFLAMMATION.

Immunity in plants is due to structural features, such as a waxy surface preventing wetting and consequent development of pathogens, thick cuticles preventing entry of germ-tubes of fungal spores; or immunity may be protoplasmic, the protoplast being an unfavourable environment for further development of the pathogen (see PHYTO-ALEXINS); or it might be acquired immunity (in context of viral diseases) when the plant recovers from an acute disease, or when resistance to virulent strains is conferred by presence of avirulent ones. In the latter (non-sterile) cases, active virus persists in the recovered or protected plant. Freedom from a second attack of an acute disease, or protection from the effects of virulent strains, persists only as long as the plants are infected. Plants are not known to produce antibodies.

IMMUNIZATION. Process rendering an animal less susceptible to infection by pathogens, to toxins, etc. May involve use of VACCINE, or passive IMMUNITY through injection of appropriate antibodies. See INOCULATION.

IMMUNOELECTROPHORESIS. See ELECTROPHORESIS.

IMMUNOFLUORESCENCE. Use of antibodies, with a fluorescent marker dye attached, in order to detect whereabouts of specific antigens (e.g. enzymes, glycoproteins) by formation of antibody–antigen complexes which show up on appropriate illumination.

IMMUNOGLOBULIN (Ig). Member of one of five major classes of globin protein with antibody activity. See IgA, IgD, IgE, IgG, IgM.

IMMUNOLOGICAL MEMORY. See B-CELL, MEMORY CELLS.

IMMUNOLOGICAL TOLERANCE. See IMMUNE TOLERANCE.

IMPLANTATION (NIDATION). Attachment of mammalian BLASTOCYST to wall of uterus (endometrium) prior to further development, placenta formation, etc. In humans the blastocyst is small and penetrates the endometrium, passing into the subepithelial connective tissue (*interstitial implantation*). This involves breaking of JUNCTIONAL COMPLEXES between endometrial cells, and proteolytic enzymes may be secreted by the TROPHOBLAST to achieve this.

IMPRINTING. (1) Form of LEARNING, often restricted to a specific *sensitive period* of an animal's development, when a complex stimulus may appear to elicit no marked response at the time of reception but nonetheless comes to form a model whose later presentation (or something appropriately similar) elicits a highly significant response. Particularly prevalent in birds. *Filial imprinting* involves narrowing of preferences in social companion (e.g. to mother, or to artificial object, in ducklings); *sexual imprinting* involves the preferential directing of sexual behaviour towards individuals similar to those encountered early in life. (2) See GENOMIC IMPRINTING.

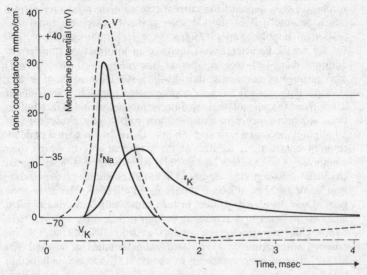

Fig. 35. The action potential curve (V) resulting from changes in sodium and potassium conductances (^{g}Na, ^{g}K) across an axon membrane at a point on its surface during propagation of an impulse.

IMPULSE. An ALL-OR-NONE RESPONSE comprising an ACTION PO-
TENTIAL propagated along the plasmalemma of an excitable tissue
cell, such as a nerve axon (between NODES OF RANVIER in myelinated axons) or muscle fibre. Impulses are initiated at synapses by
depolarizations of the postsynaptic membrane's resting potential
(usually internally negative by some 70 mv), generally brought about
by release of an excitatory neurotransmitter molecule (for general
details see ACETYLCHOLINE). This opens up Na^+- and K^+- *ligand-
gated channels* in the postsynaptic membrane and allows influx of
Na^+ and efflux of K^+ along their electrochemical gradients; Cl^--
channels remain closed if the transmitter is excitatory. Enzymic
degradation of the transmitter restores these channels to their closed
state, but current resulting from ion flow opens *voltage-gated* Na^+-
and K^+- channels in the adjacent membrane, and the flow of ions
which results causes depolarization, further current flow along the
membrane and further depolarization.

Voltage-gated channels close again when depolarization in their
region reaches its peak. These combined causes and effects result in
propagation of an action potential away from the site of original
depolarization. Because the K^+-channels open later than the Na^+-
channels and stay open longer, the action potential has the characteristics shown in Fig. 35.

There is a period of less than 1 ms, when the Na$^+$-channels are closing and the K$^+$-channels are open, when the membrane is unresponsive to a depolarizing current (the *absolute refractory period*) which because it is so short enables nerves to carry the rapid succession of impulses (up to 2500 s^{-1} for large diameter fibres, 250 s^{-1} for small diameter fibres) involved in information transfer. In addition there is a recovery period (the *relative refractory period*) after passage of an impulse during which stimuli must be of greater strength than normal to cause a propagated impulse. It lasts about 2 ms from the end of the absolute refractory period. On stimulation, individual nerve or muscle fibres respectively either conduct an impulse or contract, or they do not. There are no partial conductions or contractions because of the statistical way in which their membrane ion channels open: only when sufficient ligand-gated channels are open (the threshold level) does sufficient depolarization occur to open adjacent voltage-gated channels (the *all-or-none rule*). Their depolarizing effect in turn causes adjacent voltage-gated channels to open in a reiterated fashion (*accelerating positive feedback*). Unmyelinated nerves conduct at from 0.5–100 m. s^{-1}, increasing with diameter; myelinated axons conduct at around 120 m. s^{-1}. Rise in temperature up to about 40°C increases conduction rate.

Depolarization of muscle *sarcolemma* occurs through release of acetylcholine at NEUROMUSCULAR JUNCTIONS. Impulses are then propagated along the sarcolemma in just the same way as along nerves, but are carried inwardly to myofibrils by transverse tubules. See RESTING POTENTIAL, MUSCLE CONTRACTION, GATED CHANNELS.

INBREEDING. Sexual reproduction involving fertilization between gametes from closely related individuals, or in its most extreme form between gametes from the same (usually haploid or diploid) individual or genotype. Such *selfing* is not uncommon, even obligatory, in some plants. One end of a continuum, with OUTBREEDING at the other (see BREEDING SYSTEM). The process tends to produce homozygosity at loci (at all loci instantaneously in haploid selfing), with expected disadvantages from the expression of deleterious alleles and reduction in the level of genetic variance among offspring (see GENETIC VARIATION). However many plant populations which both outbreed and inbreed (e.g. *Viola*, violets, and gynodioecious species) may, by regular exposure to selection of rare alleles with recessive deleterious effects, be 'purged' of two such alleles for each death resulting from their expression. See ASSORTATIVE MATING, GENETIC LOAD, INBREEDING DEPRESSION.

INBREEDING DEPRESSION. Increase in proportion of debilitated or inviable offspring consequent upon INBREEDING.

INCISOR. Chisel-edged tooth of most mammals, occurring at the front of the dentary. Primitively three on each side, of both upper and lower jaws. Gnawing teeth of rodents (which grow continuously) and tusks of elephants are modified examples. Used for nipping, gnawing, cutting and pulling. See DENTAL FORMULA.

INCLUSION GRANULE. Microscopically visible bodies produced in the cytoplasms of many plant and animal cells, sometimes in the nucleus, as a result of viral infection. Often consist largely of virions, which may form crystals.

INCLUSIVE FITNESS. See FITNESS.

INCOMPATIBILITY. (Bot.) (1) In flowering plants, the failure to set seed (i.e. failure of fertilization and subsequent embryo development) after either self- or cross-pollination has occurred. It is due to the inability of the pollen tubes to grow down the style. (2) In physiologically heterothallic organisms, it is the failure to reproduce sexually in single or mixed cultures of the same mating type. Genetically determined in both cases, it prevents the fusion of nuclei alike with respect to alleles at one or more loci, thus preventing inbreeding. It is analogous to negative ASSORTATIVE MATING. (3) In horticulture, inability of the scion to make a successful union with the stock. (Zool.) The cause of rejection of a graft by the host organism through an immune response. See IMMUNITY.

INCOMPLETE DOMINANCE. See DOMINANCE.

INCOMPLETE FLOWER. Flower which lacks one or more of the kinds of floral parts, i.e. lacking sepals, petals, stamens or carpels.

INCUS. One of the mammalian EAR OSSICLES, homologous with the quadrate bone of other vertebrates.

INDEHISCENT. (Of fruits) not opening spontaneously to liberate their seeds; e.g. hazel nuts.

INDEPENDENT ASSORTMENT. (1) See MENDEL'S LAWS. (2) Events occurring in normal diploid MEIOSIS which cause one representative from each non-homologous chromosome pair to pass together into any gamete randomly, irrespective of the eventual genetic composition of the gamete. Results in random RECOMBINATION and is an important source of GENETIC VARIATION in eukaryotic populations. See ABERRANT CHROMOSOME BEHAVIOUR (*meiotic drive*).

INDETERMINATE GROWTH. Unrestricted or unlimited growth; continues indefinitely.

INDETERMINATE HEAD. Flat-topped INFLORESCENCE possessing sterile flowers with the youngest flowers in the centre.

INDICATOR SPECIES. Species whose ecological requirements are well understood and which, when encountered in an area, can provide valuable information about it. In palaeolimnology, for example, certain diatoms are invaluable indicator species enabling inferences to be made about past lake environments. Absence of an indicator species (e.g. a lichen) from an area where it might be expected to occur could be symptomatic of pollution or some other environmental impoverishment.

INDIGENOUS. Indicating an organism native to a particular locality or habitat.

INDIVIDUAL DISTANCE. In some vertebrates (particularly birds) and arthropods, the distance from its body within which an individual animal will not tolerate another of its own (and commonly of any other) species. Often results from a compromise between attraction towards other individuals and repulsion from them at short distances. See TERRITORY.

INDOLEACETIC ACID. See IAA.

INDUCIBLE ENZYME. Enzyme synthesized only when its substrate is present. See ENZYME, GENE REGULATION, JACOB-MONOD THEORY.

INDUCTION. In embryology, the process resulting from combined effects of EVOCATION and competence (see COMPETENT); results in production by one tissue (the *inducing tissue*) of a new cellular property in a dependently differentiating second tissue where the inducing tissue neither exhibits the resulting property nor alters its developmental properties as a result of the interaction. *Primary induction* events take place early in development; *secondary inductions* take place later in development.

INDUSIUM. Membranous outgrowth from undersurface of leaves of some ferns, covering and protecting a group of developing sporangia (a sorus).

INDUSTRIAL MELANISM. Occurrence, common in insects, of high frequencies of dark (*melanic*) forms of species in regions with high industrial pollution, where surfaces on which to rest are darkened by soot and where atmospheric SO_2 levels are high enough to prevent crustose lichen growth. A mutation darkening an individual will tend to be selected for (and hence come to predominate) in polluted regions since it will decrease the bearer's risk of falling prey to a visual predator; but in non-polluted parts of the species range the non-melanic form will be advantageous and occur with higher frequency. In the peppered moth, *Biston betularia*, heterozygous mutants collected in the mid-nineteenth century were paler than they are today, providing evidence in support of the theory that DOMINANCE is an evolving property of characters in populations of species.

Industrial melanism provides one of the best examples of evolution within species and of selection resulting in POLYMORPHISM; but not all melanism is necessarily adaptive against visual predation. *Thermal melanism* has been suggested in one ladybird (*Adalia bipunctata*), in which dark forms absorb more energy in regions where atmospheric soot lowers levels of incident solar radiation. They warm up earlier in the season, and gain a reproductive benefit by being mobile sooner than non-melanics. The precise roles of migration and predation on gene frequencies in industrial melanism have yet to be elucidated.

INFLAMMATION. Local response to injury in vertebrates; also involved in ALLERGY. Involves vasodilation and increased permeability of capillaries in damaged area due largely to release of HISTAMINE and serotonin from MAST CELLS. White blood cells, nutrients and fibrinogen enter and neutrophils are followed into the area by monocytes which become transformed into wandering MACROPHAGES for engulfing dead tissue, dead neutrophils, bacteria, etc. Fibrin forms from fibrinogen leaked into the tissues from blood, creating an insoluble network localizing and trapping invading pathogens, forming a fibrin clot preventing haemorrhage while isolating any infected region. *Pus* usually results following inflammation and comprises dead and living white blood cells and cell remains from damaged tissues.

INFLORESCENCE. Collective term for specific arrangement of flowers on an axis, grouped according to the method of branching, into: a) indefinite, or racemose; b) definite, or cymose. In a), branching is monopodial, inflorescences consisting of a main axis which increases in length by growth at its tip, giving rise to lateral flower-bearing branches. These open in succession from below upwards or, if the inflorescence axis is short and flattened, from the outside inwards. The following are recognized (see Fig. 36): RACEME, whose main axis bears stalked flowers; PANICLE, compound raceme, such as oat; CORYMB, raceme with flowers borne at the same level due to elongation of the stalk (pedicel) of lower flowers, e.g. candytuft; SPIKE, raceme with sessile flowers, e.g. plantain; SPADIX, spike with fleshy axis, e.g. cuckoo pint; CATKIN, spike of unisexual, reduced and often pendulous flowers, e.g. hazel, birch; UMBEL, raceme in which the axis has not lengthened, the flowers arising at the same point to form a head with the oldest outside and youngest at the centre, e.g. carrot, cow parsley; CAPITULUM, where the axis of the inflorescence is flattened and laterally expanded, with growing point in centre, and bearing closely crowded sessile flowers (florets), the oldest at the margin and youngest at the centre, e.g. dandelion.

In b), branching is sympodial and the main axis ends in a flower, further development taking place by growth of lateral branches, each behaving in the same way. The CYME is described as a MONO-CHASIUM when each branch of the inflorescence bears one other

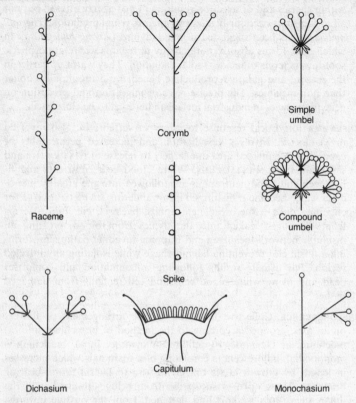

Fig. 36. *Diagrams of different types of inflorescence.*

branch (e.g. iris), and as a DICHASIUM when each branch produces two other branches (e.g. stitchwort). Inflorescences are often *mixed* (part indefinite, part definite): a raceme of cymes.

INFRASPECIFIC VARIATION. Variation within SPECIES. It takes several different forms. Clarification depends on population structure, BREEDING SYSTEM and effectiveness of gene flow of the particular case. *Clines* refer to variable phenotypic characters whose distributions display gradients mappable geographically on to gradients in environmental conditions. Morph ratio clines occur when the ratios of different morphs change in a similarly graded way. Phenotypic plasticity may be the cause but where genetic fixation is involved the variation may be closer to ecotypic. *Ecotypes* involve adaptation of populations to local edaphic, climatic or biotic influences. A *form* in botany is the category within the species generally applied to members showing trivial variations from normal (e.g. in petal colour). In zoology the term is often synonymous with morph (see below), or else a seasonal variant, or used as a neutral term when it is unclear whether a species, subspecies or lesser category is appropriate. A *morph* is one form of a polymorphic species population (see POLY-MORPHISM).

A *race* is a non-formal category used chiefly in zoological contexts. Geographical races are practically synonymous with subspecies (see later). Host races are those species populations with the same favoured hosts (if parasites) or food plants (when egg-laying, feeding, etc.); such preferences may involve various genetic and non-genetic influences. A *subspecies* is a formal taxonomic category used to denote various forms (types), commonly geographically restricted, of a polytypic species. It should ideally be used of evolutionary lineages rather than mere phenetic subdivisions of a species. Most easily applied when a population is geographically isolated from other populations of the species (e.g. on an island, mountain top). Sub-specific status is often conferred on populations which are really part of a clinal series for the characters used but where intermediate populations have not been studied. In some groups (e.g. Diptera) taxonomists have dispensed with the category. A *variety* is a formal category in botany below the level of subspecies and is used of groups which differ, for various reasons, from other varieties within the same subspecies.

INFUNDIBULUM. (1) Outpushing, or stalk, from floor of vertebrate forebrain attaching the PITUITARY to the hypothalamus. Its terminal swelling produces the posterior pituitary (neurohypophysis); its tissues combine with those growing up from the embryonic mouth to form the combined pituitary organ. (2) Anterior end of ciliated funnel of vertebrate oviduct.

INFUSORIA. Term formerly applied to rotifers, protozoa, bacteria, etc., found in cooled suspensions of boiled hay, etc.

INGUINAL. Relating to the groin.

INHIBITION. (Nervous). Prevention of activation of an effector through action of nerve impulses. Some inhibitory NEURO-TRANSMITTER molecules hyperpolarize rather than depolarize postsynaptic membranes at synapses thus reducing the probability of a propagated action potential at a synapse (see INHIBITORY POST-SYNAPTIC POTENTIAL). Alternatively, an inhibitory neurone may, by its activity, reduce the amount of excitatory neurotransmitter released by another neurone stimulating an effector. See SUMMATION.

INHIBITORY POSTSYNAPTIC POTENTIAL (IPSP). Hyperpolarization of a postsynaptic membrane at a synapse; brought about usually by release of a NEUROTRANSMITTER from the presynaptic membrane which fails to open ligand-gated Na^+- or K^+-channels, but instead opens Cl^--channels, making the inside of the cell more negative (polarized) than it was during the resting potential. Tends to inhibit formation of an ACTION POTENTIAL at a synapse. See SUMMATION.

INITIAL(S). (Bot.) Cell, or cells, from which tissues develop by division of differentiation, as in APICAL MERISTEMS; or a cell from which an antheridium develops in bryophytes.

INNATE IMMUNE RESPONSE. See IMMUNITY.

INNER CELL MASS. Group of cells formed after sinking inwardly from outer layer of the mammalian morula (blastocyst); determined by the 64-cell stage to become the future embryo rather than trophoblast.

INNER EAR. See EAR, INNER.

INNERVATION. Nerve supply to an organ.

INNOMINATE ARTERY. Short artery arising from aorta of many birds and mammals and giving rise to right subclavian artery (to fore-limb) and right carotid artery (to head).

INNOMINATE BONE. Each lateral half of the PELVIC GIRDLE when pubis, ilium and ischium are fused into a single bone as in adult reptiles, birds and mammals.

INOCULATION. Injection of living or otherwise mildly infective pathogen into a person or domestic animal followed usually by a mild but non-fatal infection which results in the patient's immunity to the virulent pathogen. Nowadays rarely used, immunization by non-infective agents being preferred. See IMMUNITY.

INOPERCULATE. Opening of a SPORANGIUM or ASCUS by an irregular tear or plug to liberate spores.

INOSINIC ACID (IMP). Purine nucleotide precursor of AMP and GMP. Also a rare monomer in nucleic acids where, being similar to guanine, it normally pairs with cytosine. Where it occurs at the 5′-end of an anticodon it may pair with adenine, uracil or cytosine in the 3′-end of the codon. See WOBBLE HYPOTHESIS.

INOSITOL. Water-soluble carbohydrate (a sugar alcohol) required in larger amounts than vitamins for growth by some organisms.

INOSITOL 1,4,5-TRIPHOSPHATE (InsP$_3$). A SECOND MESSENGER produced by phospholipase C activity as a breakdown product of the minor cell membrane phospholipid phosphatidylinositol. Hydrophilic, InsP$_3$ diffuses into the cytosol and initiates calcium ion (Ca^{++}) release from the endoplasmic reticulum (see CALMODULIN). Another second mesenger, diacylglycerol, is a product of the same phospholipase C activity.

INQUILINISM. See SYMBIOSIS.

INSECTA (HEXAPODA). Class of ARTHROPODA whose members have a body with distinct head, thorax and abdomen (see TAGMOSIS). Head bears one pair of antennae and paired mouthparts (mandibles, maxillae and a single fused labium); thorax bears three pairs of legs and frequently either one or two pairs of wings on second and/or third segments; abdomen bears no legs but other appendages may be present (e.g. see CERCI). Found as fossils from Devonian onwards. Most have a tracheal system with spiracles for gaseous exchange, and excretion by means of MALPIGHIAN TUBULES. METAMORPHOSIS either effectively lacking (APTERYGOTA), partial (EXOPTERYGOTA) or complete (ENDOPTERYGOTA). More numerous in terms of species and individuals than any other metazoan class.

INSECTIVORA. Order of placental mammals (e.g. moles, shrews, hedgehogs); a primitive insect-eating or omnivorous group resembling and probably phylogenetically close to Cretaceous ancestors of all placentals. Have small, relatively unspecialized teeth (but incisors tweezer-like). Tree shrews and elephant shrews tend nowadays to be placed in separate orders, Macroscelidia and Scandentia respectively.

INSERTION SEQUENCE (IS, I. ELEMENT). One sort of TRANSPOSABLE ELEMENT capable of inserting into bacterial chromosomes using enzymes they encode. Their ends form INVERTED REPEAT SEQUENCES. IS's may also occur in PLASMIDS (e.g. F FACTOR). Can mediate integration of plasmids into main bacterial chromosome by recombination, but cannot self-replicate; are therefore only inherited when integrated into other genomes which do have DNA replication origins. Can mediate a variety of DELETIONS, INVERSION and self-excision. See AUXOTROPH.

INSTAR. Stage between two ecdyses in insect development or the final adult stage.

INSTINCT. Behaviour which comprises a stereotyped pattern or sequence of patterns; typically remains unaltered by experience, appears in response to a restricted range of stimuli and without prior opportunity for practice. Distinction between this and learnt behaviour has been blurred by research in the last two decades: attention has focused upon developmental pathways of different behavioural responses. Even learnt behaviour presumably has some heritable component; the HERITABILITY of some behavioural patterns is high, and such behaviour tends still to be termed instinctive.

INSULIN. Protein hormone comprising 51 amino acids in two chains held together by disulphide bridges. Secreted by beta-cells of vertebrate pancreas in response to high blood glucose levels (e.g. after a meal) as monitored by the beta-cells themselves. Promotes uptake by body cells (especially muscle, liver) of free glucose; promotes enzymes converting glucose to glycogen (glycogenesis) and fatty acids (lipogenesis), and prevents breakdown of glycogen (glycogenolysis). Is thus a *hypoglycaemic* hormone – the only one in most vertebrates – reducing blood glucose. Opposed in its action by GLUCAGON, the two hormones together regulating and maintaining blood glucose at appropriate levels (about 100 mg glucose/100 cm³ blood in humans) through negative feedback via the pancreas. See DIABETES.

INTEGRATION. See NERVOUS INTEGRATION.

INTEGUMENT. (Bot.) (Of seed plants) outer cell layer or layers of ovule covering nucellus (megasporangium) and ultimately forming the seed coat. Most flowering plants have two integuments, an inner and an outer. (Zool.) Outer protective covering of an animal, such as skin, cuticle.

INTERCALARY. (Of a meristem) situated between regions of permanent tissue, such as at bases of nodes and leaves in many monocotyledons, or at junctures of stipes and blade in some brown algae.

INTERCALATED DISC. See CARDIAC MUSCLE.

INTERCELLULAR. Occurring between cells. Often applied to the matrix or ground substance secreted by cells of a tissue, as in CONNECTIVE TISSUES. For *intercellular fluid*, see TISSUE FLUID. See EXTRACELLULAR, INTERSTITIAL, INTRACELLULAR.

INTERCELLULAR JUNCTION. Any of a variety of cell–cell adhesion mechanisms, particularly abundant between animal epithelial cells, the three commonest of which are (1) DESMOSOMES, which are principally adhesive, (2) *gap junctions*, involved in intercellular communication, and (3) *tight junctions*, occluding the intercellular space, thereby restricting movements of solutes. Gap junctions consist of cylindrical channel proteins with a channel diameter of 1.5 nm, coupling cells electrically (as in electrical synapses and cardiac muscle) and in all probability metabolically (see PLASMODESMATA). Tight junctions perform a selective barrier function in cell sheets, preventing diffusion of ions, etc., from one side of the sheet to the other through intercellular spaces. This is essential to proper functioning of epithelia such as intestinal mucosae and proximal convoluted tubules of vertebrate kidneys. See Fig. 37.

INTERCELLULAR SPACE. (Bot.) In plants, air-filled cavities between walls of neighbouring cells (e.g. in cortex and pith) forming internal aerating system. Spaces may be large, making tissue light and spongy as in AERENCHYMA, occasionally harbouring algae, particularly blue-green algae.

INTERCOSTAL MUSCLES. Muscles between ribs of tetrapods which work in conjunction with the DIAPHRAGM during VENTILATION. *External intercostals* elevate ribs in quiet breathing; *internal intercostals* depress ribs aiding forced expiration.

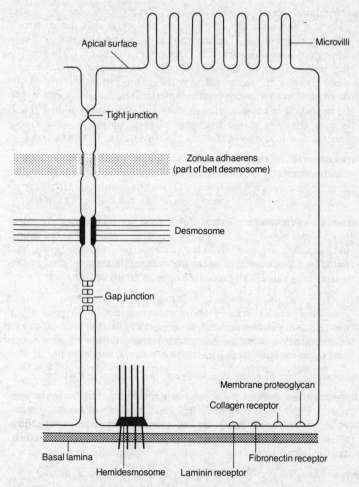

Fig. 37. *Some of the cell–cell adhesion mechanisms of epithelia.*

INTERFASCICULAR CAMBIUM. Vascular cambium arising between vascular bundles. Compare FASCICULAR CAMBIUM.

INTERFERONS. Proteins produced by virally infected animal cells and some lymphocytes, often in response to presence of double-stranded RNA in the cell. Prevent viral replication where antibodies do not penetrate. Extracellularly (as INTERLEUKINS) some can activate NATURAL KILLER CELLS and induce viral resistance in uninfected tissue.

INTERGRADATION ZONE. See HYBRID ZONE.

INTERLEUKINS (LYMPHOKINES). Soluble factors produced by helper T-CELLS and a variety of other non-leucocytic cells (hence term a misnomer); involved (1) in recognition of foreign antigens by T-cells while in contact with ANTIGEN-PRESENTING CELLS, (2) in amplifying proliferation of activated T-cells, (3) rendering macrophages more effective at phagocytosing pathogens, (4) in promoting HAEMOPOIESIS. B-CELLS can produce interleukins, but this seems not to be important in cell-mediated immune response. Interleukin-2 (IL-2) is used clinically in the treatment of melanoma and some kidney cancers. See COLONY-STIMULATING FACTOR, INTERFERONS.

INTERMEDIATE FILAMENTS. Insoluble, tough protein fibres appearing in eukaryotic cells where mechanical stress is applied. Diameters intermediate between actin filaments and microtubules. Help keep Z-discs of adjacent muscle sarcomeres in line. See CYTOSKELETON.

INTERNAL ENVIRONMENT. Extracellular fluid of metazoans, including tissue fluid, blood plasma and lymph. Its constancy of composition, pH, pressure, temperature, etc., particularly in homeotherms, was noted by Claude Bernard in the mid-nineteenth century; mechanisms maintaining this afford exquisite examples of HOMEOSTASIS.

INTERNEURONE (INTERNUNCIAL NEURONE, RELAY NEURONE). Neurone synapsing between sensory and motor neurones in a typical spinal REFLEX arc. Vertebrate interneurones are confined to GREY MATTER of central nervous system. Afford cross-connections with other neural pathways, enabling INTEGRATION of reflexes, and learning.

INTERNODE. Part of plant stem between two successive NODES.

INTEROCEPTOR. Receptor detecting stimuli within the body, in contrast with exteroceptor. Include BARORECEPTORS, PROPRIOCEPTORS, pH-receptors and receptors sensitive to concentrations of dissolved O_2 or CO_2 (see CAROTID BODY, CAROTID SINUS).

INTERPHASE. Interval between successive nuclear divisions, usually mitotic but also preceding or occasionally following meiosis. Somewhat misleading term suggesting a quiescent or resting interval in the CELL CYCLE (indeed the nucleus is often termed a resting nucleus during it). On the contrary, it is the period during which most components of the cell are continuously made. Cell mass generally doubles between successive mitoses.

INTERSEX. Individuals, often sterile and usually intermediate between males and females in appearance; sometimes hermaphrodite; resulting from failure of the mechanism of SEX DETERMINATION, often through chromosomal imbalance. A FREEMARTIN is an example where hormonal causes are involved. The discovery of intersexes in

Drosophila led to an understanding of its balance mode of sex determination. See GYNANDROMORPH, TESTICULAR FEMINIZATION.

INTERSPECIFIC. Between species; as in *interspecific* COMPETITION.

INTERSTITIAL. Lying in the spaces (interstices) between other structures, *interstitial cells* of vertebrate gonads lie either between the ovarian follicles or between the seminiferous tubules of the testis; the latter cells secrete the hormone TESTOSTERONE. See LUTEINIZING HORMONE.

INTERSTITIAL CELL STIMULATING HORMONE (ICSH). See LUTEINIZING HORMONE.

INTERSTITIAL FLUID. See TISSUE FLUID.

INTESTINE. That part of the ALIMENTARY CANAL between stomach and anus or cloaca. Responsible for most of the digestion and absorption of food and (usually) formation of dry faeces. In vertebrates, former role is often performed by the anterior *small intestine* (see DUODENUM, JEJUNUM, ILEUM), which commonly has a huge surface area brought about by a combination of (i) folds of its inner wall, (ii) VILLI, (iii) BRUSH BORDERS to the epithelial mucosal cells, (iv) SPIRAL VALVES, if present, (e.g. elasmobranchs) and (v), in herbivores especially, its considerable length. The more posterior *large intestine*, or *colon*, is usually shorter and produces dry faeces by water reabsorption. The junction between the two intestines is marked in amniotes by a valve, and often a CAECUM. Products of digestion are absorbed either into capillaries of the submucosa or, in CHYLOMICRONS, into lacteals of the lymphatic system. Such digestion as occurs in the caecum and large intestine is largely bacterial.

INTRACELLULAR. Occurring within a cell, which generally means within and including the volume limited by the plasma membrane. Contents of food vacuoles and endocytotic vesicles, although geographically within the cell, are not strictly intracellular until they have passed through the vacuole or vesicle membrane and into the cytosol. See ENDOSYMBIOSIS, GLYCOCALYX.

INTRASPECIFIC. Within a species. See DEME, INFRASPECIFIC VARIATION.

INTROGRESSION (INTROGRESSIVE HYBRIDIZATION). Infiltration (or diffusion) of genes of one species population into the gene pool of another; may occur when such populations come into contact and hybridize under conditions favouring one or the other, the hybrids and their offspring backcrossing with the favoured species population. See HYBRID, HYBRID SWARM.

INTROMITTANT ORGAN. Organ used to transfer semen and sperm into a female's reproductive tract. Include CLASPERS, PENIS.

INTRON. DNA sequence lying within a coding sequence (or its RNA transcript) and resulting in so-called 'split-genes'; common in enkaryotic genes but only a few examples in prokaryotes. Almost invariably they do not encode functional cell products and must be spliced out from RNA (pre-mRNA) during RNA PROCESSING to avoid translation into missense protein. Many of the vertebrate genes cloned to date contain several or many introns. DNA from T4 bacteriophage and from mitochondria of yeast and other fungi have *self-splicing* introns, but eukaryote introns are not self-splicing. It has been suggested that the progenote ancestor of all living groups of organisms had introns in its DNA, for while they have been found in ARCHAEBACTERIA, eubacteria appear to have lost them completely. See EXON, GENE DUPLICATION.

INTRORSE. (Of anther dehiscence) towards the centre of a flower, promoting self-pollination. Compare EXTRORSE.

INTUSSUSCEPTION. (Bot.) Insertion of new cellulose fibres and other material into an existing and expanding CELL WALL, increasing its surface area. Cellulose microfibrils are interwoven among those already existing, as opposed to being deposited on top of them. Compare APPOSITION.

INULIN. Soluble polysaccharide, composed of polymerized fructose molecules, occurring as stored food material in many plants, such as members of the Compositae and in dahlia tubers. Absent from animals.

INVAGINATION. Intucking of a layer of cells to form a pocket opening on to the original surface. Common in animal development, as during GASTRULATION.

INVERSION. (1) A type of chromosome MUTATION in which a section of chromosome is cut out, turned through 180° and rejoined, or spliced back, to the chromosome upside down. If long enough, it results in an *inversion loop* in POLYTENE or meiotic cells heterozygous for the inversion. The inverted region is stretched in order to pair up homologously with its partner, as shown in Fig. 38.

If CROSSING-OVER occurs within the inversion loop then the chromosomes which result are nearly always abnormal, having either deletions, duplications, too many centromeres or none at all (acentric). This normally results in reduced fertility. See SUPERGENE. (2) Hydrolysis of sucrose by INVERTASE to equimolar concentrations of glucose and fructose.

INVERT SUGAR. Equimolar mixture of glucose and fructose, usually resulting from digestion of sucrose by INVERTASE.

INVERTASE (SUCRASE). See SUCROSE.

INVERTEBRATE. Term designating any organism that is not a member of the VERTEBRATA. There are many invertebrate chordates.

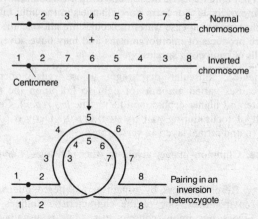

Fig. 38. *Diagram showing the effect of a large inversion upon pairing of a bivalent during meiotic prophase. Cross-overs within such an inversion hybrid (heterozygous for the inversion) results in acentric fragments and reduced fertility. Numbers indicate positions of loci.*

INVERTED REPEAT SEQUENCE. DNA sequences, often lying on either side of TRANSPOSABLE ELEMENTS and, when single-stranded, run in opposite directions along the chromosome (i.e. are palindromic) and may form a hairpin loop by folding back and base-pairing. Double-stranded inverted repeats are also found.

IN VITRO. 'In glass' (Latin). Biological process occurring, usually under experimental conditions, outside the cell or organism; e.g. in a test-tube.

IN VIVO. Biological process occurring within a living situation, e.g. in a cell or organism.

INVOLUCRE. A protective investment. (1) In thalloid liverworts, scale-like upgrowth of the thallus overarching the archegonia; (2) in leafy liverworts and mosses, groups of leaves surrounding the sex organs; (3) in many flowering plants (e.g. Compositae), group of bracts enveloping the young inflorescence.

INVOLUTION. (1) Decrease in size of an organ, e.g. thymus and other lymphoid tissue after puberty, contrasting with hyperplasia and hypertrophy. See ATROPHY. (2) Rolling over of cells during GASTRULATION, from the surface towards the interior of the developing gastrula. (3) Production of abnormal bacteria, yeasts, etc. (e.g. in old cultures).

IONOPHORE. One of a range of small organic molecules facilitating ion movement across a cell membrane (usually the plasma membrane). They either enclose the ion and diffuse through the membrane (e.g. valinomycin-K$^+$) or form pore-channels in the lipid bilayer (e.g. gramicidin), in which case water molecules are allowed through too. Some are products of microorganisms and may have adverse effects upon cells of competing species.

IRIS. Pigmented, muscular diaphragm whose reflex opening and closing causes varied amounts of light to fall upon the retinas of vertebrate and higher cephalopod EYES (the *iris reflex*). Contributes to depth of focus during ACCOMMODATION. Derived from fused CHOROID and retinal layers in vertebrates.

IRISH MOSS. Common name given to the red alga *Chondrus crispus*.

IRRITABILITY. Responsiveness by organism to altered internal and/or external environment: one of the characteristic abilities of living systems. Does not imply consciousness, but is usually a purely mechanical response, as with reverse of ciliary beat on contact with an object by *Paramecium*.

ISCHIUM. Ventral, back-projecting, paired bones of vertebrate PELVIC GIRDLE. They bear the weight of a sitting primate.

ISIDIUM. Rigid protuberance of upper part of a lichen thallus which may break off and serve for vegetative reproduction.

ISLETS OF LANGERHANS. Groups of endocrine cells scattered throughout the vertebrate PANCREAS; some of these cells secrete INSULIN, some GLUCAGON.

ISOANTIGEN. See ALLOANTIGEN.

ISOBILATERAL. (Of leaves) having the same structure on both sides. Characteristic of leaves of monocotyledons (e.g. irises), where leaf-blade is more or less vertical and the two sides are equally exposed. Compare DORSIVENTRAL.

ISODIAMETRIC. Having equal diameters; used to describe cell shape when length and width are essentially equal.

ISOELECTRIC POINT. The pH of solution at which a given protein is least soluble and therefore tends to precipitate most readily. At this pH the net charge on each of the protein molecules has the highest probability of being zero, and as a result they repel each other least in solution. They also tend not to move in an electric field, e.g. during ELECTROPHORESIS.

ISOENZYMES (ISOZYMES). Variants of a given enzyme within an organism, each with the same substrate specificity but often different substrate affinities (see MICHAELIS CONSTANT); separable by methods such as ELECTROPHORESIS. E.g. lactate dehydrogenase occurs in five different forms in vertebrate tissues, the relative amounts varying from tissue to tissue.

ISOGAMY. Fusion of gametes which do not differ morphologically, i.e. are not differentiated into macro- and microgametes. Compare ANISOGAMY.

ISOGENEIC (SYNGENEIC). Having the same genotype.

ISOGRAFT (SYNGRAFT). Graft between isogeneic individuals, such as identical twins, or mice of the same pure inbred line. Unlikely to be rejected. See GRAFT.

ISOKONT. (Bot.) Motile cell or spore possessing two flagella of equal length. Compare HETEROKONT.

ISOLATING MECHANISMS. Mechanisms restricting gene flow between species populations, sometimes of the same but usually of different species. Sometimes classified into *prezygotic mechanisms*, including any process (including behaviour) tending to prevent fertilization between gametes from members of the two populations, and *postzygotic mechanisms*, which prevent development of the zygote to maturity or render it partially or completely sterile. These mechanisms are likely to arise during geographical isolation (a non-biological isolating mechanism) of populations of the same species but they may be reinforced by selection during subsequent sympatry. Their role in sympatric speciation is under investigation. See SPECIATION.

ISOMERASE. Any enzyme converting a molecule to one of its isomers, commonly a structural isomer.

ISOMORPHIC. Used of ALTERNATION OF GENERATIONS, particularly in algae, where the generations are vegetatively identical. Compare HETEROMORPHIC.

ISOPODA. Order of the crustacean subclass MALACOSTRACA, containing such forms as aquatic waterlice (e.g. *Asellus*) and terrestrial woodlice (e.g. *Oniscus*). No carapace; body usually dorso-ventrally flattened. Females have a brood pouch in which young develop directly. Little division of labour between appendages. Many parasitic forms. About 4000 species.

ISOPTERA. Termites (white ants). Order of EUSOCIAL exopterygote insects, with an elaborate system of CASTES, each colony founded by a winged male and female; wings very similar, elongated, membranous and capable of being shed by basal fractures. Numerous apterous forms.

ISOTONIC. Of solutions having equal solute concentration (indicated by their *osmotic pressure*). If separated solely by selectively permeable membranes (e.g. cell membranes) there will be no net passage of water in either direction since they will have the same WATER

POTENTIAL. In general, whether or not water moves from one isotonic cell to another will depend upon their respective water potentials. See HYPERTONIC, HYPOTONIC.

ISOTYPE. (1) See ANTIBODY DIVERSITY. (2) Duplicate of type specimen, or HOLOTYPE.

ISOZYME. See ISOENZYME.

J

JACOB-MONOD THEORY. An influential theory of prokaryotic GENE EXPRESSION, of value in the explanation of eukaryotic gene expression. Its basic concept is that of the *operon*, a unit of TRANSLATION, comprising a group of adjacent structural genes (CISTRONS) on the chromosome, headed by a non-coding DNA sequence (the *operator*) whose configuration binds a protein (the *repressor*, or *regulator*) encoded by a regulator gene elsewhere on the chromosome (see Fig. 39). In one model, the repressor binds the operator preventing the enzyme RNA polymerase from gaining access to an adjacent DNA sequence (the *promoter*), which it must do if any of the structural genes of the operon are to be transcribed. The repressor–operator complex is stable and only broken if another molecule, the *inducer*, binds to it, in which case the inducer–repressor complex loses its affinity for the operator and transcription of the operon's cistrons results. This serves to explain prokaryotic enzyme induction, where the presence of substrate, acting directly or indirectly as inducer, is required for an enzyme to be produced by a cell. In one model of enzyme repression, the repressor does not bind to the operator until it has itself bound to some other molecule, the *corepressor* (e.g. enzyme product, or other gene product). Only then will transcription of the operon be inhibited. Mutants in the repressor gene, affecting repressor shape so that it cannot bind the operator or the corepressor, will result in CONSTITUTIVE production of the operon's cistron products. One observation which the theory helped explain was that in the bacterium *Escherichia coli* the enzymes encoded by what is now referred to as the *lac* operon were either all produced together or not produced at all. Other regulatory genes, especially in eukaryotes, encode TRANSCRIPTION FACTORS, which bind upstream of genes and assist binding of RNA polymerase to the promotor. See NUCLEAR RECEPTORS.

JAVA MAN. See HOMO.

JAWS. Paired (upper and lower) skeletal structures of GNATHOSTOMATA almost certainly deriving from the third pair of GILL ARCHES of an ancestral jawless (agnathan) vertebrate. Upper jaw (see MAXILLA) varies in its articulation with the braincase (see AUTOSTYLIC, AMPHISTYLIC and HYOSTYLIC JAW SUSPENSION). Progressive reduction in number of skeletal elements in lower jaw (see MANDIBLE) during vertebrate evolution, only the dentaries remaining in mammals. Tooth-bearing. See DENTITION.

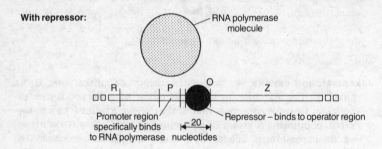

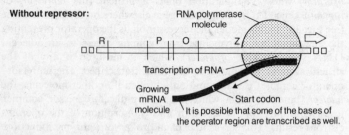

Fig. 39. Simplified diagram indicating the effect of a repressor molecule (above) in inhibiting transcription of gene Z by RNA polymerase compared with loss of this inhibition (below) when it is removed. Gene R encodes the repressor. © J. D. Watson: Molecular Biology of the Gene 3rd ed. (1976), Fig 14-12(a). Pub. Benjamin/Cummings.

JEJUNUM. Part of the mammalian small intestine succeeding the duodenum and preceding the ileum. Has larger diameter and longer villi than the rest of the small intestine, from which it is not anatomically distinct.

JELLYFISH. See SCYPHOZOA.

JOINTS. Articulations of animal endoskeletons or exoskeletons, the former definitive of arthropods, the latter characteristic of vertebrates where they may be either immovable (e.g. between skull bones), partly movable (e.g. between vertebrae) or freely movable (e.g. hinge joints, ball-and-socket joints). Commonly a feature of lever systems employing antagonistic muscles.

J-SHAPE GROWTH FORM. Type of population growth form, in which density increases rapidly in an exponential manner and then stops abruptly as environmental resistance or other limit takes effect more or less suddenly. See GROWTH CURVES, EXPONENTIAL GROWTH, S-SHAPE GROWTH FORM.

JUGULARS. Major veins in mammals and related vertebrates returning blood from the head (particularly the brain) to the superior vena(e) cava(e). Usually in the form of paired interior and exterior jugulars fusing on each side to form common jugulars before joining the subclavian veins and ultimately draining into the venae cavae.

JUNCTIONAL COMPLEX. See INTERCELLULAR JUNCTION.

JURASSIC. GEOLOGICAL PERIOD of Mesozoic era extending from about 195–135 Myr BP. A major part of the age of reptiles, during which the ARCHOSAUR radiation, begun in the Triassic, continued. Plesiosaurs and ichthyosaurs also flourished. Mammal-like reptiles (e.g. therapsids) dwindled; true mammals were scarce. Earliest fossil birds discovered (e.g. ARCHAEOPTERYX) were deposited in the Upper Jurassic.

JUVENILE HORMONE (NEOTENIN). Hormone produced by insect CORPORA ALLATA. See also ECDYSONE.

K

KAPPA PARTICLES. Gram-negative bacterial species present endosymbiotically (as commensals) within cytoplasm of ciliate *Paramecium aurelia*; their maintenance requires activity of some nuclear genes of host cell. May be transferred from one host cell to another during CONJUGATION, and may produce toxins (e.g. *paramecin*) which kill sensitive *Paramecium* strains but not the producer cells (termed *killer cells*). Inheritance of the killer trait is an instance of extrachromosomal or CYTOPLASMIC INHERITANCE.

KARYOGAMY. Fusion of nuclei or their components. Feature of eukaryotic sexual reproduction. See FERTILIZATION, HETERO-KARYOSIS. Compare PLASMOGAMY.

KARYOLYMPH. Nuclear ground substance in which chromosomes are embedded. Rarely used term.

KARYOTYPE. Characteristics of the set of chromosomes of a cell or organism (their number, sizes and shapes). A photograph or diagram of chromosomes, generally arranged in pairs and in order of size, is termed a *karyogram*.

KATABOLISM. See CATABOLISM.

KEEL (CARINA). (Zool.) Thin medial plate-like projection from sternum (breast-bone) of modern flying birds (*carinates*) and bats providing attachment for wing muscles. Absent from RATITES and many flightless birds. (Bot.) (1) Ridge alongside a fold applied to coalescent lower petals of a papilionaceous corolla of the pea family. (2) In some pennate diatoms, summit of a ridge bearing the raphe, where the valve is sharply angled at the raphe.

KELP. Common name for brown algae of the Order Laminariales.

KERATIN. Tough fibrous sulphur-rich protein of vertebrate epidermis forming resistant outermost layer of skin. See CORNIFICATION, CYTOSKELETON.

KETONE BODIES. Substances such as acetoacetate and hydroxybutyrate produced mainly in the liver from acetyl coenzyme A, itself an oxidation product of fatty acids, released for use by peripheral tissues as fuel. The metabolic pathway is termed *ketogenesis*.

KIDNEY. Major organ of nitrogenous EXCRETION and OSMOREGULA-TION in many animal groups of little or no homology. Its elements usually open directly to the exterior in invertebrates, but usually via a common excretory duct in vertebrates. Functional units in vertebrates, kidney tubules or *nephrons*, were probably originally paired in each trunk segment and drained through a pair of ducts, one on each side of the body (see WOLFFIAN DUCT). In higher vertebrates, anterior tubules (forming what remains of the *pronephros*) are embryonic and transitory, kidney function being normally dominated by the opisthonephros (*mesonephros* and *metanephros*), whose segmental organization is all but lost in the adult, a new excretory duct (the *ureter*) draining from the mass of nephrons into the bladder. The mesonephros is the functional kidney in adult fish and amphibians. In embryonic amniotes the two kidneys are initially mesonephric, lying in the trunk, their nephrons having glomeruli and coiled tubules; but the whole structure loses its urinary role during development and in males becomes invaded by the vasa efferentia. Each mesonephric duct gives off *ureteric buds* which grow into the intermediate mesoderm and develop into the collecting ducts, pelvis and ureter – in due course draining the *metanephros*. Each bud gives rise to a cluster of capillaries, a *glomerulus*, and a long tubule differentiating into *Bowman's capsule, proximal convoluted tubule, loop of Henle* and *distal convoluted tubule*, joining the collecting duct. Units developing from the cap tissue are termed *nephrons*, most of whose components lie in the kidney *cortex*, only the loops of Henle lying in the *medulla*, where they join the collecting ducts. Over a million nephrons may occur in each mammalian metanephros.

Hydrostatic pressure in the blood forces water and low molecular mass solutes (not proteins) out of the glomeruli into the Bowman's capsules. In mammals 80% of this glomerular filtrate is then reabsorbed across the cells of the proximal convoluted tubule by FACILITATED DIFFUSION and ACTIVE TRANSPORT into capillaries of the *vasa recta* draining the kidney. The descending and ascending limbs of the loop of Henle form a COUNTERCURRENT SYSTEM whose active secretion and selective permeabilities result in a high salt concentration in the interstitial fluid of the medulla, enabling water to be drawn back into the medulla osmotically from the collecting ducts if these are rendered permeable by ANTIDIURETIC HORMONE. Urea is also reabsorbed, but never against a concentration gradient, half being excreted on each journey through the kidneys. Kidneys play a major role in osmoregulation and help regulate blood pH by controlling loss of HCO_3^- and H^+. The remnants of glomerular filtrate from all collecting ducts comprise the URINE. See RENIN.

KILOBASE (Kb). 1000 bases or base pairs of nucleic acid.

KINASE. Enzyme transferring a phosphate group from a HIGH-ENERGY PHOSPHATE compound to a recipient molecule, often an enzyme, which is thereby activated and able to perform some function. Opposed by phosphatase activity, which removes the transferred phosphate group. See PHOSPHAGEN, PROTEIN KINASE, PHOS-PHORYLASE KINASE, ENTEROKINASE, CASCADE.

KINESIN. Ubiquitous and complex eukaryotic protein with the ability to bind separately to microtubules and organelles and then generate the force (through its ATPase activity) required to move the latter along the former. Kinesin-dependent organelle movement is usually centrifugal, unlike dynein-dependent movement. See CYCLOSIS, MYOSIN.

KINESIS. Movement (as opposed to growth) of an organism or cell in response to a stimulus such that rate of locomotion or turning depends upon intensity but not direction of the stimulus. Compare TAXIS, TROPISM.

KINETIN. A purine, probably not occurring naturally, but acting as a CYTOKININ in plants.

KINETOCHORE. Structure developing on CENTROMERE of chromosome, usually during late mitotic and meiotic prophases. MICROTUBULES appear to embed in it and possibly grow out from it.

KINETOPLAST. Organelle present in some flagellate protozoans (the Kinetoplastida, e.g. *Trypanosoma, Leishmania*) and containing sufficient DNA for visibility under light microscopy when suitably stained. Apparently a modified mitochondrion; commonly situated near the origin of a flagellum.

KINETOSOME. See CENTRIOLE.

KINGDOM. Taxonomic category with the greatest generality commonly employed, inclusive of phyla. There has been controversy over the number of kingdoms to employ, most nowadays favouring four: MONERA, ANIMALIA, PLANTAE, FUNGI. The term *Protista* has in the past been used to cover all unicellular organisms, but these can usually be reasonably satisfactorily housed in the kingdoms mentioned above. The term *Protoctista* has found favour with some, covering unicellular eukaryotes.

KININ. See CYTOKININ, BRADYKININ.

KINORHYNCHA. Class of minute marine ASCHELMINTHES (or a phylum in its own right) with superficial metameric appearance but without true segmentation. Share a syncytial hypodermal structure with GASTROTRICHA but unlike them lack external cilia. Cuticle covered with spines. Muscular pharynx similar to that of nematode worms. Nervous system a ring around the pharynx with four longitudinal nerve cords. Usually dioecious.

KIN SELECTION. Selection favouring genetic components of any behaviour (in its broadest sense) by one organism having beneficial consequences for another, and whose strength is proportional to the

RELATEDNESS of the two organisms. Contrast GROUP SELECTION. See HAMILTON'S RULE.

KLINEFELTER'S SYNDROME. Syndrome occurring in men with an extra X-chromosome giving genetic constitution XXY. Usually results from NON-DISJUNCTION and expresses itself by small penis, sparse pubic hair, absence of body hair, some breast development (gynaeco-mastia), small testes lacking spermatogenesis. Long bones often longer than normal. See TURNER'S SYNDROME, TESTICULAR FEMINIZATION.

K_M-VALUE. See MICHAELIS CONSTANT.

KRANZ ANATOMY (K. MORPHOLOGY). Wreath-like arrangement (*Kranz* being German for wreath) of palisade mesophyll cells around a layer of bundle-sheath cells, forming two concentric CHLOROPLAST-containing layers around the vascular bundles; typically found in leaves of C_4 plants, such as maize and other important cereals. See PHOTOSYNTHESIS.

KREBS CYCLE (CITRIC ACID CYCLE, TRICARBOXYLIC ACID CYCLE, TCA CYCLE). Cyclical biochemical pathway (see Fig. 40) of central impor-tance in all aerobic organisms, prokaryotic and eukaryotic. The reac-tions themselves contribute little or no energy to the cell, but de-hydrogenations involved are a source of electrons for ELECTRON TRANSPORT SYSTEMS via which ATP is produced from ADP and inorganic phosphate. In addition, some GTP is produced directly during the cycle. Cells with mitochondria perform the cycle by means of enzymes within the matrix bounded by the inner mito-chondrial membrane; in prokaryotes these enzymes are free in the cytoplasm. Bulk of substrate for the cycle is acetate bound to coenzyme A as acetyl CoA (see PANTOTHENIC ACID), but inter-mediates of the cycle can act as substrates. Acetate is usually derived from pyruvate produced by GLYCOLYSIS, amino acid oxidation or fatty acid oxidation, a decarboxylation and oxidation within the mitochondrion then generating carbon dioxide, acetate and reduced NAD. See ATP.

KRUMMHOLZ. Region between the alpine and tree lines, where trees are dwarfed and deformed owing to severe environmental conditions, particularly wind.

K-SELECTION. Selection for those characteristics which enable an organism to maximize its FITNESS by contributing significant num-bers of offspring to a population which remains close to its CARRY-ING CAPACITY, *K*. Such populations are characterized by intensely competitive and density-dependent interactions among adults, with few opportunities for recruitment by young. Adults invest heavily in growth and maintenance, have a small reproductive commitment and

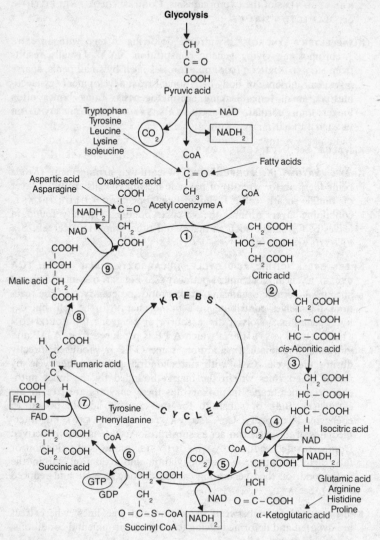

Fig. 40. Diagram of the Krebs cycle indicating where amino acids and fatty acids can enter the process.

generally have delayed maturity. See DENSITY-DEPENDENCE, *r*-SELECTION.

KUPFFER CELL. See RETICULO-ENDOTHELIAL SYSTEM.

L

LABELLING. Variety of indispensable techniques for detecting presence and/or movement of certain isotopic atoms both *in vitro* and *in vivo*. Isotopes used are either radioactive (RADIOISOTOPES) or differ in their atomic masses without being radioactive. Commonly used radioactive isotopes include ^{14}C, ^{35}S, ^{32}P and ^{3}H; non-radioactive 'heavy' isotopes include ^{15}N and ^{18}O. Material containing the unusual isotope is often administered briefly (pulse-labelling), and 'chased' by unlabelled material. The time taken for label to pass through the system and the route it takes contribute to our understanding of the dynamics of biological systems and processes (e.g. cell membranes, photosynthesis, aerobic respiration, DNA and protein synthesis) as well as of molecular structure. See AUTORADIOGRAPHY, SOUTHERN BLOT TECHNIQUE, DNA HYBRIDIZATION.

LABIAL. Referring to LABIUM, or to lip-like structure.

LABIATE PROCESS. Tube or opening through the valve wall of a diatom (BACILLARIOPHYTA) with an internal flattened tube or longitudinal slit often surrounded by two lips.

LABIUM. (Bot.) (1) Lower lip of flowers of the family Labiatae. (2) The lip subtending the ligule in the lycopod *Isoetes*. (Zool.) (1) An insect mouthpart forming the lower lip and comprising a single structure formed from a pair of fused second maxillae and bearing a pair of palps distally. Compare crustacean MAXILLIPED. (2) One member of two pairs of skin folds (*labia*) in female mammals protecting the exterior opening of the vagina.

LABRUM. Plate of exoskeleton hinged to front of the head in some arthropod groups and serving to enclose a space (cibarium) in front of the actual mouth. Found in insects and crustaceans. Trilobites had a similar structure, but placed ventrally on the head.

LABYRINTHODONTIA. Subclass of extinct amphibians found as fossils from Upper Devonian to Upper Triassic (see GEOLOGICAL PERIODS). Generally believed to have had rhipidistian ancestors.

LABYRINTHULALES. Net slime moulds. Order of colonial organisms; cells naked, spindle-shaped, secreting an extracellular membrane-limited matrix in the form of a network of tubes (slimeways) through which they move. Aquatic; mostly marine. Affinities uncertain.

lac OPERON. DNA sequence in the genome of the bacterium *Escherichia coli* comprising an *operator, promoter*, and three structural genes, transcribed into a single mRNA, encoding enzymes involved in lactate uptake and metabolism. See JACOB-MONOD THEORY.

LACTATION. Process of milk production by mammary glands. Involves hormone activity, notably of PROLACTIN, itself released by hypothalamic *prolactin releasing factor* reflexly secreted during sucking at nipples. Oestrogens and progesterone promote prolactin secretion but inhibit milk secretion; however after delivery the female sex hormone levels in the blood drop abruptly and this inhibition is removed. Sucking at the nipples reflexly releases OXYTOCIN from the posterior pituitary causing muscle contraction in the breast alveoli and milk letdown. See HUMAN PLACENTAL LACTOGEN.

LACTEAL. Lymph vessel draining a VILLUS of vertebrate small intestine. After digestion, reconstituted fats are released into the lacteal as CHYLOMICRONS.

LACTIC ACID. A carboxylic acid, $CH_3CHOHCOOH$, produced by reduction of pyruvate during anaerobic respiration (see GLYCOLYSIS). Many vertebrate cells can produce lactate, notably muscle and red blood cells. After transport in the blood it may be converted back by the liver to glucose during GLUCONEOGENESIS. Probably responsible for sensation of muscle fatigue. See OXYGEN DEBT. Also formed in metabolism of many bacteria (e.g. from lactose in souring of milk), as well as in fungi (e.g. Deuteromycotina), commonly in association with alcohol or acetaldehyde.

LACTOGENIC HORMONE. See PROLACTIN.

LACTOSE. Disaccharide found in mammalian milk. Comprises galactose linked to glucose by a $\beta.1$–4 glycosidic bond. A reducing sugar.

LACUNA. (Zool.) Any small cavity, such as those containing bone or cartilage cells.

LAGOMORPHA. Order of placental mammals including pikas, rabbits and hares. Gnawing herbivores differing from rodents chiefly in having two pairs of incisors as opposed to one pair in the upper jaw, the second pair being small and functionless.

LAG PHASE. See GROWTH CURVES.

LAMARCK, LAMARCKISM. Jean Baptiste de Lamarck (1744–1829) was a French natural philosopher who united a wide range of scientific interests under general principles. Coined the term *biology* in 1802, and worked in the newly-created National Museum of Natural History as professor of the zoology of the lower animals, although his previous work had been largely botanical. First to classify animals into invertebrates and vertebrates and to use such taxonomic categories as Crustacea and Arachnida. Lamarck's evolutionary views owed something to BUFFON's interest in modification of organisms by changes in their environments – changes which for Lamarck altered an organism's needs and habits. He held (as did Geoffroy St. Hilaire) that continuous spontaneous generation was required to restock the lowest life forms which had evolved into more complex ones.

For Lamarck the mechanism of evolutionary change was environmental: organs which assist an organism in its altered conditions are strengthened (e.g. the giraffe has acquired its high shoulders and long neck by straining to reach higher and higher into trees for leaves); others progressively atrophy through disuse. Such acquired characters, he thought, were then inherited. But there were and are no clear examples of such inheritance; nor does this theory account satisfactorily for evolutionary stability (stasis). It would require a theory of inheritance completely at variance with that receiving experimental support today. See CUVIER, DARWIN, WEISMANN, NEO-LAMARCKISM, MUTATION.

LAMELLA. Any thin layer or plate-like structure. (Bot.) (1) One of the spore-bearing gills in the fruiting body of a mushroom or related fungus, attached to the underside of the cap (pileus) and radiating from centre to margin. (2) One of a series of double membranes (thylakoids) within a chloroplast which bear photosynthetic pigments. (3) In bryophytes, a thin sheet of flap-like plates of tissue on dorsal surface of the thallus or leaves. (Zool.) One of the concentric layers of hard calcified material of compact bone forming part of a HAVERSIAN SYSTEM.

LAMELLIBRANCHIA. See BIVALVIA.

LAMELLIPODIUM. Sheet-like extension, or flowing pseudopodium, of the leading edges of many vertebrate cells during locomotion, some forming attachments with the substratum, others carried back in waves (ruffling, producing a *ruffled border*). Many give rise to microspikes. See CELL LOCOMOTION, LECTIN.

LAMINA. (1) Sheet or plate; flat expanded portion of a leaf or petal. (2) In brown algae (Phaeophyta), expanded leaf-like portion of the thallus.

LAMINA PROPRIA. Loose connective tissue of a MUCOUS MEMBRANE (e.g. of gut mucosa), binding the epithelium to underlying structures and holding blood vessels serving the epithelium.

LAMINARIN. Storage polysaccharide product of brown algae (PHAEOPHYTA); composed of $\beta.1$–3 linked glucans, containing 16 to 31 residues. May also be different degrees of branching. Occurs as an oil-like liquid outside chloroplasts, in a vesicle surrounding the pyrenoid.

LAMINAR PLACENTATION. Attachment of ovules over surface of carpel.

LAMPBRUSH CHROMOSOME. BIVALENTS during diplotene in some vertebrate (notably amphibian) oocytes in which long chromatin loops, which are transcriptionally very active, form at right angles to the chromosome axis and become covered with newly transcribed RNA. Not certain that much of this RNA acts as mRNA in protein synthesis; but if it is functional, the high transcriptional activity is presumably an adaptation to serving a relatively large cell from a single nucleus. See CHROMOSOME.

LAMPREYS. See CYCLOSTOMES.

LAMP SHELLS. See BRACHIOPODA.

LARGE INTESTINE. See COLON.

LARVA. Pre-adult form in which many animals hatch from the egg and spend some time during development; capable of independent existence but normally sexually immature (see PAEDOGENESIS, PROGENESIS). Often markedly different in form from adult, into which it may develop gradually or by a more or less rapid METAMORPHOSIS. Often dispersive, especially in aquatic forms. In insects especially, the phase of greatest growth in the life cycle. Examples include AMMOCOETE, CATERPILLAR, LEPTOCEPHALUS, NAUPLIUS, TADPOLE, TROCHOPHORE, VELIGER.

LARYNX. Dilated region at upper end of tetrapod trachea at its junction with the pharynx. 'Adam's Apple' of humans. Plates of cartilage in its walls are moved by muscles and open and close glottis. In some tetrapods, and most mammals, a dorso-ventral and membranous fold (vocal cord) within the pharynx projects from each side wall, vibrations of these producing sounds. Movement of the cartilage plates alters the stretch of the cords and alters pitch of sound. See SYRINX.

LASSO CELL. Cell type characteristic of CTENOPHORA, whose tentacles are armoured with these sticky thread-cells for capturing prey. Do not penetrate prey, unlike nematocysts of cnidarians.

LATE WOOD (SUMMER WOOD). Last part of the growth increment formed in growing season, containing smaller cells; more dense than early wood.

LATENT PERIOD (REACTION TIME). Time between application of a stimulus and first detectable response in an irritable tissue.

LATERAL LINE SYSTEM (ACOUSTICO-LATERALIS SYSTEM). Sensory system of fish and aquatic and larval amphibians whose receptors are clusters of sensory cells (*neuromast organs*) derived from ectoderm, found locally in the skin or within a series of canals, or grooves on head and body. Neuromasts resemble cristae of the VESTIBULAR APPARATUS of higher vertebrates, having a gelatinous cupula but lacking otoliths; they are probably homologous structures. Pressure waves in the surrounding water appear to distort the neuromasts, sending impulses via the vagus nerves on either side, where they associate with the *lateral lines* themselves – an especially pronounced pair of these sensory canals, one running the length of each flank. The head canals are served by the facial nerves.

LATERAL MERISTEMS. Meristems giving rise to secondary tissue; the vascular cambium and cork cambium.

LATERAL PLATE. See MESODERM.

LATEX. Fluid product of several flowering plants, characteristically exuding from cut surfaces as a milky juice (e.g. in dandelions, lettuce). Contains several substances, including sugars, proteins, mineral salts, alkaloids, oils, caoutchouc, etc.; rapidly coagulates on exposure to air. Function not clearly understood, but may be concerned in nutrition and protection, as well as in healing wounds. Latex of several species is collected and used in manufacture of several commercial products, the most important being rubber.

LATIMERIA. See COELACANTHINI.

LAURASIA. One of two great Upper Carboniferous land masses, the other being GONDWANALAND, formed by the breakup of Pangaea. Comprised what are now North America, Greenland, Europe and Asia. Originally straddling the Equator, it gradually moved northwards by continental drift. For much of the Jurassic and Cretaceous much of Europe was covered by the Tethys and Turgai Seas, the latter only drying up to link Europe and Asia about 45 Myr BP. Separation of Gondwanaland and Laurasia was completed by early Cretaceous (130 Myr BP) with the result that much of the later radiation of dinosaurs took place in Laurasian continents but not in Gondwanaland.

LAVER. General name given to the edible dried preparation made from the red alga *Porphyra* (Rhodophyta).

LDL (LOW-DENSITY LIPOPROTEIN). See CHOLESTEROL, LIPOPROTEIN.

L-DOPA (l-DIHYDROXYPHENYLALANINE). Intermediate in the pathway from phenylalanine to noradrenaline and immediate precursor of the brain neurotransmitter DOPAMINE. Dopamine is deficient from the caudate nuclei of the brain in patients with Parkinson's disease. While neither oral nor intravenous dopamine reaches the brain, *l*-dopa does so and is of widespread clinical use in treating parkinsonism, at least in the short term. Transplantation of embryonic adrenal medulla (production site for *l*-dopa) into the caudate nuclei of sufferers from parkinsonism is under clinical test but raises ethical issues.

LEAF. Major photosynthesizing and transpiring organ of bryophytes and vascular plants; those of the former are simpler, non-vascular and not homologous with those of the latter, which consist usually of a leaf stalk (*petiole*), attached to stem by LEAF BASE, the leaf blade typically lying flattened on either side of the main vascular strand, or *midrib*. The lamina is often lobed or toothed, possibly reducing the mean distance water travels from main veins to sites of evaporation, helping to cool the leaf. Hairy leaves probably reduce insect damage, as may latex channels, resin ducts, essential oils, tannins and calcium oxalate crystals. Some leaves produce hydrogen cyanide when damaged. During evolution, leaves have become modified to serve in reproduction, either sexually (as SPOR-OPHYLLS), or asexually (as producers of propagating buds). Leaves of higher plants usually have a bud in their axils and are important producers of GROWTH SUBSTANCES. See COTYLEDON, CUTICLE, LEAF BLADE, PHYLLOTAXIS.

LEAF BASE (PHYLLOPODIUM). Usually the expanded portion of the leaf, attached to the stem.

LEAF BLADE (LAMINA). Thin, flattened and flexible portion of a LEAF; major site of PHOTOSYNTHESIS and TRANSPIRATION, for which it is admirably adapted. May be simple (comprising one piece) or compound (divided into separate parts, leaflets, each attached by a stalk to the petiole). A typical dicotyledon leaf presents a large surface area, photosynthesizing cells being arranged immediately below the upper epidermis allowing maximum solar energy absorption. The blade is provided with a system of supporting veins bringing water and mineral nutrients and removing photosynthetic products. Has a system of intercellular spaces opening to the atmosphere through STOMATA, permitting regulated gaseous exchange and loss of water vapour; its thinness reduces diffusion distances for gases, keeping their concentration gradients steep, so increasing diffusion rates. Its large surface area and transpiration rate help coding. Evergreen leaves generally have blades twice as thick as deciduous ones, but neither can offer too great a wind resistance. See KRANZ ANATOMY, MESOPHYLL.

LEAF GAP. Localized region in vascular cylinder of the stem immediately above point of departure of LEAF TRACE (leaf trace bundle) where the parenchyma rather than vascular tissue is differentiated. In some plants where there are several leaf traces (leaf trace bundles) to a leaf, these are associated with a single leaf gap. Leaf gaps are characteristic of ferns, gymnophytes and flowering plants (anthophytes).

LEAFLET. Leaf-like part of a compound leaf.

LEAF SCAR. Scar marking where a leaf was formerly attached to the stem.

LEAF SHEATH. Base of a modified leaf, forming a sheath around the stem (e.g. in grasses, sedges).

LEAF TRACE. (1) Vascular bundle extending between vascular system of stem and leaf base; where more than one occurs, each constitutes a leaf trace. (2) Vascular supply extending between vascular system of stem and leaf base, consisting of one or more vascular bundles, each known as a *leaf trace bundle*.

LEARNING. Acquisition by individual animal of behaviour patterns, not just as an expression of a maturation process but as a direct response to changes experienced in its environment. Various forms of learning include CONDITIONING, HABITUATION and IMPRINTING. Insight learning, in which an animal uses a familiar object in a new way to solve a problem creatively, may be a form of instrumental conditioning: actions may come to be selected because their consequences form part of a route to obtaining a goal. Once clearly contrasted with INSTINCT, but it is now realized that these are not mutually exclusive categories of process. Learning ability is a clear example of adaptability, and of the adaptiveness of behaviour. See MELANOCYTE-STIMULATING HORMONE.

LECITHIN. See PHOSPHOLIPID.

LECTIN. Proteins and glycoproteins cross-linking cell-surface carbohydrates and other antigens, often causing cell clumping (see AGGLUTINATION). May act as antigens themselves, as do the mitogenic lymphocyte-stimulators *phytohaemagglutinin* (*PHA*) and *Concanavalin A* (*ConA*). In plants, may provide toxic properties of seeds. Are also involved in artificial CAPPING of specific cell surface components.

LECTOTYPE. Specimen or other component of original material, selected to serve as a nomenclatural type when no HOLOTYPE was designated at the time of publication, or as long as it is missing. See ISOTYPE, NEOTYPE, SYNTYPE.

LEGHAEMOGLOBIN. See HAEMOGLOBIN.

LEGUME. (1) A pod; fruit of members of the Family Leguminoseae (peas, beans, clovers, vetches, gorse, etc.). Dry fruit formed from single carpel that liberates its seeds by splitting open along sutures into two parts. (2) Used by agriculturists for a particular group of fodder plants (clovers, alfalfa, etc.) belonging to the Leguminoseae. Important in crop rotation, having symbiotic nitrogen-fixing root nodule bacteria.

LEMMA. Lower member of pair of BRACTS surrounding grass flower, enclosing not only the flower but also the other bract (PALEA).

LEMURS. See DERMOPTERA (flying lemurs); and PRIMATES.

LENS. Transparent, usually crystalline, biconvex structure in many types of eye, serving to focus light on to light-sensitive cells. In vertebrate eyes, constructed of numerous layers of fibres of the protein *crystallin*, arranged like layers of an onion. Normally enclosed in connective tissue capsule and held in position by suspensory ligaments. See ACCOMMODATION, EYE.

LENTIC. (Of freshwaters) where there is no continuous flow of water, as in ponds, lakes. Compare LOTIC.

LENTICEL. Small raised pore, usually elliptical, developing in woody stems when the epidermis is replaced by cork; packed with loosely arranged cells and allowing gaseous exchange between interior of stem and the atmosphere.

LEPIDOPTERA. Butterflies and moths. Endopterygote insect order. Two pairs of large membranous wings, with few cross-veins, and covered with scales; larva a CATERPILLAR, usually herbivorous and sometimes a defoliator of economic importance. Adults feed on nectar using highly specialized and often coiled proboscis formed from grooved and interlocked maxillae. There is no simple way to distinguish all moths from all butterflies, but in Europe any lepidopteran with club-tipped antennae, flying in the day, and capable of folding its wings vertically over its back, is a butterfly.

LEPIDOSAURIA. Dominant subclass of living reptiles; possess DIAPSID skulls and overlapping horny scales covering the body. Includes the orders Rhyncocephalia (*Sphenodon*, the tuatara) and Squamata (lizards, snakes, amphisbaenids).

LEPTOCEPHALUS. Oceanic larva of European eel. Migrates over 2000 miles across Atlantic from breeding site near West Indies (Sargasso Sea) to European fresh waters, where it becomes adult. Transparent.

LEPTOMA. Thin area in wall of gymnophyte pollen, through which the pollen tube emerges.

LEPTOME (LEPTOID). Photosynthate-conducting cell, approaching phloem in structure and function, found in bryophytes (esp. Bryales).

LEPTOSPORANGIATE. (Of vascular plant sporangia) arising from a single parent cell and possessing a wall of one layer of cells. Spore production is low in comparison with EUSPORANGIATE type.

LEPTOTENE. Stage in first prophase of MEIOSIS during which chromosomes are thin and attached at both ends to the nuclear membrane. DNA has already replicated but each chromosome appears as one thread, sister chromatids being closely apposed.

LEUCOCYTE (WHITE BLOOD CELL). Nucleated blood corpuscle lacking haemoglobin. Includes *granulocytes* (neutrophils, eosinophils, baso-

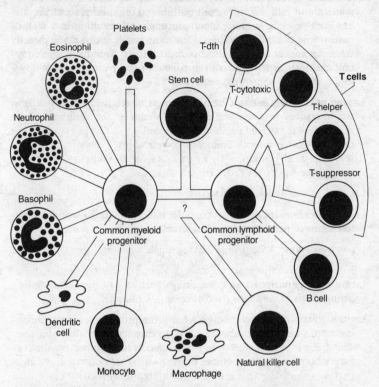

Fig. 41. *Diagram showing the variety of human leucocytes and their origins. Platelets are not leucocytes but are derived from a myeloid cell line.* Adapted from N. Staines, J. Brostoff and K. James, *Introducing Immunology,* Gower Medical Publishing, 1985, courtesy of the authors and publishers.

phils), with granules in their cytoplasm, developing typically from MYELOID TISSUE, and *agranulocytes* (lymphocytes and monocytes), lacking cytoplasmic granules and developing typically from LYMPHOID TISSUE. Lymphocytes are of two kinds: T-cells and B-cells (see also IMMUNITY). The monocytes in tissue fluid are called *wandering macrophages*, and these and neutrophils are the major phagocytic leucocytes. Natural killer cells may also be of lymphoid origin. See specific cell types, and Fig. 41.

LEUCOPLAST. Colourless plastid found in cells of plant tissues not normally exposed to light. Includes AMYLOPLASTS, storing starch, ELAIOPLASTS, storing oil, and ALEUROPLASTS, storing protein.

LEUCOSIN. See CHRYSOLAMINARIN.

LEUKAEMIA. Malignant overproduction by MYELOID TISSUE of white blood cells, crowding out normal red cell- and platelet-producing lines (leading to poor blood clotting) and resulting in a lack of mature and normal white cells. Death can result not so much directly from the cancer as from these indirect effects. Treatment by X-rays and chemotherapy may result in partial or complete remission; myeloid transplants sometimes required.

LEYDIG CELLS (INTERSTITIAL CELLS). Groups of relatively scanty cells in the interstices between seminiferous tubules of the vertebrate testis responsible for steroid production (especially TESTOSTERONE). Have unusually large smooth endoplasmic reticulum. Testosterone output is synergistically enhanced by PROLACTIN and LUTEINIZING HORMONE. See MATURATION OF GAMETES.

LH. See LUTEINIZING HORMONE.

LIANA. Climbing plant of tropical forests, with long, woody, rope-like stems of peculiar anatomical structure.

LIBRIFORM FIBRE. Xylem fibre having thick walls and greatly reduced pits.

LICE. See SIPHUNCULATA (sucking lice), MALLOPHAGA (biting lice), and PSOCOPTERA (book lice).

LICHEN. Symbiotic association between a fungus and an alga, developing into a unique morphological form distinct from either partner. The fungal partner (mycobiont) is usually a member of the ASCOMYCOTINA, and sometimes of the BASIDIOMYCOTINA; the algal partner (phycobiont) is either a green alga (CHLOROPHYTA), or a blue-green alga (CYANOBACTERIA). Provide dominant flora in large areas of mountain and arctic regions, where few other plants can exist, and play an important role in the primary colonization of bare areas. Lichens may be *crustose*, forming a thin, flat crust on the substratum; *foliose*, flat with leaf-like lobes; or *fruticose*, upright branched forms. Very slow-growing, they vary greatly in size, from a millimetre to several metres across; they consist of tissues, the most characteristic comprising loosely interwoven, branched hyphae forming a net-like structure. Algal cells may be evenly distributed among the hyphae, but more often occur in a thin layer. Anatomically, some of the foliose lichens are the most complex. Non-sexual reproduction occurs through fragmentation of the thallus as well as through *soredia* production – discrete structures comprising a few algal cells surrounded by hyphae and INSIDIA. The fungus may also reproduce non-sexually by conidia which form in the PYCNIDIA. Sexual reproduction is confined to the fungal component, structures formed commonly being ascocarps (apothecia or perithecia). Ascospores are formed in asci and discharged. On germination of the ascospores, new lichen individuals are formed if the algal partner happens to be present; but in its absence the fungus dies. In arctic regions, certain lichens are a valuable food source (e.g. Iceland moss, reindeer moss). Others provide dyes (e.g. *Roccella* provides litmus), as well as being sources of medicines, poisons, cosmetics and perfumes. Taxonomically, lichens form a GRADE, rather than a CLADE.

LIFE. Complex physico-chemical systems whose two main peculiarities are (1) storage and replication of molecular information in the form of nucleic acid, and (2) the presence of (or in viruses perhaps merely the potential for) enzyme catalysis. Without enzyme catalysis a system is inert, not alive; however, such systems may still count as biological (e.g. all viruses away from their hosts). Other familiar properties of living systems such as nutrition, respiration, reproduction, excretion, irritability, locomotion, etc., are all dependent in some way upon their exhibiting the two above-mentioned properties.

Living systems also have an evolutionary history. Whatever the ORIGIN OF LIFE may have been, all *existing* life forms derive from living antecedents. The earliest living system would have been very different from any modern life form, particularly so in their genetic systems (modes of storage and implementation of molecular information).

LIFE CYCLE. Progressive series of changes undergone by an organism or a lineal succession of organisms from fertilization to death of the stage producing the gametes beginning an identical series of changes. Life cycles, as much as phenotypes, display adaptation to the external environment. Even within a species there may be more than one way in which a generation can be completed. There may in particular be considerable variation in methods of reproduction. Sometimes this involves an alternation of generations (*metagenesis*) between distinct sexual and asexual individuals. The simplest form of life cycle is purely ASEXUAL, involving repeated binary fission and/or budding. In eukaryotes sexual life cycles may be classified according to whether or not haploid mitosis and/or diploid mitosis occur (see Fig. 42). Such life cycles may be either (a) haplontic, in which only haploid cells can divide by mitosis, the diplophase being represented by a single nucleus (e.g. *Spirogyra, Chlamydomonas, Mucor*), (b) diplontic, in which only diploid cells can divide by mitosis, all haploid cells being immediate products of meiosis (e.g. most animals), or (c) haplodiplontic (or diplohaplontic), in which both haploid and diploid mitoses can occur. Haplophase and diplophase may be more or less equally prominent (e.g. the alga *Ulva*), or one may be dominant (haplophase in mosses and liverworts, diplophase in ferns and seed plants). In most haplodiplontic cases (yeasts are somewhat variable) gametes are produced by mitosis, and a mitotic product of the zygote undergoes meiosis.

It is usual with plants to refer to the haplophase as the *gametophyte*, and to the diplophase as the *sporophyte*. Some plants, such as the

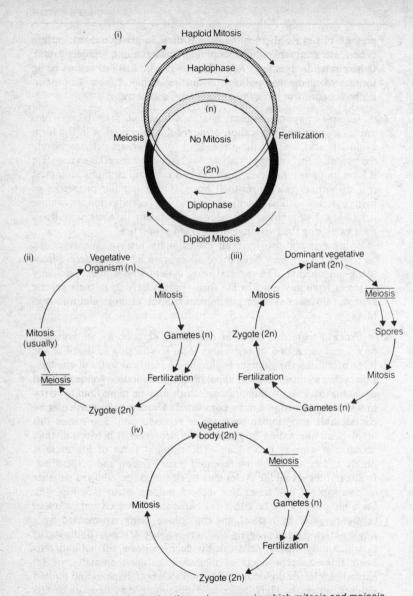

Fig. 42. (i) Scheme indicating the various ways in which mitosis and meiosis can be included within a eukaryotic sexual life cycle. Both fertilization and meiosis may or may not be followed by mitosis. The situation in which no mitosis at all occurs (the central route) is rarely if ever found: cell number would be increased solely by meiosis. (ii) The typical haplontic life cycle (e.g. Mucor, Spirogyra, Chlamydomonas). (iii) The typical haplodiplontic life cycle (e.g. mosses, liverworts, ferns, conifers, flowering plants). Spores may be produced homosporously or heterosporously. (iv) The typical diplontic life cycle (e.g. Fucus, most animals).

alga *Ulva*, have distinct haploid and diploid individuals in the life cycle (clear alternation of generations), and even in mosses and ferns these phases are practically distinct individuals. But during evolution, the gametophyte has become increasingly restricted, represented in flowering plants only by the microspore, megaspore and their greatly reduced mitotic products. Although development of gametophytic heterospory, begun in ferns, has been taken far further in flowering plants, it is the sporophyte that has become the dominant phase from ferns to seed plants, probably indicating a greater robustness and adaptability of the diploid plant in terrestrialization. Most animal life cycles are diplontic, but in some forms (e.g. polychaetes) it may be that products of meiosis undergo mitosis to produce gametes. In most cases of male haploidy, sperm are produced mitotically, although in bees they are formed by unipolar meiosis. Life cycles of animal parasites often involve POLYEMBRYONY as well as two or more hosts, while those of periodical cicadas (Hemiptera) may take 17 years to complete since nymphs take this long to reach maturity. See ALTERNATION OF GENERATIONS, PARTHENOGENESIS.

LIGAMENT. Form of vertebrate CONNECTIVE TISSUE joining bone to bone. *Yellow elastic ligaments* consist primarily of elastic fibres and form relatively extensible ligaments joining vertebrae, and true vocal cords; *collagenous ligaments* by contrast consist largely of parallel bundles of collagen and resist extension.

LIGASE. ENZYME catalysing condensation of two molecules and involving hydrolysis of ATP or another such triphosphate. DNA LIGASE is much used in GENE MANIPULATION, as well as forming part of DNA REPAIR MECHANISMS.

LIGNIN. Complex polymeric molecule composed of phenylpropanoid units associated with CELLULOSE (as *lignocellulose*) in cell walls of sclerenchyma, xylem vessels and tracheids, making them strong and rigid. After cellulose, lignin is the most abundant plant polymer, forming 20–30% of the wood of trees.

LIGNOCELLULOSE. Major chemical component of wood. Valuable resource, not least of energy as when converted to methane or alcohol in techniques of BIOTECHNOLOGY. See CELLULOSE, LIGNIN.

LIGULE. (1) Membranous outgrowth arising (a) from junction of leaf blade with leaf sheath in many grasses, (b) from base of leaves of certain lycopods. (2) Flattened corolla of ray flower in Compositae.

LIMITED GROWTH. Growth with fixed or definite limit; determinate growth.

LIMITING FACTOR. Any independent variable, increase in whose value leads to increase in the value of a dependent variable. Ideally, values of other independent variables should be held constant while this relationship is examined. In plots of dependent against independent variables, the latter are *limiting* only while there is a linear or near-linear relationship to the plot. Thus in the plot of initial velocity of an enzyme reaction (V_o) against substrate concentration, S, the latter is limiting only until the enzyme begins to be saturated with substrate (see ENZYME). In ecology, the term applies to any variable factor of the environment whose particular level is at a given time limiting some activity of an organism or population of organisms; e.g. temperature may limit photosynthesis when other conditions would favour a higher rate; growth of planktonic diatoms may become limited by depletion of dissolved silica.

LIMNOLOGY. Study of fresh waters and their biota.

LIMULUS. See MEROSTOMATA.

LINKAGE. Two or more loci (see LOCUS), and their representative genes, are said to be linked if they occur on the same chromosome. Such loci normally occur in the same linear sequence on all homologous chromosomes, so that chromosomes form *linkage groups*.

Contrary to the second of MENDEL'S LAWS of inheritance, eukaryotic genes which are linked (forming part of a chromosome) will tend to pass together into nuclei produced by MEIOSIS and are randomly assorted. This is very important, because whereas there is an equal statistical probability of all combinations of unlinked genes ending up in a given nucleus after meiosis (simply through the behaviour of non-homologous chromosome pairs), genes which are linked can only be prevented from passing together into nuclei by CROSSING-OVER, or by some chromosome MUTATION such as translocation, which separates them (see LINKAGE DISEQUILIBRIUM). The events of meiosis create new linkage groups and thereby new sets of phenotypic characters in organisms upon which SELECTION can act. Without linkage (i.e. without chromosomes) selection could never alter the probabilities with which characters appear together in different individuals of a population, and adaptation and evolution in so far as we understand them could not have occurred. Linkage makes possible the selection of *coadapted combinations* of alleles of different gene loci, which will tend to be transmitted together. Because rates of crossing-over between loci can be altered by selection (see MAP DISTANCE), adaptive combinations of genes, and of their resulting character combinations, can be preserved rather than disrupted by meiosis. See CHROMOSOME MAPPING, RECOMBINATION, SEX LINKAGE, SUPERGENE.

LINKAGE DISEQUILIBRIUM. Occurrence in a population of two or more loci so tightly linked that few of the theoretically possible gene combinations are found. E.g., if loci A and B are represented by just two alleles each in the population (*A* and *a*; *B* and *b*) yet only individuals with genotypes *AABB* and *aabb* are found in any frequency, then there is linkage disequilibrium between the two loci. Without crossing-over, and without selection for and against any of the genotypes, one expects linkage equilibrium ($AB + ab = Ab + aB$) to be achieved eventually. See LINKAGE, SUPERGENE.

LINKAGE MAP. See CHROMOSOME MAP.

LINNAEUS, CAROLUS. (Adj. *Linnean*) Swedish naturalist (1707–88), physician and originator of modern system of BINOMIAL NOMENCLATURE. In his *Systema Naturae Fundamenta Botanica* he assigned to every known product of nature (including minerals) its place on one great system of classification; indeed, discovery of the natural method of such classification was for him the naturalist's task. He thought he had achieved this for species and genera but only partially for classes and orders. Nature was for him imbued with divine *economy*: nothing happened unnecessarily, nor did any created form ever become extinct, although fossil evidence clearly forced him to consider this possibility. He was a little perplexed by the discovery of hybridization between species, and in later works considered that new species might arise in this way; but fundamentally his thinking was by today's standards conservative and non-evolutionary. His chief critic was BUFFON. See GREAT CHAIN OF BEING.

LINNEAN HIERARCHY. Arrangements of organisms into TAXA; forming nested sets, with KINGDOM the highest taxonomic category and SPECIES the lowest.

LINOLEIC ACID. ESSENTIAL FATTY ACID of mammals, making up to 20% of the total fatty acid content in their trigylcerides.

LIPASE. Enzyme hydrolysing fatty acid esters; it converts triglycerides (fats) into fatty acids and glycerol. Present in pancreatic juice and succus entericus of vertebrates. *Lipoprotein lipase* on surfaces of adipose tissue cells is largely responsible for hydrolysis of triglycerides within CHYLOMICRONS in blood plasma.

LIPIDS. Wide and heterogeneous assemblage of organic compounds having in common their solubility in organic solvents such as alcohol, benzene, diethyl ether, etc. Include fats and waxes, steroids and sterols (e.g. cholesterol), glycolipids, phospholipids, terpenes, fat-soluble vitamins, prostaglandins, carotenes and chlorophylls. See MICELLE.

LIPOGENESIS. FATTY ACID synthesis.

LIPOLYSIS. Hydrolysis of lipids, particularly of triglycerides. See ADIPOSE TISSUE, LIPASE.

LIPOPROTEIN. Micellar complex of protein and lipids. *Apolipoproteins*, protein components of these complexes, transport otherwise insoluble lecithin, triglyceride and cholesterol in vertebrate blood plasma. The transport lipoprotein particle comprises an outer lipid bilayer with specific conjugated protein components, within which the transported molecules are either free or esterified to bilayer fatty acids. There is a complex turnover and interaction of these particles, CHYLOMI-CRONS, low density lipoprotein (LDL), high-density (HDL) and very high-density (VHDL) forms increasing relative to the amount of protein in the particle. See CHOLESTEROL, LIPASE, ADIPOSE TISSUE, LIVER, COATED PIT.

LIPOSOME. Artificially produced spherical lipid bilayers, 25 nm or more in diameter, which can be induced to segregate out of aqueous media. Their selective permeabilities to organic solutes and the subsequent reactions which can occur within them have led some to propose a similar structure as a cell prototype. See COACERVATE, CHYLOMICRON.

LIST. Cellulose extension of cell wall in some armoured dino-flagellates, usually extending out from the cingulum and/or sulcus.

LITHOPHYTE. Plant found growing on rocks.

LITHOSERE. SERE originating on exposed rock surface.

LITHOTROPH. Autotrophic bacterium (e.g. nitrifying or sulphur bac-terium) obtaining its energy from oxidation of inorganic substances (e.g. sulphur, iron) by inorganic oxidants: the terminal hydrogen acceptor in respiration is always inorganic.

LITTORAL. Inhabiting the bottom of sea or lake, near the shore. From shore to 200 m in the sea; to 6–10 m in lakes, depending upon the extent of rooted vegetation.

LIVER. Gland, usually endodermal in origin and arising as a diver-ticulum of gut. Livers in different phyla are not homologous. In vertebrates its main glandular function is production of BILE, which leaves via the hepatic duct for storage in the gall bladder. Has a wide capability enzymatically, much of which is inducible. Structural unit is the *lobule*, a roughly hexagonal block of cuboidal cells (*hepatocytes*) supplied at its corners with products of digestion via factors of the hepatic portal vein, and with oxygen by branches of the hepatic artery. Blood leaves the lobule at its centre through a vessel leading to the hepatic vein. Bile leaves through *bile canaliculi*, also at corners of lobules. As a homeostatic organ it is second to none, involved in production of GLYCOGEN from monosaccharides and fat and its subsequent storage; absorption and release of blood glucose, absorption of CHYLOMICRONS, deamination of amino acids, urea production, GLUCONEOGENESIS, raising of blood temperature, production of prothrombin, fibrinogen, albumin and other plasma proteins, storage of vitamins A, D, E and K, red blood cell breakdown, detoxification of some poisons and steroid hormone conversion to cholesterol.

LIVERWORTS. See HEPATICOPSIDA.

LOCULE. Compartment, cavity, or chamber; (a) in ASCOMYCOTINA, chambers containing ASCI; (b) in flowering plants, cavity of ovary where ovules occur; (c) in diatoms, a chamber within the frustule, having a constricted opening on one side and a VELUM on opposite side.

LOCULICIDAL. (Bot.) Describing dehiscence of multilocular capsule by longitudinal splitting along a dorsal suture (midrib) of each carpel; e.g. iris. Compare SEPTICIDAL.

LOCUS (GENE LOCUS). Position on homologous chromosomes occupied, normally throughout a species population, by those genes which determine the state of a particular phenotypic character (see GENE). E.g., a locus for eye colour in *Drosophila*. This position may be determined by CHROMOSOME MAPPING, various occupants of a locus in a given population being referred to as ALLELES. Relative positions of loci within an individual's cells may be altered by some chromosome MUTATIONS. See LINKAGE.

LODICULES. Reduced perianth of grass flowers; two small scale-like structures below ovary which at time of flowering swell up, forcing open enclosing bracts (pales), exposing stamens and pistil.

LOG PHASE (LOGARITHMIC PHASE). See GROWTH CURVES.

LOMASOME. Membranous configuration associated with cell (or hyphal) wall of fungi, occurring singly or in groups, and situated between the plasmalemma and the wall; comprises membranous tubules, vesicles or parallel sheets lying in a matrix. May play a role in normal development of either the plasmalemma or the wall.

LOMENTUM. Type of leguminous fruit, constricted between seeds and breaking into one-seeded portions when ripe; e.g. in bird's foot trefoil.

LONG-DAY PLANTS. Plants that must be exposed to light for periods longer than some critical length for flowering to occur (and consequently flowering primarily during the summer). See PHOTO-PERIODISM, SHORT-DAY PLANTS, DAY-NEUTRAL PLANTS, PHYTOCHROME.

LONG SHOOT. Main branch in some gymnophytes (gymnosperms) bearing short dwarf shoots; e.g. in *Pinus, Ginkgo*.

LOOPED DOMAIN. See CHROMOSOME.

LOOP OF HENLE. See KIDNEY.

LOPHOPHORE. Hollow ring of ciliary feeding tentacles surrounding the mouth and (strictly) containing an extension of the coelom, as in Ectoprocta, Brachiopoda and Phoronidea. Applied loosely to any ring of oral tentacles, as in some polychaete worms and entoproctans.

LORICA. Envelope surrounding the protoplast, but not attached to it as the wall is. Seen in various algae, e.g. *Trachelomonas, Dinobryon*.

LOTIC. Of freshwaters where there is continuous flow of water e.g. streams, rivers. Compare LENTIC.

LUCIFERASE, LUCIFERIN. See BIOLUMINESCENCE.

LUMBAR VERTEBRAE. Bones of the lower back region, lacking rib attachments and situated between thoracic and sacral vertebrae.

LUMEN. (1) Cavity within tube (e.g. within blood vessel, gut) or within sac. (2) Cavity within cell wall of plant cell, from which the protoplast has been lost.

LUNG. Sac-like organ of gaseous exchange, invariably with moist inner surface. (1) In vertebrates, they arise as a diverticulum of the pharynx (see GAS BLADDER) and were present in fish prior to the rise of amphibians, serving probably as an adaptation to drought and/or poorly oxygenated water (see DIPNOI). Generally paired, and in most higher tetrapods internally subdivided into bronchi, bronchioles and alveoli, they lie surrounded by coelomic membranes that in mammals will later form fluid-lubricated pleural cavities separating lungs from thorax (see MEDIASTINUM). Lungs are relatively small in birds, where AIR SACS are the major respiratory surfaces. Lungs lack muscles of their own and are ventilated by rib muscles of the trunk and by the diaphragm (in mammals). Only in birds does air actually circulate through lungs; elsewhere air is tidal and terminates in richly vascularized alveoli, providing a huge surface area for gaseous exchange. The lung's (pulmonary) blood circulation has evolved in tetrapods towards the DOUBLE CIRCULATION of birds and mammals. Stretch receptors in a lung's connective tissue walls feed back to RESPIRATORY CENTRES controlling ventilation. (2) In molluscs, the lung is most advanced in pulmonate gastropods where it is a specialization of the mantle cavity and opens to the air via a valved pneumostome. Muscles ventilate the chamber.

LUNG BOOK. Organ or gaseous exchange in some air-breathing arachnids (e.g. in scorpions and, in conjunction with tracheae, some spiders). Consists of leaf-like projections sunk into pits opening through a narrow pneumostome to the air. Four pairs occur in scorpions, two pairs or one pair in some spiders, none in others. Gaseous exchange is by diffusion.

LUNGFISH. See DIPNOI.

LUTEIN. A xanthophyll pigment.

LUTEINIZING HORMONE (LH, INTERSTITIAL CELL-STIMULATING HORMONE,

ICSH). A glycoprotein GONADOTROPHIN, produced by the anterior PITUITARY under influence from hypothalamic releasing factor (GnRF). Its main function in males is to stimulate interstitial cells of the testis to produce TESTOSTERONE, which in turn shuts off GnRF release. In females, brings about ovulation (see MENSTRUAL CYCLE), and transforms the ruptured Graafian follicle into a corpus luteum with subsequent progesterone production.

LYCOPHYTA. Club mosses. There are five living genera: *Lycopodium, Selaginella, Isoetes, Phylloglossum,* and *Stylites,* giving about 1000 living species, representatives of an evolutionary lineage extending back to the Devonian period. Lycopods split very early on into two major groups. The first remained herbaceous and is still represented in today's flora; the second, the lepidodendrids (or tree ferns), became woody and tree-like and were the dominant plants of coal-forming forests of Carboniferous period (e.g. *Lepidodendron*). Became extinct in the Permian period (about 280 Myr BP). Living lycopods are small, evergreen plants with upright or trailing stems that bear numerous small leaves. Sporangia are borne singly in axils of leaves, sporophylls occurring in groups at intervals along the stem or forming terminal cones. They are either homosporous (e.g. *Lycopodium*), with small mycotrophic prothalli, wholly or partly subterranean; or heterosporous (e.g. *Selaginella*), with reduced prothalli remaining largely enclosed by the spore wall. See LIFE CYCLE.

LYMPH. Clear fluid within vessels of LYMPHATIC SYSTEM and derived from TISSUE FLUID and resembling it in composition, except in those lymph vessels between the gut and venous system after a meal, when it is generally 'milky' due to presence of CHYLOMICRONS. It returns proteins from the tissue fluid to the blood. Lymphocytes enter as it passes through LYMPH NODES.

LYMPHATIC (LYMPHOID) SYSTEM. System of LYMPH-containing vessels and the organs producing and accumulating lymphocytes linked by these vessels in vertebrates (see Fig. 43). The 'second' circulatory system in these animals. Comprises a blindly-ending meshwork of highly permeable endothelial *lymph capillaries* (resembling blood capillaries, but having non-return valves) permeating most body tissues (not the nervous system) and joining to form larger vessels (usually not larger than 2–3 mm diameter), resembling veins, with non-return valves but thinner connective tissue walls. Finally joins the venous system, usually near the heart. TISSUE FLUID drains into lymph capillaries and is slightly concentrated by loss of some water and electrolytes as it passes along the system, under the same forces as achieve venous return. Most of the large *lymph trunks* enter the left *thoracic duct* which opens into the venous system close to or at the left subclavian vein, but some also enter from a right thoracic duct near the right subclavian vein. See LYMPH NODE, LYMPHOID TISSUE.

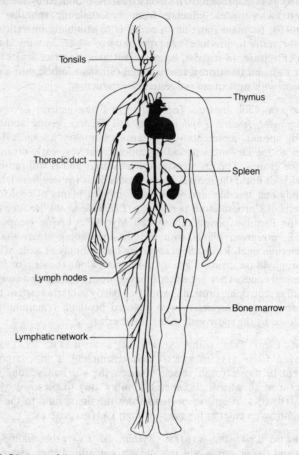

Tonsils

Thymus

Thoracic duct

Spleen

Lymph nodes

Bone marrow

Lymphatic network

Fig. 43. *Diagram of the human lymphatic system and its associated lymphoid organs and tissues.*
Adapted from N. Staines, J. Brostoff and K. James, *Introducing Immunology*, Gower Medical Publishing, 1985, courtesy of the authors and publishers.

LYMPH HEART. Enlarged part of lymphatic vessel with muscular pulsating wall. Present in many vertebrates, but not in birds or mammals. Pumps lymph.

LYMPH NODE. Ovoid structures on lymph vessels of mammals and to a lesser extent of birds (but not other vertebrates); up to 25 mm long in humans, comprising connective tissue framework of capsule and inner extensions (trabeculae) supporting successive internal parenchyma tissues: (a) the cortex of B-CELLS, (b) the paracortex of T-CELLS and (c) the medulla of T- and B-cells. Afferent lymphatic vessels join the capsule, where valves ensure that lymph passes progressively towards the medulla, where an efferent lymph vessel rejoins the lymphatic circulation. As it passes through the node, lymph is processed by fixed macrophages of the RETICULO-ENDOTHELIAL SYSTEM lining the sinuses, while lymphocytes respond to antigens from adjacent ANTIGEN-PRESENTING CELLS. Each node has its own blood supply, and both B-cells and T-cells respond by clonal expansion in a way dependent upon the type of antigenic stimulation. See LYMPHOMA.

LYMPHOCYTE. One of two kinds of vertebrate white blood cell (LEUCOCYTE), confined to the blood system. Most are *small lymphocytes* (diameter 6–8 μm), agranular with high nuclear:cytoplasmic ratio; but *large lymphocytes* are larger (diameter 8–10 μm) and granular, with lower nuclear:cytoplasmic ratio and with azurophilic granules (staining with azure dyes) in the cytoplasm; some may serve as NATURAL KILLER CELLS. Small lymphocytes (B-CELLS and T-CELLS) are involved in both humoral and cell-mediated IMMUNITY. Lymphocytes can move by amoeboid locomotion but most are non-phagocytic. They develop in LYMPHOID TISSUE.

LYMPHOID TISSUE. Vertebrate tissue in which LYMPHOCYTES develop (e.g. Bursa of Fabricius in birds, thymus in mammals). Lymphocytes are produced in *primary lymphoid tissue* (thymus, embryonic liver, adult bone marrow) and migrate to *secondary lymphoid tissue* (spleen, lymph nodes, unencapsulated lymphoid regions of gut submucosa, respiratory and urogenital regions), where ANTIGEN-PRESENTING CELLS and mature T-CELLS and B-CELLS occur. The kidney is a major secondary lymphoid organ in lower vertebrates. Tissues in animals without close chordate affinities may nonetheless form regional aggregates of phagocytes which are loosely termed lymphoid. See MYELOID TISSUE.

LYMPHOKINES. See INTERLEUKINS.

LYMPHOMA. Any tumour composed of lymph tissue. *Hodgkin's disease* is a malignant lymphoma of reticulo-endothelial cells in lymph nodes and other lymphoid tissues.

LYSIGENOUS. (Of secretory cavities in plants) originating by dissolution of secreting cells; e.g. oil-containing cavities in orange peel. Compare SCHIZOGENOUS.

LYSIS. Destruction of cells through damage to or rupture of plasma membrane, e.g. by osmotic shock. In bacteria, may be brought about by infection by BACTERIOPHAGE. See HYDROLYSIS.

LYSOGENIC BACTERIA, LYSOGENY. See BACTERIOPHAGE.

LYSOSOMES. Diverse membrane-bound vacuolar organelles, forming integral part of eukaryotic intracellular digestive system (see Fig. 44). Contain large variety of hydrolytic enzymes. Substances for digestion

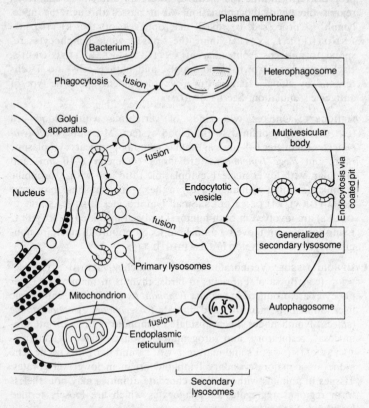

Fig. 44. *Diagram showing the source of primary lysosomes and their fusions with other organelles to form secondary lysosomes. Ladder-like rings indicate coated vesicles.*

entering the cell's geography by endocytosis (heterophagy) become part of the lysosomal system when they fuse with *primary lysosomes* (0.5 μm diameter), which have not been involved in hydrolytic activity, budded from the GOLGI APPARATUS. The vacuoles (*heterophagosomes*) formed by this fusion are sites of digestion analogous to the gut, and similar digestion products diffuse into the cytosol through the lysosomal membrane. Undigested remains may persist in 'residual bodies' for some time. Autophagy often involves organelles being wrapped up in a membranous vacuole, probably originating from smooth endoplasmic reticulum, and its subsequent fusion with primary lysosomes to form *autophagosomes*. These are characteristic of cells involved in hormone-related developmental reorganization. Autophagosomes and heterophagosomes are types of *secondary lysosome*, and may be considerably larger than primary lysosomes. Some

extracellular enzyme secretion results from fusion of primary lysosomes directly with the plasma membrane.

Lysozyme. Class of ENZYME catalysing hydrolysis of (glycosaminoglycan) walls of bacteria, leading to rupture and death of
remaining protoplast. Secreted by skin and mucous membranes and
found in tears, saliva and other body fluids of mammals; also in egg
white. Provides one innate immune response to bacteria. Lysozyme
was among the first proteins whose three-dimensional molecular
structure was elucidated.

M

MACROCYST. Cyst developed by aggregation of myxamoebae in some slime moulds (see MYXOMYCOTA).

MACROGAMETE (MEGAGAMETE). See GAMETE, ANISOGAMY.

MACROMOLECULE. Molecule of very high molecular weight, characteristic of biological systems, e.g. proteins, nucleic acids, polysaccharides, and complexes of these.

MACRONUCLEUS (MEGANUCLEUS). See NUCLEUS.

MACRONUTRIENTS. Substances required in large amounts for plant growth; e.g. nitrates, phosphates, sulphates, potassium, calcium and magnesium.

MACROPHAGE. Phagocytic cell of vertebrate connective tissue but not typically of the blood itself. Included here are *wandering macrophages* derived from monocytes (see LEUCOCYTE), and more static macrophages (*histiocytes*) dispersed throughout connective tissue but capable of migrating towards a site of infection. Phagocytes of the RETICULO-ENDOTHELIAL SYSTEM are rather more specialized macrophages.

MACROSPORANGIUM. See MEGASPORANGIUM.

MACROSPORE. See MEGASPORE.

MACROSPOROPHYLL. See MEGASPOROPHYLL.

MACULA. (1) See FOVEA. (2) Small receptor in walls of the utricle and saccule of tetrapod VESTIBULAR APPARATUS providing information about position of the head in relation to gravity when the animal is moving or at rest. Lying in planes perpendicular to one another, they comprise an epithelium containing support cells and HAIR CELLS whose stereocilia are embedded in a gelatinous layer on which lie *otoliths* of calcium carbonate which move the gelatinous layer, so pulling on the stereocilia. Impulses generated travel along CRANIAL NERVE VIII to the MEDULLA. Otoliths have an inertia which makes them stay at relative rest or movement as the body respectively moves or comes to rest. See STATOCYST.

MAJOR HISTOCOMPATIBILITY COMPLEX (MHC). Mammalian gene complex of several highly polymorphic linked loci encoding glycoproteins involved in many aspects of immunological recognition, both between lymphoid cells and between lymphocytes and antigen-presenting cells. Initially detected as the region encoding antigens involved in graft rejection, the most important region in humans (the HL-A system) is located on the sixth chromosome pair; in mice the comparable region (the H-2 complex) is located on the seventeenth chromosome pair. Some encoded glycoproteins bind intracellular peptides and take them to the plasma membrane, exposing the cell to cytotoxic T-CELL surveillance. See IMMUNITY.

MALACOSTRACA. Largest subclass of CRUSTACEA, including prawns and shrimps (Amphipoda), crabs and lobsters (Decapoda) and wood-lice (Isopoda). Thorax with eight segments, covered by carapace (not in isopods or amphipods); abdomen usually with six. Abdominal appendages on all segments, usually functioning in swimming. Thoracic appendages serve for locomotion and sometimes for feeding, and often bear gills. Compound eyes stalked.

MALARIA. A widespread and debilitating human disease, caused by a protozoan parasite (*Plasmodium* spp.) injected by mosquitoes of the genus *Anopheles*. Of many forms, the most commonly fatal is due to *P. falciparum*. Parasite life cycle includes asexual multiplicative stages in human liver and erythrocytes, fertilization in the mosquito gut lumen, with subsequent asexual multiplication in its wall. Work on a vaccine has been held up by ANTIGENIC VARIATION of the parasite. Mosquitoes are now resistant to many insecticides, as is the parasite to prophylactic drugs (e.g. chloroquine); but drug combinations interfering with plasmodial folic acid metabolism can be effective. Short-lasting oil on surfaces of mosquito breeding pools may be successful. A vaccine is the ultimate answer, and there are hopes that one may soon be available. See DNA PROBE, SICKLE-CELL ANAEMIA.

MALE HAPLOIDY. Arrhenotoky. Condition in which males arise from unfertilized eggs, females from fertilized ones. Males have no father; females have two parents, but only three grandparents. Males are cytologically haploid, producing gametes by mitosis (see LIFE CYCLE). Universal in hymenopteran insects; found also in some scale insects (Coccoidea) and mites, among several other animal groups. Form of PARTHENOGENESIS. See HAMILTON'S RULE, THELY-TOKY.

MALE STERILITY. See CYTOPLASMIC MALE STERILITY.

MALLEUS. See EAR OSSICLES.

MALLOPHAGA. Biting lice; bird lice, feather lice. Exopterygotan insects; ectoparasitic on birds and occasionally on mammals. Flattened, with tarsal claws and reduced eyes. Cannot pierce skin (unlike SIPHUNCULATA), but bite small particles of FEATHERS or hair. Include hen louse, *Menopon*.

MALPIGHIAN CORPUSCLE. In vertebrate KIDNEY, a glomerulus and its associated Bowman's capsule. See NEPHRON.

MALPIGHIAN LAYER. Innermost layer of epidermis of vertebrate skin, next to dermis. An active region of MITOSIS, non-stem cells being pushed to the surface, ageing and becoming cornified. Its cells often contain the pigment melanin, darkening the skin as a protection against ultraviolet light.

MALPIGHIAN TUBULES. Long, blind-ending, slender tubes lying in the haemocoeles of most myriapods and terrestrial insects and arachnids (where independently evolved), but not Onychophora. Involved in excretion and osmoregulation. In insects, open into intestine near junction of hindgut and midgut. Water, nitrogenous waste and salts enter tubule lumen, probably by active secretion. The hindgut then selectively reabsorbs water and useful metabolites, leaving a precipitated concentrate rich in uric acid salts which then leaves with the faeces. See COXAL GLANDS.

MALTOSE. Disaccharide sugar composed of two glucose molecules bonded in an $\alpha[1,4]$glycosidic linkage. Formed by the enzymatic degradation of starch by AMYLASES. Occurs e.g. in germinating seeds such as barley, and during digestion in animals.

MAMMALIA. Class of vertebrates evolving in the Triassic, with present forms distinguished from all others by presence of body hair and secretion of milk, generally via mammary glands. Homoiothermic with a diaphragm used in ventilation of lungs. Lower JAW is composed of a single pair of bones (dentaries), thus differing from their therapsid reptilian ancestors, presumed also to have been homoiothermic. Only the left systemic arch remains (see AORTA, DORSAL). Three bones form the EAR OSSICLES in each middle ear. Mostly of small size until the Tertiary. Three subclasses: oviparous PROTOTHERIA (monotremes, e.g. spiny anteaters, duck-billed platypus); viviparous METATHERIA (marsupials, e.g. opossums, wombats, koalas, kangaroos) and EUTHERIA (placentals, in 16 extant orders). Several extinct orders. See MAMMAL-LIKE REPTILES.

MAMMAL-LIKE REPTILES (THERAPSIDA). Reptilian subclass radiating from the mid-Permian and becoming extinct at the end of the Triassic. Sluggish herbivores and active carnivores, the latter with elbow and knee swung towards (tucked into) the body allowing rapid four-footed gait. The herbivores may have formed herds. Small Mesozoic therapsids gave rise to the earliest mammals, contemporaries of the ruling archosaurs for several million years. See REPTILIA.

MAMMARY GLANDS. Milk-producing glands, peculiar to the ventral surfaces of female viviparous mammals (similar structures are present in monotremes). Rudimentary in males, unless abnormal hormonally. Develop from epidermis and resemble sweat glands, consisting of a branching series of ducts terminating in secretory alveoli during pregnancy. See MILK, PROLACTIN.

MANDIBLE. (1) Of vertebrates, the lower JAW. (2) Of insects, crustacea and myriapods, one of the first pair of MOUTHPARTS ('jaws'), usually involved in biting and crushing food and heavily sclerotized. See MAXILLA.

MANDIBULAR ARCH. See VISCERAL ARCHES.

MANNAN. Polysaccharide, composed of mannose units and occurring in cell walls of some algae and yeasts.

MANNITOL. A 6-carbon sugar alcohol, oxidised to mannose. Component of the polysaccharide laminarin.

MANNOGLYCERATE. Saccharide storage product in some red algae (RHODOPHYTA).

MANTLE. (Bot.) (1) A dense mass of fungal hyphae surrounding a root. (2) That part of the diatom valve that bends away at 90°. (Zool.) Surface layer of visceral hump of MOLLUSCA, secreting shell. Flaps of the mantle enclose the mantle cavity. Similar structure is found in brachiopods.

MAP DISTANCE. A measure of the frequency of CROSS-OVER between two linked chromosome marker loci, equal to the percentage recombination in meiotic products, provided there are no double or multiple cross-overs between the markers. The further apart geographically two loci are on a chromosome the more will their apparent distance (as derived from cross-over values) be foreshortened by double and multiple cross-overs between them. Map distances between loci are reduced (underestimated) to the extent that such crossovers occur between them, so that long map distances are best derived from summation of short distances, using intervening markers. Suppression of all crossing-over between two loci (see SUPPRESSOR MUTATION) would effectively render them indistinguishable from a single complex locus. This might have selective advantages. See SUPERGENE.

MARGINAL MERISTEM. MERISTEM located along margin of leaf primordium and forming the blade.

MARGINAL PLACENTATION. Attachment of ovules to margin of a carpel.

MARSUPIAL. See METATHERIA.

MARSUPIUM. Pouch of many marsupials and spiny anteater (*Echidna*, PROTOTHERIA). Fold of skin supported by epipubic bone of pelvic girdle, forming pouch containing mammary glands or similar structures, into which newborn (or eggs in echidnas) are placed. Young marsupials attach there to teats, or lick milk in the case of teatless *Echidna*.

MASS FLOW. Hypothesis for explaining TRANSLOCATION of materials in phloem of vascular plants, relying on passive processes which do not demand energy expenditure. SIEVE-TUBES and their contents are regarded as forming an osmotic system continuous throughout the plant and connected, via PLASMODESMATA, with other tissues. A gradient of osmotic, and hence turgor, pressure exists within the sieve-tube system as a result of entry into it of synthesized materials (photosynthates) from the leaf mesophyll, increasing pressure. Their removal for consumption in areas of growth and storage lowers this pressure. The pressure gradient thus brought about results in mass flow of water and dissolved solutes through the system from regions of high to regions of low pressure. Circulation of water is completed via the xylem, with which the phloem is intimately linked. This explanation of phloem translocation has been challenged by those who believe that active, or accelerated, diffusion plays a role.

MAST CELL. Granular LEUCOCYTE derived from MYELOID TISSUE, often associated with mucosal epithelia. When not in connective tissue, dependent upon T-CELLS for reproduction. Granules in cytoplasm contain SEROTONIN, HEPARIN and HISTAMINE; released when allergen cross-links the IgE molecules bound to plasma membrane. Involved in many allergies, but also in immunity to parasites. See BASOPHIL.

MASTIGONEME. Exceedingly fine lateral projection (tinsel or flimmer) of certain flagella. Often numerous. Probably increase surface area and frictional resistance during movement.

MASTIGOPHORA. Often employed as a class of PROTOZOA, but including both holophytic, holozoic and facultative forms. Usually refers to flagellated protozoans, both *zoomastigophorans* such as trypanosomes and *phytomastigophorans* such as euglenoids. Close affinities of flagellated and pseudopodial protozoans have resulted in some authorities uniting them in the Class Sarcomastigophora. Likely that all these are grades rather than clades. See FLAGELLUM.

MASTOID PROCESS. Part of mammalian auditory capsule (periotic bone) containing air spaces communicating with middle ear.

MATERNAL EFFECT (M. INFLUENCE). Instances in which some aspect of phenotype of normal sexually produced offspring is controlled more by the genotype of maternal parent than by its own. Distribution of cytoplasmic molecules in the egg, under the control of maternal genes, is one notable example in that it may have a marked effect upon CLEAVAGE and later development. For instance, during oogenesis of the *Drosophila* egg, maternal mRNA products of the BICOID GENE are retained at the anterior pole of the egg, and the eventual protein translation product disperses some way along the egg, forming the kind of gradient required of a positional signal in the POSITIONAL INFORMATION theory of development. See MATER-NAL INHERITANCE, OSKAR GENES.

MATERNAL GENE. Term sometimes used specifically for a gene responsible for a MATERNAL EFFECT.

MATERNAL INHERITANCE. Characters inherited through the female line only. Not all genes are transmitted via nuclear chromosomes (see cpDNA, mtDNA, PLASMIDS). Evolution of anisogamy means that the female parent may contribute more genetically to its offspring than does the male parent. See MATERNAL EFFECT, CYTOPLASMIC INHERITANCE.

MATING TYPE. In, e.g. algae and fungi, used to designate a particular genotype with respect to compatibility in sexual reproduction. Gametes are identical in appearance and referred to as plus (+) and minus (−), rather than as male and female. See HETEROTHALLISM, SEX.

MATURATION OF GERM CELLS. Processes normally taken to include the cell division and differentiation involved in the production of functional gametes from their germ mother cells.

During *oogenesis* in humans, oocyte development is arrested at first meiotic prophase in the newborn female, most oocytes being surrounded by follicle cells, some already organized in developing follicles; but these degenerate before puberty. At puberty, FOLLICLE STIMULATING HORMONE restarts follicle development and a monthly mid-cycle surge of pituitary LUTEINIZING HORMONE causes just one follicle to develop further. Follicle cells connect to the oocyte by GAP JUNCTIONS and contribute small molecules from which it synthesizes macromolecules. The primary oocyte completes its first meiotic division; the follicle enlarges and ruptures, releasing a secondary oocyte. This will only complete MEIOSIS when fertilized (see Fig. 45a).

Spermatogenesis in the TESTIS (Fig. 45b) is also under pituitary control, LH (= ICSH) being the more important hormone, but acting synergistically with FSH. ICSH causes LEYDIG CELLS to release testosterone, some of which probably stimulates the germinal epithelium of the seminiferous tubules to divide. ICSH also affects the SERTOLI CELLS, influencing the rate at which spermatozoa are released into their tubule lumen. Spermatogonia remain attached to each other after mitosis, as do their primary and secondary spermatocytes and spermatids (forming a *syncytium*). Meiosis takes about 24 days to complete in man and most sperm differentiation (*spermiogenesis*, or *spermateliosis*) occurs after meiosis is completed, spermatids losing excess cytoplasm and developing an ACROSOME and, from one of the two centrioles, a flagellum − all the while surrounded and protected from immune attack by their

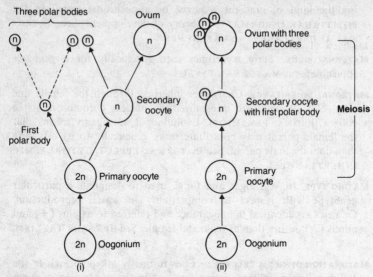

Fig. 45a. Diagram illustrating vertebrate oogenesis (i) with polar bodies shown distinct from oocyte; (ii) with polar bodies remaining attached to oocyte, as normally occurs.

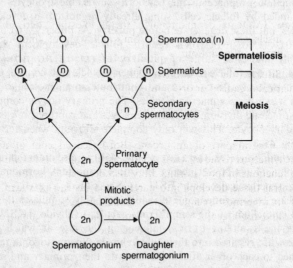

Fig. 45b. Diagram illustrating vertebrate spermatogenesis.

Sertoli cell (gametes are antigenically unlike body cells). Spermiogenesis takes about 9 weeks in man. Sperm normally cannot fertilize an ovum until they have been in the epididymis, nor until they have undergone CAPACITATION in the female reproductive tract. See MENSTRUAL CYCLE.

MAXILLA. (1) One of the DERMAL BONES of the vertebrate upper jaw (and of the face in man) carrying all the upper teeth, except incisors. (2) Sometimes used for the whole vertebrate upper jaw. See PRE-MAXILLA. (3) One of a pair of MOUTHPARTS in insects, crustaceans and myriapods, lying beneath the mandibles and articulating with the head capsule by a ball-and-socket joint. Sclerotized. In insects, bears a jointed sensory palp. In crustaceans, tends to be a phyllopodium, and may serve in feeding and gaseous exchange (see BIRAMOUS APPENDAGES).

MAXILLULE. Paired crustacean mouthpart, lying behind the mandibles and in front of the maxillae. Usually a phyllopodium (see BIRAMOUS APPENDAGES), passing food to the mouth.

MEATUS. A passage; e.g. external auditory meatus (see EAR, EXTERNAL).

MECKEL'S CARTILAGE. Paired bar of cartilage, one forming each side of lower jaw of gnathostome embryo, and of adult elasmobranch. In most other vertebrates it becomes reduced, and partly ossified, as the *articular bone* (see EAR OSSICLES), which in non-mammals forms hinge of lower jaw. Represents part of third VISCERAL ARCH.

MECONIUM. Contents of mammalian foetal intestine; derived from glands discharging into gut, and from swallowing of amniotic fluid during process of excretion.

MEDIAN. Situated in or towards the plane dividing a bilaterally symmetrical organism or organ into right and left halves.

MEDIASTINUM. Space between the two pleural cavities containing the heart in its pericardium, aorta, trachea, oesophagus, thymus, etc.

MEDULLA. Central part of an organ. (Bot.) Central core of usually parenchymatous tissue in stems where vascular tissue is in cylindrical form; functions in food storage. May also occur in some roots where the central tissue develops into parenchyma instead of xylem. Also refers to innermost region of the thallus in lichens and in some brown and red algae. (Zool.) (1) Central part of an animal organ, typically where outer part is termed the *cortex*. See ADRENAL GLAND, mammalian KIDNEY, LYMPH NODE. (2) Abbreviation of MEDULLA OBLONGATA.

MEDULLA OBLONGATA (MEDULLA). Most posterior part of vertebrate brain. See BRAINSTEM.

MEDULLARY RAY. Thin vertical plate of parenchyma tissue, one to several cells wide, running radially through the STELE. May be either *primary*, passing from pith (medulla) to cortex between primary vascular bundles, or *secondary*, formed from cambium during secondary thickening, ending blindly in secondary xylem and phloem. Since these latter have no connection with pith (medulla), they are sometimes termed *vascular* (*phloem* or *xylem*) rays. The function of medullary rays is in storage and radial translocation of synthesized (organic) materials.

MEDULLATED NERVE FIBRE. A myelinated NERVE FIBRE.

MEDUSA. Free-swimming form of the CNIDARIA. Shaped like bell or umbrella, swimming by rhythmic contractions of circular muscle in the rim and/or subumbrella producing a jet-propulsion effect through the water and involving concentrations of nerve cells (ganglia) and formation of nerve tracts or rings (two in jellyfish). Relatively thick mesogloea. Produced by budding, they themselves are sexual. Absent from life cycles of corals, sea anemones and some hydrozoans (e.g. *Hydra*).

MEGAGAMETE. See GAMETE, ANISOGAMY.

MEGAGAMETOPHYTE. In heterosporous plants, the female gametophyte; located within the ovule of seed plants.

MEGAKARYOCYTE. Large, highly polyploid bone marrow cell giving rise to PLATELETS by a kind of pinching-off, or cellular autotomy.

MEGANUCLEUS (MACRONUCLEUS). See NUCLEUS.

MEGAPHYLL. Type of leaf possessing a branched system of veins and with a LEAF TRACE generally associated with one or more LEAF GAPS in STELE of stem. Usually associated with siphonostelic vascular system in stem. Characteristic of ferns and seed plants. Compare MICROPHYLL.

MEGASPORANGIUM (MACROSPORANGIUM). SPORANGIUM in which MEGASPORES are formed; i.e. a meiosporangium of heterosporous plants, producing usually one to four megaspores. In seed plants, an OVULE.

MEGASPORE (MACROSPORE). Larger of two kinds of spore (meiospore) produced by heterosporous ferns; the first cell of the female gametophyte generation of these and of seed plants. It becomes the EMBRYO SAC in flowering plants (ANTHOPHYTA).

MEGASPORE MOTHER CELL. Diploid cell (meiosporocyte) in which MEIOSIS will occur, contained within the megasporangium (within nucellus of ovule); produces one or more megaspores. See DOUBLE FERTILIZATION.

MEGASPOROPHYLL (MACROSPOROPHYLL). Leaf or modified leaf, bearing MEGASPORANGIA. In flowering plants, the carpel. Compare MICROSPOROPHYLL.

Meiosis (reduction division). Process whereby a nucleus divides by
two divisions into four nuclei, each containing half the original
number of chromosomes, in most cases forming a genetically non-
uniform haploid set (see Fig. 46). A necessary aspect of eukaryotic
sexual reproduction, for without it fertilization would usually double
the chromosome number every generation. Meiosis ensures that all
gamete nuclei from diploid parents contain a haploid set of chromosomes. It also ensures, in sexually outbred populations at least, wide
GENETIC VARIATION between offspring. This results from genetic
RECOMBINATION, both random and non-random.

The first meiotic division is initiated by DNA replication (S
phase), so that each chromosome comes to comprise two sister
chromatids, as in mitosis; but at the start of prophase the sister
chromatids of each chromosome remain tightly together, their early
appearance (*leptotene*) giving the impression of unreplicated chromosomes. They remain attached to the nuclear membrane. Chromosomes
then pair up homologously (*zygotene*), as a synaptonemal complex
develops between them holding them together (*synapsis*). Each such
pair is termed a *bivalent*; but where there is little homology (e.g.
between sex chromosomes as in the heterogametic sex of most vertebrates) synapsis is only partial. Completion of synapsis is followed by
shortening and thickening of the bivalents (*pachytene*), during which
interval, often lasting for days, CROSSING-OVER occurs between
non-sister chromatids, producing recombination between homologous
chromosomes. The synaptonemal complex then dissolves (*diplotene*)
and the two homologous chromosomes fall apart, except where
crossing-over holds them together at one or more visible *chiasmata*.
Chromosomes begin to unwind (decondense) and may commence
RNA transcription again. At *diakinesis*, any RNA synthesis ceases
(but see LAMPBRUSH CHROMOSOMES); bivalents shorten, thicken
and detach from the nuclear membrane. For the first time they
appear as four distinct chromatids, linked at their centromeres and
chiasmata.

After first meiotic prophase, the nuclear membrane disintegrates
and at *first metaphase* bivalents lie on the midline of the cell, between
the two poles. To commence *first anaphase* the spindle fibres, attached
to kinetochores of the centromeres, then pull the two members of
each bivalent to opposite poles, the sister chromatids appearing
somewhat more 'splayed out' than at mitotic anaphase. Chance
governs which pole each chromosome of a bivalent moves to, ensuring
random recombination between non-homologous chromosomes.
Each of the two sets of chromosomes produced is HAPLOID, although
each dyad has chromatids of mixed parental origin. A short *first
telophase* and *interphase* may follow; or the second meiotic division
proceeds at once. Nuclear membranes normally reform, chromosomes
decondensing, then recondensing, at a brief *second prophase*. The
nuclear envelope breaks down, a spindle forming either parallel (e.g.

plant megaspores) at right angles to the first. *Second metaphase*, *anaphase* and *telophase* pass quickly, resembling mitotic phases except that non-identical sister *chromatids* separate, lying on the metaphase plate until the centromeres holding them together separate at anaphase. Again it is a matter of chance which of two poles a chromatid moves to. Nuclear membranes form around the four haploid chromosome sets to complete meiosis. See LIFE CYCLE, SPERMATOGENESIS, OOGENESIS and MATURATION OF GERM CELLS for vertebrate gametogenesis. See COST OF MEIOSIS, POLYPLOID.

MEIOSPORE. Spore produced by meiosis.

MEIOTIC DRIVE. See ABERRANT CHROMOSOME BEHAVIOUR.

MELANIN. Dark brown pigment of many animals and product of tyrosine metabolism, giving brown and yellow colouration to skin, hair, etc. Often located in melanophores. See CHROMATOPHORE, MALPIGHIAN LAYER.

MELANISM. See INDUSTRIAL MELANISM.

MELANOCYTE-STIMULATING HORMONE. Peptide hormone of the *pars intermedia* of the vertebrate anterior PITUITARY GLAND causing darkening of the skin in lower vertebrates by contraction of melanin in melanophores (see CHROMATOPHORE). In humans it appears to raise the general excitability of neurones in the central nervous system, probably affecting learning. Release inhibited by hypothalamic factor.

MELANOPHORE. See CHROMATOPHORE.

MELATONIN. See PINEAL GLAND.

MEMBRANE. See CELL MEMBRANES.

MEMBRANE BONE. See DERMAL BONE.

MEMBRANE POTENTIAL. Electrical potential across a cell membrane. All plasma membranes have such voltage gradients, the inside negative with respect to the outside. Most pronounced in animal excitable tissues. See RESTING POTENTIAL.

MEMBRANOUS LABYRINTH. Vertebrate inner ear, comprising the VESTIBULAR APPARATUS and, in higher vertebrates, the COCHLEA.

MEMORY CELL. A kind of mature B-CELL capable of clonal expansion when appropriately triggered (*sensitized*) by antigen, typically after initial antigen-presentation. See ANTIGEN-PRESENTING CELL.

MENDEL, JOHANN (GREGOR). Austrian experimental biologist (1822–84) of peasant stock. After study at Olomouc University, he entered Brno monastery in 1843 as it provided opportunities for continued academic work. His experimental work involved production of pea

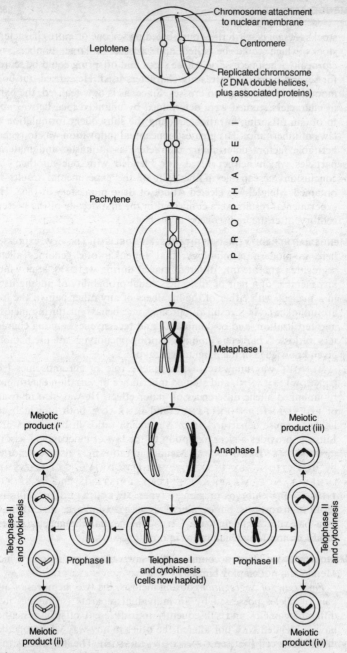

Fig. 46. The diagram shows the behaviour of one pair of homologous chromosomes during meiosis. One cross-over results in a chiasma, visible at metaphase I. From this stage onwards the two chromosomes are shown black or white to indicate origins of segments.

stocks (*Pisum* is self-fertile) pure-breeding for one or more characters; stocks which could be selfed and crossed in large numbers, reciprocally when necessary, whose seeds and offspring could be scored for several pairs of contrasting characters used. He carried out both monohybrid and dihybrid crosses. Since, as is now realized, the pairs of characters studied were determined by unlinked loci, he was able to obtain offspring in ratios enabling the subsequent formulation of laws of inheritance. His greatest conceptual innovation was to regard heritable factors determining characters as atomistic and material particles which neither fused nor blended with one another – a conclusion inescapable in the light of the experimental results he obtained. Mendel was elected Abbot of Brno monastery in 1868. His experimental results were confirmed in the first decade of the present century, after the 'rediscovery' of his laws.

MENDELIAN HEREDITY (M. INHERITANCE, MENDELISM). The view, expressed here in modern terminology, that in eukaryotic genomes alleles segregate (separate into different nuclei) during MEIOSIS, after which any member of a pair of alleles has equal probability of finding itself in a nucleus with either of the members of any other pair (if the loci are unlinked). As a result of chromosome behaviour during meiosis and fertilization, and of dominance and recessiveness among characters, ratios of characters among offspring phenotypes are predictable, given knowledge of the parental genotypes.

MENDEL was unaware of the genetic role of chromosomes (see GENE, WEISMANN), and studied inheritance of variation determined by unlinked allelic differences of major effect. He was also unaware of POLYGENIC INHERITANCE and LINKAGE, both of which are liable to cause departures from Mendelian ratios in breeding work. Linkage provides a clear exception to the law of independent assortment (see MENDEL'S LAWS). Mendelian ratios may also be distorted by MUTATION, SEX-LINKAGE, MEIOTIC DRIVE, CYTOPLASMIC INHERITANCE, MATERNAL EFFECT, EPISTASIS and by SELECTION among embryos or gamete types. MALE HAPLOIDY will also result in distortions; but even here, as with sex-linkage, alleles behave in a basically Mendelian way. It is simply that the genetic system produces non-Mendelian ratios in breeding work.

MENDEL'S LAWS. This account of the laws uses terminology which Mendel did not employ himself.

First Law, of Segregation: during meiosis, the two members of any pair of alleles possessed by an individual separate (segregate) into different gametes and subsequently into different offspring, neither having blended with nor altered the other in any way while together in the same cell (but see GENE CONVERSION). The law asserts that alleles retain their integrities (barring mutation) during replication from generation to generation. None of the cells normally produced by meiosis contains two alleles from any locus.

Second Law, of Independent Assortment: asserts that during meiosis all combinations of alleles are distributed to daughter nuclei with equal probability, distribution of members of one pair having no influence on the distribution of members of any other pair. It holds, in effect, that during meiosis random reassortment occurs between alleles at different loci. Thus if one locus is represented by the two alleles *A* and *a*, while another locus is represented by alleles B and b, then all four haploid nuclei *AB*, *aB*, *Ab* and *ab* will be formed in equal frequency by meiosis. Mendel's second law is refuted by LINK- AGE. Genes at linked loci tend to retain their linear sequences on chromosomes during meiosis and therefore tend to be inherited as blocks (but see CROSS-OVER VALUE).

Mendel's first law is a consequence of the behaviour of all chromosomes during meiosis. His second law is a consequence of the independent behaviour of non-homologous chromosomes during meiosis. See MENDELIAN HEREDITY.

MENINGES. Three membranous coverings of the vertebrate brain, spinal cord, and spinal nerves as far as their exits from between the vertebrae. Innermost is the vascular *pia mater*, separated from the *arachnoid* by the fluid-filled arachnoid spaces in which a fine web of fibres and villi reabsorb the CEREBROSPINAL FLUID back into the venous system. Outermost membrane, the *dura mater*, is a thick, fibrous membrane lining the skull and separated from the arachnoid below it by the dural sinus draining blood from the brain. The pia mater supplies capillaries to the ventricles of the brain.

MENSTRUAL CYCLE. See Fig. 47. Modified oestrous cycle of catarrhine primates; characterized by sudden breakdown of uterine endometrium, producing bleeding (*menstruation*), and by absence of a period of heat (oestrous) in which the animal is particularly sexually receptive. Controlled and coordinated nervously and hormonally, in particular by the hypothalamus and pituitary gland and the gonads. A mature human cycle has the following main sequence of events: a hypothalamic releasing factor (FSH-RF) causes release of FOLLICLE-STIMULATING HORMONE from the pituitary, initiating growth of one GRAAFIAN FOLLICLE in the ovaries and its consequent increased production of oestrogens, especially oestradiol. High oestrogen levels eventually inhibit release of both hypothalamic FSH-RF and LH-RF, decreasing FSH and LH output; but about two days prior to *ovulation* a marked rise in output of pituitary LUTEINIZING HORMONE occurs. Through a positive feedback mechanism, rising levels of oestrogens induce LH release (via LHRF) from the pituitary, which has increased sensitivity at this time. A mid-cycle surge of LH induces rupture of the Graafian follicle (ovulation) with release of the egg and development of the Graafian follicle into a CORPUS LUTEUM, which secretes oestradiol and increasing amounts of progesterone, which maintains the vascularization of the

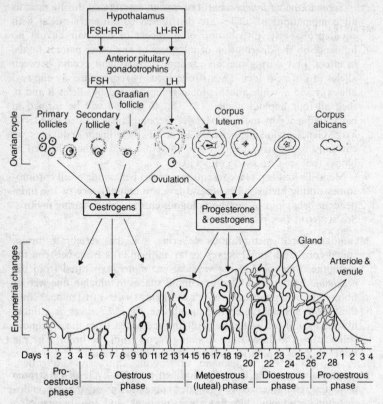

Fig. 47. *Diagram illustrating the hormonal and endometrial changes in the human menstrual cycle when no pregnancy occurs.*

uterine endometrium begun by oestrogens. Progesterone is secreted only so long as small amounts of pituitary LH maintain the corpus luteum, about 10–12 days in women unless pregnancy occurs. With atrophy of the corpus luteum, progesterone and oestradiol levels drop sharply, causing *menstruation* – the shedding of the uterine lining each month in the absence of pregnancy. If there is no pregnancy, the cycle repeats immediately as hypothalamic FSH-RF is released again, having been inhibited by the negative feedback effect of steroid sex hormones. If pregnancy occurs, maintenance of the corpus luteum is sustained for a short period by LH, then by HCG (HUMAN CHORIONIC GONADOTROPHIN) secreted by the implanted blastocyst and later by the placenta. See CONTRACEPTIVE PILL, MATURATION OF GERM CELLS.

MERICARP. Single-seeded portion of SCHIZOCARP.

MERISTELE. Individual vascular unit of DICTYOSTELE.

MERISTEM. Localized region of active mitotic cell division in plants, from which permanent tissue is derived. New cells formed by activity of a meristem become variously modified to form characteristic tissues of the adult (e.g. epidermis, cortex, vascular tissue, etc.). A meristem may have its origin in a single cell (e.g. in ferns), or in a group of cells (e.g. in flowering plants). The principal meristems in latter group occur at tips of stems and roots (APICAL MERISTEMS, or growing points), between xylem and phloem of vascular bundles (CAMBIUM) in cortex (CORK CAMBIUM), in young leaves and (e.g. in many grasses) at bases of internodes (intercalary meristems). Meristems may also arise in response to wounding.

MERISTEMATIC ACTIVITY. State of active mitosis in a MERISTEM.

MERISTODERM. Outer meristematic cell layer in certain brown algae.

MEROBLASTIC. See CLEAVAGE.

MEROCRINE GLAND. Gland whose cells secrete their product while remaining intact: no portion is pinched off (see APOCRINE GLAND), nor do cells have to disintegrate in order to release their product (see HOLOCRINE GLAND). Examples include vertebrate salivary glands and exocrine cells of the pancreas.

MEROPLANKTONIC. Term describing organisms that spend part of their life cycle in the plankton and part in the benthos as a resting stage.

MEROSTOMATA. Class of aquatic arthropods (formerly the order Xiphosura of the class Arachnida). Includes the extinct EURYPTERIDA. The only living forms (king crabs, e.g. *Limulus*) have a broad cephalothorax (prosoma) covered dorsally by carapace which covers limbs and in which is located a pair of eyes. Chelicerae chelate. Pedipalps resemble walking limbs. Gnathobases (spiny basal segments of legs) function as mandibles, which are absent. Opisthosoma represented by fused tergites forming a single dorsal plate, with a long caudal spine hinged to its posterior border. Gill books serve for gaseous exchange. The trilobite larva of *Limulus* has chelicerae and lacks antennae; however the chelicerate arthropods could have been derived from a pro-trilobite. See ARTHROPODA.

-merous. As a suffix, referring to the number of parts, e.g. corolla pentamerous, consisting of five petals.

MESARCH. Type of maturation of primary xylem from a central point outwards; i.e. the oldest xylem elements (protoxylem) are surrounded by later-forming metaxylem.

MESENCEPHALON. See MIDBRAIN.

MESENCHYME. Embryonic mesoderm comprising widely scattered tissue giving rise to connective tissue, blood, cartilage, bone, etc.

MESENTERY. (1) Double-layered extension of the peritoneum attaching stomach and intestines to the dorsal body wall. Contains blood, lymph and nerve supply to these organs. See OMENTUM. (2) Vertical partitions of body wall of anthozoans (sea anemones) forming compartments within the enteron.

MESOCARP. Middle layer of mature ovary wall, or pericarp; between the exocarp and endocarp. See FRUIT.

MESODERM. Middle germ layer of triploblastic animals, coming to lie between ectoderm and endoderm from gastrulation onwards (see GERM LAYER for mesodermal derivatives). In animals with a large coelom it is separable into an inner *splanchnic mesoderm* forming outer covering of digestive tract and its diverticula, and an outer *somatic mesoderm* from which, when present, develop the somatic skeleton and musculature, kidneys and gonads. Part of the mesoderm commonly becomes divided during development into a series of blocks, or *somites*, which form the basic developmental units of segmented coelomates. In vertebrates the somites, which are dorsal and on either side of the neural tube, each undergo differentiation into lateral *dermatome* and medial *myotome* with *sclerotome* around the neural tube and notochord. These will give rise respectively to dermis, striated muscle and vertebral column and adjacent rib components. Vertebrate mesoderm undivided into somites forms unsegmented *lateral plate mesoderm*, giving rise to splanchnic and somatic mesoderm. The *extraembryonic mesoderm* of amniotes is mesoderm that has spread out to cover the trophoblast and forms the CHORION.

MESOGENE DEVELOPMENT. Development of stomata where GUARD CELLS and subsidiary cells share a common parental cell.

MESOGLOEA. Layer of jelly-like material between ectoderm and endoderm of coelenterates. Merely a non-cellular collagenous membrane in hydrozoans such as *Hydra*, but much enlarged, thickened and fibrous in jellyfish (scyphozoans) and in these and some other groups comes to contain cells derived from the two tissue layers. It is not itself a germ layer.

MESOKARYOTE. Term proposed for the dinoflagellate (PYRRO-PHYTA) nucleus, where the chromosomes persist in a condensed state at all times; used also to denote the dinoflagellate evolutionary position, between prokaryotes and eukaryotes.

MESONEPHROS. See KIDNEY.

MESOPHILIC. (Of microorganisms) with optimal temperature for growth between 20°–45°C.

MESOPHYLL. Internal tissue of LEAF BLADE; differentiated into upper *palisade* and lower *spongy* mesophyll, the latter cells with generally fewer chloroplasts and separated by large air spaces. See LEAF.

MESOPHYTE. Plant found growing under average conditions of water supply. Compare HYDROPHYTE, XEROPHYTE.

MESOSOME. Infolded region of plasma membrane of bacterial cell, containing electron transport system, attaching to the circular chromosome and involved in initiation (and termination?) of chromosomal replication. Generally located near newly-forming cell wall in binary fission. Very probably homologous with mitochondrial cristae. See LOMASOME.

MESOTHELIUM. Epithelium-like layer covering vertebrate SEROUS MEMBRANES; flattened squamous cells derived from mesoderm, and therefore not strictly epithelial.

MESOZOIC. Geological era extending from 225–70 Myr BP; includes Triassic, Jurassic and Cretaceous periods. See GEOLOGICAL PERIODS.

MESSENGER RNA (mRNA). Single-stranded RNA molecule, translated on ribosomes into a polypeptide. Produced (transcribed) by RNA polymerase, only one of the two DNA strands being read by the polymerase and acting as template (see PROTEIN SYNTHESIS). The transcription product (hnRNA in eukaryotes) generally contains some base sequences not coding for any part of the eventual polypeptide, these being cut out enzymatically (see RNA PROCESSING, INTRON) before the processed mRNA is attached to a ribosome. The amino acid sequence of the polypeptide reflects the mRNA nucleotide sequence, as dictated by the GENETIC CODE. Most mRNA molecules are short-lived in the cell, but some eukaryotic cells lacking nuclei have relatively long-lived mRNA, as do some egg cells.

METABASIDIUM. Cell in which MEIOSIS occurs in members of the BASIDIOMYCOTINA. Compare PROBASIDIUM.

METABOLA. See PTERYGOTA.

METABOLISM. Sum of the physical and chemical processes occurring within a living organism; often intended to refer only to its enzymic reactions. May be regarded as comprising *anabolism* (build-up of molecules) and *catabolism* (breakdown of molecules). Frequently used in context of a particular class of compounds within an organism, e.g. fat metabolism. The term *metabolic rate* is often used rather loosely, more or less synonymously with respiratory rate, the level of oxygen consumption in aerobes being an indicator of the general metabolic activity of the organism.

METABOLITE. Substance participating in METABOLISM. Some are intermediary compounds of biochemical pathways; others are taken in from the environment.

METACARPAL BONES. Rod-like bones of fore-foot of tetrapods articulating with carpals (wrist-bones) proximally and finger bones (phalanges) distally. Usually one corresponding to each digit. Compare METATARSAL BONES. See PENTADACTYL LIMB.

METACARPUS. Region of tetrapod fore-foot containing metacarpal bones. Palm region of man.

METACENTRIC. See CENTROMERE.

METACHROMATIC GRANULE (POLYPHOSPHATE BODY, VOLUTIN GRANULE). Spherical granule containing stored phosphate found in cells of blue-green algae (CYANOBACTERIA). Absent from young cells or cells grown in phosphate-deficient medium; prominent in older cells.

METACHROMATIC STAIN. See STAINING.

METACHRONAL RHYTHM. Pattern of beating adopted by groups of cilia, segmented parapodia of polychaetes and arthropod limbs, in which each unit (cilium, parapodium, limb) is at a slightly different stage in the beat cycle from those on either side of it. Result is a smooth progression of waves of beating along the units. Often occurs in forms of locomotion, microphagous feeding and exchange of water over a respiratory surface.

METAGENESIS. See ALTERATION OF GENERATIONS.

METAMERIC SEGMENTATION (METAMERISM). See SEGMENTATION.

METAMORPHOSIS. Process during, and as a result of, which an animal undergoes a comparatively rapid change from larval to adult form. Under hormonal control, it is most notable in the life histories of many marine invertebrates, the majority of insects (especially ENDOPTERYGOTA) and of AMPHIBIA. Often requires destruction of much larval tissue (see LYSOSOMES) and changes in gene expression. Of enormous evolutionary and ecological significance. See CORPORA ALLATA, THYROXINE.

METANEPHROS. See KIDNEY.

METAPHASE. Stage in MITOSIS and MEIOSIS.

METAPHLOEM. Primary PHLOEM formed, after the protophloem.

METAPHYTON. Algae present between epiphytic algae, occurring as a loose collection of non-motile or weakly motile forms, lacking ready means of attachment to a substratum.

METAPLASIA. Transformation of one kind of adult cell into another not usually found in that part of the body. May occur during tumour formation.

METARTERIOLES. Minute branches of arterioles (8–18 μm in diameter). Generally give rise to capillary beds which may be opened and closed by precapillary sphincters. See CAPILLARY.

METASTASIS. Movement of infectious microorganisms or malignant CANCER CELLS, usually via blood stream or lymph, from one focus of growth to another part of the body where they set up further foci, in the case of cancer cells often by attachment to the endothelium of a capillary.

METATARSAL BONES. Rod-like bones of tetrapod hind-foot, usually one corresponding to each digit, articulating with ankle bone (tarsal) proximally and toe bones (phalanges) distally. See PENTADACTYL LIMB.

METATARSUS. Region of tetrapod hind-foot containing metatarsal bones.

METATHERIA. Marsupials. Subclass or infraclass of MAMMALIA containing just one order (Marsupialia). Originated in early Cretaceous; short gestation period and very undeveloped offspring at birth; a protective pouch (marsupium) in most forms investing the mammary glands of the female and housing the young while they complete development. Restricted to Australian region (including New Guinea) and the New World; fossil marsupials appear to be absent from Europe and Mongolia. Decline in diversity during late Cretaceous. Many resemble placental mammals of various orders, exhibiting CONVERGENCE. See PLACENTA

METAXYLEM. Primary xylem formed after the protoxylem. Cell elongation is generally complete, or nearly so, by the time of its production.

METAZOA. Multicellular animals; a subkingdom of the ANIMALIA. Characterized by differentiation of cells enabling division of labour, organization of at least tissue grade, development including recognizable larva and intercellular communication. Sponges (PARAZOA) are sometimes regarded as non-metazoan since they lack a nervous system providing communication between cells; but this is achieved in other remarkable ways.

METHANOGENS. Particular bacteria occurring in a reducing environment devoid of oxygen, and able to produce methane (marsh gas) from carbon dioxide and hydrogen.

MHC. See MAJOR HISTOCOMPATIBILITY COMPLEX.

MICELLE. Particle of colloidal size, normally spherical, which in aqueous medium has a hydrophilic exterior and hydrophobic interior. Commonly consists of a phospholipid monolayer, sometimes with associated protein, within which non-polar materials may be housed and transported. Several storage polysaccharides (e.g. amylose, amylopectin) form hydrated micelles rather than true solutions within cells. See BILE, CHYLOMICRON, LIPOPROTEIN, LIPOSOME.

MICHAELIS CONSTANT (MICHAELIS-MENTEN CONSTANT). For a given enzyme–substrate reaction, the value (K_M) of the substrate concentration at which the initial velocity of the reaction (V_O) is half maximal. A low value indicates high affinity of enzyme for substrate. Allosteric enzymes do not strictly have K_M values. See ENZYME.

MICROBE. Microscopic organism of any type, whether prokaryotic or eukaryotic. Many are pathogenic.

MICROBIOLOGY. Branch of biology dealing with MICRO-ORGANISMS.

MICROBODIES. Term often used to indicate PEROXISOMES and glyoxysomes (see GLYOXYLATE CYCLE).

MICRODISSECTION. Technique used in operating upon small organisms whereby the specimen is dissected while viewed through a microscope. Instruments are normally manipulated by mechanical means. Micromanipulation is often used where living cells or their nuclei are removed or inserted.

MICROFILAMENT. See ACTIN, CYTOSKELETON.

MICROFOSSIL. Microscopic fossil; includes spores, pollen grains, algae, tracheids, pieces of plant cuticle, animals, etc.

MICROGAMETE. See GAMETE, ANISOGAMY.

MICROGAMETOPHYTE. One of two types of gametophyte produced by HETEROSPOROUS plants; develops from microspore. In ferns, comprises a *prothallus* (confined to the microspore in clubmosses like *Selaginella*); in seed plants, comprises microspore within wall of pollen grain, plus its mitotic products, including pollen tube and its nuclei.

MICROGRAPH. Photograph of an image obtained during either light or electron microscopy.

MICROMETRE (MICRON). Unit of length often used in microscopy. Symbol μm; 10^{-3} mm; 10^{-6} m. Many bacterial cells are approximately 1 μm in length. 1 μm = 1000 nm = 10 000 Å.

MICRON. See MICROMETRE.

MICRONUCLEUS. See NUCLEUS.

MICRONUTRIENT. Substance required by an organism from its environment for healthy growth, but only in minute amounts. Includes TRACE ELEMENTS and VITAMINS.

MICROORGANISM. Microscopically small organism. Includes unicellular plants and animals, bacteria and many fungi. Viruses are commonly included as well.

MICROPHAGY. Methods employed by animals in feeding upon particles which are small in relation to their own size. Such feeding tends to occur continually, frequently by sieving of particles from water by such devices as baleen plates (baleen whales), by ciliary-mucus devices in pharyngeal gill slits (urochordates, cephalochordates) and other gills (bivalve molluscs); by lophophores; by trapping particles in bristle-fringed trunk limbs (many crustacea), or even cytoplasmic nets (heliozoans, foraminiferans).

MICROPHYLL. Type of leaf usually but not always small, possessing very simple vascular system comprising single branched vein. LEAF TRACE is not associated with a LEAF GAP in stele of stem. Associated with PROTOSTELIC vascular system. Characteristic of club mosses (LYCOPHYTA), horsetails (SPHENOPHYTA) and related forms. Compare MEGAPHYLL.

MICROPYLE. Canal formed by extension of integument(s) of ovule beyond apex of nucellus; recognizable in a mature seed as a minute pore in seed coat through which water enters at start of germination.

MICROSCOPE, MICROSCOPY. Microscopes are instruments employing lenses to produce magnified images, and hence fine detail, of objects too small to observe clearly with the naked eye. Earliest *light microscopes* (visible light as the transmitted medium) appeared in about 1590 in Holland and had two glass lenses. In these *compound microscopes* one short focal length lens, the eye-piece lens, produces a magnified apparent image of a real image produced by a second short focal length lens, the objective lens, placed closer to the object. Total magnification is the product of the magnifications produced by these two lenses. Hooke (1665) first recorded cells in cork, and Leeuwenhoek first noted infusoria (1676) and bacteria (1683), which were first stained by von Gleichen (1778). In the 1870s E. Abbe and C. Zeiss produced the first oil-immersion objective lens enabling a good image of up to ×1500 magnification. Achromatic and aplanatic objective lenses (the finest available to this day) were developed by J. J. Lister, those of 1886 from Zeiss being of a very high quality.

The thinner the material being observed, the greater the clarity of image; hence the improvement of embedding, cutting and sectioning methods to match the evolution of the microscope. Light reflected from a point object cannot be recombined again to form another true point, but only a disc of light. When discs representing adjacent object points overlap detail is lost. The *resolving power* of a microscope, its ability to distinguish fine detail, is proportional to the wavelength of the transmitted medium. Visible light has a wavelength of about 0.5 μm and the best resolving power (even using visible light of the shortest wavelength) is about 0.45 μm. Objects closer together will not be resolved as more than one object. Where an object (e.g. a typical cell) is transparent, with features differing only in refractive index, light rays will emerge with different phase relations depending on the paths they have taken. The resulting image has uniform brightness, but the technique of *phase contrast microscopy* makes use of its phase differences so as to produce the image that would have been seen had these been amplitude differences.

Electron microscopes use wave properties of electrons fired through the object held in a vacuum. In transmission electron microscopes, electrons are accelerated through the microscope by a large voltage (up to 1000 kV), those passing through the object hitting a screen, which fluoresces giving an image. The electrons are focused by electron magnets. The wavelength of electrons is inversely related to the voltage used, but even in low voltage apparatus is about 0.005 nm – four orders of magnitude (10^4) less than that of a light wave. Resolving power of the order of 1 nm for biological material (less for crystals) is achievable. This is about 200 times better than in light microscopy. Materials for observation are commonly first fixed and then embedded in Araldite® prior to sectioning by an ultramicrotome, giving sections 20–100 nm thick. The dangers of so generating artefacts are well known and are avoided as far as possible. Biological material contains few heavy atoms and consequently its electron-scattering ability is poor, but can be improved by soaking the object in, or spraying it with, a salt of a heavy metal ('staining' it). In *scanning electron microscopy* the electron beam causes the object to emit its own electrons. These can be used to produce an image which is built up as the electron beam scans the specimen (rather in the way a television picture is produced). Micrographs produced have a three-dimensional quality, but lower resolution than in transmission microscopy. One technique often employed for looking at hydrophobic interiors of membranes is *freeze-fracture electron microscopy*, in which cells are frozen in ice to the temperature of liquid nitrogen ($-196\,°C$) and then fractured, the plane of fracture tending to pass through the middles of lipid bilayers. Exposed fracture surfaces are then shadowed with platinum and carbon followed by digestion of the organic content (*freeze-etching*), leaving a platinum 'replica' which can be examined with the electron microscope. This is particularly useful for detecting where proteins are located in membranes. In electron microscopy, the specimen is inevitably in a vacuum and exposed to high temperature electron bombardment; hence it is impossible to view living material, specimens requiring cooling during viewing.

MICROSOMES. Products of homogenization of endoplasmic reticulum. Rough microsomes are closed vesicles with ribosomes attached to their outer surfaces; smooth microsomes lack ribosomes and may derive from plasma membrane as well as from smooth endoplasmic reticulum.

MICROSPIKE. Small FILOPODIUM.

MICROSPORANGIUM. Sporangium in which microspores are formed; borne on a microsporophyll. In seed plants, a *pollen sac.*

MICROSPORE. Smaller of two kinds of spore produced by heterosporous plants (e.g. ferns, seed plants); first cell of microgametophyte generation in these plants, developing after meiosis from a *microspore mother cell* (microsporocyte). In seed plants, becomes a pollen grain. Compare MEGASPORE.

MICROSPOROPHYLL. Leaf, or modified leaf, that bears microsporangia. In flowering plants, a STAMEN. Compare MEGASPOROPHYLL.

MICROTOME. Machine for cutting extremely thin sections of tissue, etc. For light microscopy the material is first fixed and either frozen or embedded in paraffin wax, sections cut with a steel knife usually being 3–20 μm thick; for electron microscopy, fixing is followed by embedding in a resin such as Araldite®, sections cut with the glass or diamond knife of an *ultramicrotome* being 20–100 nm thick.

MICROTUBULAR ROOT. Microtubules attaching to BASAL BODY of flagellum within a motile cell.

MICROTUBULE. One of the essential protein filaments of the CYTO-SKELETONS of probably all eukaryotic cells, and of their cilia, flagella, basal bodies, centrioles and mitotic and meiotic spindles. Each microtubule consists of a hollow cylinder, about 25 nm in diameter, made up of 13 protofilaments of the protein TUBULIN. Each protofilament consists of globular tubulin molecules polymerized together. Like ACTIN filaments, microtubules grow and depolymerize at different rates at their two ends. Microtubule roles include guiding organelle and chromosome movement in the cell, causing cell elongation by their own elongation and involvement in ciliary and flagellar beating. Various associated proteins play modifying roles in the behaviour of these microtubules. See SPINDLE.

MICROVILLUS. Minute finger-like projection from the surfaces of many eukaryotic cells, particularly animal epithelia involved in active uptake (e.g. small intestine, kidney proximal convoluted tubule) where, several thousand strong, they constitute the *brush borders* observed in electron micrographs. Each microvillus is about 1μm long and 0.1 μm in diameter. About 40 ACTIN microfilaments run along its length, supported by accessory proteins (e.g. alpha-actinin, fimbrin). They may be capable of retraction and extension, possibly through sliding of actin over myosin (see MUSCLE CONTRACTION). Brush borders may increase area of plasma membrane available for absorption 25-fold. *Stereocilia* are long, thick microvilli, about 4 μm in length (see

HAIR CELL), and the rhabdomeres of arthropod compound eyes are also composed of microvilli (see EYE). See INTERCELLULAR JUNCTIONS for diagram.

MIDBRAIN. See BRAIN.

MIDDLE EAR. See EAR, MIDDLE.

MIDDLE LAMELLA. Layer of intercellular material cementing together the primary walls of adjacent cells. Contains calcium pectate. See CELL WALL.

MIDGUT. (1) In vertebrate development, a somewhat arbitrary division of the endodermal layer in contact with remnants of the yolk sac and giving rise to part of duodenum, rest of small intestine and upper part of large intestine. (2) Cylinder of endodermal epithelial cells of the arthropod gut (mesenteron). Not lined by cuticle; secretes a *peritrophic membrane*, enclosing food and separating and protecting the epithelium from abrasion. Most digestion occurs here, with glycogen storage and secretion of urate (see MALPIGHIAN TUBULES). Commonly bears one or more pairs of diverticula (midgut or hepatic caeca), increasing its surface area.

MILDEW. (1) Plant disease caused by a fungus, producing superficial, powdery or downy growth on host surface. (2) Fungus causing such a disease. (3) Often used synonymously with MOULD.

MILK. (1) Complex aqueous secretion of MAMMARY GLANDS, with which mammals suckle their young. Composition varies, but usually rich in suspended fat, in protein (mostly *casein*) and sugar (mainly *lactose*). In addition, minerals, vitamins and antibodies (including antibacterial IgA in human milk, but not in cow's milk) and (again in humans but not in cows) the important iron-binding protein *lactoferrin* inhibiting bacterial growth in the baby's intestine. Milk production is under PROLACTIN control, and its ejection involves OXYTOCIN release during the *sucking reflex*. See PARTURITION. (2) See CROP MILK.

MILK TEETH. See DECIDUOUS TEETH.

MILLIPEDES. See DIPLOPODA.

MILLON'S TEST. Common test for proteins and phenols. Proteins give positive results because of presence of the amino acid tyrosine. A few cm^3 of reagent ($150 \, g \cdot dm^{-3}$ aqueous mercuric sulphate in 15% v/v sulphuric acid) is added to test solution in a test tube and boiled. After cooling and addition of a few drops of sodium nitrite solution ($10 \, g \cdot dm^{-3}$) a brick red colour indicates a positive result. The BIURET REACTION is nowadays preferred in schools since Millon's reagent is potentially harmful.

MIMICRY. Relational term, indicating that the signal component of some mutually beneficial evolved signal–response pairing between two organisms (one signalling, the other responding) has been simulated or employed by a third party to its own advantage, often to the detriment of one or both of the original parties.

In *Müllerian mimicry* (after F. Müller) two or more organisms independently derive protection from predation, for example by tasting repellent; but they also benefit mutually through convergent evolution of similar warning (aposematic) colouration and pattern, predators learning by association to avoid both after tasting one. In *Batesian mimicry* (after H. W. Bates) an edible or otherwise relatively defenceless organism (the *mimic*) gains protection from predation by resemblance to a distasteful, poisonous or harmful organism, termed the *model*. In this case, the benefit gained will depend in part upon the ratio of availabilities of models and mimics, predators tending to require reinforcement of the learnt association of colour/pattern and disagreeableness. Some species (e.g. certain butterflies) may be Batesian mimics of more than one model (see POLYMORPHISM). In *aggressive mimicry*, the mimic feeds on one of the original parties in the original signal–response pairing. Some instances of CRYPSIS (e.g. stick or leaf insects) may appear to overlap the criteria of mimicry given above; but it can be argued that crypsis does not involve manipulation of a true signal (as opposed to background noise) in the service of a new function. Many flowering plants, notably several orchids, achieve pollination through mimicry of signals (e.g. visual, olfactory) attracting specific insect pollinators. Mimicry has provided much support for Darwinian evolution.

MINIMAL MEDIUM. Medium for growing microorganisms, containing ample quantities of all minimal nutritional requirements of the wild-type organism. See PROTOTROPH, AUXOTROPH.

MIOCENE. Geological epoch; subdivision of Tertiary period. See GEOLOGICAL PERIODS.

MIRACIDIUM. Ciliated larval stage of endoparasitic trematode platyhelminths (flukes). Commonly bores into snail, where it develops into a sporocyst. See TREMATODA.

MITES. See ACARI.

MITOCHONDRION (CHONDRIOSOME). Cytoplasmic organelle of all eukaryotic cells engaging in aerobic respiration, and the source of most ATP in those cells. Vary in number from just one to several thousand per cell. Most electron micrographs of mitochondria show them as either cylindrical (up to 10 μm long and 0.2. μm in diameter) or roughly spherical (0.5–5 μm in diameter). Have two membranes, the outer smooth and generally featureless, the inner invaginating to produce *cristae*, generally at right angles to long axis of the mitochondrion. Mitochondria from very active tissues have large numbers of tightly-packed cristae, those from relatively anoxic cells have few cristae. The matrix within the inner membrane is often granular and contains ribosomes (of prokaryotic type), several copies of a circular DNA molecule (see mtDNA) and other protein-synthesizing components, several enzymes (e.g. those for the KREBS CYCLE, part of the UREA cycle and for FATTY ACID OXIDATION) and variable amounts of calcium and phosphate ions. Enzymes for haem synthesis (both for cytochromes and haemoglobin) occur within the mitochondrial matrix.

Inner membrane is the site of the ELECTRON TRANSPORT SYSTEM, the ATP-synthesizing complex and enzymes involved in fatty acid synthesis. Enzymes in outer membrane include those involved in oxidation of adrenaline and serotonin, and others engaged in phospholipid metabolism. The two membranes differ in their permeabilities: outer freely permeable to small and medium-sized molecules and ions; inner permeable only to small uncharged molecules such as O_2 and undissociated H_2O and impermeable to glucose, but with various TRANSPORT PROTEINS permitting exchange of ATP and ADP, Krebs cycle intermediates and the accumulation of pyruvate, calcium and phosphate ions against concentration gradients, all coupled to electron transport (see ACTIVE TRANSPORT). The ATP synthetase complex ($F_0F_1ATPase$) is also embedded in inner membrane, comprising at least nine different polypeptides in two separable units: five make a spherical head ($F_1ATPase$) which can catalyse ATP synthesis; remainder comprise a proton channel (F_0) through which protons re-enter the mitochondrion after ejection by the electron transport system (see BACTERIORHODOPSIN and Fig. 14a).

New mitochondria arise by growth and division of existing mitochondria (resembling binary fission in bacteria). Many of their proteins are encoded by genes in the nuclear genome, in particular the enzymes of the Krebs cycle and fatty acid oxidation, much of the electron transport system, the outer membrane proteins and DNA and RNA polymerases. Mitochondrial GENETIC CODE differs from both 'nuclear' and bacterial codes in that the triplet UGA codes for tryptophan and is not a stop codon. Other codons may also have different meanings, even between mitochondria from different organisms.

Mitochondria of most C_3 plants of temperate latitudes do not engage in aerobic respiration during daylight, this being restricted to the hours of darkness. Instead these plants engage in the seemingly wasteful process of PHOTORESPIRATION, which is greatly reduced or absent in C_4 plants of tropical origin.

Mitochondria are generally regarded as having originated by ENDOSYMBIOSIS from a purple photosynthetic bacterium.

MITOSIS (KARYOKINESIS). Method of nuclear division which produces two daughter nuclei, genetically identical to each other and to the original parent nucleus. Commonly accompanied by division of parental cytoplasm (cytokinesis) around the daughter nuclei to produce two daughter cells (cell division). It is the usual method by which nuclei are replicated during the growth, development and repair of multicellular organisms and coenocytic fungi. Prokaryotes lack nuclei and have no mitosis, replicating and segregating their DNA by a different process. Body cells (and nuclei) not undergoing mitosis are typically in INTERPHASE. The first stage of mitosis (*prophase*) commences with the first appearance of condensed chromosomes, each consisting of two identical sister chromatids held together by a CENTROMERE. In contrast to their behaviour in first meiotic prophase, homologous mitotic chromosomes do not normally pair up (see Fig. 48).

During *prophase* a spindle of MICROTUBULES assembles outside the nucleus. If centrioles are present, they have replicated in S-phase (see CELL CYCLE) and form foci for origins of spindle fibres. The two ASTERS get pushed apart by growth of microtubules. In *metaphase* the nuclear membrane disintegrates and microtubules growing from the KINETOCHORES of the chromosome centromeres attach to those of the spindle, the resulting agitation making all chromosomes come to lie in one plane (the metaphase plate) half way between the two spindle poles. Each chromosome lies with its long axis at right angles to the axis of the spindle. Metaphase may be quite lengthy. *Anaphase* begins when kinetochores separate, the two sister chromatids of a chromosome being pulled towards the opposite spindle poles, which move further apart. Each chromatid is now a chromosome in its own right. A few minutes later, once chromosomes have reached the poles, a new nuclear membrane appears around each group of chromosomes, which decondense. Nucleoli reappear, constituting *telophase*. Mitosis is now complete and two daughter cells are normally produced by cytokinesis. Mitosis normally takes between a half and three hours. See FISSION, CLEAVAGE. Compare MEIOSIS.

MITOSPORE. Spore produced as a result of mitosis. Contrast MEIOSPORE.

MITRAL VALVE. Valve comprising two membranous flaps and their supporting tendons between the atrium and ventricle on left side of the hearts of birds and mammals. See HEART.

MIXED TISSUE. Tissue containing more than one cell type, all originating from same group of stem cells in development. XYLEM and PHLOEM are examples.

MIXOTROPH. PHOTOAUTOTROPH capable of utilizing organic compounds in the environment; *facultative heterotroph* or *facultative parasite* may also refer to a mixotroph.

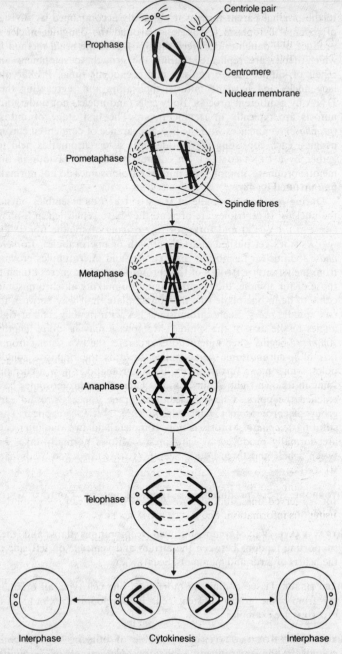

Fig. 48. *Diagrams showing the behaviour of two pairs of chromosomes in animal mitosis.*

Labels on figure:

Prophase — Centriole pair, Centromere, Nuclear membrane

Prometaphase — Spindle fibres

Metaphase

Anaphase

Telophase

Interphase — Cytokinesis — Interphase

MODIFICATION. Methylase activity preventing or reversing host restriction (see PHAGE RESTRICTION). Specific bases of the phage DNA within target site of the restriction endonuclease are methylated. Non-heritable change.

MODIFIER. Gene capable of altering expression of another gene at a different locus in the genome. Selection for such modifiers can establish a genetic background within which a new mutation at a locus will consequently be expressed to a greater or lesser extent. Modifiers are thus strong candidates for control of DOMINANCE and recessiveness of characters. See DOSAGE COMPENSATION, EPISTASIS.

MODULATOR. (Of enzyme). See allosteric ENZYME, ATP.

MOLAR. Crushing back tooth of mammal, without milk tooth predecessor (unlike premolar). Usually has several roots and complicated pattern of ridges and projections on grinding surface. See DENTITION.

MOLE. See INSECTIVORA.

MOLECULAR ACTIVITY (TURNOVER NUMBER). Number of substrate molecules catalysed per minute by a single enzyme molecule (or active site). Carbonic anhydrase has one of the highest molecular activities: 36×10^6 min^{-1} molecule^{-1}.

MOLECULAR CLOCK. The more important it is for the amino acid sequence of a protein to be conserved, the more slowly does that sequence evolve (the more conservative it is) through evolutionary time. For instance, the discarded peptide fragments formed when fibrinogen is converted to fibrin have no biological import and their sequence evolves rapidly, whereas haemoglobin and especially HISTONES are very conservative. Can be used to estimate the time since any two evolutionary lineages diverged, using proteins from living species as representatives of those lineages; however, it has to be remembered that the 'clock' runs faster for neutral sequences than for conserved ones. Indeed, evolutionary lineages can be established using this information. See MUTATION RATE

MOLLUSCA. Phylum of bilaterally symmetrical, unsegmented invertebrates. Largely aquatic; coelom restricted to spaces around heart and within kidneys and gonads. Characteristically soft-bodied, with anterior head (rudimentary in Bivalvia); commonly with rasping tongue (radula), large muscular ventral foot (modified to 'arms' in Cephalopoda); with viscera usually in a hump dorsal to the foot. Outer layer of visceral hump called the *mantle*, usually covered by a shell. Mantle typically enclosing a *mantle cavity* in which usually lie the gills (ctenidia), anus and opening of kidney and reproductive ducts. Heart typically present; circulation often includes both open (haemocoele) and closed components. Nervous system of cords and ganglia (see CENTRAL NERVOUS SYSTEM). Development mostly by spiral CLEAVAGE, often with trochophore larva. Includes Monoplacophora, Amphineura (chitons), Gastropoda (snails, slugs), Scaphopoda, Bivalvia (clams, oysters, etc.) and Cephalopoda (octopus, cuttlefish, squid). Enormous radiation among the phylum. About 80 000 species living; numerous fossils back to the Cambrian.

MONADELPHOUS. (Of stamens) united by their filaments to form tube surrounding the style; e.g. in lupin, hollyhock. Compare DIADELPHOUS, POLYADELPHOUS.

MONERA. Taxon sometimes used in classification to include all, and only, prokaryotic organisms (BACTERIA, CYANOBACTERIA, ARCHAEBACTERIA, MYCOPLASMAS). Commonly has the status of a KINGDOM. See PROKARYOTE.

MONGOLISM. See DOWN'S SYNDROME.

MONOCARPIC. (Of a plant) flowering once during its life.

MONOCHASIUM. See INFLORESCENCE.

MONOCHLAMYDEOUS (HAPLOCHLAMYDEOUS). (Of flowers) having only one whorl of perianth segments.

MONOCLONAL ANTIBODY. See ANTIBODY.

MONOCOLPATE. Referring to a pollen grain with one furrow or groove through which pollen tube will emerge.

MONOCOTYLEDONAE. Smaller of the two classes into which flowering plants (ANTHOPHYTA) are divided; distinguished from the DICOTYLEDONAE by the presence of a single seed leaf (cotyledon) in the embryo, and by other structural features, such as parallel-veined leaves, stem vascular tissue in the form of scattered closed bundles, flower parts usually in threes or multiples of three. A few monocotyledonous plants are large (e.g. palm trees), but majority are small. Includes many important food plants; e.g. cereals, fodder grasses, bananas, palms, also ornamentals, e.g. orchids, lilies, tulips.

MONOCYTE. Largest of the kinds of vertebrate LEUCOCYTE. See also MYELOID TISSUE.

MONOECIOUS. (1) (Of plants), having both male and female reproductive organs on the same individual; in flowering plants, having unisexual, male and female, flowers on the same plant, e.g. hazel. See DIOECIOUS, HERMAPHRODITE. (2) (Of animals) see HERMAPHRODITE.

MONOKARYON (MONOCARYON). A fungal cell, hypha or mycelium in which there are one or more genetically identical haploid nuclei. Characteristic of early phase in life cycles of many of the BASIDIOMYCOTINA. Compare DIKARYON, HETEROKARYOSIS.

MONOPHYLETIC. Of a taxon or taxa originating from and including a single stem species (either known or hypothesized) and either including the whole CLADE so derived (a holophyletic taxon), or else excluding one or more smaller clades nested within it (a paraphyletic taxon).

MONOPLANETISM. In some fungi (Oomycetes), where a zoospore of only one type is formed. Contrast DIPLANETISM.

MONOPODIUM. An axis produced and increasing in length by apical growth; e.g. trunk of a pine tree. Compare SYMPODIUM.

MONOSACCATE. Refers to a type of pollen grain with a single air bladder.

MONOSACCHARIDES. Simple sugars, with molecules often containing either five carbon atoms (pentoses such as ribose, $C_5H_{10}O_5$) or six (hexoses such as glucose and fructose, $C_6H_{12}O_6$). Some, notably glucose and its amino derivatives, are monomers of biologically important POLYSACCHARIDES and GLYCOSAMINOGLYCANS. As carbohydrates their empirical formula approximates to $C_x(H_2O)_x$. Trioses (3-carbon sugars) such as phosphoglyceraldehyde and phosphoglyceric acid are important intermediates of many biochemical pathways (e.g. see GLYCOLYSIS, PHOTOSYNTHESIS).

MONOSOMY. Abnormal chromosome complement, one chromosome pair in an otherwise diploid nucleus being represented by just a single chromosome. An instance of ANEUPLOIDY. May arise by NON-DISJUNCTION. The norm in the heterogametic sex of those animals with the XO/XX mode of SEX DETERMINATION.

MONOSTROMATIC. Refers to a THALLUS one cell in thickness.

MONOSULCATE. Refers to a pollen grain with one furrow or groove on distal surface.

MONOTREMATA. Monotremes. The only order of the mammalian subclass PROTOTHERIA.

MONOZYGOTIC TWINS. Twins developing from two genetically identical cells, cleavage products of a single fertilized egg which become completely separate (failure to do so may give rise to Siamese twins). They will be of the same sex. The nine-banded armadillo regularly produces a genetically identical litter and this amounts to a form of asexual reproduction. See FRATERNAL TWINS, FREEMARTIN, POLYEMBRYONY.

MORPH. One form of INFRASPECIFIC VARIATION; one of the *forms* of a species population exhibiting POLYMORPHISM, generally of a markedly obvious kind.

MORPHACTINS. Group of synthetic compounds derived from fluorenecarboxylic acid, that reduce and modify plant growth. Internally, the orientation of the SPINDLE axes of dividing cells is altered; most striking external effect is development of dwarf, bushy habit due to shortening of the internodes and loss of apical dominance (see AUXINS). Other effects include inhibition of phototropism and geotropism, of seed germination, and of lateral root development.

MORPHOGEN. Substance responsible for some aspect of MORPHOGENESIS. Controversy surrounding their putative existence seems to be resolving in their favour, largely through work on the development of the chick limb bud (see RETINOIC ACID) and on developmental genetics in *Drosophila* and other arthropods (see BICOID GENE and references cited there). It seems to be clear that at least some of the molecules with morphogenic activity are TRANSCRIPTION FACTORS.

MORPHOGENESIS. The generation of form and structure during development of an individual organism. *Morphogenetic movements* involve displacements and migrations of large numbers of cells during ontogeny, being particularly pronounced during GASTRULATION. One aspect of such movements, if they are to result in appropriate cell location, is ADHESION between cells.

Considerable interest surrounds the theory that an animal's form develops through a hierarchical series of decisions determining the fates of cells and clones derived from them as cell number increases. The cues enabling these decisions to be made are believed to result in PATTERN FORMATION, and take the form of molecules (*morphogens*) activating or suppressing gene expression. See COMPARTMENT, POSITIONAL INFORMATION.

MORPHOLOGY. (1) Study of the form, or appearance, of organisms (both internal and external). Anatomy is one aspect of morphology. *Comparative morphology* is important in evolutionary study. (2) The actual 'appearance' of an organism, including such diverse features as behaviour, enzyme constitution and chromosome structure. Any and all detectable features of an organism.

MORULA. Animal embryo during CLEAVAGE, at the stage when it is a solid mass of cells (blastomeres). Stage prior to the BLASTULA. Human embryos implant at this stage.

MOSAIC. (1) Organism comprising clones of cells with different genotypes derived, however, from the same zygote (unlike CHIMAERAS). See HETEROCHROMATIN, OVOTESTIS. (2) (Bot.) Symptom of many virus diseases of plants; patchy variation of normal green colour, e.g. tobacco mosaic.

MOSAIC DEVELOPMENT (DETERMINATE DEVELOPMENT). Development in which CLEAVAGE of the egg produces blastomeres lacking the capacity to develop into entire embryos when isolated, even under favourable conditions, specific regions of the embryos being absent. Opposite of REGULATIVE DEVELOPMENT. Tunicates and echinoderms are among those animals with clear mosaic development. Those cleavage divisions which set up the mosaic condition are termed determinate cleavage divisions. May sometimes result from nonrandom distributions of cytoplasm in the egg (see MATERNAL EFFECT).

MOSSES. See BRYOPSIDA.

MOTOR END-PLATE. See NEUROMUSCULAR JUNCTION.

MOTOR NEURONE (MOTOR NERVE). Neurone carrying impulses away from central nervous system to an effector (usually a muscle or gland). Efferent neurone. In vertebrates, motor neurones leave spinal cord via the ventral horn of a spinal nerve.

MOTOR UNIT. One motor neurone, its nerve terminals, and the skeletal muscle fibres innervated by them. There may be 200 or more muscle fibres in each vertebrate motor unit in large muscles of the leg or trunk.

MOULD. (1) Any superficial growth of fungal mycelium. Frequently found on decaying fruit or bread. (2) Popular name for many fungi.

MOULTING. (1) For moulting in arthropods, see ECDYSIS. (2) The periodic shedding of FEATHERS in birds, and of hair in mammals.

MOUTHPARTS. Paired appendages of arthropod head segments, surrounding the mouth and concerned with feeding and sensory information. Evolved from walking limbs. Similarities in mouthparts between different arthropod groups are often due to convergence rather than homology. Arachnids have a pair of CHELICERAE and PEDIPALPS but no mandibles; crustaceans a pair each of MANDIBLES, MAXILLULES, MAXILLAE and MAXILLIPEDS; insects a pair of mandibles and maxillae, a LABIUM and a LABRUM; centipedes a pair of mandibles and two pairs of maxillae; millipedes a pair of mandibles and maxillae. Mouthparts show most adaptive radiation in insects, commonly being modified for piercing and/or sucking (see PROBOSCIS). See Fig. 49.

mtDNA. Mitochondrial DNA. The mitochondrial genome almost always consists of a circular duplex, always less than 100 kilobases in length (16.5 kb in humans). In animals, generally contains less than 20 structural genes, including some for mitochondrial ribosomal RNAs, transfer RNAs and perhaps for some subunits of a few of the respiratory enzymes. Mitochondrial DNA evolves rapidly, with a mutation rate up to 10 times that of nuclear DNA. See MITOCHONDRION, PLASMID, CYTOPLASMIC MALE STERILITY, cpDNA.

MUCIFEROUS BODY. Organelle of some algal cells (e.g. Euglenophyta,

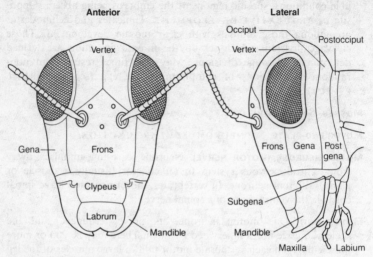

Fig. 49. *Anterior and lateral views of a locust head showing positions of mouthparts and head areas.*

Pyrrophyta) which can be ejected. They lack complex structure of TRICHOCYSTS.

MUCILAGE. Slimy fluid containing complex carbohydrates; secreted by many plants and animals. Swells in water, and is often involved in water retention.

MUCILAGE CANAL. Elongate cells in cortex of some brown algae (Order Laminariales) and cycads that may be involved in conduction of mucilage.

MUCILAGE HAIRS. Specialized mucilage-producing hairs of bryophytes, occurring most commonly near leaf axils and growing points of gametophores.

MUCINS (MUCOPROTEINS). Jelly-like, sticky or slippery GLYCO-PROTEINS. Formed by complexing a GLYCOSAMINOGLYCAN to a protein, the former contributing most of the mass. Some provide an intercellular bonding material, others lubrication. See PEP-TIDOGLYCAN.

MUCOPEPTIDES. Major constituents of cell walls of blue-green algae (CYANOBACTERIA) and bacteria, constituting up to 50% of the dry weight. They are GLYCOSAMINOGLYCANS, with chains of alternating N-acetylglucosamine and N-acetylmuramic acid residues linked by peptides sometimes including diaminopimelic acid; not found in eukaryotes.

MUCOPOLYSACCHARIDE. See GLYCOSAMINOGLYCAN.

MUCOPROTEINS. See MUCINS.

MUCOSA. See MUCOUS MEMBRANE.

MUCOUS MEMBRANE. General name for a moist epithelium and, in vertebrates, its immediately underlying connective tissue (*lamina propria*). Applied particularly in context of linings of gut and urogenital ducts. Epithelium is usually simple or stratified, often ciliated; usually contains goblet cells secreting mucus.

MUCUS. Slimy secretion containing MUCINS, secreted by goblet cells of vertebrate mucous membranes. Some invertebrates produce similar sticky or slimy fluids.

MÜLLERIAN DUCT. Oviduct of jawed vertebrates (i.e. excluding Agnatha). Generally paired, except in birds. A mesodermal tube, arising embryologically from mesothelium of coelom in close association with WOLFFIAN DUCT, and opening at one end by a ciliated funnel into coelom, joining the cloaca (or its remnant in placental mammals) at the other. Muscular and ciliary movements pass eggs down the tube; where fertilization is internal, sperms pass up it. In marsupials and placentals, gives rise to *Fallopian tube*, *uterus* and *vagina*. In all placentals, posterior parts of the pair of tubes fuse to produce a single, median, vagina and, as in humans, a single median uterus. The duct is just a remnant at most in males.

MÜLLERIAN MIMICRY. See MIMICRY.

MULTIAXIAL. Having an axis with several apical cells that give rise to nearly parallel filaments; found in many red algae (RHODOPHYTA).

MULTICELLULARITY. Occurrence of organisms composed not merely of more than one cell, but of cells between which there is (a) considerable division of labour brought about by cell DIFFERENTIA-TION, and (b) a fairly high level of intercellular communication. Increase in cell number must be accompanied by intercellular ADHES-ION, and by itself is of relatively little evolutionary significance. However it provides opportunities for division of labour and specialization between cells, which may enable further increase in size, providing scope for improved competitiveness in a variety of respects, and exploitation of fresh niches. Sponges (PORIFERA) display considerable communication between cells (see AMOEBOCYTES), although this does not take the form of a nervous system. Occasionally used rather loosely of groups of cells between which there is little differentiation and for which the terms colonial (e.g. *Volvox*) or filamentous (e.g. *Spirogyra*) are more appropriate. Social organization of prokaryotic MYXOBACTERIA and eukaryotic MYXOMYCOTA may well be termed multicellular. See ACELLULAR, COLONY, COEND-BIUM.

MULTIENZYME COMPLEX. Complex and compound molecule usually containing several enzyme subunits performing different stages in a biochemical pathway along with intrinsic cofactors (prosthetic groups). The DNA replicating enzymes (DNA *polymerase complex*) of cells, including those coded by viral genomes, form such complexes, as do *pyruvate dehydrogenases* of mitochondria. Often of sufficient size to be visible in electron micrographs, as are RIBOSOMES – justifiably considered to be multienzyme complexes.

MULTIFACTORIAL INHERITANCE. Any pattern of inheritance where variation in a particular aspect of phenotype is dependent upon more than one gene locus (in this sense synonymous with *polygenic*) but often also, especially in some human clinical disorders, upon particular environmental conditions which tip the balance in favour of expression: enough genes may predispose one to having the disease, but exposure to some environmental factor may be necessary before actual symptoms of the disease are presented. Allelic differences involved are usually of varying but small individual effect. Sex differences may also be involved.

MULTIFINGER LOOP. See TRANCRIPTION FACTORS.

MULTIGENE FAMILIES. genes with considerable base sequences in common and thought to have descended from a single ancestral gene through GENE DUPLICATION and modification. The common base sequences are believed to be homologous and conserved. It is thought such multigene families are represented by genes encoding globins, immunoglobulins and NUCLEAR RECEPTORS, by HOMOEOBOX- and PAIRED BOX-containing genes and by genes whose products contain MULTIFINGER LOOPS.

MULTINUCLEATE. Cell or hypha containing many nuclei. See ACELLULAR, COENOCYTE, SYNCYTIUM.

MULTIPLE ALLELISM. Simultaneous occurrence within a species population of more than two alleles (*multiple alleles*) at a given gene locus. A common phenomenon, contributing not only to continuous variation of a character in the population but sometimes also, as in inheritance of human BLOOD GROUPS, to discontinuous variation. Common at loci responsibility for INCOMPATIBILITY in plants. See SUPERGENE.

MULTITUBERCULATA. Extinct mammalian order within the PROTOTHERIA. Mainly of late Jurassic–Cretaceous times, but by extending into the Cenozoic its span far exceeds that of any other mammalian order. Had multicusped teeth (as opposed to the three or fewer of therian mammals). Probably the first herbivorous mammals; somewhat rodent-like in form and dentition. Some were medium-sized. See PANTOTHERIA.

MULTIVALENTS. Associations of more than two chromosomes, joined by chiasmata, at meiosis in POLYPLOIDS (seen most clearly at first metaphase). Where homology extends to more than just pairs of chromosomes (as it does particularly in AUTOPOLYPLOID cells) synapsis will probably result in crossing-over between any homologous regions present. When such multiply chiasmate chromosomes move apart at anaphase the result may be a chain or ring of chromosomes, and since multivalents do not always disjoin regularly, autoployploids often have a proportion of meiotic products with unbalanced (aneuploid) chromosome numbers giving loss of fertility. See ALLOPOLYPLOID.

MUREINS. Group of MUCINS found in bacterial cell walls.

MUSCARINE. An ALKALOID which, like nicotine, mimics acetylcholine action at certain cholinergic junctions (*muscarinic junctions*). Affects target organs of parasympathetic nervous system (e.g. vagus nerve); its effects being blocked by atropine, those of nicotine by curare.

MUSCI. See BRYOPSIDA.

MUSCLE. Any of a spectrum of animal tissues of mesenchymal origin, from fibroblast-like cells through to skeletal muscle fibres. Smooth and striated muscle are found in all animal phyla from the Coelenterata upwards. Contractile role of muscle is attributable to a proliferation of protein microfilaments from cell cortex to cell interior. See STRIATED MUSCLE, SMOOTH MUSCLE, CARDIAC MUSCLE.

MUSCLE CONTRACTION. The force-generating response of muscle to stimulation. May be *isotonic* (when muscle shortens during contraction) or *isometric* (when there is no change in muscle length during contraction). Molecular mechanism of contraction is probably similar in smooth, cardiac and striated muscle but is best understood in the last. Myofibrils in a resting striated muscle fibre consist of very precise arrangements of filaments of proteins ACTIN and MYOSIN (see STRIATED MUSCLE). In order for contraction to occur, several conditions must be met. Currently accepted account (*sliding filament hypothesis*) holds that force generation can only occur when actin and myosin filaments make contact and form the complex ACTOMYOSIN. This they can only do in the presence of intracellular calcium ions (Ca^{++}). But the tubular sarcoplasmic reticulum has CALCIUM PUMPS which accumulate Ca^{++}, keeping it scarce. These pumps are shut off when ACTION POTENTIALS reach the terminal cisternae of the sarcoplasmic reticulum where they make contact with inwardly folded transverse tubules (T-tubules) of the fibre's outer membrane (see Fig. 61). Only then does Ca^{++} flow from the cisternae, surrounding the myofilaments. These ions then bind to *troponin* molecules attached to actin filaments and in turn cause a shift in positions of

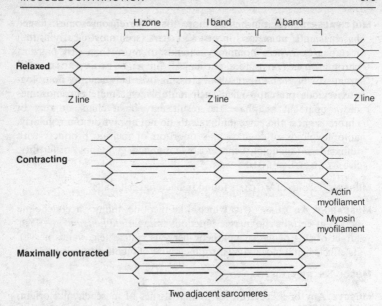

Fig. 50. *Diagram illustrating the sliding filament hypothesis of striated muscle contraction.*

tropomyosin molecules also attached to actin. These unmask sites on actin to which myosin can bind, forming actomyosin. But the 'heads' on myosin filaments will only bind to actin after they have been primed to do so by hydrolysis of ATP, which each head catalyses through its own ATPase activity. Only when ATP has been hydrolysed (i.e. before contraction occurs) is the myosin head ready to attach to an actin filament, changing its own shape (allosterically) so as to pull the actin filament past it. It can then release its bound ADP and inorganic phosphate, hydrolyse another ATP molecule (which requires magnesium ions, Mg^{++}) and swing back ready for another power stroke. Repeated cycles of this sequence result in the actin filament sliding along relative to myosin; and since in muscle cells actin filaments are bound to Z-discs, the H-zone in the middle of the contracting sarcomere will narrow and darken and the I-bands disappear: the sarcomere has then fully contracted.

Striated muscle contracts with an ALL-OR-NONE RESPONSE. It relaxes only if ATP is available to free myosin from actin and if Ca^{++} ions are actively removed from sarcoplasm by terminal cisternae, which they are in the absence of action potentials along the sarcolemma (and transverse tubules). In the absence of ATP the actomyosin complex persists and *rigor mortis* occurs.

There is limited availability of ATP in sarcoplasm, and the immediate source of energy for its resynthesis is another HIGH-ENERGY

PHOSPHATE compound, the PHOSPHAGEN *creatine phosphate* (*phosphocreatine*). If muscle is forced to contract repeatedly the amplitude of contractions will decrease and ultimately fail altogether when energy resources are expended. Muscle, like nerve, has brief absolute and relative *refractory periods* (see IMPULSE) after an action potential has passed a point on its membrane. During these the membrane becomes repolarized. The relative refractory period of striated muscle is much shorter than that of cardiac muscle, but longer than that of nerve. See TETANUS.

Striated muscle will respond to a succession of rapid stimuli. In moderate activity oxygen supply to muscles is adequate to prevent build-up of lactic acid, but if work rate exceeds oxygen supply an OXYGEN DEBT will arise. Prior to this, however, the oxygen stored in MYOGLOBIN will have been used up. *Muscle tone* (*tonus*) is the sustained low-level reflexly controlled background muscle contraction without which muscle flabbiness and poor responsiveness result. MUSCLE SPINDLES are responsible for maintaining this tone, which is variable. The effect of *training* on muscles is mainly to increase capillary circulation and hence diffusion rates of materials to and from muscle fibres. Endurance training results in hypertrophy of muscle tissue.

Smooth muscle contractions are either *tonic* or *rhythmic*. Some smooth muscle is self-exciting (see MYOGENIC). In tonic contractions slow waves of electrical activity pass across adjacent cells, which are often fused to form a syncytium. Bursts of action potentials then cause the muscle to contract in functional blocks when the waves reach a sufficient intensity. Parasympathetic stimulation is responsible for tonus in much of the vertebrate gut. Rhythmic contractions of smooth muscle in walls of tubular organs (*peristaltic* contractions) are also due to rapid spikes of action potentials; but the initiating signal here is usually stretching of the tube wall, which elicits contraction either directly or reflexly via local nerve plexuses (see AUERBACH'S PLEXUS).

Cardiac muscle contracts and relaxes rapidly and continuously, with a rhythm dictated by its intrinsic pacemaker and the neural and endocrine influences upon it. It has a longer refractory period than striated muscle.

MUSCLE SPINDLE. Stretch receptor (proprioceptor) of vertebrate muscle. When muscle is stretched, sensory nerve endings in its spindles fire and send impulses via the spinal cord to the cerebellum and cerebral cortex. When muscle contracts these impulses are inhibited. Responsiveness of the spindle can be altered by central control from the reticular formation of the midbrain: each spindle can be made to contract even when its muscle is not contracting, thus enabling regulation of the rate of muscle contraction in response to different loads. Increased spindle firing due to muscle stretch will result in reflex muscle contraction and reduced spindle firing. The nerve circuitry is quite complex and involves inhibitory and excitatory loops.

MUSHROOM. Popular name for edible fruiting bodies of fungi belonging to the Agaricaceae (BASIDIOMYCOTINA), especially species of *Agaricus*.

MUTAGEN. Any influence capable of increasing MUTATION rate. Usual effect is chemical alteration, addition, substitution or dimerization of one or more bases or nucleotides of the genetic material (DNA or RNA). Mutagens include ALKYLATING AGENTS, acridine dyes, ultraviolet light, X-rays, beta and gamma radiation. See CARCINOGEN, AMES TEST.

MUTANT. An individual, stock or population expressing a MUTATION.

MUTATION. Alteration in the arrangement, or amount, of genetic material of a cell or virus. They may be classified as either *point mutations*, involving minor changes in the genetic material (often single base-pair substitutions), or *macromutations* (e.g. deletions), involving larger sections of chromosome.

The effects of macromutations on chromosome structure are often visible during mitosis and meiosis and include INVERSION, TRANSLOCATION, DELETION, DUPLICATION (see TRANSPOSABLE ELEMENT) and POLYPLOIDY. NON-DISJUNCTION produces chromosome imbalance (e.g. aneuploidy). Whereas point mutations commonly cause amino acid substitutions in the polypeptide encoded by the gene and often have minor effects on gene function, macromutations (especially if they involve deletions) commonly lead to syndromes of abnormalities which seriously reduce FITNESS, and are more often lethal. Many point mutations fail to result in amino acid substitutions because the GENETIC CODE is degenerate. Others, though altering amino acid sequence, either have little effect or only partially inactivate gene function (as in *leaky* auxotrophs of fungi and bacteria). But a single amino acid replacement in a critical position can abolish an enzyme's activity. One type of radical effect that can follow from a single base-pair substitution is the creation of STOP CODON within the open reading frame, usually rendering the translation product useless. Mutations creating stop codons (*nonsense* mutations) can often be phenotypically suppressed by compensating mutations in the anticodon sequence of tRNA molecules (*suppressor mutations*). *Frameshift mutations* are nucleotide additions and deletions not involving a multiple of three base-pairs, which move the 'reading frame' of tRNA to the left or right during the translation phase of PROTEIN SYNTHESIS. These usually have drastic effects on gene function.

A distinction may be made in some multicellular organisms between mutations occurring in body cells (somatic mutations), and those occurring in GERM LINE cells. Only the latter can normally be inherited and play a part in evolution (but see POLYPLOIDY).

By themselves, mutations do not normally direct the path of evolution: there is no 'mutation pressure'; rather, they provide the heritable variation upon which selection may act. It is usually accepted that mutations are random with respect to requirements of the cells or organisms in which they arise; but debate is developing as to whether or not some bacterial populations can respond to selection pressure by increasing the frequency of mutation rates for favourable genes. See MUTAGEN, GENE CONVERSION, GENETIC VARIA-TION, MOLECULAR CLOCK.

MUTATION RATE. The frequency with which MUTATIONS arise in populations of organisms or in tissue culture. Experiments suggest that an average of about one base-pair changes 'spontaneously' per 10^9 base-pair replications. If proteins are, on average, encoded by about 10^3 base-pairs then it would take about 10^6 cell generations before the protein contained a mutation. See MOLECULAR CLOCK, MUTAGEN.

MUTUALISM. See SYMBIOSIS.

MYCELIUM. Collective term for mass of fungal hyphae constituting vegetative phase of the fungus; e.g. mushroom spawn.

MYCOBIONT. Fungal partner of a LICHEN; usually member of ASCOMYCOTINA. Compare PHYCOBIONT.

MYCOLOGY. Study of fungi.

MYCOPHAGE. Fungal PHAGE.

MYCOPHYTA. See EUMYCOTA.

MYCOPLASMAS. General name for group of prokaryotes distinguished by possession of the smallest known cells, as little as 0.1 μm in diameter. Cells lack cell walls, but possess ribosomes and all other protein-synthesizing machinery, encoded by a very small genome of fewer than 650 genes (about 20% of a bacterial genome). Some are saprotrophic; others are pathogens of plants and animals, including humans. Known by a variety of names, including *pleuropneumonia-like organisms* (PPLOs), after the disease caused in cattle by the first member to be described. Six genera to date, about 50 of the 60 species belonging to *Mycoplasma*. Their ribosomal RNA sequences suggest they form a natural group, probably of bacterial origin. *Mycoplasma-like organisms* (MLOs) have been isolated from over 200 plant species and have been implicated in more than 50 plant diseases, often with symptoms of yellowing or stunting; they appear to be confined to sieve-tubes and to be passed passively from one sieve-tube member to another through sieve-plate pores.

MYCORRHIZA. Symbiotic association between a fungus and the root of a higher plant. Mycorrhizas are of common occurrence. Two main types exist: (a) *endotrophic*, in which the fungus is within the cortex cells of the root, e.g. of orchids; and (b) *ectotrophic*, where it is external, forming a mantle that completely invests the smaller roots, e.g. of pine trees. Mycorrhizal plants benefit from the association, in ectotrophic forms particularly by the fungal transport of ions (e.g. phosphate), which are otherwise relatively immobile in the soil, into the plant roots. Distributions of some plants are limited by distribution of the appropriate mycorrhizal fungus, vital as it may be for establishment and growth of seedling trees of several species (e.g. pines). Association of the fungus with the tree is necessary for development and reproduction of the fungus. Plants grown in soils which have had excess artificial fertilizer added, or which have been sprayed regularly with some herbicides, have been shown to have poorer mycorrhizae.

MYCOSIS. Disease of animals caused by fungal infection; e.g. ringworm.

MYCOTROPHIC. (Of plants) having MYCORRHIZAS.

MYELIN SHEATH. Many layers of membrane of SCHWANN CELL (in peripheral nerves) or of oligodendrocyte (central nervous system) wrapped in a tight spiral round a nerve axon forming a sheath preventing leakage of current across the surrounded axon membrane except at *nodes of Ranvier*. Cell membranes comprising the myelin sheath contain large amounts of the glycolipid galactocerebroside. Such *myelinated neurones* conduct faster than equivalent non-myelinated ones as current 'jumps' from one node to the next. See IMPULSE.

MYELOID TISSUE. Major site of vertebrate HAEMOPOIESIS, restricted except in embryo to red bone marrow (e.g. in ribs, sternum, cranium, vertebrae, pelvic girdle). Mesodermal in origin. Myeloid stem cells give rise to cell lineages distinct from those of LYMPHOID TISSUE. Products include MONOCYTES and descendant MACROPHAGES, MAST CELLS, BASOPHILS, NEUTROPHILS, EOSINOPHILS, MEGA-KARYOCYTES and their PLATELETS. In embryos, liver and spleen are principal myeloid sites.

MYELOMA. Malignant cancers of MYELOID TISSUE; causing anaemia, especially in the middle-aged and elderly. See CANCER CELL.

MYOBLAST. Precursor cells of vertebrate SKELETAL MUSCLE fibres. Can divide but eventually fuse to form typical multinucleate syncytia of differentiated skeletal muscle tissue.

MYOFIBRIL. Structural unit of STRIATED MUSCLE fibres, several to each fibre.

MYOGENIC. (Of muscle tissue) capable of rhythmic contraction independently of external nervous stimulation as a result of presence of PACEMAKER. CARDIAC MUSCLE always has such a pacemaker and SMOOTH MUSCLE may have.

MYOGLOBIN. Conjugated protein of vertebrate muscle fibres. Single polypeptide, whose iron-haem prosthetic group binds molecular oxygen. Its oxygen equilibrium curve lies well to the left of HAEMO-GLOBIN's, enabling it to load and unload its oxygen at lower oxygen partial pressures. It gives up its oxygen only when the muscle is under anoxic conditions. See Fig. 31.

MYOSIN. A protein found in the majority of eukaryotic cells. At least two classes of myosin exist, a single-headed tail-less variety (*myosin I*) involved in CELL LOCOMOTION, and a two-headed, tailed variety (*myosin II*) involved in MUSCLE CONTRACTION. Each filament of myosin II is thicker than those of ACTIN, comprising a tail composed of two heavy α-helices of about 134 nm in length by which it interacts with other myosin filaments, and two pairs of light chains forming a pair of knob-like heads at each end containing actin-binding ATPase sites essential for ACTOMYOSIN formation, as in muscle contraction. Its presence in STRIATED MUSCLE is responsible for A-bands. Myosin I molecules, on the other hand, are single-headed and instead of the tail have a non-α-helical domain housing a second actin-binding site which is ATP-independent (i.e. lacks ATPase activity). Besides existing freely in the cytosol, myosin I appears capable of interacting with phospholipid bilayers of membranes and may mediate such membrane-dependent cell movements as occur in the absence of myosin II. Single-headed myosins have also been located within the microvilli of vertebrate gut epithelia and of *Drosophila* eye rhabdomes.

MYOTOME. See MESODERM.

MYRIAPODA. Class of ARTHROPODA including two important sub-classes: carnivorous Chilopoda (centipedes), and herbivorous DI-PLOPODA (millipedes). Have long bodies of many segments, and distinct heads bearing one pair of antennae, mandibles and at least one pair of maxillae (see MOUTHPARTS). Terrestrial, with tracheal system. Centipedes have one pair of legs per segment and flattened bodies; millipedes have two pairs of legs per segment (the result of segment fusion) and cylindrical bodies.

MYXAMOEBA. Naked cell characteristic of vegetative phase of slime fungi (MYXOMYCOTA) and some simple fungi; capable of amoeboid locomotion.

MYXOBACTERIA. Small group of rod-shaped bacteria, distinguished by gliding movement in contact with a solid surface and by delicate, flexible, cell wall. In many, vegetative cells mass together (particularly when starved) to form minute fruiting bodies in which cells dif-ferentiate into durable spores. Compare eukaryotic MYXOMYCOTA.

MYXOMYCETES. See MYXOMYCOTA.

MYXOMYCOTA. Slime moulds; myxamoebae. Group of simple eukaryotes with plant- and animal-like characteristics. Affinities uncertain: possibly a fungal division (Myxomycetes), or a protozoan class (Mycetozoa). All have naked, motile assimilitative phases and all produce spores or spore-like structures. Four distinctive groups, of uncertain relationships. Vegetative phase of largest group comprises naked, multicellular masses of protoplasm known as *plasmodia*. These move and ingest food while internal differentiation produces a fruiting body (sporangium) held upright on a stalk and foot plate. Stalk and spore cells become covered in gelatinous cellulose walls, and spores germinate under appropriate conditions to produce either single-celled myxamoebae, or biflagellate swarm cells which commonly behave as gametes and copulate in pairs. Some slime moulds, including dictyostelids, produce *macrocysts*: aggregates of cells within a thin primary wall. A central cell enlarges by ingesting other myxamoebae phagotrophically, after which a thick secondary wall is deposited within the primary layer. At germination, myxamoebae are released as the wall breaks. Meiosis is thought to occur prior to spore formation. Widely distributed in damp conditions. Large aggregates of *Fuligo septica* occur on tanner's bark.

MYXOPHYCEAN STARCH. Storage polysaccharide of the CYANOBACTERIA, similar to glycogen.

MYXOPHYTA. See CYANOBACTERIA.

N

NAD **(NICOTINAMIDE ADENINE DINUCLEOTIDE, COENZYME I)**. See Fig. 51.
Dinucleotide COENZYME, derivative of NICOTINIC ACID, required
in small amounts in many REDOX REACTIONS where oxidoreductase
enzymes transfer hydrogen (i.e. carry electrons), as in KREBS CYCLE
and ELECTRON TRANSPORT SYSTEM (ETS) in respiration.

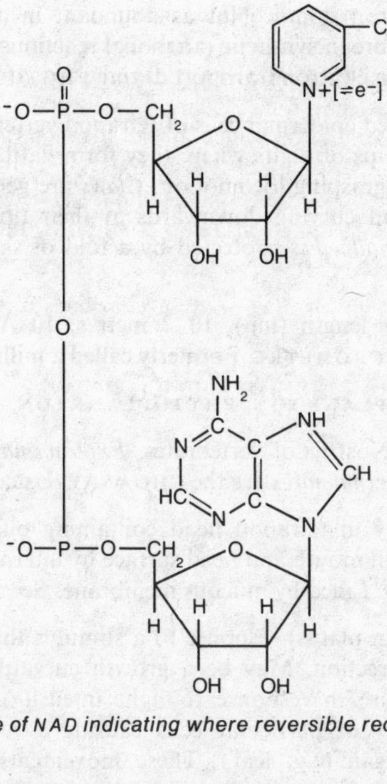

Fig. 51. Structure of NAD indicating where reversible reducton occurs, giving NADH.

In the Krebs cycle the oxidized form of the coenzyme (NAD^+)
receives one hydrogen atom from isocitrate (under the influence of
isocitrate dehydrogenase) to become NADH, while another hydrogen atom from the isocitrate becomes a proton:

isocitrate + NAD^+ = ketoglutarate + CO_2 + NADH + H^+

NAD dehydrogenase, which contains the flavoprotein FMN and is an
important part of mitochondrial ETS, then transfers both the proton

and the hydrogen from reduced N A D to its F M N component, releasing N A D⁺ for re-use:

$$NADH + H^+ + FMN = NAD^+ + FMNH_2$$

By this means, electrons can be collected as N A D H from a variety of sources and funnelled into the E T S. So N A D H is an energy source for A T P synthesis, through its link to the E T S, and N A D H produced in glycolysis can enter mitochondria and provide this service.

NADP (NICOTINAMIDE ADENINE DINUCLEOTIDE PHOSPHATE, COENZYME II). A dinucleotide coenzyme; a phosphorylated N A D, and like it involved in several dehydrogenase-linked redox reactions, but usually acting as an electron donor. Not as abundant in animal cells as N A D. Involved more in synthetic (anabolic) reactions than in breakdown. Major role in electron transport during PHOTOSYNTHESIS.

NAIL. Hard, keratinized epidermal cells of tetrapod vertebrates covering upper surfaces of tips of digits where they form flattened structures, often for arboreal grasping locomotion. *Claws* are generally tougher still, narrowing and curving downwards at their tips. The mitotic epithelium of the *nailbed* is protected by a fold of skin. See CYTO-SKELETON.

NANOMETRE. Unit of length (nm). 10^{-9} metres; 10 Ångstroms; one thousandth of a MICROMETRE. Formerly called a millimicron.

NANOPLANKTON. See PLANKTON, PHYTOPLANKTON.

NARES. (Sing. *naris*) Nostrils of vertebrates. *External nares* open on to surface of head; *internal nares* are the CHOANAE. Usually paired.

NASAL CAVITY. Cavity in tetrapod head containing olfactory organs, communicating with mouth and head surface by internal and external NARES respectively. Lined by mucous membrane. See PALATE.

NASTIC MOVEMENT. (In plants) response to a stimulus that is independent of stimulus direction. May be a growth curvature (e.g. flower opening and closure in response to light intensity), or a sudden change in turgidity of particular cells causing a rapid change in position of an organ (e.g. leaf). These movements are classified according to nature of the stimulus; e.g. *photonasty* is a response to alteration in light intensity, *thermonasty* to change in heat intensity, *seismonasty*, to shock. Compare TROPISM.

NATURAL CLASSIFICATION. CLASSIFICATION which groups organisms according to their degree of relatedness (i.e. phylogenetically). It provides much implicit information about organisms, giving high predictive value. See NATURAL KIND, NATURAL TAXON.

NATURAL KILLER CELL (NK CELL). Type of LEUCOCYTE of higher vertebrates capable of recognizing alterations to surfaces of virally infected and cancerous cells, of binding to them and then killing them. Activated by INTERFERON. Component of the INNATE IMMUNE SYSTEM.

NATURAL KIND. A term filling a need experienced by some who seek to justify a classificatory system on objective, non-arbitrary grounds. Such people hold that objects would 'sort themselves' into categories, or kinds, if all facts about them were known. At such a time one could predict the kind to which an organism belonged given knowledge only of those facts about it that *materially caused* it to be a member of that kind. This predictability presupposes a degree of theoretical sophistication which might be unattainable given the epigenetic nature of biological systems. So, even if natural kinds do exist in biology, we might never be in a position to know which they are. See CLASSIFICATION.

NATURAL ORDER. Linnean category, equivalent to FAMILY. See LINNAEUS.

NATURAL SELECTION (SELECTION). Most widely accepted theory concerning the principal causal mechanism of evolutionary change ('descent with modification'); propounded by Charles DARWIN and Alfred Russel WALLACE. The theory asserts that, given diversity (both genetic and phenotypic) among individuals making up a species population, not all individuals in the population at time t_o will contribute equally to the make-up of the population at a subsequent time $t_{,}$. To the extent that this is due to the effects of heritable differences upon individuals, natural selection has acted.

Confusion arises over the use of Herbert Spencer's phrase 'the survival of the fittest'. Individual organisms do not survive through geological time (unlike some evolutionary lineages), but what they inherit and pass on does: that is, GENES (see UNIT OF SELECTION). The theory of natural selection asserts that the genetic composition of an evolutionary lineage will change through time by non-random transmission of genes from one parental generation to the next, a non-randomness ('selection') due solely to the fact that not all gene combinations are equally suited to a given environment, and that consequently individuals differ in their biological (Darwinian) FITNESS. Constraints upon phenotype from the environment, which produce this differential gene transmission, are termed *'selection pressure'*. It is commonly assumed that all regular components of a species' phenotype have been favoured by natural selection, but evolution may sometimes result from causes other than natural selection (see GENETIC DRIFT).

When a character, especially a POLYGENIC one, is under *directional* selection (*orthoselection*) in a population, it undergoes a progressive (incremental) shift in its mean value with time. When alternative phenotypes (e.g. red-eye and white-eye) at one character mode (here eye-colour) are each favoured in the same population, selection is said to be *disruptive* and the population may become polymorphic for that character mode. When a particular (mean) character state is favoured, selection is said to be *stabilizing*. See SEXUAL SELECTION.

NATURAL TAXON. Taxon comprising organisms more closely related to one another than to organisms in any other taxon at the same taxonomic level.

NAUPLIUS. Larval form of many crustaceans. Oval, unsegmented, bearing three pairs of appendages. It approximates to the 'head' of the eventual adult, successive segments being added in an anterior direction from the rear as it develops.

NAUTILOID. Designating those cephalopod molluscs resembling and including the pearly nautilus, *Nautilus*. Numerous fossil forms, first appearing in the Cambrian. Have chambered shell, either coiled or straight.

NEANDERTHAL MAN. See *Homo*.

NEARCTIC. ZOOGEOGRAPHICAL REGION consisting of Greenland, and North America southwards to mid-Mexico.

NECRIDIUM (SEPARATION DISC). Cell in a blue-green alga filament (CYANOBACTERIA), whose death results in formation of a HORMOGONIUM.

NECROSIS. Death of cells or tissues, especially through disease.

NECROTROPHIC. (Of organisms) feeding on dead cells and tissues.

NECTARY. Fluid-secreting gland of flowers. *Nectar* contains sugars, amino acids and other nutrients attractive to insects, especially in insect-pollinated flowers.

NEGATIVE FEEDBACK. See HOMEOSTASIS.

NEGATIVE STAINING. Those methods employed in both light and electron MICROSCOPY by which only the background is stained, the unstained specimen showing up against it.

NEKTON. Swimming animals of pelagic zone of the sea or lake. Includes fishes and whales. See BENTHOS, PLANKTON, PELAGIC.

NEMATOCYST. (Bot.) Structurally complex ejectile organelle produced in a few dinoflagellate genera (*Nematodinium* and *Polykrikos*) bearing a remarkable resemblance to comparable structures found in the Cnidaria. (Zool.) Inert stinging capsule produced by a cnidarian CNIDOBLAST.

NEMATODA. Roundworms (eelworms, threadworms). Abundant and ubiquitous animal phylum (or class of the ASCHELMINTHES). Unsegmented, triploblastic and pseudocoelomate. Circular in cross-section, with a characteristic undulating, or thrashing movement. Elastic CUTICLE of collagen acts as antagonist of the unique longitudinal muscles (no circular muscle) during swimming. Muscle cells have contractile bases adjacent to cuticle and non-contractile 'tails' lying in the vacuolated parenchymatous pseudocoelom. Gut with suctorial pharynx; anus terminal. Nervous system of simple nerve cords, ganglia and anterior nerve ring. Excretory system intracellular, consisting of two longitudinal canals. Cilia absent; sperm amoeboid. Sexes usually separate. Includes free-living and parasitic forms. FILARIAL WORMS of man and domestic animals, and root eelworms (e.g. of potato) are of great economic importance. Each species has its own number of cells, and there are usually four moults.

NEMATOMORPHA. Small phylum (or aschelminth class) of thin, elongated worms resembling nematodes but lacking excretory canals and with a brain linked to a single ventral nerve cord. Their larvae bore into insects. Like nematodes, have only longitudinal muscles in body wall. See ASCHELMINTHES for more general properties.

NEMERTINA (NEMERTEA, RHYNCHOCOELA). Ribbon, or proboscis, worms. Small phylum of mostly marine worms with platyhelminth-like characteristics. Differ in having tube-like gut with mouth and anus (gut entire), peculiar proboscis, simpler reproductive system and a circulatory system.

NEO-DARWINISM. Brand of DARWINISM, current since early decades of 20th century, which combines Darwin's theory of evolution by natural selection with MENDELIAN HEREDITY and post-Mendelian genetic theory. Accounts, more successfully than Darwin was able, for the origin and maintenance of variation within populations. The combination has to some extent resolved problems surrounding the nature and origins of species (see SPECIATION).

NEOGEA. See NEOTROPICAL REGION.

NEOGENE. Collective term for the Miocene, Pliocene and Pleistocene epochs. See GEOLOGICAL PERIODS.

NEOGNATHAE. Largest superorder of NEORNITHES, including all non-fossil and non-ratite species. Many orders. See PALAEOGNATHAE.

NEO-LAMARCKISM. View, generally discredited, that acquired characters may be inherited. Notoriously espoused by the Stalinist biologist T. D. Lysenko (mainly for crops and domesticated animals). In another episode this century, Austrian biologist P. Kammerer tried to demonstrate the phenomenon in midwife toads and sea squirts. More recently still, genetic transfer of acquired immunity in rats has been alleged. Experimental support for the view has generally been inconclusive, and sometimes even fabricated. See LAMARCK, GENE-TIC ASSIMILATION, MUTATION.

NEOLITHIC (NEW STONE AGE). Phase of human history, commencing approximately 10 000 years BP, and succeeding the PALAEOLITHIC, during which domestication of animals and cultivation of plants first occurred. Production of sophisticated stone tools (sometimes by mass-production), arrowheads, fine bone ornaments, etc., took place. Farming began to supersede hunter-gathering ecologically. First detected in Mesopotamia, from which it spread.

NEOPALLIUM. See CEREBRAL CORTEX.

NEOPLASM. Tumour, or cancerous growth. Malignant if invasion or METASTASIS occurs, or is likely to occur; otherwise *benign*. See CANCER CELL.

NEORNITHES. See AVES.

NEOTENY. Retardation of somatic development. Form of HETEROCHRONY, often confused with PROGENESIS. Involves a slowing in rate of growth and development of specific parts of the body relative to (especially) the reproductive organs, although frequently accompanied by delayed onset of sexual maturity. Usually results in retention of juvenile features by otherwise adult animal. Formerly used in broad sense of PAEDOMORPHOSIS. Examples include the amphibian *Ambystoma* (axolotl); development of combat and display structures in some social mammals (e.g. mountain sheep, giraffe, African buffalo); increased brain size in slowly-growing mammals with small litter sizes and intense parental care, e.g. some cat species and, arguably, humans. Correlation observed between presence of neotenic attributes and conditions favouring K-SELECTION.

NEOTROPICAL REGION. ZOOGEOGRAPHICAL REGION consisting of South and most of Central America. Compare Nearctic.

NEOTYPE. Specimen chosen as a replacement of the HOLOTYPE when that is lost or destroyed.

NEPHRIDIUM. Tubular organ present in several invertebrate groups (platyhelminths, nemerteans, rotifers, annelids, some molluscan larvae and *Amphioxus*), developing independently of the COELOM as an intucking of the ectoderm. Lumen often formed by intracellular hollowing-out of nephridial cells, closing internally (*protonephridium*) or acquiring an opening, the nephrostome, into the coelom (*metanephridium*). Protonephridia end internally either in FLAME CELLS or SOLENOCYTES. Adult annelids usually have metanephridia, often replacing larval protonephridia, and sometimes these have ciliated funnels and resemble COELOMODUCTS (which they are not). Nephridia open either to outside of the body (via *nephridiopores*), or into the gut. Carry excretory products, wafted by cilia or flagella; may have an osmoregulatory role; occasionally carry gametes.

NEPHRON. Functional unit of vertebrate KIDNEY.

NERITIC. Inhabiting the sea over the continental shelf, arbitrarily taken to be where it is shallower than 200 metres. Compare OCEA-NIC.

NERVE. (Bot.) Narrow thickened strip of tissue found running the length of the middle of a moss leaf (costa). (Zool.) Bundle of motor and/or sensory NEURONES and GLIAL CELLS, with accompanying connective tissue, blood vessels, etc., in a common connective tissue sheath, or *perineurium*. Each neurone conducts independently of its neighbours. *Mixed nerves* contain both sensory and motor neurones. Nerves may be nearly as long as the whole animal and contain thousands of (usually myelinated) neurones. See CRANIAL NERVE, SPINAL NERVE, NERVOUS SYSTEM.

NERVE CELL. See NEURONE.

NERVE CORD. Rod-like axis of nervous tissues forming, usually, a longitudinal through-conduction pathway and integration centre for both sensory and motor information and forming a major element of the CENTRAL NERVOUS SYSTEM. Usually linked anteriorly to the BRAIN. In segmented animals especially, a major route for REFLEX ARCS. In invertebrates the nerve cord or cords are often *nerve chains*, composed of linked ganglia. In vertebrates it forms the SPINAL CORD. Nerve cords in annelids and arthropods are usually ventral and paired; in chordates they are single, dorsal and usually hollow (see NEURULATION).

NERVE ENDING. Structure forming either the sensory or motor end of a peripheral neurone. If the former, may comprise free nerve endings or a RECEPTOR end organ; if the latter, usually consists of a motor end-plate (see NEUROMUSCULAR JUNCTION).

NERVE FIBRE. Axon of a NEURONE, and its MYELIN SHEATH if present. Diameters vary from $1-20$ μm in vertebrates up to 1 mm in GIANT FIBRES of some invertebrates. Nerve fibres commonly branch towards their termini into small-diameter 'twigs'.

NERVE IMPULSE. See IMPULSE.

NERVE NET. Network of neurones, often diffusely distributed through tissues, making up all or most of the nervous system of coelenterates and echinoderms and a large component of peripheral nervous systems of hemichordates. Cells may fuse to form a syncytium or form specialized synapses capable of transmitting in both directions. In some coelenterates, especially motile ones, there is often division of labour between two nerve nets, one (*through-conduction net*) conducting faster and more unidirectionally than the other. Nerve nets characteristically conduct away slowly in all directions from a point of stimulation, temporal SUMMATION and FACILITA-

TION involving an increasing area of excitation with increased stimulus strength. Found in gut walls of some arthropods, molluscs and vertebrates.

NERVE PLEXUS. Diffuse network of neurones and/or ganglia. In vertebrates, *brachial plexi* and *sacral plexi* are associated with limb movements and consist of anastomosing spinal nerves. *Solar plexus* is the collective term for a number of ganglia in the coeliac/anterior mesenteric region connected to the sympathetic nerve chain by the splanchnic nerves.

NERVOUS INTEGRATION. Process whereby sensory inputs, often from more than one source and modality, either give rise to unified motor responses or are stored under some principle of association. SYNAPSES are the basic physical units of integration, providing for SUMMATION of excitatory and inhibitory potentials. Non-synaptic membranes of the postsynaptic cell may integrate through ADAPTATION, ACCOMMODATION or by their refractory periods (see IMPULSE). Ganglia, nerve 'nuclei' and brains all depend on synaptic connexions for integration. Reflex arcs are major sites of nervous integration, *relay neurones* often communicating sensory information to the brain where even quite complex behaviour is often reflexly coordinated. Quite simple muscular activity often requires fairly complex nervous integration, in vertebrates employing feedback from MUSCLE SPINDLES, after which the CEREBELLUM coordinates this information and initiates appropriate motor outputs. In segmented invertebrates this role is often performed by segmental ganglia, largely independently of the brain. All forms of LEARNING involve integration of nervous information, as do long- and short-term memory. See HYPOTHALAMUS, NEUROENDOCRINE COORDINATION.

NERVOUS SYSTEM. Complement of nervous tissue (NEURONES, NERVES, RECEPTORS and GLIAL CELLS) serving to detect, relay and coordinate information about an animal's internal and external environments and to initiate and integrate its effector responses and activities. Present in all multicellular animals except sponges, developing from the ectodermal GERM LAYER. Characteristic mode of information carriage lies in patterns of nerve IMPULSES transmitted along neurones, NEUROTRANSMITTERS relaying the impulse pattern at SYNAPSES. Presence of synapses enables some nervous integration to occur; this is more marked the more advanced the nervous system. Nervous systems may be relatively simple, as in NERVE NETS, but even here there is a tendency towards *through-conduction pathways* enabling rapid transfer of impulses from one region to specific, and often distant, parts. Thus invertebrate and vertebrate NERVE CORDS mediate local reflexes, via *segmental* and *spinal nerves* respectively. Further, ganglia often serve as integration centres. The nerve cord(s) and BRAIN make up the CENTRAL NERVOUS SYSTEM (CNS), the *peripheral nervous system* comprising most of the nerves conducting impulses towards and away from it (in vertebrates, these include the spinal and CRANIAL NERVES). The AUTONOMIC NERVOUS SYSTEM forms an additional visceral motor circuit in vertebrates, integrated anatomically and functionally with the CNS.

Compared with ENDOCRINE SYSTEMS, nervous systems provide for *reception* of more specific environmental information and, through integration centres simultaneously receptive to inputs from a wide variety of sources (see SYNAPSE), can elicit responses (generally more complex, preprogrammed and integrated) far more quickly. There are no structures equivalent to sense organs or integration centres in endocrine systems, which rely upon unidirectional transport of solutes via the blood system (see HORMONES). It is the distribution patterns of receptor sites on the membranes of target cells which are instrumental in responses to hormonal messages, diffusely broadcast as they are when compared to the pin-point accuracy of nerve impulses. Nervous systems function via rapid, multi-directional and highly integrated through-conduction pathways. Not surprisingly, rapid responses to environmental changes are integrated via the nervous system, whereas seasonal and circadian responses often involve the endocrine system. See NERVOUS INTEGRATION, NEUROENDOCRINE COORDINATION.

NEURAL ARCH. Arch of bone resting on centrum of each VERTEBRA, forming tunnel (neural canal) through which spinal cord runs.

NEURAL CREST. Band of embryonic vertebrate ectoderm on both sides of the developing neural tube, giving rise to dorsal root ganglia, CHROMAFFIN CELLS, SCHWANN CELLS, and other cell types indicated in Fig. 22 for GERM LAYER. Neural crest cells often attain their final positions after lengthy migrations.

NEURAL PLATE. Flat expanse of chordate ectodermal tissue, the first-formed embryonic rudiment of the nervous system. Will sink and round up to form neural tube.

NEURAL SPINE. Spine of bone which may be produced from top of NEURAL ARCH, running up between dorsal muscles of each side and serving for their attachment. Successive spines are usually bound by ligaments.

NEURAL TUBE. Hollow dorsal tube of chordate embryonic nerve tissue formed by rolling up of neural plate, *neural folds* so produced fusing in the mid-dorsal line, a process termed *neurulation* (see DESMOSOMES). The epidermis then fuses above the neural tube. Failure to roll up and fuse gives rise to *spina bifida* and, in its most extreme form, *anencephaly*. The tube expands in front to form the brain and its ventricles, the narrower more posterior part forming the spinal cord; thereby produces the central nervous system and peripheral motor neurones.

NEUROBLAST. Embryonic and presumptive nervous tissue cell.

NEUROCRANIUM. The part of the skull surrounding brain and inner ear, as distinct from the part composing JAWS and their attachments, the SPLANCHNOCRANIUM.

NEUROENDOCRINE COORDINATION. Combined and integrated activities of the nervous and endocrine systems; involved in many physiological and behavioural responses to internal and external signals in multicellular animals.

Moulting (ECDYSIS) in some insects is initiated nervously but is mainly hormonally controlled; copulation and ovulation in many female mammals are integrated so that hormonal influences bring about OESTROUS behaviour (involving nervous control); the ensuing cervical stimulation initiates nervous release of gonadotrophic releasing factors (GnRFs), bringing about ovulation and increasing the probability of fertilization. The vertebrate hypothalamus and pituitary illustrate the close anatomical and physiological links between the nervous and endocrine systems. For instance, a sudden fall in air temperature around a mammal induces shivering. This involves sensory input from skin to the hypothalamus, whose output (along vagus nerve) causes adrenaline release from adrenals followed by rapid rhythmic contractions in appropriate body muscles and rise in body temperature. See NEUROHAEMAL ORGAN, NEUROSECRETORY CELLS.

NEUROGLIA. See GLIAL CELL.

NEUROHAEMAL ORGAN. Organ lying outside the nervous system and storing secretion from numbers of NEUROSECRETORY CELLS, releasing it into the blood. Particularly widespread in arthropods. Insect CORPORA CARDIACA, and *sinus glands* in the eyestalks of some crustaceans are examples.

NEUROHUMOUR. See NEUROTRANSMITTER.

NEUROHYPOPHYSIS. See PITUITARY GLAND.

NEUROMAST ORGAN. See LATERAL LINE SYSTEM.

NEUROMUSCULAR JUNCTION. Area of membrane between a motor neurone and the muscle cell membrane forming a SYNAPSE between them. The area of muscle membrane under the nerve is the *motor endplate*. Each SKELETAL MUSCLE fibre commonly receives just one terminal branch of a neurone. Each nerve impulse releases a 'jet' of acetylcholine, and resulting small depolarizations of the endplate summate in a graded way to a critical threshold level (about -50 mV internal negativity) at which an action potential is generated and travels along the muscle (see IMPULSE, MUSCLE CONTRACTION).

NEURONE (NEURON, NERVE CELL). Major cell type of nervous tissue, specialized for transmission of information in the form of patterns of IMPULSES. Nucleus and surrounding cytoplasm comprise the *cell body* (*perikaryon*, or *soma*), which may be the site of multiple synaptic connexions with other nerve fibres. In non-receptor cells numerous projections from the cell body, *dendrites*, provide a large surface area for synaptic connexions with other neurones, while one or more other regions of the cell body (*axon hillocks*) extend into long thin *axons* carrying impulses away from the cell body to other neurones or to effectors, making contacts via SYNAPSES and usually through secretion of NEUROTRANSMITTERS.

The membrane of the axon hillock region commonly has the lowest threshold for production of ACTION POTENTIALS and is often the site of their origin. Most neurones have one axon (*monopolar*); others have two (*bipolar*), or several (*multipolar*). Axons are also termed *nerve fibres*, and may or may not be covered in a MYELIN SHEATH. Myelinated fibres conduct faster than unmyelinated ones. Unmyelinated fibres, lacking the electrical insulation of myelinated fibres, are normally located in association centres of the nervous system (see GREY MATTER).

NEUROPHYSIN. See NEUROSECRETORY CELLS.

NEUROPIL(E). Mass of interwoven axons and dendrites within which neurone cell bodies of the central nervous system are embedded.

NEUROPTERA. Order of ENDOPTERYGOTE insects including alderflies and lacewings. Two similar pairs of membranous wings which, when at rest, are held up over body. Biting mouthparts. Larvae carnivorous, often taking insect pests (lacewing larvae eat aphids). Mostly terrestrial; some aquatic (e.g. alderfly larvae). Alderflies (Megaloptera) include some of the most primitive endopterygotes.

NEUROSECRETORY CELLS. Cells present in most nervous systems, combining ability to conduct nerve impulses with terminal secretion of hormones which travel via the blood and act on target cells. Do not transmit impulses to other cells. Distinction between these and ordinary neurones is blurred as the latter secrete NEUROTRANSMITTERS, although not into the blood. Examples include cells in the brains, CORPORA CARDIACA, CORPORA ALLATA and THORACIC GLANDS of insects, and cells in the vertebrate HYPOTHALAMUS. Hormones produced are transported inside the axon, and then usually in the blood, by proteins termed *neurophysins*. See NEUROHAEMAL ORGAN, CHROMAFFIN CELL.

NEUROTRANSMITTER (NEUROHUMOUR). Low molecular mass substances released in minute amounts at interneural, neuromuscular and neuroglandular SYNAPSES. May be excitatory (depolarizing postsynaptic membrane) or inhibitory (hyperpolarizing postsynaptic membrane). Most widespread is ACETYLCHOLINE, but others are NORADRENALINE, DOPAMINE, SEROTONIN, glutamic and aspartic acids, glycine, GABA, ENKEPHALINS and ENDORPHINS. See IMPULSE.

NEURULA. Stage of vertebrate embryogenesis after most gastrulation movements have ceased, manifested externally by presence of NEURAL PLATE. Stage ends when NEURAL TUBE is complete.

NEUTER. Organism lacking sex organs, but otherwise normal.

NEUTROPHIL. Type of LEUCOCYTE. Phagocytic POLYMORPH.

NEWT. See URODELA.

NICHE (ECOLOGICAL NICHE). Originally (J. Grinnell, 1914) considered to be the spatial and dietary conditions, biotic and abiotic, within which a particular type of organism is found. It was appreciated these were complex and possibly different for different populations of the same species. A later approach (C. Elton, 1927) was to regard a niche as a role within a community enacted equivalently by different species in different communities and defined for animals largely by feeding habits and size. Both approaches view a niche as an immutable 'place' (in a broad sense) within a community, and neither identifies it with the organism occupying that place. A much more abstract approach (G. E. Hutchinson, 1944) took a niche to be the totality of environmental factors (the n-dimensional hyperspace) acting on a species (or species population). If temporal considerations are included (e.g. nocturnality/diurnality), then the COMPETITIVE EXCLUSION PRINCIPLE may be expressed thus: realized niches do not intersect. This approach defines a niche strictly with respect to (in terms of) its occupant, and with respect to a set of continuous (as opposed to discrete) axes, i.e. axes upon which all other niches are also defined. Modern niche theory is concerned particularly with resource competition between species. See SPECIES.

NICOTINAMIDE ADENINE DINUCLEOTIDE. See NAD.

NICOTINAMIDE ADENINE DINUCLEOTIDE PHOSPHATE. See NADP.

NICOTINIC ACID (NIACIN). VITAMIN of the B-complex, lack of which is part of the cause of *pellagra* in man (symptoms being dermatitis and diarrhoea). Yeast, fortified white bread and liver are good sources, but it can also be synthesized in the body from the amino acid tryptophan, of which milk, cheese and eggs are good dietary sources. Contributes structural components to coenzymes NAD and NADP. Synthesized by many microorganisms.

NICTITATING MEMBRANE. Transparent membranous skinfold (third eyelid) of many amphibians, reptiles, birds and mammals, lying deeper than the other two; often very mobile. Moves rapidly over the cornea independently of the other two, if these move at all, cleaning it and keeping it moist. Often used under water.

NIDATION. See IMPLANTATION.

NIDICOLOUS. Of those birds which hatch in a relatively undeveloped state (naked, blind) and stay in the nest, being tended, for some time after hatching. Compare NIDIFUGOUS.

NIDIFUGOUS. Of those birds which hatch in a relatively advanced and mobile state and are capable of leaving the nest immediately and of searching for food, often assisted by one (usually) or both parents. Compare NIDICOLOUS.

NIT. Egg of human louse; cemented to hair.

NITRIFICATION. Conversion of ammonium ions (NH_4^+) to nitrite and nitrate ions (NO_2^-, NO_3^-) by chemotrophic soil bacteria. Major genera involved are, respectively, *Nitrosomonas* and *Nitrobacter*. Resulting nitrate ions are main nitrogen source of plants but are easily leached from soil. Free oxygen is essential for the process and low soil temperature and pH greatly reduce its rate. See NITROGEN CYCLE.

NITROGEN CYCLE. Circulation of nitrogen atoms, brought about mainly by living organisms. Inorganic nitrogenous compounds (chiefly nitrates) are absorbed by autotrophic plants from soil or water and synthesized into organic compounds. These autotrophs die and decay or are eaten by animals, and the nitrogen, still in the form of organic compounds (e.g. proteins, nucleic acids), returns to the soil or water via excretion or by death and decay. Ammonifying and nitrifying bacteria then convert them to inorganic compounds (see AMMONIFICATION, NITRIFICATION). Some nitrogen is lost to the atmosphere as nitrogen gas by DENITRIFICATION. A great deal (about 140–700 $mg.m^{-2}.yr^{-1}$, more in fertile areas) is extracted from the atmosphere by N-fixing bacteria and blue-green algae (see NITROGEN FIXATION). Lightning causes oxygen and nitrogen to react, producing oxides of nitrogen which react with water to form nitrate ions, adding on average approx. 35 $mg.m^{-2}.yr^{-1}$ of nitrogen to the soil.

NITROGEN FIXATION. Incorporation of atmospheric nitrogen (N_2) to form nitrogenous organic compounds. Majority (about 10^8 tonnes globally per year) produced by free-living nitrogen-fixing prokaryotes, both bacteria and blue-green algae (CYANOBACTERIA), the energy required coming either from photosynthesis (e.g. in *Chlorobium*) or soil organic compounds (e.g. in *Clostridium*). Symbiotic nitrogen fixation occurs in root nodules of leguminous plants (esp. Family Fabaceae), where bacteria of genus *Rhizobium*, otherwise free-living in the soil, infect roots and colonize cortical cells. The bacteria use carbohydrate supplied via host phloem, providing in turn nitrogenous products via host xylem. Other N-fixing symbioses occur between the fungus *Actinomyces* and alders (*Alnus*), and between blue-green algae and liverworts, ferns and gymnophytes. The enzyme involved, *nitrogenase*, is highly energy-demanding (using ATP and electron donors from host) and produces ammonia only under anaerobic conditions. See NITROGEN CYCLE.

NK CELL. See NATURAL KILLER CELL.

NODAL BRACT. Modified leaf-like appendage emanating from a NODE on a stem, cone or fruiting apex.

NODE. (Bot.) Part of plant stem where one or more leaves arise.

NODE OF RANVIER. Exposed region of axon of NERVE FIBRE, where the MYELIN SHEATH is absent between Schwann cells. Only here does current flow through the membrane during passage of an IMPULSE along the myelinated nerve, so that these axons have much faster impulse transmission than non-myelinated axons. The conduction is sometimes referred to as saltatory, since it leaps from one node to the next.

NOMINALISM. Approach to CLASSIFICATION which denies the reality of categories employed, holding instead that they are artificial constructs employed for convenience or for naturalistic reasons. Compare ESSENTIALISM.

NON-DISJUNCTION. Chromosomal MUTATION resulting in failure of either a) the two members of a bivalent to separate during the first meiotic anaphase, or b) the two sister chromatids of a chromosome to separate at second meiotic anaphase. Results in aneuploidy: resulting cells have either one too many or one too few chromosomes. Trisomies and monosomies may arise this way. In man, may cause DOWN'S SYNDROME.

NORADRENALINE. (In USA, NOREPINEPHRINE.) Generally excitatory NEUROTRANSMITTER of sympathetic nervous system, produced at postsynaptic nerve termini. Like ADRENALINE, a derivative of the amino acid tyrosine (an amino group replacing adrenaline's methyl group), and like it secreted by the ADRENAL GLAND, but in smaller quantities. Implicated in maintaining AROUSAL in the human brain, in dreaming and in generation of mood.

NOSTRILS. See NARES.

NOTOCHORD. Rod of vacuolated tissue enclosed by firm sheath and lying along long axis of chordate body, between central nervous system and gut. Present at some stage in all chordates. In most vertebrates, occurs complete only in embryo (or larva) but remnants may persist between the vertebrae, which obliterate it. Found in larval and adult cephalochordates, larval urochordates, and perhaps adult hemichordates. In these latter forms it acts skeletally, antagonizing the myotomes. Usually regarded as mesodermal, and with it forming the *chorda-mesoderm*.

NOTOGEA. Australian ZOOGEOGRAPHICAL REGION.

NUCELLUS. Tissue surrounding the megasporangium (ovule) in a seed of a gymnophyte or flowering plant; part of the sporophyte; surrounded by one or two integuments. Generally provides nutrition for megaspore.

NUCLEAR PORE. See NUCLEUS.

NUCLEAR PROTEINS, NUCLEAR RECEPTORS. Proteins which, once bound by an attachment site to an intracellular ligand, are then able to bind to a specific nuclear chromatin region and initiate/enhance transcription of target genes. Unbound receptor molecules differ in their location, some being nuclear, some cytoplasmic. Different receptor molecules may show considerable sequence homology; but even so they may bind a wide variety of biologically and biosynthetically unrelated ligands. Such nuclear receptor families include the *steroid hormone receptor* family and *thyroid hormone-retinoic acid-vitamin D_3 receptor* family. Some nuclear receptors may have oncogenic properties if their amino acid sequence becomes altered through mutation in the encoding gene. See ENHANCER, TRANSCRIPTION FACTORS.

NUCLEASE. Enzyme degrading a nucleic acid. See DNase, RNase, RESTRICTION ENDONUCLEASE. Some occur within LYSOSOMES.

NUCLEIC ACID. A polynucleotide. There are two naturally occurring forms, DNA and RNA, polymers formed by condensation of nucleotides by, respectively, DNA and RNA polymerase. Both forms occur in all living cells, but only one of them in a given virus. DNA (except in RNA viruses) carries the store of molecular information (the genetic material, a major component of CHROMOSOMES); RNA is involved in deciphering of this information into cell product during transcription and translation. See GENE, NUCLEOPROTEIN, PROTEIN SYNTHESIS.

NUCLEIC ACID HYBRIDIZATION. See DNA HYBRIDIZATION.

NUCLEOHISTONE. See CHROMATIN, CHROMOSOME.

NUCLEOID. DNA-containing region of prokaryotic cell, where the main chromosome and any PLASMIDS are housed; not membranebound. DNA highly condensed.

NUCLEOLUS. Spherical body, occupying up to 25% of nuclear volume, one or more of which stain with basic dyes in interphase nuclei. Its size reflects its level of activity. Consists of decondensed chromatin containing the chromosome-specific and highly amplified tandem gene sequences for ribosomal RNA (*nuclear organizers*), along with their transcription products (ribosomal RNA sequences), certain RNA-binding proteins and associated ribosomal proteins, all involved in the early phase of ribosome construction. Nucleoli usually dwindle and disappear early in nuclear division, but reappear as RNA synthesis recommences in telophase. Much of the nucleolar RNA and protein seem to be carried by chromosomes while the nucleolus itself is disassembled. See GENE AMPLIFICATION.

NUCLEOPROTEIN. Complex of nucleic acid and associated proteins forming a CHROMOSOME. In eukaryotes the proteins include histones. See CHROMATIN.

NUCLEOSIDES. The base-ribose moieties of NUCLEOTIDES. Common naturally-occurring forms in RNA are adenosine, guanosine, cytidine and uridine, while *deoxy* forms (deoxyribose instead of ribose) of all but uridine occur in DNA, deoxythymidine replacing uridine.

NUCLEOSIDE TRIPHOSPHATE (NUCLEOTIDE DIPHOSPHATE). Molecule comprising a NUCLEOSIDE to which three phosphate groups are bound, usually at the 5′ hydroxyl of the pentose. ATP, GTP, CTP and UTP are substrates in RNA synthesis, while the *deoxy* forms dATP, dGTP, dCTP and dTTP are substrates in DNA synthesis (see DNA REPLICATION).

NUCLEOSOME. Fundamental packing unit of a eukaryotic CHROMOSOME, comprising an octomeric HISTONE core and 146 base pairs of DNA wound around it. Transcription of eukaryotic RNA occurs on DNA organized as nucleosomes.

NUCLEOTIDE. A molecule comprising either a purine or pyrimidine base bonded to either a phosphorylated ribose or (in the case of a deoxynucleotide) deoxyribose moiety. A phosphorylated nucleoside. Nucleotides and deoxynucleotides comprise the monomers of RNA and DNA respectively, in which they are linked by phosphodiester bonds. In *cyclic nucleotides* such as cyclic AMP (see AMP), the phosphate moiety forms a diester bond with the 3′ and 5′ hydroxyls of the ribose. Nucleotides are incorporated into nucleic acids by polymerases which use nucleoside or deoxynucleoside triphosphates as substrates. See DNA REPLICATION.

NUCLEUS. (1) The organelle in eukaryotic cells making about 10% of the cell volume and containing the cell's CHROMOSOMES. Together with the CYTOPLASM of the cell it comprises the cell PROTOPLASM. A eukaryotic cell may have no nucleus, or one or more; but each such cell ultimately derives from one which was nucleated.

Nuclei vary in size, even growing and diminishing within the same cell, tending to be largest in actively synthetic (e.g. secretory) cells.

Each nucleus has an *envelope* of two membranes punctuated by *pore complexes* of 30–100 nm internal diameter (nuclear pores) where the inner and outer membranes are continuous. Pore complexes are involved in active transport of molecules in to and out from the *nucleoplasm* within the envelope. The inner nuclear membrane has attachment sites for INTERMEDIATE FILAMENTS (chiefly *lamins*) constituting the thin *nuclear lamina*, which support the CHROMATIN fibres during interphase. Other intermediate filaments surround the cytoplasmic surface of the outer membrane. Phosphorylation of the lamins at mitotic and meiotic prophase in higher eukaryotes (e.g. not diatoms, dinoflagellates or fungi) is associated with disintegration of the nuclear membrane, while their dephosphorylation at anaphase causes vesicles of nuclear membrane to assemble around chromosomes and gradually fuse to reform the nuclear envelope. The outer membrane of the envelope is usually continuous in places with the ENDOPLASMIC RETICULUM. Nucleoplasm contains enzymes, ribosomes and other proteins, all of which originated on ribosomes in the cytosol and subsequently passed into the nucleus (having appropriate peptide sorting signals). Some of these are involved in transcription of ribosomal RNA from multiple copies of rRNA genes located on particular chromosomes (often tandemly, as *nucleolar organizers*). The decondensed chromatin involved, RNA transcripts and protein-RNA complexes which form, together constitute a NUCLEOLUS, of which there is usually at least one per nucleus.

If properly stained, chromosomes are visible in interphase nuclei as thin grains of chromatin. Interphase nuclei contain relatively de-condensed chromatin. This condenses (shortens and thickens by coiling) during nuclear division to give chromosomes visible in light microscopy. Ciliates and some other protozoa contain two nuclear classes: a *macronucleus* concerned with the vegetative 'day-to-day' running of the cell, and one or more *micronuclei* involved in CONJUGATION. The macronucleus, which contains much of the genome in multiple and often fragmented copies, divides amitotically (see ABERRANT CHROMOSOME BEHAVIOUR), the micronucleus by mitosis. Mature eukaryotic cells in which the nucleus has disintegrated (e.g. mammalian red blood cells) may rely on 'long-lived' messenger RNA for protein synthesis. Nuclear transplantation work has revealed much about cell differentiation, not least that the nucleus commonly only releases information to the cytoplasm (in the form of mRNA) when 'instructed' by it to do so, for example by NUCLEAR PROTEINS and some steroid hormones.

(2) One of the several anatomically distinct aggregations of nerve cell bodies (ganglia) in the vertebrate BRAIN. The *lateral geniculate nuclei* within the THALAMUS, for example, contain cell bodies of neurones of the two optic nerves. (3) The central part of an atom of a chemical element, containing at least one proton and usually at least one neutron.

NUDE MICE. Developmentally abnormal mice, lacking thymus (*athymic*). Used in immunological work.

NULL HYPOTHESIS. Hypothesis constructed in order to derive from it a hypothetical set of experimental results (*expected results*) which would obtain under specified experimental conditions were the hypothesis true, and which may be compared with *observed results* derived from an actual experiment operating under those conditions. If there is a significant discrepancy between the two sets of results (see CHI-SQUARED TEST), then it is probable that the null hypothesis is incorrect. See EXPERIMENT.

NULLISOMIC. Abnormal chromosome complement in which both members of a chromosome pair are absent from an otherwise diploid nucleus. Type of ANEUPLOIDY; viable only in a polyploid (e.g. wheat, which is hexaploid).

NUMERICAL PHYLETICS. Erection of an inferred phylogeny of living organisms through numerical analysis of characteristics. See PHYLETICS.

NUMERICAL TAXONOMY. Erection of classifications by numerical analysis of characteristics. See CLASSIFICATION.

NURSE CELLS. (Bot.) In some liverworts, sterile cells among the spores lacking any special wall thickening and frequently disintegrating before spores mature. (Zool.) For Sertoli cells, see TESTIS.

NUT. Dry indehiscent, single-seeded fruit, somewhat similar to an achene but the product of more than one carpel and usually larger, with a hard woody wall; e.g. hazel nut. A small nut is called a nutlet.

NUTATION (CIRCUMNUTATION). Spiral course pursued by apex of plant organ during growth due to continuous change in position of its most rapidly growing region; most pronounced in stems, but also occurs in tendrils, roots, flower stalks, and sporangiophores of some fungi.

NYCTINASTY. Response of plants to periodic alternation of day and night; e.g. opening and closing of many flowers, 'sleep movements' of leaves. Related to changes in temperature and light intensity. See PHOTONASTY, THERMONASTY.

NYMPH. Larval stage of exopterygote insect. Resembles adult (e.g. in mouthparts, compound eyes), but is sexually immature. Wings absent or undeveloped.

O

OBLIGATE. Term indicating some type of restriction in an organism's way of life, from which it cannot depart and survive. Used particularly in the contexts of parasites, aerobes and anaerobes. Contrast FACULTATIVE.

OCCIPITAL CONDYLE. Bony knob at back of skull, articulating with first vertebra (the atlas). Absent in most fish, whose skulls do not articulate with vertebral column. Single in reptiles and birds, double in amphibians and mammals.

OCCIPUT (OCCIPITAL REGION). (1) In vertebrates, an arbitrarily delimited region of the skull and head in the neighbourhood of the occipital condyle. (2) In insects, a plate of exoskeleton forming back of the head.

OCEANIC. Inhabiting the sea where it is deeper than 200 metres. Compare NERITIC.

OCELLUS. General term for several types of *simple eye*, as found in some annelids, insects (the only eyes of larval endopterygotans), arachnids and other arthropods. Incapable of image-formation. See EYE.

OCHRE MUTATION. A 'stop' mutation; UAA triplet of messenger RNA. The ochre codon forms the normal terminus of the coding regions of many codons. See GENETIC CODON, AMBER MUTATION.

OCULOMOTOR NERVE. Third CRANIAL NERVE of vertebrates. Supplies four of the extrinsic eye muscles and, by neurones of parasympathetic system, via the ciliary ganglion, intrinsic eye muscles of ACCOMMODATION and pupil constriction. A ventral root.

ODONATA. Dragonflies and damselflies. Order of exopterygote insects with aquatic nymphs. Carnivorous as nymphs and adults, with biting mouthparts, hinged on an extensible 'mask' in the nymph. Large, compound eyes; two pairs of similar wings, folded over the back in damselflies (Zygoptera) and horizontally in dragonflies (Anisoptera). Some fossil (Carboniferous) forms had wingspans in excess of half a metre.

ODONTOBLASTS. Cells lying in pulp cavities of vertebrate teeth, sending processes into adjacent dentine, which they help form.

ODONTOID PROCESS. See ATLAS.

OEDEMA. Swelling of tissue through increase of its TISSUE FLUID volume. Can occur during INFLAMMATION, through blockage of vessels of LYMPHATIC SYSTEM returning tissue fluid to venous system (see FILARIAL WORM) and through low levels of blood proteins causing poor return of tissue fluid through CAPILLARIES, as in protein malnutrition (kwashiorkor).

OESOPHAGUS. Anterior part of gut, between pharynx and stomach. Usually concerned simply in peristaltic passage of food to stomach. May contain a CROP.

OESTRADIOL, β-OESTRADIOL. Most potent OESTROGEN in female mammals.

OESTROGENS. Group of STEROIDS, both naturally-produced ovarian hormones and artificial homologues of them. Synergists of GROWTH HORMONE. High blood levels inhibit release of gonadotrophic releasing factors (GnRFs) from the HYPOTHALAMUS. The most potent natural oestrogen in humans is β-oestradiol; two others are oestrone and oestriol. Produced by theca cells of developing GRAAFIAN FOLLICLES of the ovary and, after ovulation, by granulosa cells; adrenal cortices also produce small amounts.

Main roles of oestrogens in female mammals are: development of MAMMARY GLANDS (synergistically with PROGESTERONE), regulation of fat deposition, softening and smoothing of skin texture (in humans), regulation of uterine endometrium, again with progesterone (see MENSTRUAL CYCLE), promotion of flattening and widening of PELVIC GIRDLE and GROWTH spurt at puberty (all *secondary sexual characteristics*). Effects on sexual behaviour depend on the species involved, but may cause 'maternal behaviour' if injected into males.

Oestrogens enter target cell nuclei in combination with specific cytoplasmic receptors, annealing to specific chromosome regions and initiating mRNA transcription (see GENE EXPRESSION). They are major components of the CONTRACEPTIVE PILL, and some synthetic forms (e.g. stilboestrol) have been shown to be carcinogenic. See NUCLEAR RECEPTORS.

OESTROUS CYCLE. Reproductive cycle of short duration (usually 5–60 days) occurring in sexually mature females of many non-human mammals, in the absence of pregnancy. Regulated by endocrine (often neuroendocrine) interactions involving the HYPOTHALAMUS, pituitary and ovaries. The phases in the cycle include: *oestrus*, *metoestrus*, *dioestrus* and *pro-oestrus*. Compare human MENSTRUAL CYCLE.

OESTRUS. Follicular phase of the OESTROUS CYCLE during which sexual desire and attractions of the female may be heightened, leading to copulation (see NEUROENDOCRINE COORDINATION); less so in some primates. In cats and dogs, periods of oestrus (when females are 'on heat') occur two or three times per year, separated by long periods of *anoestrus*.

OLECRANON PROCESS. Bony process on mammalian ulna, extending below elbow joint and employed in attachment of muscles straightening limb.

OLFACTORY NERVE. First CRANIAL NERVE of vertebrates, running from olfactory area of cerebral cortex to olfactory bulb (see OLFACTORY ORGANS).

OLFACTORY ORGANS. Organs of smell. In vertebrates, consist of sensory epithelia in nasal cavities, whose cells respond to molecules dissolved in their moist mucous membranes. *Olfactory cells* are bipolar neurones whose dendrites synapse with branches of the olfactory nerve in a mass of grey matter termed an *olfactory bulb* within the braincase. The olfactory nerve travels to olfactory regions of cerebral cortex.

OLIGOCENE. Geological epoch, subdivision of the Tertiary period, lasting from 38–26 Myr BP. See GEOLOGICAL PERIODS.

OLIGOCHAETA. Class of annelid worms which includes earthworms. See ANNELIDA.

OLIGOPEPTIDE. See PEPTIDE.

OLIGOSACCHARIDE. Carbohydrate formed by joining together monosaccharide units (4–20) in a chain by links termed *glycosidic bonds*, formed enzymatically. Are intermediate digestion products of POLYSACCHARIDES, and some form side chains of GLYCOPROTEINS and GLYCOLIPIDS. See GLYCOSYLATION.

OLIGOTROPHIC. Term describing lakes and rivers that are unproductive in terms of organic matter formed; nutrient status of the water and nutrient supply are very low. Compare EUTROPHIC, MESOTROPHIC.

OMMATIDIUM. See EYE.

OMNIVORE. Animal eating both plant and animal material in its diet. See CARNIVORE, HERBIVORE.

ONCHOCERIASIS. 'River blindness'. See FILARIAL WORMS.

ONCHOSPHERE (HEXACANTH). Six-hooked embryo of tapeworms (CESTODA). Develops from egg, usually while still in proglottis. Will bore through gut wall of secondary host and is then carried in blood to host tissues, in which it lodges and develops into the CYSTICERCUS.

ONCOGENE. Genetic locus of tumour virus (*oncogenic virus*) involved in the conversion of host cell into a CANCER CELL (*neoplasia*). These loci often encode for PROTEIN KINASES which phosphorylate many target proteins, some of them components of the host plasma membrane. Some such modification of the membrane is thought to alter the cell's response to *growth factors* in its environment, causing it to divide when previously it was unresponsive. Most retrovirus oncogenes originate from the host cell genome and get into the viral genome by a transduction-like event. Such cellular genes with the potential to become oncogenes are termed *proto-oncogenes* or *cellular oncogenes* (c-oncs). Some oncogenes may become expressed when TRANSPOSABLE ELEMENTS insert adjacently to them in the genome. See SUPPRESSOR MUTATION.

ONCOGENIC VIRUS. See ONCOGENE, VIRUS.

ONE GENE–ONE ENZYME HYPOTHESIS. See GENE.

ONTOGENY. The whole course of an individual's development, and life history. Compare PHYLOGENY. See RECAPITULATION.

ONYCHOPHORA. Small group of animals (e.g. *Peripatus*) with annelid and arthropod affinities, sometimes regarded as a distinct phylum, but here placed as a class within the ARTHROPODA. Represent an early stage of *arthropodization*. Body segmented with soft, unjointed cuticle 1μm thick and permeable to water; body wall with longitudinal and circular smooth muscles; head not demarcated, bearing a pair of long, mobile *pre-antennae*, simple jaws and oral papillae. Limbs fleshy, unsegmented, with terminal claws; probably evolved from parapodial appendages. The heart has ostia. Coelom replaced in adult by haemocoele. Gaseous exchange by tracheae and spiracles. Excretory system by segmentally repeated *coxal glands*, similar to those of arachnids and crustaceans but with ciliated excretory ducts. The pair of ventral nerve cords lack segmental ganglia but have numerous connectives. Eyes simple. Forest-dwelling, generally nocturnal predators. About 70 species. Some aquatic Cambrian fossils, such as *Aysheaia* and *Opabinia*, show affinities. See TARDIGRADA.

OOCYTE. Cell undergoing MEIOSIS during OOGENESIS. *Primary oocytes* undergo the first meiotic division, *secondary oocytes* undergo the second meiotic division. See MATURATION OF GERM CELLS.

OOGAMY. Form of sexual reproduction involving production of large non-motile gametes, which are fertilized by smaller motile gametes. Extreme form of ANISOGAMY, occurring in all metazoans and some plants.

OOGENESIS. Production of ova, involving usually both meiosis and maturation. See MATURATION OF GERM CELLS.

OOGONIUM. (1) (Bot.) Female sex organ of certain algae and fungi, containing one or several eggs (OOSPHERES). (2) (Zool.) Cell in an animal ovary which undergoes repeated mitosis, giving rise to oocytes. Compare SPERMATOGONIUM.

Oosphere. (Bot.) Large, naked, spherical non-motile macrogamete (egg); formed within an OOGONIUM.

Oospore. Thick-walled resting spore formed from a fertilized OO-SPHERE.

Opal mutation. A 'stop' mutation. A UGA mRNA triplet. See AMBER MUTATION.

Open bundle. Vascular bundle in which a vascular cambium develops.

Operator. See JACOB-MONOD THEORY.

Operculum. (Bot.) Lid or cover; in certain fungi, part of cell wall; in mosses, a multicellular apical lid that opens the sporangium. (Zool.) (1) Cover of gill slits of holocephalan and osteichthyan fishes, and of larval amphibia. Contains skeletal support. (2) Horny or calcareous plate borne on back of foot in many gastropod molluscs, brought in over the body when animal withdraws into its shell. (3) A second EAR OSSICLE of many urodele and anuran amphibians; flat plate fitting into oval window of inner ear. May replace the stapes.

Operon. See JACOB-MONOD THEORY.

Ophidia. Snakes. Suborder of the Order SQUAMATA (sometimes a separate order). Limbless reptiles, with exceptionally wide jaw gape due to mobility of bones. Eyelids immovable, nictitating membrane fused over cornea; no ear drums.

Ophiuroidea. Brittlestars. Class of ECHINODERMATA; star-shaped, with long, sinuous ambulatory arms radiating from clearly delineated central disc; mouth downwards; well-developed skeleton of articulating plates; tube feet without suckers; no pedicillariae; easily break up by autotomy; scavengers.

Ophthalmic. Alternative adjective to optic.

Opiates. See ENDORPHINS, ENKEPHALINS.

Opisthobranchia. Subclass of gastropod molluscs, in which shell and mantle cavity are reduced or lost. Body bilateral and slug-like. Sea hares (e.g. *Aplysia*), other nudibranchs, and the actively swimming pteropods.

Opisthosoma. One of the TAGMATA of chelicerate arthropods (Merostomata and Arachnids). Comprises the trunk segments, devoid of walking limbs; often (e.g. in scorpions) separable into an anterior *mesosoma* and posterior *metasoma*.

Opsonins. Proteins coating foreign particles, such as surfaces of pathogenic microorganisms, rendering them more susceptible to ingestion by phagocytic leucocytes (*opsonization*). Many are ANTI-BODIES.

OPTIC CHIASMA. Structure formed beneath vertebrate forebrain by those nerve fibres of the right optic nerve crossing to left side of brain and those of left side crossing to the right side. In most vertebrates all the fibres cross; in mammals about 50% on each side do so. See BINOCULAR VISION, DECUSSATION.

OPTIC NERVE. Second CRANIAL NERVE of vertebrates; really part of brain wall. See RETINA.

OPTIMIZATION THEORY (OPTIMALITY THEORY). Very generally the theory that, through natural selection, the behaviours of organisms are such that they tend to the most cost-effective use of time and available resources, given environmental circumstances. Despite charges of being vacuous and truistical, and although similar statements may be deducible from some formulations of Darwinian theory, their heuristic value has been great, prompting detailed and testing studies which have greatly enhanced understanding in such diverse fields as foraging behaviour, diet choice, habitat selection and competition in animals, and reproductive behaviour more generally. See GAME THEORY.

ORBIT. Cavity or depression in vertebrate skull, housing eyeball.

ORDER. Taxonomic category; inclusive of one or more families, but itself included by the class of which it is a member. See CLASSIFICATION.

ORDOVICIAN. GEOLOGICAL PERIOD, second of Palaeozoic era. Lasted approximately from 500–440 Myr BP.

ORGAN. Functional and anatomical unit of most multicellular organisms, consisting of at least two tissue types (often several) integrated in such a way as to perform one or more recognizable functions in the organism. Examples in plants are roots, stems and leaves; in animals, liver, kidney and skin. Organs may be integrated into functional SYSTEMS. Sometimes it is debatable where the limits of organs are; thus the stomach is often regarded as an organ, but on occasions so is the whole alimentary canal.

ORGAN CULTURE. By partial immersion in nutrient fluid, the growth and maintenance *in vitro* of an organ after removal from the body. The organ must be small (usually embryonic) since nutrients must diffuse into it from outside. See TISSUE CULTURE.

ORGAN OF CORTI. See COCHLEA.

ORGANELLE. Structural and functional part of a cell, distinguished from the cytosol. Often membrane-bound (e.g. nucleus, mitochondria, chloroplasts, endoplasmic reticulum); sometimes not (e.g. ribosomes). The plasma membrane is itself an organelle. Those in cytoplasm are termed *cytoplasmic organelles*, in contrast to the nucleus. See CELL for diagram (Fig. 3).

ORGANIZER. Any part of an embryo which can induce another part to differentiate. Classic example is dorsal lip of the BLASTOPORE. See EVOCATION, INDUCTION, POSITIONAL INFORMATION.

ORGANOTROPHIC. See HETEROTROPHIC.

ORIENTAL. ZOOGEOGRAPHICAL REGION comprising India and Indo-China south to WALLACE'S LINE.

ORIGIN OF LIFE. Major hurdles to the origin of life-forms resembling even the simplest known cells are the production of a) a genetic system which is b) sufficiently discrete from other such systems to enable its collection of properties to be a unit in reproductive competition with them. These two requirements would seem to put a premium on simple genetic (information) systems engaging in, at least initially, autocatalysis, with subsequent expansion of the stored information to code for molecules which do not themselves store information but instead form part of the structure of the system.

Even the simplest living systems today are far too complicated to resemble at all closely the earliest forms of life. Three major features of present living systems (construction from complex, often polymeric, organic compounds; catalysis by proteins; storage of molecular information in the form of nucleic acid) are likely to be the end results of evolution by natural selection between systems which were originally far simpler and lacked all these features. The following account is highly conjectural, but raises some of the issues being debated.

For long it was thought that the early Earth's atmosphere consisted of reducing molecules (ammonia, methane, hydrogen and water vapour), and attempts to create likely precursor molecules of life (amino acids, bases) by passing electrical sparks through such gaseous mixtures were rewarding. However it is probable that four thousand million years ago the sun was weaker in terms of total thermal output than it is now, and more active in terms of ultraviolet output. Implications for the Earth's early atmosphere and the origin of life are that: a) without an atmospheric GREENHOUSE EFFECT, the surface temperature of the Earth would have been too low to avoid ocean freezing, and that b) incident ultraviolet radiation on the Earth's atmosphere would have resulted in photolysis of methane and ammonia, releasing free hydrogen out of the atmosphere, with probable retention of carbon dioxide and atomic and molecular nitrogen. The carbon dioxide could have provided the greenhouse effect: the early oceans were not frozen.

One of the precursors of adenine (a base present in the nucleic acid RNA) may well have been hydrogen cyanide (HCN). RNA has become a lively contender in the 'origin' debate on account of its recently discovered catalytic ability (see ENZYME). It is now clear that, in principle at least, RNA is capable of self-catalysis (*autocatalysis*). HCN may have been formed from methane present as a trace gas in the early mantle. Hydrothermal vents (undersea hot springs) provide reducing compounds which, by permeating rock, could form complex molecules, cooled sufficiently by the water to reduce the likelihood of their dissociation. These, in association with the self-assembling crystalline organization provided by such clays as *kaolinites* and *illites*, could come to form polymers supported by the lattice of the crystal.

The first membranes might have arisen by self-assembly, forming spherical vesicles in much the same way that some phospholipids do when mixed with water. Such lipid vesicles might then have accumulated any organic molecules soluble in their membranes and if these polymerized inside they might not have been able to escape. Continued polymerization within might have put the vesicles under strain, promoting their enlargement by incorporation of further membrane components, as can happen with artificial LIPOSOMES. Rupture of the spheres followed by renewed growth of the rupture products would have achieved a simple form of reproduction.

In the past, RNA may have had wider catalytic ability than it has now (see TELOMERE). If RNA were among early polymers, formed perhaps on clay crystals trapped within membrane spheres, and were it to catalyse amino acid polymerization, then protein synthesis might gain independence from clay crystals, RNA also supporting the protein. If some of these proteins were to associate with the membrane in such a way as to stabilize it and even promote entry of building materials, selection could have favoured those systems which in replicating retained catalytic production of the membrane proteins by RNA; that is, which achieved the reproducible link between genotype and phenotype. Something ike a simple cell would have been produced. DNA could eventually have been synthesized off the RNA (retroviruses do this today), the resulting duplex providing a more stable store of information.

ORNITHINE CYCLE (UREA CYCLE). See UREA.

ORNITHISCIA. Extinct order of ARCHOSAURS ('ruling reptiles'), lasting throughout Jurassic to late Cretaceous. 'Bird-hipped' and herbivorous dinosaurs. Some early forms were bipedal. Both the pubes and ischia of the PELVIC GIRDLE pointed downwards and backwards. Included stegosaurs (e.g. *Stegosaurus*), ankylosaurs and ceratopsids (e.g. *Triceratops*). Compare SAURISCHIA.

ORNITHOLOGY. Study of birds.

ORNITHOPHILY. Pollination by birds.

ORTHOGENESIS. Theory, prevalent in early decades of 20th century, that the evolutionary path of a lineage can acquire an 'impetus', or 'inexorable trend' carrying it in a direction independently of selective constraints imposed by the environment. It was once suggested that the genetic material itself might somehow be responsible. Now discredited.

ORTHOPTERA. Large order of exopterygote insects, including locusts, crickets and grasshoppers. Medium or large insects, usually with two pairs of wings (sometimes absent), front pair narrow and hardened, hind pair membranous; hind legs usually large and modified for jumping. Stridulatory organs; usually involving wings being rubbed together (*alary*), or inner face of hind femur being rubbed against a hardened vein on forewing (*femoro-alary*). Paired auditory organs (*tympana*) either on anterior abdomen or tibias of front legs.

ORTHOSELECTION. Directional selection. See NATURAL SELECTION.

OSCULUM. Large opening through which water leaves the body of a sponge (PORIFERA), having entered through *ostia*.

OSKAR GENES. (*osk.*) A group of MATERNAL EFFECT genes in *Drosophila*, whose mRNA transcription products become localized in the posterior pole of the developing oocyte. Embryos from females mutant for any of these genes develop normal heads and thoraxes, but lack abdomens. The *osk* protein product appears to activate transcription of the *nannos* (*nos*) gene, whose product in turn seems to activate *knirps* transcription (see BICOID GENE).

OSMIUM TETROXIDE. Substance (OsO_4) which, in aqueous solution, blackens fat and is often used to demonstrate myelin sheaths of neurones. Used as a fixative in light and electron microscopy. Sometimes called osmic acid.

OSMOPHILIC BODY. Lipid-containing granules, appearing dark when treated with osmium fixation.

OSMOREGULATION. Any mechanism in animals regulating a) the concentration of solutes within its cells or body fluids, and/or b) the total volume of water within its body.

Each major inhabited environment, freshwater, marine and terrestrial, poses its own osmotic problems for organisms. Cells approximate to sea water in WATER POTENTIAL, but have lower water potentials than freshwater and far higher water potentials than air at normal humidities. Freshwater protozoans and sponges use CONTRACTILE VACUOLES to expel osmotic water from their cells; the SODIUM PUMPS of most animal cell membranes are important in limiting the cell's internal ion concentration. Protection from osmotic uptake of water and from desiccation is achieved by a body surface covered with an impermeable CUTICLE or SKIN, but there are usually soft parts (e.g. gut, gills) through which water enters. Adaptive behaviour may reduce osmotic dangers; thus limpets adhere closely to rocks at low tide; marine mussels and periwinkles retract into shells. *Euryhaline* animals tolerate wide fluctuations in water potentials of their surroundings, possibly through a limited ability to swell in hypotonic conditions and by a reduction in intracellular levels of some organic compounds, or inorganic ions such as sodium and chloride. Some engage in active ion uptake (e.g. of sodium) if the external medium becomes dilute enough to cause sodium loss by diffusion. Unless urine can be made hypotonic to body fluids, inorganic ions will be lost by this route, and a kidney which reabsorbs ions from the excretory fluid will reduce this loss. ANADROMOUS animals, such as many cyclostomes and some fish, automatically take up water in freshwater but produce a copious hypotonic urine, active ion uptake by the gills replacing urinary loss. On return to marine conditions they lose water to the sea and correct this by drinking sea water: sodium, potassium and chloride are absorbed by the intestine and excreted across the gills (see SODIUM PUMP).

Restriction of water loss on land is associated with evolutionary change from excretion of ammonia (ammonotely) to urea (ureotely) and uric acid (uricotely). Amphibian metamorphosis includes a change from ammonotely to ureotely; land reptiles produce a urine hypotonic to body fluids, but because of uricotely (uric acid being insoluble requires only incidental water in its removal) the total volume produced is not great. Evolution of the AMNIOTIC EGG would have been less likely without the uricotely of reptilian and bird embryos. Marine birds and reptiles do not regularly drink sea water, but they absorb much salt from their food and excrete it through *salt glands* above the orbits of the eyes.

Vertebrate KIDNEYS, especially mammalian, can produce a urine which is variably copious and hypotonic to body fluids or sparse and hypertonic, depending upon the water balance of the body. The hypothalamus monitors blood concentration, and any increase above the homeostatic norm results in release of ANTIDIURETIC HORMONE by the posterior PITUITARY. This causes uptake of water by the collecting ducts and reduced volume of hypertonic urine. Moreover, the hypothalamic *thirst centre* now initiates drinking, restoring blood concentration. Sudden loss of blood volume (or pressure) causes *aldosterone* release from the ADRENAL cortex (see ANGIOTENSINS). This increases potassium excretion and sodium retention by the kidney, tending to make extracellular (tissue) fluid more concentrated than body cell fluid, lowering its water potential and drawing water out from the cells to help restore blood volume. After a blood meal, the bug *Rhodnius* disposes of excess fluid by release of a *diuretic hormone* from its thoracic ganglia. This accelerates active secretion of sodium ions from the gut, chloride and water following them into the haemocoele, where they are dealt with by MALPIGHIAN TUBULES. Osmoregulation may therefore involve structural, physiological and behavioural adaptations.

Osmosis. The net diffusion of water across a selectively permeable membrane (permeable in both directions to water, but varyingly permeable to solutes) from one solution into another of lower WATER POTENTIAL. The *osmotic pressure* of a solution is the pressure which must be exerted upon it to prevent passage of distilled water into it across a semipermeable membrane (one impermeable to all solutes, but freely permeable to solvent), and is usually measured in pascals, Pa (1 Pa = 1 Newton/m²). The plasmalemma is selectively (i.e. not semi-) permeable and permits selective passage of solutes (see CELL MEMBRANES), as does the *tonoplast* in plant cells.

The osmotic pressure of a solution depends upon the ratio between the number of solute and solvent particles present in a given volume: 1 mole of an undissociated (non-electrolytic) substance dissolved in 1 dm³ of water at 0 °C has an osmotic pressure of 2.26 MPa (megapascals). But a solution of 0.01 mole of sodium chloride per dm³ of water has almost twice the osmotic pressure of a solution of 0.01 mole of glucose per dm³ of water. This is because sodium chloride in water dissociates almost completely into Na^+ and Cl^- ions (giving twice the particle number) whereas glucose does not dissociate in water. Solutions of equal osmotic pressure are *isotonic*; one with a lower osmotic pressure than another is *hypotonic* to it, the latter being *hypertonic* to the former.

When cells produce polymers from component monomers they may dramatically reduce the osmotic pressure of their cytoplasm; hence polymers (e.g. starch, glycogen) make good storage compounds because they are *osmotically inactive*. Glucose is far more *osmotically active* than is starch.

Osmosis is a physical process of great importance to all organisms, affecting relations with their environments as well as between their component cells. If an animal cell takes up water osmotically its plasma membrane will eventually rupture (*osmotic shock*); but the SODIUM PUMP reduces this problem under normal physiological conditions. Cells with cell walls are normally prevented from osmotic rupture. Those which lose water osmotically shrink and become *plasmolysed* (see PLASMOLYSIS). Water relations of cells and organisms are generally discussed in terms of water potential. See TURGOR, DIFFUSION, OSMOREGULATION.

Osmotic Activity and Osmotic Pressure. See OSMOSIS.

Ossicle. A very small bone, or calcified nodule.

Ossification. Formation of bone. May occur by replacement of CARTILAGE (*endochondral ossification*) or by differentiation of non-skeletal mesenchyme (*intramembranous ossification*), forming DERMAL BONES. In the former, cartilage first calcifies and hardens to form *ossification centres*. The cartilage cells die and those in the periochondrium develop into *osteoblasts* and lay down a ring of bone to form the *periosteum*, which enlarges as blood vessels invade and osteoblasts enter vacant lacunae to develop into *osteocytes*. The bone lengthens and thickens, finally becoming mineralized. Cartilage remains as *articular cartilage* forming JOINTS at the bone ends, and in the *growth plate* in the bone EPIPHYSIS. See DIAPHYSIS, GROWTH.

OSTEICHTHYES. Bony fishes. Largest vertebrate class. Some fossils come from the Upper Silurian but a considerable radiation had occurred by the Lower to Middle DEVONIAN. Includes subclasses CHOANICHTHYES (see DIPNOI and CROSSOPTERYGII) and ACTINOPTERYGII (which itself includes the TELEOSTEI). Other fish groups (e.g. agnathans, placoderms, acanthodians) contain bone; but the osteichthyans are the only jawed fishes with bony vertebrae and gill arches and with paired fins, but lacking bony plates on head and body. ACANTHODII are sometimes regarded as a subclass of the Osteichthyes, and do not really fall outside the above criteria. Compare CHONDRICHTHYES.

OSTEOBLAST. Cell type responsible for formation of calcified intercellular matrix of BONE.

OSTEOCLAST. Multinucleate cell which breaks down the calcified intercellular matrix of BONE. Remodelling of bone shape by such activity accompanies bone GROWTH.

OSTEOSTRACI. Extinct order of AGNATHA (Monorhina: single external nostril). Head covered by strong bony shield. Heterocercal tail. See OSTRACODERMI.

OSTIOLE. Pore in fruit bodies of certain fungi, and in conceptacles of brown algae through which, respectively, spores or gametes are discharged.

OSTIUM. (1) One of the many inhalant pores of sponges. See PORIFERA. (2) Opening in the arthropod heart, into which blood flows from the haemocoele.

OSTRACODA. Subclass of CRUSTACEA. Most are a few mm long. Aquatic. CARAPACE bivalved, closable by adductor muscles, and completely covering body. FILTER FEEDING. Trunk limbs few and small; food gathered by head appendages. About 2000 species.

OSTRACODERMI. Group of fossil AGNATHA (Diplorhina: external nostrils paired) found from Upper Silurian to mid-Devonian, but probably present from Cambrian. Covered in bony armour; up to 50 cm in length. Lacked paired appendages. Similarities with cyclostomes. Some may have had mouthparts, but not jaws. Vertebrae so far absent from fossils; notochord apparently the skeletal support. See OSTEOSTRACI.

OTIC. Concerning the ear. See reference under AUDITORY.

OTOCYST. See STATOCYST.

OTOLITH. Granule of calcium carbonate in the vertebrate MACULA. See also STATOCYST.

OUTBREEDING. Effect of any mechanism which tends to ensure that gametes which fuse at fertilization produce zygotes with a higher degree of heterozygosity than they otherwise would. Darwin realized that many plants (e.g. orchids, primroses) are structurally adapted to cross-pollination, and subsequent work on the genetics of incompatibility mechanisms, preventing self-fertilization, has shown that selection must favour such plants under certain conditions. Mechanisms promoting outbreeding include having separate sexes (see DIOECIOUS), DICHOGAMY, HETEROSTYLY, INCOMPATIBILITY mechanisms and, commonly in animals, the social structure of the breeding group (e.g. incest taboos in humans). Advantage of outbreeding in sexual populations is that it serves to free genetic variability in the population to selection and thereby reduces evolutionary stagnation of the population (see GENETIC VARIATION). Compare INBREEDING.

OUTCROSSING. (Bot.) Pollination between (normally genetically) different plants of same species.

OVARIAN FOLLICLE. See GRAAFIAN FOLLICLE.

OVARY. (1) (Bot.) Hollow basal region of a carpel, containing one or more ovules. In a flower which possesses two or more united carpels the ovaries are united to form a single compound ovary. (2) (Zool.) Main reproductive organ in female animals, producing *ova*. In invertebrate development eventual GERM CELLS often become distinct at early cleavage. In vertebrates a pair of ovaries develops from mesoderm in the roof of the abdominal coelom; they become invaded by cells from the endoderm, but it is debatable whether these or mesodermal cells developing *in situ* become the *primordial germ cells*; however, the ovary epithelium (*germinal epithelium*) invaginates and some of its cells begin to undergo *oogenesis* (see MATURATION OF GERM CELLS). Each *oogonial* cell becomes surrounded by other cells to form a *follicle*. The ovaries produce various sex hormones, for details of which in mammals see MENSTRUAL CYCLE. See TESTIS.

OVERTURN. Complete mixing of a body of water, from surface to bottom; the breakdown of STRATIFICATION. Results from various external factors.

OVIDUCT. (Zool.) Tube carrying ova away from ovary (or from the coelom into which ova are shed) to the exterior. See MÜLLERIAN DUCT.

OVIPARITY. Laying of eggs in which the embryos have developed little, if at all. Many invertebrates are *oviparous*, as are the majority of vertebrates, including the prototherian mammals. See OVOVIVI-PARITY, VIVIPARITY.

OVIPOSITOR. Organ formed from modified paired appendages at hind end of abdomens of female insects, through which eggs are laid (*oviposition*). Consists of several interlocking parts. Frequently long (e.g. in ichneumon wasps), and capable of piercing animals or plants, permitting eggs to be laid in otherwise inaccessible places. Stings of bees and wasps are modified ovipositors.

OVOTESTIS. Organ of some hermaphrodite animals (e.g. the garden snail, *Helix*; sea bass, Serranidae) serving as both ovary and testis. In the citrus tree pest *Icerya purchasi* (cottony cushion scale) diploid ('female') individuals transform into functional self-fertilizing hermaphrodites, possessing an ovotestis in which the centre is haploid and testicular, the cortex diploid and ovarian. They are thus chromosomal MOSAICS. In order to achieve the haploid interior, one set of chromosomes is eliminated in some of the early ovotestis cells (see ABERRANT CHROMOSOME BEHAVIOUR (3)).

OVOVIVIPARITY. Development of embryos within the mother, from which they may derive nutrition, but from which they are separated for most, or all, of development by persistent EGG MEMBRANES. Examples include many insects, snails, fish, lizards and snakes. See OVIPARITY, VIVIPARITY.

OVULATION. Release of ovum or oocyte from mature follicle of vertebrate ovary. See MENSTRUAL CYCLE for hormonal details.

OVULE. Structure found in seed plants which develops into a SEED after fertilization of an egg cell within it (see DOUBLE FERTILIZATION). In gymnophytes (gymnosperms), ovules are unprotected; in flowering plants (anthophytes), they are protected by the MEGASPOROPHYLL, which forms a closed structure (CARPEL) within which they are formed singly or in numbers. Each ovule is attached to carpel wall by stalk (funicle) which arises from its base (chalaza). A mature flowering plant ovule comprises a central mass of tissue (the NUCELLUS) surrounded by one or two protective layers (integuments) from which the seed coat is ultimately formed. Integuments enclose the nucellus except at the apex, where a small passage (the micropyle) permits entry of water and oxygen during germination. Within the nucellus is a large oval cell, the EMBRYO SAC, developed from the megaspore and containing the egg cell.

OVUM. Unfertilized, non-motile, egg cell. In many animals it is an OOCYTE. Product of the OVARY.

OXIDATIVE PHOSPHORYLATION. Process by which energy released during electron transfer in aerobic RESPIRATION is coupled to production of ATP. See BACTERIORHODOPSIN, NAD, MITOCHONDRION.

OXIDOREDUCTASES. Major group of ENZYMES; catalyse REDOX REACTIONS.

OXYGEN DEBT. Amount of oxygen needed to oxidize the LACTIC ACID produced in the anaerobic work done by muscle during vigorous exercise, and to resynthesize *creatine phosphate* used (see PHOSPHAGEN). Until then, oxygen intake remains above normal while lactic acid remains in the muscle and blood. May be oxidized to pyruvic acid, converted to carbohydrate by the liver in GLUCONEOGENESIS, or neutralized by blood bicarbonate BUFFER and excreted as *sodium lactate* by the kidneys.

OXYGEN DISSOCIATION CURVES (OXYGEN EQUILIBRIUM CURVES). See HAEMOGLOBIN.

OXYGEN QUOTIENT. See Q_{O_2}.

OXYHAEMOGLOBIN. See HAEMOGLOBIN.

OXYTOCIN. Oligopeptide hormone secreted by *pars nervosa* of PITUITARY glands of birds and mammals. Involved in contraction of alveoli of mammary glands during expression of milk, and in promoting smooth muscle contraction of uterus during coitus, and during PARTURITION.

P

P₁. Parental generation in breeding work. Their offspring constitute the F_1 generation.

PACEMAKER. (1) Group of modified cardiac muscle cells in *sinus venosus* of vertebrate HEART. In mammals and birds, forms *sinuauricular* (*sinoatrial*) *node* and lies in right atrial wall near superior vena cava. Cells have wandering membrane potentials, with a built-in tendency to depolarize (the so-called pacemaker potential), and the cell with the fastest intrinsic rate of depolarization leads the rest to fire with it simultaneously. The electrical current generated spreads to adjacent *Purkinje fibres*, and from there to atrial muscle cells, which contract together. The intrinsic pacemaker discharge is increased by sympathetic nerves from the thorax and decreased by branches of the vagus (parasympathetic system). See CARDIO-ACCELERATORY and CARDIO-INHIBITORY CENTRES. (2) A *myogenic pacemaker* occurs in the smooth muscles of vertebrate gut; longitudinal and circular layers each have one.

PACHYTENE. See MEIOSIS.

PAEDOGENESIS. Form of HETEROCHRONY, in which reproductive organs undergo accelerated development relative to rest of body, giving larval maturity. The alternative term PROGENESIS has been advocated. See NEOTENY, PAEDOMORPHOSIS, PARTHENO-GENESIS.

PAEDOMORPHOSIS. Evolutionary displacement of ancestral features to later stages of development in descendant organisms, either through PROGENESIS or NEOTENY. See HETEROCHRONY.

PAIR-RULE GENES. A class of at least eight segmentation genes, particularly studied in *Drosophila*, mutants in which bring about repetitive deletion of specific parts of alternating segments. The fact that they are expressed in seven or eight stripes during cellularization of the *Drosophila* BLASTODERM is a key event in the PATTERN FORMATION process. The earliest to be expressed in development (in a manner dependent on the prior expression of GAP GENES) are *runt* (*runt*) and *hairy* (*h*), whose protein products regulate each other's transcriptions as with gap genes, producing narrowed and alternating *runt* and *h* domains. The products encoded by *runt* and *h* act as negative regulators of the pair-rule genes *fushi tarazu* (*ftz*) and *even-skipped* (*eve*), whose expression patterns end up as seven alternating stripes each. Expression of a third pair-rule gene, *paired* (*prd*), follows and produces fourteen stripes of product. The result of this mutually interactive sequence of gene activation and repression is that blastoderm cells emerge with different combinations of gene product that can now serve as reference points for further pattern formation, particularly in defining the domains of the SEGMENT-POLARITY GENES *engrailed* (*en*) and *wingless* (*wg*).

PALAEARCTIC. ZOOGEOGRAPHICAL REGION, consisting of Europe, north Africa, and Asia south to Himalayas and Red Sea.

PALAEOBOTANY (PALEOBOTANY). Study of fossil plants.

PALAEOCENE. Geological epoch; earliest division of Tertiary period. See GEOLOGICAL PERIODS.

PALAEOECOLOGY. Study of relationships between fossil organisms and their environments. Largely concerned with reconstruction of past ecosystems through inferences from fossils and their sediments.

PALAEOGENE. Collective term for Palaeocene, Eocene and Oligocene epochs. See GEOLOGICAL PERIODS.

PALAEOGNATHAE. Ratites. Superorder of the subclass NEORNITHES. Includes birds with a reduced breastbone, and which are therefore secondarily flightless. Examples are: ostrich, rheas, cassowaries, tinamous, kiwis, and the extinct moas and elephant birds. Their feathers lack barbs. Probably represent a GRADE rather than a CLADE. See NEOGNATHAE.

PALAEOLITHIC. The Old Stone Age of human prehistory. Lasted from about 1.8 Myr–10 000 yr BP. Its chronology commences with the beds at Olduvai housing *Homo habilis* remains. See HOMO, NEOLITHIC.

PALAEONTOLOGY. Study of FOSSILS and evolutionary relationships and ecologies of organisms which formed them.

PALAEOZOIC. Earliest major geological era. See GEOLOGICAL PERIODS.

PALATE. Roof of the vertebrate mouth. In mammals and crocodiles the roof is not homologous with that of other vertebrates; a new (false) palate has developed beneath original palate, by bony shelves projecting inwards from bones of upper jaw. In mammals, bony part of false palate (*hard palate*) is continued backwards by a fold of mucous membrane and connective tissue, the *soft palate*.

PALATOQUADRATE (PTERYGOQUADRATE). Paired cartilage or cartilage bone forming primitive upper jaw (as in CHONDRICHTHYES and embryo tetrapods). See AUTOSTYLIC JAW SUSPENSION, HYOSTYLIC.

PALEA (SUPERIOR PALEA, PALE). Glume-like bract of grass spikelet on axis of individual flower which, with the LEMMA, it encloses.

PALINDROMIC. Reading the same forwards and backwards. Some DNA sequences are palindromic. See INVERTED REPEAT SEQUENCE, INSERTION SEQUENCE.

PALISADE. See MESOPHYLL.

PALLIUM. See CEREBRAL CORTEX.

PALMELLOID. (Of algae) describing an algal colony comprising indefinite number of single, non-motile cells, embedded in mucilaginous matrix.

PALPS. Paired appendages of many invertebrates, on the head or around the mouth. In polychaete annelids, tactile (on head); in bivalve molluscs, ciliated flaps around mouth generating feeding currents; in crustaceans, distal parts of appendages carrying mandibles (locomotory or feeding); of insects, parts of first and second MAXILLAE, sometimes olfactory.

PALYNOLOGY. See POLLEN ANALYSIS, PALAEOECOLOGY.

PANCREAS. Compound gland of vertebrates, in mesentery adjacent to duodenum; endocrine and exocrine functions. Secretes *pancreatic amylase*, *lipase*, *trypsinogen* and *nucleases* from its ACINAR CELLS in an alkaline medium of sodium hydrogen carbonate, promoted by CHOLECYSTOKININ and SECRETIN, and by the vagus (cranial nerve X), when the gastric phase of digestion is complete. The two major pancreatic hormones are INSULIN and GLUCAGON, secreted respectively from β and α cells of the Islets of Langerhans. See DIGESTION.

PANCREATIN. Extract of pancreas containing digestive enzymes.

PANCREOZYMIN. See CHOLECYSTOKININ.

PANGAEA. Ancient land mass persisting over 200 Myr, until end of JURASSIC. Began to break up in late Triassic, eventually forming LAURASIA and GONDWANALAND.

PANGENESIS. Theory adopted by Charles DARWIN to provide for the genetic variation his theory of natural selection required. Basically Lamarckist, it supposed that every part of an organism produced 'gemmules' ('pangenes') which passed to the sex organs and, incorporated in the reproductive cells, were passed to the next generation. Modification of the body, as through use and disuse, would result in appropriately modified gemmules being passed to offspring. Severely criticized by Darwin's cousin, Francis Galton. Empedocles held a similar view, but Aristotle rejected it.

PANICLE. Type of INFLORESCENCE.

PANMIXIS. Result of interbreeding between members of a species population, no important barriers to gene flow occurring within it.

The whole population represents one GENE POOL. Panmictic animal populations tend to have high VAGILITY.

PANTOTHENIC ACID. VITAMIN of the B-complex; precursor of COENZYME A (CoA), a large molecule comprising a nucleotide bound to the vitamin. See KREBS CYCLE.

PANTOTHERIA. Extinct order of primitive therian Jurassic mammals. Small and insectivorous. Have therian features: large alisphenoid on wall of braincase, and triangular molar teeth. Contemporaries of MULTITUBERCULATA.

PAPILLA. Projection from various animal tissues and organs. *Dermal papillae* project from dermis into epidermis of vertebrates, providing contact (and finger-print patterns); in feather and hair follicles, papillae provide blood vessels. *Tongue papillae* increase surface area for taste buds in mammals; they are cornified for rasping in cats, etc.

PAPPUS. Ring of fine, sometimes feathery, hairs developing from calyx and crowning fruits of the Family Compositae (e.g. dandelion). Act as parachute in wing dispersal of fruit.

PARABIOSIS. Surgical joining together of two animals so that their blood circulations are continuous. Each member of the pair is termed a *parabiont*. Often employed to monitor humoral influences in behaviour and development (e.g. in insect moulting).

PARADERMAL SECTION. Section cut parallel to surface of a flat structure, e.g. a leaf.

PARALLEL EVOLUTION. Possession in common by two or more taxa of one or more characteristics, attributable to their having similar ecological requirements and a shared genotype inherited from a common ancestor: the common characteristics would be HOMOLOGOUS. In CLADISTICS, no distinction is made between parallel evolution and CONVERGENCE.

PARALLEL VENATION. Pattern of leaf venation, where principal veins are parallel or nearly so; characteristic of MONOCOTYLEDONAE.

PARAMYLON. Storage polysaccharide composed of $\beta[1,3]$-linked glucose units; characteristic of EUGLENOPHYTA.

PARAPATRY. Where the ranges of two populations of the same or different species overlap they are SYMPATRIC; but if the ranges are contiguous, i.e. if they abut for a considerable part of their length but do not overlap, the distributions show parapatry. Distributions of several organisms formerly regarded as single species have been shown to consist of several SIBLING SPECIES or SEMISPECIES with parapatric distributions, as in frogs of *Rana pipiens* species group.

PARAPHYLETIC. Term describing taxon or taxa originating from and including a single stem species (known or hypothetical) but excluding one or more smaller CLADES nested within it. E.g. if, as is commonly accepted, flowering plants arose from a gymnophyte ancestor, then gymnophytes are a *paraphyletic group* since the group does not include all descendants of a common ancestor. Such taxa are not permitted in CLADISTICS but are much used by adherents of EVOLUTIONARY TAXONOMY. See CLADE, MONOPHYLETIC.

PARAPHYSIS. (Bot.) Sterile filament, numbers of which occur in mosses and certain algae, interspersed among the sex organs, and in the hymenia of certain fungi (ASCOMYCOTINA, BASIDIOMYCOTINA).

PARAPODIUM. Paired metameric fleshy appendage projecting laterally from the body of many polychaete annelid worms (especially errant polychaetes). Usually comprises a more dorsal *notopodium* and a ventral *neuropodium*, each with bundles of chaetae and endowed with a supporting chitinous internal *aciculum* to which muscles moving the parapodium are attached.

PARASEGMENT. Unit in insect development which corresponds not to a morphological segment but to the posterior COMPARTMENT of one segment and the anterior compartment of the next most posterior segment. See HOMOEOBOX.

PARASEXUAL CYCLE. See PARASEXUALITY.

PARASEXUALITY. Fungal LIFE CYCLE (*parasexual cycle*) which includes the following: occasional fusion of two haploid heterokaryotic nuclei in the mycelium to form a diploid heterozygous nucleus; mitotic division of this nucleus during which crossing-over (*mitotic crossing-over*) occurs, then restoration of haploidy to the nucleus by either mitotic NON-DISJUNCTION, or some form of *chromosome extrusion* removing a haploid set of chromosomes (see ABERRANT CHROMOSOME BEHAVIOUR). The non-sexual spores produced differ genetically from the parent mycelium. It accounts for the variation of pathogenicity in certain fungal plant pathogens, and has enabled genetical studies to be made using members of the DEUTEROMYCOTINA.

PARASITE. One kind of symbiont. Organism living in (*endoparasite*) or on (*ectoparasite*) another organism, its *host*, obtaining nourishment at the latter's expense. Metabolically dependent upon their hosts, as are carnivores, herbivores, etc. Distinction between herbivorous caterpillar and ectoparasitic fluke is not clear-cut. Endoparasites generally display more, and more specialized, adaptations to parasitism than do ectoparasites, and often include both primary and secondary HOSTS in the life cycle. *Obligate parasites* cannot survive independently of their hosts; *facultative parasites* may do so. *Partial parasites* (e.g. mistletoe,

Viscum) are plants which photosynthesize and also parasitize a host. Sometimes relationships between members of the same species are parasitic (e.g. males of some angler fishes live attached to the female and suck her blood). Placental reproduction shares features with parasitism, as do forms of viviparity in which young emerge causing death of the parent (e.g. in the midge *Miastor* and many aphids and water fleas). See SYMBIOSIS, MALARIAL PARASITE, PARASITOID.

PARASITOID. Insects, and some other animals, which introduce their eggs into another animal, in which they grow and develop in a slow and controlled manner using the host's resources without killing it. At maturation they emerge and usually do cause death of the host (unlike most PARASITES).

PARASTICHY. See PHYLLOTAXY.

PARASYMPATHETIC NERVOUS SYSTEM. See AUTONOMIC NERVOUS SYSTEM.

PARATHYROID GLANDS. Endocrine glands of tetrapod vertebrates, usually paired, lying near or within THYROID depending on species. Arise from embryonic gill pouches, and produce PARATHYROID HORMONE. Removal produces abnormal muscular convulsions within a few hours.

PARATHYROID HORMONE (PTH, PARATHORMONE). Polypeptide hormone of parathyroid glands operating with vitamin D and CALCITONIN in control of blood calcium levels. Injection releases calcium from bone and raises blood Ca^{++} level, inhibiting further parathyroid hormone release, apparently via direct negative feedback on parathyroid glands. Also reduces Ca^{++} excretion by the kidneys. Deficiency produces muscle spasms.

PARATYPE. Any specimen, other than the type specimen (HOLOTYPE) or duplicates of this, cited with the original taxonomic description and naming of an organism.

PARAZOA. Subkingdom of the ANIMALIA, containing the phylum PORIFERA (sponges).

PARENCHYMA. (Bot.) Tissue comprising living, thin-walled cells, often almost as broad as long, and permeated by a system of intercellular spaces containing air. Cortex and pith are typically composed of parenchyma. (Zool.) Loose tissue of irregularly shaped vacuolated cells within gelatinous matrix and forming a large part of the bodies of some invertebrate groups, notably platyhelminths and nematodes.

PARIETAL. (Bot.) Referring to peripheral position, as in chloroplasts of some algal cells located near the cell's periphery. (Zool.) *Parietal bones* lie one on each side of the vertebrate skull, behind and between the eye orbits. See also PINEAL GLAND.

PARIETAL EYE. See PINEAL EYE.

PARIETAL PLACENTATION. (Bot.) Attachment of ovules in longitudinal rows on carpel wall.

PARTHENOCARPY. (Bot.) Development of fruits without prior fertilization. Occurs regularly in banana and pineapple (which are therefore seedless). Can be induced by certain auxins in unfertilized flowers, e.g. those of tomato.

PARTHENOGENESIS. Development of an unfertilized gamete (commonly an egg cell) into a new individual. One of a spectrum of forms of uniparental SEXUAL REPRODUCTION.

A parthenogenetic egg cell (or nucleus) may become diploid either through nuclear fusion, or through a restitution division (see RESTITUTION NUCLEUS). Sometimes cleavage products of a haploid egg may undergo fusion, producing a diploid embryo. Phenomenon includes non-gametic forms of AUTOMIXIS and in animals is a common cause of MALE HAPLOIDY. In animals, THELYTOKY (absence of males) enables rapid production of offspring without food competition from males. *Cyclical parthenogenesis* (as in some aphids and flukes) involves a combination of thelytoky and bisexual fertilization. In some aphids thelytoky prevails in summer, males only appearing in autumn or winter when fertilization occurs. In the midges *Miastor* and *Heteropeza*, larvae possess functional ovaries enabling progenetic reproduction by automixis, adults not appearing for generations; some larval flukes are progenetic. Some instances of thelytoky (automictic, or meiotic, thelytoky) involve meiotic egg-production, and two of the four meiotic products sometimes fuse to restore diploidy; in others (apomictic, or ameiotic, thelytoky), mitosis produces the egg cells. In some cases diploidy may be restored by ENDOMITOSIS after meiosis.

In the protozoan *Paramecium*, fusion may occur of two of the micronuclei produced meiotically from the cell's parent micronucleus (*autogamy*) but no new individual is produced.

Since development of unfertilized eggs can occur, on rare occasions, in many species (e.g. *Drosophila* and grasshoppers), and be induced artificially in many others (see CENTRIOLE), it is still surprisingly rare, especially since it avoids the COST OF MEIOSIS. Thelytokous forms seem to be liable to early extinction compared with their bisexual relatives, probably through progressive homozygosity (see GENETIC VARIATION).

Haploid egg cells have been recorded developing parthenogenetically in plants, but this seems to have had little evolutionary impact. Unreduced (diploid) gametophytes may arise either from an unreduced megaspore (*diplospory*) or from an ordinary unreduced somatic cell of the sporophyte (*apospory*). Both are genetically equivalent to apomictic (ameiotic) parthenogenesis in animals. In dandelions

(*Taraxacum*) the megaspore mother cell undergoes meiosis, the first division producing a restitution nucleus, the second producing two cells, each unreduced, from one of which the 8–nucleate embryo sac is produced. See GYNOGENESIS, PARTHENOCARPY.

PARTURITION. Expulsion of foetus from uterus at end of pregnancy in therian mammals. Involves OXYTOCIN and PROSTAGLANDINS. Smooth muscle contractions of the uterine wall (labour) force the offspring out.

PASSAGE CELLS. Cells of the ENDODERMIS, typically of older monocot roots, opposite protoxylem groups of stele, remaining unthickened and with CASPARIAN STRIPS only after thickening of all other endodermis cells. Allow transfer of material between cortex and vascular cylinder.

PASSERINES. Members of largest avian order, the Passeriformes. Perching birds, characterized by having large first toe directed back, the other three forward. Includes most of the common inland birds. See NEOGNATHAE.

PASTEUR, LOUIS (1822–95). French chemist and microbiologist; professor of chemistry at the Sorbonne, but worked mostly at the Ecole Normale in Paris. Became director of the Institut Pasteur, Paris, in 1888. Championed the view that fermentation was a *vital* rather than a simple *chemical* process, as against the chemical theory of Liebig and Berzelius. Already aware, with others, that yeasts were associated with alcoholic fermentations, he demonstrated presence of microorganisms in other fermentations. In 1858 he demonstrated fermentation in the absence of organic nitrogen, destroying the chemical theory. Pasteur's many experiments supported the germ theory of fermentation, as against the theory of spontaneous generation, and its implications were appreciated by Joseph Lister in his work on antisepsis in the 1860s. Pasteur came to accept the role of microorganisms in disease, showing that attenuated forms of bacteria produced by serial culture could be used in inoculation to immunize the host. His vaccines against anthrax and rabies, like Jenner's earlier ones against smallpox, were instrumental in establishing the germ theory of disease. See PASTEURIZATION, VIRCHOV.

PASTEUR EFFECT. Phenomenon whereby onset of aerobic respiration inhibits glucose consumption and lactate accumulation in all facultatively aerobic cells, conserving substrates. Depends upon the allosteric inhibition of glycolytic enzymes by high intracellular ATP to ADP ratio. See ATP, GLYCOLYSIS.

PASTEURIZATION. Method of partial sterilization, after Louis PASTEUR, who discovered that heating wine at a temperature well below its boiling point destroyed the bacteria causing spoilage without affecting its flavour. Widely used to kill some disease-causing bacteria in food, e.g. tubercle bacteria in milk (by heating at 62°C for 30 min, or at 72°C for 15 secs followed by rapid cooling), delaying its fermentation. Compare TYNDALLIZATION.

PATELLA. Kneecap. Bone (sesamoid bone) over front of knee joint in tendon of extensor muscles straightening hind limb. Present in most mammals, some birds and reptiles.

PATHOGEN. Disease-causing parasite, usually a microorganism.

PATHOLOGY. Study of diseases, or diseased tissue.

PATRISTIC. (Of similarity) due to common ancestry. See CLADISTICS.

PATTERN, PATTERN FORMATION. Describing phenomena whereby cells in different parts of embryo become locked into different developmental pathways, coordinated in such a way as to produce a viable multicellular system. Early animal embryos engage in a hierarchy of decisions involving progressive steps in *regional specification*, in which POSITIONAL INFORMATION is imparted to cells, which then respond to it. Cells of a particular histological type may have arrived at their condition via alternative, *non-equivalent* routes. Parts of an embryo acquire different DETERMINED states through regionalization processes. See COMPARTMENT, PAIR-RULE GENES.

PEAT. Accumulated dead plant material which has remained incompletely decomposed owing, principally, to lack of oxygen. Occurs in moorland, bogs and fens, where land is more or less completely waterlogged; often forms a layer several metres deep. Of local value as fuel for burning.

PECK ORDER. See DOMINANCE (2).

PECTIC COMPOUNDS. Acid polysaccharide carbohydrates, present in CELL WALLS of unlignified plant tissues; comprise pectic acid and pectates, pectose (propectin) and pectin. Form gels under certain conditions. Principal components are galacturonic acid, galactose, arabinose and methanol. Form the basis of fruit jellies.

PECTORAL FIN. See FINS.

PECTORAL GIRDLE (SHOULDER GIRDLE). Skeletal support of vertebrate trunk for attachment of fins or forelimbs; in fish, attaches to skull. Primitively, a curved bar of cartilage or bone on each side of body, fusing ventrally to form a hoop transverse to long axis, incomplete dorsally. Each bar bears a joint with fin or limb (see GLENOID CAVITY). Components are: *scapulae* dorsal to the joint, *coracoids* ventrally. Additional dermal bones are the *cleithra* in fish and primitive tetrapods, and *clavicles*, usually on the ventral side. In mammals each clavicle joins a scapula at a process of the coracoid, the *acromion*. Scapulae do not articulate with the vertebral column or ribs (compare

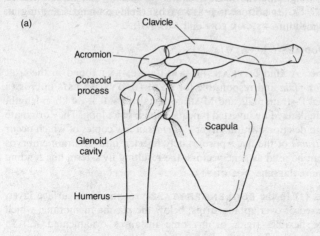

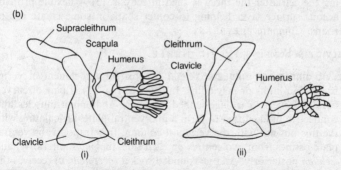

Fig. 52. (a) *Human shoulder region, viewed from the front.* (b) *Pectoral girdle and fin of* (i) *crossopterygian fish and* (ii) *early fossil amphibian. One side only shown in each case.*

PELVIC GIRDLE). In tetrapods, coracoids and clavicles join mid-ventrally to the sternum.

PEDICEL. (Bot.) Stalk of individual flower of an inflorescence. (Zool.) Narrow tube-like 'waist' of many hymenopteran insects.

PEDIPALPS. Second pair of head appendages of ARACHNIDA. See MOUTHPARTS.

PEDUNCLE. Stalk of an INFLORESCENCE.

PEKING MAN. Early form of *Homo erectus* from China; formerly termed *Sinanthropus*. See HOMO.

PELAGIC. Inhabiting the mass of water of lake or sea, in contrast to the lake or sea bottom (see BENTHOS). Pelagic animals and plants are divided into PLANKTON and NEKTON.

PELECYPODIA. See BIVALVIA.

P ELEMENT. A kind of TRANSPOSABLE ELEMENT found in the fruit fly *Drosophila* and responsible for HYBRID DYSGENESIS in crosses between P-strain male and M-strain female flies. 0.5–1.4 kb in length, they are flanked by inverted repeats 31 base pairs long. They originate through deletions within larger *P factors*, a few copies of which occur in *P strains* of the fly. Appropriately injected into M-strain embryos they can be used as gene vectors, the resulting fly's germ line tending to acquire the gene. See COPIA.

PELLICLE. (1) In the EUGLENOPHYTA, a proteinaceous surface layer, composed of overlapping strips, below the plasma membrane, which can be flexible, rigid, or in some instances ornamented. (2) In armoured Dinoflagellates, that portion of the cell covering surrounding the cell after the theca is shed in ecdysis. (3) A flexible proteinaceous surface layer helping to confer shape in some ciliate protozoans. Compare PERIPLAST.

PELVIC FIN. See FINS, PELVIC GIRDLE.

PELVIC GIRDLE (HIP GIRDLE). Skeletal support for attachment of vertebrate hind-limbs or pelvic fins. In fish (see Fig. 53), a pair of curved bars of bone or cartilage embedded in the abdominal muscles and connective tissue, fused to form a mid-ventral plate, articulating with the fins but not with the vertebral column. In tetrapods, the ventral plate ossifies from two centres on each side: the *pubis* anteriorly, and *ischium* posteriorly. A large rounded socket (*acetabulum*) receives the head of the femur on each side where these two bones join a third and dorsal element, the *ilium*, which unites with one or more sacral vertebrae to form a complete girdle around this region of the trunk, giving rigid support to hind-limbs for locomotion. Pelvic girdle structure varies in tetrapod classes. In mammals, the ilium extends anteriorly towards the SACRUM, while the pubis and ischium have moved posteriorly, hardly reaching the acetabulum. In most mammals, many reptiles and *Archaeopteryx* the pubes articulate or fuse mid-ventrally to form the *pubic symphysis*, consisting in humans of fibrocartilage between the two *coxal bones* (fused ilium, pubis and ischium on each side). In monotremes and marsupials a pair of *prepubes* reaches forward from the pubes to form a body wall support. Compare PECTORAL GIRDLE.

PELVIS. (1) The PELVIC GIRDLE. (2) Lower part of the vertebrate abdomen, bounded by the pelvic girdle. (3) The *renal pelvis*. See KIDNEY.

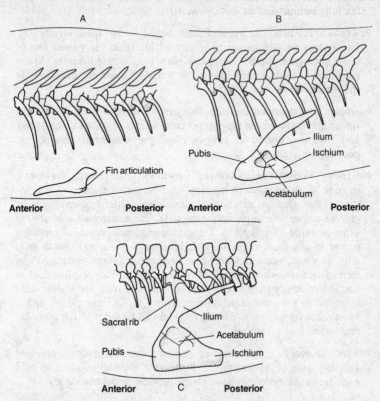

Fig. 53. *Pelvic girdles of (A) fish, (B) early tetrapod and (C) later tetrapod with attachment of girdle to enlarged sacral rib.*

PENETRANCE. A dominant character determined by an ALLELE is either always expressed in any individual where it occurs (*completely penetrant*), or is expressed in some individuals but not in others (*incompletely penetrant*). Once a character finds expression, it may be expressed to varying degrees in different individuals (*variable expressivity*). Possible factors affecting penetrance and expressivity include the genetic background (see MODIFIER) and environmental influences during development. See EPIGENESIS.

PENICILLINS. See ANTIBIOTIC.

PENIS. Unpaired intromittant organ of male mammals, some reptiles and a few birds (especially of those mating on water). In mammals, contains the terminal part of the urethra.

PENNATE DIATOM. Diatoms (BACILLARIOPHYTA) that are bilaterally symmetrical in valve view.

PENTADACTYL LIMB. The type of limb found in tetrapod vertebrates. Evolved as an adaptation to terrestrial life from the paired fins of crossopterygian fishes. The basic plan is illustrated below. Many modifications occur through loss or fusion of elements, especially in the terminal parts. See Fig. 54.

PENTOSE. Monosaccharide with five carbon atoms in molecule, e.g. ribose and deoxyribose (important constituents of NUCLEIC ACIDS), and ribulose. Found in various plant polysaccharide chains, e.g. pectin, gum arabic.

PENTOSE PHOSPHATE PATHWAY (PHOSPHOGLUCONATE PATHWAY, HEXOSE MONOPHOSPHATE SHUNT). An alternative route to GLYCO-LYSIS for glucose catabolism, involving initially conversion of glucose-phosphate to phosphogluconate. Some intermediates are the same as those of glycolysis. Generates extramitochondrial reducing power in the form of NADPH, important in several tissues (e.g. adipose tissue, mammary gland, adrenal cortex) where fatty acid and steroid synthesis occur from acetyl coenzyme A. Ribose-phosphate is one intermediary in the pathway, and may be used for nucleic acid synthesis or for glucose production from CO_2 in some plants. Other monosaccharides are intermediates, and can be fed into the pathway and linked with glycolysis.

PEPSIN. Proteolytic enzyme, secreted as its precursor *pepsinogen* by *chief cells* of gastric pits of vertebrate stomach. Active in acid conditions. An endopeptidase, giving peptides and amino acids.

PEPTIDASE. One of a group of enzymes hydrolysing peptides into component amino acids. Also sometimes used of proteolytic enzymes in general. See ENDOPEPTIDASE, EXOPEPTIDASE.

PEPTIDE. (1) A compound of two or more amino acids (strictly, amino acid residues), condensation between them producing PEP-TIDE BONDS. Shorter in length than a polypeptide. Besides being intermediates in protein digestion and synthesis, many peptides are biologically active. Some are hormones (e.g. OXYTOCIN, ADH, MELANOCYTE-STIMULATING HORMONE); some are vasodilatory. (2) Sometimes employed in the context of any chain of amino acids, however long.

PEPTIDE BOND. Bond between two amino acids resulting from combination of an amino group ($-NH_2$) attached to the α-carbon of one amino acid and the carboxyl group ($-COOH$) attached to the α-carbon of another amino acid. The bond formed ($-NH-CO-$) involves the elimination of one molecule of water, and is a *condensation reaction*. See PROTEIN, AMINO ACID.

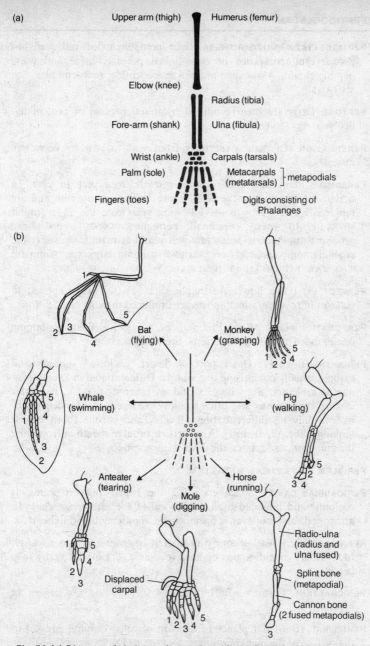

Fig. 54. (a) *Diagram of skeleton of pentadactyl limb of vertebrate, giving names of parts of fore-limb, and (in brackets) of hind-limb. On left, common name of whole part; on right, names of bones.* (b) *Diagram indicating the adaptive radiation of vertebrate limb from a basic archetype.*

PEPTIDOGLYCAN (MUCOPROTEIN). Rigid component of cell walls in prokaryotes, composed of cross-linked polysaccharide and polypeptide chains. Also called murein, mucopeptide, mucocomplex. See MUCINS.

PEPTONE. Large fragment produced by initial process of protein hydrolysis.

PERENNATION. (Of plants) surviving from year to year by vegetative means.

PERENNIAL. Plant that continues its growth from year to year. In herbaceous perennials, the aerial parts die back in autumn and are replaced by new growth the following year from the underground structure. In woody perennials, permanent woody stems above ground form starting point for each year's growth, a characteristic enabling some species (trees, shrubs) to attain large size. Compare ANNUAL, BIENNIAL, EPHEMERAL.

PERFECT. (1) (Of a flower) hermaphrodite. (2) (Of fungi) applied to stage of life cycle in which spores are produced through fertilization.

PERFORATION PLATE. (Bot.) End wall of a vessel element containing one or more holes or perforations. See XYLEM.

PERIANTH. (Bot.) (1) Outer part of flower, enclosing stamens and carpels; usually comprising two whorls. Differentiated in Dicotyledonae as an outer green CALYX and an inner COROLLA, the latter usually conspicuous and often brightly coloured. In Monocotyledonae, usually no differentiation of calyx and corolla, both whorls looking alike. (2) In leafy liverworts, a tubular sheath surrounding the archegonia and, later, the developing sporophyte.

PERIBLEM. See APICAL MERISTEM.

PERICARDIAL CAVITY. Space enclosing the heart. In vertebrates, is coelomic and bounded by a double-walled sac, the *pericardium*. In arthropods and molluscs, is haemocoelic, supplying blood to heart.

PERICARP. (Bot.) Wall of an ovary after it has matured into a FRUIT. May be dry, membranous or hard (e.g. achene, nut), or fleshy (e.g. berry).

PERICHAETIUM. Distinct whorl of leaves surrounding sex organs in mosses.

PERICLINAL. (Bot.) (Of planes of division of cells) running parallel to the surface of the plant. Compare ANTICLINAL.

PERICYCLE. (Bot.) Tissue of the vascular cylinder lying immediately within the ENDODERMIS, comprising parenchyma cells and sometimes fibres.

PERIDERM. (Bot.) Cork cambium (phellogen) and its products; i.e. cork and secondary cortex (phelloderm).

PERIGENE DEVELOPMENT. (Bot.) Development of GUARD CELLS in which the guard cell mother cell does not give rise to subsidiary cells.

PERIGONIUM. (Bot.) The antheridium, together with associated perigonial leaves or bracts (surrounding the antheridium), borne on a specialized branch (perigonial branch), in mosses and liverworts.

PERIGYNOUS. See RECEPTACLE.

PERIKARYON. Alternative term for CELL BODY of neurone.

PERILYMPH. See VESTIBULAR APPARATUS.

PERINEUM. Region between anus and urogenital openings of placental mammals.

PERIOSTEUM. Sheath of connective tissue investing vertebrate bones, and to which tendons attach. Contains osteoblasts, and white and elastic fibres.

PERIPATUS. See ONYCHOPHORA.

PERIPHERAL NERVOUS SYSTEM. See NERVOUS SYSTEM. Compare CENTRAL NERVOUS SYSTEM, AUTONOMIC NERVOUS SYSTEM.

PERIPLAST. A cell covering found in the algal division CRYPTOPHYTA, comprising cell membrane with an underlying layer of plates or membranes and an overlying layer of granular material; not as elaborate as a PELLICLE.

PERISPERM. Nutritive tissue surrounding the embryo in some seeds; derived from the NUCELLUS. Compare ENDOSPERM.

PERISSODACTYLA. Order of eutherian mammals containing the odd-toed ungulates (e.g. horses, tapirs, rhinoceroses). Walk on hoofed toes, the weight-bearing axis of foot lying along the third toe, which is usually larger than the others (some of which may have disappeared): horses have very large third, but minute second and fourth, toes. Tapirs have four toes on front feet, three on hind feet. Rhinoceroses have three toes on each foot. See ARTIODACTYLA.

PERISTALSIS. Waves of contraction of smooth muscle, passing along tubular organs such as the intestines. Serve to move material from one end of the tube to the other.

PERISTOME. (Bot.) Fringe of pointed appendages (teeth) around the opening of dehiscent moss capsule, concerned with spore liberation. (Zool.) Spirally twisted groove leading to cytostome in some ciliate protozoans.

PERITHECIUM. Rounded, or flask-shaped, structure of some ascomycotinans and lichens; with internal hymenium of asci and paraphyses and an apical pore (ostiole) through which ascospores are discharged.

PERITONEUM. Epithelium (serous membrane) lining posterior coelomic cavity (*abdominal, perivisceral* and *peritoneal cavities*) of vertebrates, around the gut. That covering the gut and other viscera is termed *visceral*; that lining the wall of the cavity is *parietal*. The MESENTERY carrying blood vessels to and from the viscera is also peritoneal.

PERIVISCERAL CAVITY. The vertebrate coelomic cavity lined by PERITONEUM.

PERMANENT TEETH. Second of the two successive sets of teeth of most mammals, replacing DECIDUOUS TEETH. See DENTITION.

PERMEASE. TRANSPORT PROTEIN, or carrier molecule, assisting in transport across cell membranes without being permanently altered in the process.

PERMIAN. Last period of Palaeozoic era, 270–225 Myr BP in duration. In Britain, Permian and Triassic marine faunas are often united as New Red Sandstone (*Permo-Trias*). Among vertebrates, palaeoniscoid fish dominated, and holosteans appeared; the LABYRINTHODONTIA survived among amphibians, but amphibians were much reduced in the later Permian; reptiles included pelycosaurs and their replacements the THRAPSIDA, but there are few representatives of the dominant Mesozoic reptiles. Flora included *Glossopteris* and *Gangamopteris*. At its close there was a considerable mass EXTINCTION of marine fauna. See GEOLOGICAL PERIODS.

PEROXISOME. MICROBODY containing CATALASE, especially in vertebrate liver and kidney cells, and those plant cells involved in PHOTO-RESPIRATION. Catalase uses hydrogen peroxide (itself produced by oxidative enzymes using molecular oxygen inside the organelle) to remove hydrogen atoms from substrates, so oxidizing (detoxifying) phenols, formic acid, formaldehyde and ethanol (oxidized to acetaldehyde). See also GLYOXYLATE CYCLE.

PEST CONTROL. Five major strategies of pest control are employed, each dependent for its effectiveness upon the ecological 'strategy' of the pest organism (see PESTS). They are: pesticide control; BIOLOGICAL CONTROL; cultural control (where agricultural or other practices are used to change the pest's habitat); breeding for pest resistance in cultivated organisms, and sterile mating control, where pest populations are variously sterilized to reduce their reproductive rates. These main approaches, and the types of pest against which they are employed, are indicated in Table 5.

PESTS. Species whose existence conflicts with human profit, conveni-

	r-pests	Intermediate pests	K-pests
Pesticides	Early widescale applications based on forecasting	Selective pesticides	Precisely targeted applications based on monitoring
Biological control		Introduction or enhancement of natural enemies	
Cultural control	Timing, cultivation sanitation and rotation →	←	Change in agronomic practice, destruction of alternative hosts
Resistance	General, polygenic resistance →	←	Specific, monogenic resistance
Genetic control			Sterile mating technique

Table 5. *Principal control methods appropriate for different pest strategies.*

ence, or welfare. Some cause serious nuisance; their injuriousness is well established and their control is either a social or economic necessity. Pest status commonly arises from (a) entry of species into previously uncolonized regions; (b) some change in the properties of previously unproblematic species; (c) changes in human activities, bringing contact with species to which there was previous indifference; (d) increase in abundance of a species, with resulting nuisance value.

Pest status may be interpreted in terms of ecological strategies wrought by different selection pressures to which pest species are exposed; thus: *r-pests*, where r-SELECTION influences are uppermost; *K-pests*, where K-SELECTION influences dominate; and *intermediate pests*, lying somewhere between these two. See PEST CONTROL, WEEDS.

PETAL. One of the parts forming corolla of flower; often brightly coloured and conspicuous. See FLOWER.

PETIOLE. Stalk of a LEAF.

PEYER'S PATCHES. Patches of secondary LYMPHOID TISSUE in submucosa of amniote intestines.

pH. A quantitative expression denoting the relative proton (hydrogen ion, H^+) concentration in a solution. The pH scale ranges from 0–14: the higher the pH value, the lower the acidity. A pH of 7 indicates a neutral solution. Defined as $-\log [H^+]$, where $[H^+]$ = concentration of protons, expressed as $g.dm^{-3}$. In consequence, a solution of pH 6 has ten times the H^+ concentration of a solution of pH 7. It is important that the pH of cells and body fluids is kept within acceptable values, one reason being the effect of pH change on the shapes of globular protein molecules. See PROTEIN, BUFFER.

PHAEOPHYTA. Brown algae. Division of the Algae, almost all marine, deriving characteristic colour from large amounts of the carotenoid *fucoxanthin* in their chloroplasts, which also contain chlorophylls a, c_1 and c_2 as well as β-carotene and violaxanthin. Main storage product is the polysaccharide *laminarin*. Cell walls with cellulose as main structural component, any amorphous component being mainly *alginic acid* and *fucoidin*. Morphologically very diverse, from minute (less than 1 mm long) and filamentous, to very long (up to 60–70 m) and complex, with root-like holdfast and stem-like stipe bearing branched or unbranched leaf-like part, often provided with air bladders and relatively complex internal structure. Never unicellular or colonial.

Two types of reproductive structure: (a) unilocular single-celled sporangium releasing haploid zoospores after meiosis which will form the gametophyte generation and produce gametes; (b) plurilocular sporangium, each cell of which produces a single motile cell, functioning either as a gametangium (producing haploid gametes) if on a gametophyte, or as a sporangium (producing diploid zoospores) if on a sporophyte. Fertilization isogamous, between flagellated microgametes and non-motile macrogametes. Most attach to rocks in intertidal or subtidal regions, dominating in colder waters. A species of *Sargassum* is exceptional in being pelagic, accumulating in large quantities in the Sargasso Sea near the West Indies.

PHAGE. Viruses infecting bacteria are termed BACTERIOPHAGES; those infecting fungi are termed *mycophages*.

PHAGE CONVERSION. Phenomenon whereby new properties may be conferred upon host cells when infected by temperate phage. Each cell receiving the prophage also acquires the new property. See BAC-TERIOPHAGE.

PHAGE RESTRICTION. When phage are grown upon bacteria of one strain and then upon another, their titre may drop in the second host strain. They are then said to be *restricted* by the second host strain. Due to degradation of the phage DNA by host RESTRICTION EN-DONUCLEASES. See MODIFICATION.

PHAGOCYTE. Cell which can ingest particles from its surroundings (phagocytosis), forming vacuole composed of the plasma membrane in which the material lies. The vacuoles may then fuse with LYSO-SOMES to form *heterophagosomes*. Receptor sites on the plasma membrane may be involved in vacuole formation, as with antibody markers on surfaces of phagocytic leucocytes. Many protozoans are phagocytic, as are those vertebrate white blood cells (*neutrophils*, *monocytes*) and MACROPHAGES which engulf bacteria and clumped antigens. See LEUCOCYTE, IMMUNITY.

PHAGOCYTOSIS. Process whereby plasma membrane of a cell

encloses a particle in the external environment and traps it within a food vacuole. This is normally converted to a heterophagosome as LYSOSOMES fuse with it and enable digestion of the contents. One example of ENDOCYTOSIS; common in a variety of animal cells, all of which are thereby termed *phagocytic*.

PHAGOSOME. For *autophagosomes* and *heterophagosomes*, see LYSOSOME.

PHALANGES. Bones of vertebrate digits (fingers and toes). Each finger has 1–5 phalanges (more in whales) articulating end-to-end in a row, the proximal of each row forming a joint with a *metacarpal bone*. See PENTADACTYL LIMB.

PHANEROGAMIA. Early name for seed plants (gymnophytes, anthophytes), since reproductive organs (cones, flowers) are clearly evident.

PHANEROPHYTES. Class of RAUNKIAER'S LIFE FORMS.

PHANEROZOIC. Geological division lasting from approx. 600 Myr BP to the present. Initiated by appearance of metazoan fossils. During it atmospheric oxygen level rose from about 0.1 of present levels to that of the present.

PHARYNX. (1) The vertebrate gut between mouth (buccal cavity) and oesophagus, into which opens the glottis in tetrapods and gill slits in fish. In man and other mammals is represented by throat and back of nose; partly divided by soft palate into upper (nasal) section and lower (oral, or throat) section. Contains sensory receptors setting off swallowing reflex. The GAS BLADDER and EUSTACHIAN TUBE, where found, also open into it. See GILL POUCH. (2) Part of the gut into which gill slits open internally in urochordates and cephalochordates.

PHASE CONTRAST. See MICROSCOPE.

PHASMIDA. Order of exopterygote insects containing stick insects and leaf insects.

PHELLEM. See CORK.

PHELLODERM. Secondary cortex tissue formed by cork cambium. See CORK.

PHELLOGEN (CORK CAMBIUM). Meristematic cells producing CORK.

PHENETIC. (Of relationship or classification) see PHENETICS.

PHENETICS. Grouping of organisms into taxa on the basis of estimates of overall similarity, without any initial weighting of characters. Diagrams which result (*phenograms*) are devoid of necessary phylogenetic implications although may be interpreted phylogenetically.

Phenetics is a branch of numerical taxonomy. See CLASSIFICATION, PHYLOGENETICS.

PHENOCOPY. Environmentally induced alteration in the phenotype of an organism (or cell), commonly resulting from abnormal developmental conditions. Mimics effect of a known mutation but is nonheritable. See GENETIC ASSIMILATION. Contrast TRANS-DETERMINATION.

PHENOGRAM. See PHENETICS.

PHENOLOGY. Study of periodicity phenomena in plants, such as timing of flowering in relation to climate.

PHENOTYPE. Total appearance of an organism, determined by interaction during development between its genetic constitution (GENOTYPE) and the environment. Different phenotypes may result from identical genotypes, but it is unlikely that two organisms could share all their phenotypic characters without having identical genotypes. Shared presence of a character in two organisms does not necessitate identical genotypes with respect to that character. See DOMINANCE, PENETRANCE, ECOTYPE, PHENOTYPIC PLASTICITY.

PHENOTYPIC PLASTICITY. Extent to which phenotype may be modified by expression of a particular genotype in different environments. Its often adaptive nature is illustrated by HETEROPHYLLY in some members of *Ranunculus* (subgenus *Batrachium*) and by *Polygonum amphibium*. In animals, the phenomenon includes some cases of caste determination in insects and of SEX DETERMINATION in diverse groups. Any genotype probably has a characteristic degree of developmental plasticity (see HOMEOSTASIS). In plants, dwarf, prostrate, thorny and succulent forms are often produced when RAMETS of a cloned genotype are grown in different conditions; genetic fixation (by selection) of an altered phenotype, preadapted to the conditions inducing it, may not be uncommon in wild populations although evidence is rather sparse. In a broad sense, animal LEARNING could be included here; cell differentiation takes the range of the concept below the level of the individual. See CLINE, DEME, ECOTYPE.

PHENYLKETONURIA. Recessive human genetic disorder. The enzyme converting dietary phenylalanine to tyrosine is deficient, causing excretion of phenylpyruvate (or phenylalanine) in urine. Intellectual impairment common and epileptic attacks occur in about 25% of cases. Tendency to lighter hair and skin pigmentation than average. Can be detected soon after birth; a diet low in phenylalanine reduces symptoms.

PHEROMONES. Chemical substances which, when released into an animal's surroundings, influence the behaviour or development of other individuals of same species. Include sexual attractants in many insect species. Worker and queen bees produce several different pheromones, each with its own effect; one deer produces pheromones from at least seven sites on its body, each with a different social function.

PHLOEM. Principal food-conducting tissue in vascular plants. Mixed tissue, containing parenchyma and occasionally FIBRES, besides sieve elements, the main conducting cells (SIEVE CELLS in nonanthophytes, SIEVE-TUBE MEMBERS in anthophytes), and their COMPANION CELLS. Substances transported include sugars, amino acids, some mineral ions and growth substances. Phloem may be *primary* or *secondary* in origin, the former frequently being stretched and destroyed during elongation of the plant. Protoplasts of adjacent sieve elements connect via groups of narrow pores, the *sieve areas*, concentrated on the overlapping ends of these long, slender cells. Where the pores are large, the sieve area is termed a *sieve plate*. See MASS FLOW, VASCULAR BUNDLE, XYLEM.

PHORONIDEA. Small phylum of marine worm-like animals, unsegmented and coelomate, living in tubes (tubicolous) of chitin which they secrete. Resemble ECTOPROCTA in habit and appearance, but are only superficially colonial. Feed by ciliated LOPHOPHORES. Planktonic larva resembles a trochophore. Vascular system contains haemoglobin.

PHOSPHAGEN. One of a number of HIGH-ENERGY PHOSPHATE compounds which act as reservoirs of phosphate-bond energy in the cell, a KINASE transferring their phosphate to ADP, forming ATP. Include *phosphocreatine* and *phosphoarginine*. Nerve and muscle, especially, contain phosphagens. Creatine and arginine become phosphorylated when ATP concentration in the cell is high, the reverse occurring when the cell's ATP to ADP ratio is low. Phosphoarginine is characteristic of invertebrates, phosphocreatine of vertebrates. Both occur in echinoids and hemichordates. See CREATINE.

PHOSPHATASES. Enzymes splitting phosphate from organic compounds. Compare KINASE.

PHOSPHATE-BOND ENERGY. See HIGH-ENERGY PHOSPHATE.

PHOSPHATIDES. See PHOSPHOLIPIDS.

PHOSPHOCREATINE. See PHOSPHAGEN, CREATINE.

PHOSPHODIESTERASES. Group of enzymes capable of hydrolysing cyclic AMP to (non-cyclic) 5'-AMP. Some also hydrolyse the phosphodiester backbone (sugar-phosphate backbone) of nucleic acids.

PHOSPHOLIPIDS (PHOSPHOLIPINS, PHOSPHOLIPOIDS, PHOSPHATIDES).
Those LIPIDS bearing a polar phosphate end, commonly esterified to
a positively charged alcohol group. Major components of cell membranes and responsible for many of their properties. There are two
main groups: (a) the *phosphoglycerides*, derivatives of phosphatidic
acid, include *lecithin* (phosphatidylcholine), *cephalin* (phosphatidyleth-
anolamine) and phosphatidylinositol.

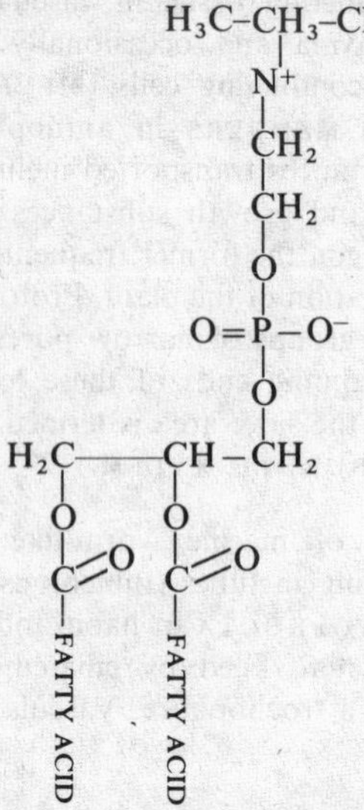

$$H_3C-CH_3-CH_3$$
$$N^+$$
$$CH_2$$
$$CH_2$$
$$O$$
$$O=P-O^-$$
$$O$$
$$H_2C \quad CH-CH_2$$
$$O \qquad O$$
$$C=O \quad C=O$$
FATTY ACID — FATTY ACID

LECITHIN (polar end in detail)

   Lecithin, an important cell membrane component (and surfactant in
vertebrate lungs), is a component of BILE, helping to render cholesterol soluble; (b) the *sphingolipids*, containing the basic *sphingosine*
instead of glycerol, forming components of plant and animal cell
membranes, e.g. *sphingomyelins* (not to be confused with *myelin*) of
liver and red blood cell membranes. See CELL MEMBRANES, CERE-
BROSIDES.

PHOSPHOPROTEINS. Proteins with one or more attached phosphate
groups. The milk protein *casein* has them attached to serine residues.
See PHOSPHORYLASE KINASE.

PHOSPHORYLASE. Enzyme which transfers a phosphate group, often
from inorganic phosphate ions, on to an organic compound which
thereby becomes *phosphorylated*. Glycogen and starch phosphorylases
are enzymes involved in the mobilization of carbohydrate reserves,
forming glucose-phosphate.

PHOSPHORYLASE KINASE. A KINASE activating a *phosphorylase* by
transfer of a phosphate group from ATP. Often cyclic AMP-depen-

dent. May exert its effect by allosteric change in the protein adjacent to the phosphorylase in a multiprotein complex. One of many regulatory enzymes in cells.

PHOSPHORYLATION. Transfer of a phosphate group by a PHOS-PHORYLASE to an organic compound. Usually ATP-dependent, this produces compounds which are highly reactive in water with other organic molecules in the presence of appropriate enzymes, when their phosphate is transferred in turn and energy made available for work. The most important energy-transfer system in metabolism. See KINASE, HIGH-ENERGY PHOSPHATE, OXIDATIVE PHOS-PHORYLATION, PHOTOPHOSPHORYLATION and ATP.

PHOTIC ZONE. Upper portion of a lake, river or sea, sufficiently illuminated for photosynthesis to occur.

PHOTOAUTOTROPH. See AUTOTROPHIC.

PHOTOAUXOTROPH. Photosynthetic organism requiring an external vitamin source.

PHOTOHETEROTROPH. See HETEROTROPHIC.

PHOTOPERIODISM. A biological response to changes in the ratio of light and dark in a 24-hour cycle. In plants, although flowering is the best known example, many other responses are photoperiodic and regulated by the photoreversible pigment PHYTOCHROME. CIRCADIAN RHYTHMS are thought to be fundamental to photoperiodism.

With respect to flowering, plants may be grouped into three categories: (a) *short-day plants*, which flower in early spring or autumn (fall) and require a light period shorter than a critical length; (b) *long-day plants*, flowering mainly during summer, requiring a light period longer than a critical length; and (c) *day-neutral plants*, where flowering is unaffected by photoperiod. Different populations within a species are often precisely adjusted to the photoperiodic regimes where they live. Some plants require only a single exposure to the critical day–night cycle in order to flower. The stimulus is perceived by the leaves and transmitted (probably by some growth substance, or *florigen*) to growing points where flowering is initiated. Photoperiodic effects in animals include initiation of mating in aphids, fish, birds and mammals. See PROXIMATE FACTOR.

PHOTOPHILE. Literally, light-loving; light-receptive phase of a CIR-CADIAN RHYTHM, lasting about 12 hours. Compare SKOTOPHILE.

PHOTOPHOSPHORYLATION. Coupling of phosphate with ADP to produce ATP, using light energy absorbed in PHOTOSYNTHESIS. See BACTERIORHODOPSIN.

PHOTOREACTIVATION. See DNA REPAIR MECHANISMS.

PHOTORECEPTOR. (Bot.) Light-sensitive region of a cell that receives stimulus in phototaxis; usually a dense area in flagellar swelling. Compare EYESPOT. (Zool.) Light-sensitive RECEPTOR; e.g. ROD CELL, EYE.

PHOTORESPIRATION. Type of very active non-mitochondrial respiration occurring in conditions of high light intensity, reduced CO_2 levels and raised O_2 levels in temperate plants carrying out C_3 PHOTOSYNTHESIS; usually absent (or low) in tropical C_4 plants. Involves oxidation of carbohydrates, takes place in PEROXISOMES, and yields neither ATP nor $NADH_2$, thus appears very wasteful. Main substrate, *glycolic acid*, is derived from oxygenation of ribulose bisphosphate (RuBP) through competitive inhibition of chloroplast ribulose bisphosphate carboxylase by molecular oxygen (see ENZYME). Glycolic acid is oxidized by molecular oxygen in peroxisomes to yield hydrogen peroxide, which is destroyed by catalase. Up to 50% of photosynthetically fixed carbon may be reoxidized to CO_2 during photorespiration, lessening the efficiency of C_3 photosynthesis.

PHOTOSYNTHESIS. The light-dependent manufacture of organic from inorganic molecules occurring in CHLOROPLASTS and the cells of blue-green algae and some bacteria, in presence of one or more types of light-trapping pigment (notably chlorophylls in plants). Almost the only carbon-fixing process on Earth, and the route by which virtually all energy enters the biosphere.

Occurs in two stages: (a) a light-trapping phase, in which light energy (*photons*) initiates photochemical reactions on pigment molecules, producing energy-rich compounds (ATP and $NADPH$) with release of molecular oxygen (plants only). This is often called the *light phase* of photosynthesis; (b) a light-independent phase, in which enzymes located off the pigment molecules use products of the light phase and incorporate (fix) carbon dioxide, using the atoms of its (inorganic) molecules to synthesize more organic molecules. Energy for this carbon-fixation comes from ATP and $NADPH$ produced momentarily earlier in the light phase. Because the carbon-fixing process is not in itself dependent upon light it is often referred to as the *dark phase* of photosynthesis. Plant chloroplasts are thought to have evolved by ENDOSYMBIOSIS from prokaryotic cells resembling blue-green algae. Photosynthesis in prokaryotes differs from that in plants in not releasing molecular oxygen. The pigment system (*photosystem*) needed to make use of hydrogen atoms in water is lacking, and instead these usually anaerobic bacteria use alternative hydrogen donors (see AUTOTROPHIC).

In chloroplasts, photons are absorbed on thylakoid membranes by chlorophylls and accessory pigments arranged in the form of ANTENNA COMPLEXES. Energy is passed by resonance transfer between the pigments until it reaches the reaction centre of one of two types of photosystem, each containing a specific form of chlorophyll a. That of photosystem II, chlorophyll a_{680}, donates a pair of excited electrons (excitons) to an organic receptor molecule. The electrons lost are replaced by a pair from a water molecule as protons and molecular oxygen is released into the thylakoid space. In the other photosystem, photosystem I, chlorophyll a_{700} donates an electron pair to another organic receptor which, assisted by a FERREDOXIN molecule, reduces NADP to release NADPH into the chloroplast stroma. Electrons lost are replaced by those from chlorophyll a_{680} after they have passed along an ELECTRON TRANSPORT SYSTEM which includes cytochromes. In this process energy is released to drive protons from the stroma across the thylakoid membrane into the thylakoid spaces, diffusing out through an ATPase in the membrane and releasing sufficient energy for synthesis of ATP from ADP and inorganic phosphate. The whole electron and proton flow sequence is called *non-cyclic photophosphorylation*. After electrons have left the chlorophyll a_{700} of photosystem I they may short-circuit back again via some of the electron transport molecules. This produces no NADPH, but proton-pumping and ATP formation do occur; this is termed *cyclic photophosphorylation*. See Fig. 146.

The dark phase of photosynthesis differs between so-called C_3 and C_4 plants. In the former (including most temperate plants), carbon dioxide is fixed by the enzyme *ribulose bisphosphate carboxylase/oxidase* (*Rubisco*), acting as a carboxylase. The two substrates are CO_2 and ribulose 1,5-bisphosphate (RuBP), and the product is the 3-carbon (hence C_3) compound 3-phosphoglyceric acid (PGA). PGA is not energetic enough for further metabolism, but is converted using ATP and NADPH from the light phase to glyceraldehyde-3-phosphate. This can be used to recycle RuBP (the CALVIN CYCLE) to synthesize starch, or sucrose. If Rubisco functions as an oxidase then RuBP serves as a substrate for PHOTORESPIRATION, and photosynthesis is less efficient.

In C_4 plants (e.g. many cereals and the rice grass *Spartina*), the first product of CO_2-fixation is 4-carbon oxaloacetate, produced by phosphoenolpyruvate carboxylase (PEP carboxylase), which uses phosphoenolpyruvate as its other substrate. In this *Hatch-Slack pathway*, oxaloacetate is further reduced to malate or changed by addition of an amino group to aspartate. These reactions occur in mesophyll cells of the leaf, and the malate (or aspartate) moves from them to bundle-sheath cells surrounding the vascular bundles (see KRANTZ ANATOMY). Here malate is decarboxylated to yield CO_2 and pyruvate, the CO_2 entering the Calvin cycle as substrate for RuBP-carboxylase while the pyruvate reacts with ATP to form more PEP molecules. Such plants can generally photosynthesize at temperatures far higher than C_3 plants, and generally have a far lower CO_2 COMPENSATION POINT. See BACTERIORHODOPSIN, CRASSULACEAN ACID METABOLISM.

PHOTOSYSTEM. See PHOTOSYNTHESIS.

PHOTOTAXIS. TAXIS in which the stimulus is light.

PHOTOTROPHIC. See AUTOTROPHIC.

PHOTOTROPISM (HELIOTROPISM). TROPISM in which light is the stimulus; e.g. the bending of the stem of an indoor plant towards a window, brought about by increased elongation of cells in the growth region on the shaded side. Has been shown to be caused by unequal distribution of the growth substance AUXIN, which migrates from the light side to the dark side of the shoot, particularly in light of wavelengths between 400–500 nm. Evidence suggests that a flavin pigment absorbing such blue light mediates the effect.

PHRAGMOPLAST. (Bot.) Spindle-shaped system of microtubules, arising in plant cells between two daughter nuclei at telophase, and within which the cell plate is formed. See PHYCOPLAST.

PHYCOBILIPROTEINS (PHYCOBILINS, BILIPROTEINS). Water-soluble blue (*phycocyanin*) or red (*phycoerythrin*) pigments present with chlorophyll in blue-green algae, Cryptophyta and Rhodophyta. Are proteins covalently conjugated to the non-protein phycobilin, which resembles a bile pigment. Function as ACCESSORY PIGMENTS. An aggregation of them on surface of a thylakoid is termed a *phycobilisome*.

PHYCOBIONT. Term referring to the algal partner in a lichen.

PHYCOCYANIN. See PHYCOBILIPROTEINS.

PHYCOERYTHRIN. See PHYCOBILIPROTEINS.

PHYCOLOGY. Study of ALGAE.

PHYCOMYCETES. In older classifications, all lower fungi; included the distantly-related Mastigomycotina and ZYGOMYCOTINA.

PHYCOPLAST. (Bot.) System of microtubules, perpendicular to former position of microtubules of mitotic spindle, along which the new wall forms in some green algae (Chlorophyta). See PHRAGMOPLAST.

PHYLLID. Flattened leaf-like appendage in bryophytes.

PHYLLOCLADE. See CLADODE.

PHYLLODE. Flat, expanded petiole replacing blade of leaf in photosynthesis.

PHYLLOTAXY (PHYLLOTAXIS). (Bot.) Arrangement of leaves on the stem: whorled, opposite or spiral. In spiral phyllotaxy, a line connecting attachment points of successive leaves forms a spiral; individual leaves regularly positioned within this spiral. In the most simple, truly alternate, arrangement, leaves are 180° apart, and passage from one half leaf to that precisely above it involves one circuit of the stem and two leaves – a phyllotaxy of 1/2. Various forms of phyllotaxy occur, such as 1/3, 2/5, 3/8, 5/13, etc. (a Fibonacci series), each fraction representing an angle made by successive leaves with the stem (looking vertically downwards). At the end of each spiral, a leaf is directly above the one at the beginning. Looking down on a stem, these points of superimposition are identified as vertical rows of leaves known as *orthostichies*. At the growing point, though leaf primordia are spirally arranged, orthostichies do not occur; but looking down at the apex, the primordia are arranged in a series of descending curves, or *parastichies*, some clockwise. Parastichy in an apex may become orthostichy in a mature shoot by straightening during elongation.

PHYLOGENETICS. Approach to biological CLASSIFICATION concerned with reconstructing PHYLOGENY and recovering the history of speciation. This is possible when speciation is coupled with, and does not proceed faster than, character modification. *Phylogenetic trees* so produced should represent the hypothetical historical course of speciation and be open to rigorous testing. The system in ascendancy today is CLADISTICS. Compare PHENETICS.

PHYLOGENY. Evolutionary history. Genealogical history of a group of organisms, in practice represented by its hypothesized ancestor–descendant relationships. Compare ONTOGENY. See RECAPITULATION.

PHYLUM. Taxonomic category often restricted to the animal kingdom; includes one or more CLASSES and is included within a KINGDOM in the taxonomic hierarchy. Corresponds to the category DIVISION in botany.

PHYSIOLOGICAL SALINE. See RINGER'S SOLUTION.

PHYSIOLOGICAL SPECIALIZATION (BIOLOGICAL SPECIALIZATION). Existence within a species of genetically different races or forms, such as the 300 or so races of the black stem rust *Puccinia graminis tritici* of wheat, which differ in their pathogenicity. See INFRASPECIFIC VARIATION.

PHYSIOLOGY. Study of processes, many either directly or indirectly homeostatic, that occur within living organisms; in multicellular organisms, includes interactions between cells, tissues and organs and all forms of intercellular communication, both energetic and metabolic.

PHYTO-. Prefix indicating a botanical context.

PHYTOALEXINS. Non-specific compounds, generally phenolic, synthesized *de novo* or in greatly increased concentration by plants in response to infection by fungi, to which they are toxic. Believed to play a primary role in disease resistance.

PHYTOCHROME. Proteinaceous plant pigment existing in two different but interconvertible forms. P_r absorbs (is *receptive* to) red light (660 nm) and P_{fr} absorbs far-red light (730 nm). When a P_r molecule absorbs a photon it is converted in milliseconds to a molecule of P_{fr}, the reverse happening when P_{fr} absorbs a photon. P_{fr} is lost from cells when the plant is placed in the dark, as it reverts to P_r or is otherwise destroyed. The active form of the pigment, P_{fr}, promotes flowering in long-day plants, inhibits flowering in short-day plants and is responsible for seed germination, for changes occurring in seedlings as they penetrate through the soil to the light, and for development of anthocyanins. The $P_r : P_{fr}$ ratio is instrumental in breaking dormancy in some seeds.

PHYTOGEOGRAPHY. See PLANT GEOGRAPHY.

PHYTOKININS. See CYTOKININS.

PHYTOPLANKTON. Plant plankton. *Euplankton* (holoplankton) comprises phytoplankton completing its life cycle suspended in open water; *meroplankton* spends part of its life cycle in bottom suspensions; *pseudoplankton* comprises casual species, occasionally in open water but derived from benthic communities. See PLANKTON.

PHYTOSOCIOLOGY. See PLANT SOCIOLOGY.

PIA MATER. The innermost of the MENINGES.

PILEUS. Cap-like part of fruiting body of fungi of Basidiomycotina (e.g. mushrooms), bearing the hymenium on its undersurface.

PILI (sing. PILUS). Proteinaceous filaments protruding through cell wall of Gram-negative bacteria, with an adhesive role. Some (*sex pili*) are involved in conjugation. See F FACTOR.

PILIFEROUS LAYER. That part of the root epidermis bearing root hairs.

PINEAL EYE. See PINEAL GLAND.

PINEAL GLAND (EPIPHYSIS CEREBRI). Small mass of nerve tissue attached to roof of third ventricle of the vertebrate midbrain; loses all nervous connection with the brain but innervated by sympathetic nervous system. In amphibia, primitive reptiles (e.g. *Sphenodon*, the tuatara) and some snakes, pineal cells form a *parietal eye* (median eye) lying within a parietal foramen of the skull, with a lens-like upper epithelium and retina-like lower part. In higher vertebrates, the pineal organ has a glandular structure, and secretes *melatonin* which inhibits gonadotrophins and their effects. Melatonin production is inhibited by exposure of the animal to light. SEROTONIN, another pineal product, and melatonin demonstrate inverse circadian rhythms in their production.

PINNA (pl. PINNAE). (Bot.) A primary division, or leaflet, of a compound leaf or frond. (Zool.) See EAR, OUTER.

PINNATE. Form of branching which occurs at uniform angles from different points along a central axis, all in one plane, as in a feather.

PINNIPEDIA. Eutherian order (or suborder of the order CARNIVORA), containing specialized and aquatic mammals: seals (Phocidae), sealions (Otariidae) and walruses (Odobenidae). Limbs are broad flippers, with webbed feet; tail very short. In true seals, hind limbs are fused with tail.

PINOCYTOSIS. Ingestion of surrounding fluid by a cell through localized invagination of its plasma membrane, which thus completely surrounds a minute drop of fluid and forms a vesicle. See ENDOCYTOSIS, LYSOSOME.

PISTIL. Either each separate carpel (in *apocarpous gynoecia*), or two or more fused carpels (in *syncarpous gynoecia*). Each typically comprises an ovary, style and stigma.

PISTILLATE. (Of flowers) naturally possessing one or more carpels but no functional stamens; also called carpellate. See STAMINATE.

PITH. Ground tissue located in the centre of stem or root, within the vascular cylinder. Usually comprises PARENCHYMA. See STELE.

PITHECANTHROPUS. See Homo.

PITS. Small, sharply defined depressions in wall of plant cell where the secondary wall is completely absent, permitting easier passage of material between adjacent cells. Usually occur over *primary pit fields* (where plasmodesmata are concentrated), but also in their absence. Often coincide with pits in the walls of adjacent cells, separated from them by a *pit membrane* comprising a middle lamella and a very thin layer of primary wall on either side, together forming a *pit pair*. Such *simple pits* connect living cells together and also occur in stone cells and some fibres. In *bordered pits*, characteristic of xylem vessels and tracheids, the pit cavity is partly enclosed by over-arching of the cell wall (the *pit border*) and the pit membrane may possess a central, thickened impermeable *torus*, closing the pit aperture if the pit membrane is displaced laterally.

PITUITARY GLAND (HYPOPHYSIS). Small but essential vertebrate gland, lying in a depression of sphenoid bone of skull and communicating with the hypothalamus (part of the diencephalon of the forebrain) by a stalk-like *infundibulum* (pituitary stalk). A composite gland, comprising an anterior lobe (*adenohypophysis*) deriving from pharyngeal ectoderm and secreting the bulk of pituitary hormones, and a posterior lobe (*neurohypophysis*) deriving from ectoderm of the hypothalamus and containing the termini of hypothalamic neurosecretory nerve axons.

The adenohypophysis releases its hormones into capillaries, drained by the hypophysial vein, under commands (*releasing factors*) from the hypothalamus which reach it via a pituitary portal system. Hormones secreted by the most anterior region (*pars distalis*) include GROWTH HORMONE, PROLACTIN, THYROID-STIMULATING HORMONE (TSH), GONADOTROPHINS and ACTH. Where present, an intermediate region (*pars intermedia*) secretes MELANOCYTE-STIMULATING HORMONE. The neurohypophysis, neurosecretory rather than endocrine in function, releases the hormones OXYTOCIN and ANTIDIURETIC HORMONE (see NEUROSECRETORY CELLS for *neurophysins*).

The pituitary forms an integral link in many homeostatic feedback circuits in the body. See HYPOTHALAMUS.

PLACENTA. (Bot.) That part of the ovary wall to which the ovules or seeds are attached. (Zool.) A temporary organ, consisting of both embryonic and maternal tissues, within the uterus of therian mammals and several other viviparous animals, which enables embryos to derive soluble metabolites by diffusion and/or active transport from the maternal blood supply. In placental mammals the organ is connected to the embryo by the *umbilical cord* and has an essential role in the immunological protection of the embryo. See Fig. 16.

In mammals (see Fig. 55), foetal components of the placenta derive initially from the TROPHOBLAST, connected with the embryonic bloodstream either through its contact with the YOLK SAC (*vitelline placentation*) or ALLANTOIS (*allantoic placentation*). In the latter, the outer covering is termed the *allantochorion*, a term also used of placentae in which blood vessels, but little else, derive from the allantois (as in humans). In both vitelline and allantoic placentation, trophoblastic villi push out from the surface of the chorion and invade the uterine lining (*endometrium*; see DECIDUA), greatly increasing the surface area (14 m^2 in humans) for exchange of solute molecules. In marsupials, vitelline placentation serves the entire, brief, intrauterine life; in eutherians it is soon replaced by allantoic placentation, and sometimes only certain restricted zones of the allantochorion (termed *cotyledons*) form villi and participate in exchanges. Functions of the mammalian placenta include: (a) allowing passage of small molecules from uterine capillaries, or the intervillous blood sinuses (depending upon the type of placentation), into the capillaries of the umbilical vein. These include O$_2$, salts, glucose, amino acids and small peptides (all by active transport), simple fats, some antibodies (see IgG) and some vitamins (A, C, D, E and K); (b) removal of embryonic excretory molecules, notably CO$_2$ and urea, by diffusion from the umbilical artery into the maternal circulation; (c) storage of glycogen, and its conversion to glucose if foetal glucose levels drop; (d) hormone production, substituting for maternal ovaries in production of the sex hormones

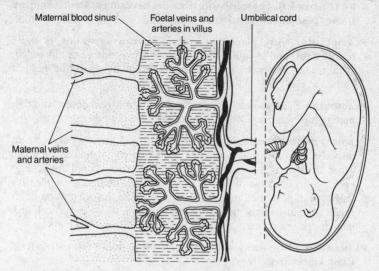

Fig. 55. Diagram of human placenta and its relationship to the foetus. Left side of diagram is magnified compared to right side.

OESTROGENS (largely oestriol) and PROGESTERONE (primates especially), and in humans producing HUMAN CHORIONIC GONADO-TROPHIN, HUMAN PLACENTAL LACTOGEN and such 'pituitary' hormones as ACTH and TSH; (e) production of ENDORPHINS, and (f) prevention of red blood cell exchange between mother and foetus, avoiding AGGLUTINATION. (The human placenta is exceptional in allowing a mutual exchange of leucocytes and blood platelets. See PARTURITION.)

PLACENTALIA. 'Placental' mammals. See EUTHERIA.

PLACENTATION. (Bot.) Manner in which the ovules are attached in an ovary. (Zool.) The type of arrangement of the PLACENTA in mammals.

PLACODERMI. Class of fossil bony fish, some large predators, mainly from the Devonian period. Bony shields on head and front part of body. AUTOSTYLIC JAW SUSPENSION; paired fins; heterocercal tail. Possibly related more to Chondrichthyes than to Osteichthyes or Acanthodii.

PLACOID SCALE (DENTICLE). Tooth-like scale, characteristic of the CHONDRICHTHYES and completely covering elasmobranch fish. Base bone-like, but made largely of dentine, with a pulp cavity and enamel-like coating (usually *vitro-dentine*). Probably represent last remnants of the bony armour covering the early vertebrate body (see

PLACODERMI). Those fish with denticles have similar teeth. Compare GANOID SCALE, COSMOID SCALE.

PLAGIOCLIMAX. Any plant community whose composition is more or less stable and in equilibrium under existing conditions but which, as a result of human intervention, has not achieved the natural CLIMAX; e.g. grassland under continuous pasture.

PLAGIOSERE. Succession of plants deflected into a new course through human intervention. Compare PRISERE.

PLAGIOTROPISM. Orientation of a plant part by growth curvature in response to gravitational stimulus, such that its axis makes an angle other than a right-angle with the line of gravitational force. E.g. exhibited by branches of a main root which makes an acute angle with the vertical. Also used in a wider sense to mean the orientation of an organ so that its axis makes a constant angle with the vertical, and includes DIAGEOTROPISM as a special type. See GEOTROPISM.

PLANARIANS. Free-living members of the PLATYHELMINTHES. Class Turbellaria.

PLANATION. Dichotomies in two or more planes, flattening out into one plane, e.g. a leaf.

PLANKTON. Organisms kept in suspension by water movements and dispersed more by those movements than by their own activities. Most are very small, occurring mainly near the water surface (the photic zone). Of great ecological and economic importance, providing food for fish and whales. See PHYTOPLANKTON, ZOOPLANKTON. Compare BENTHOS.

PLANONT (PLANOSPORE). See ZOOSPORE.

PLANTAE. The Kingdom comprising eukaryotic organisms (mostly autotrophs), usually with clearly defined cellulose-containing cell walls. Sometimes restricted to those having advanced tissue differentiation (e.g. bryophytes and vascular plants), where the diplophase (sporophyte) includes an embryo and the haplophase (gametophyte) produces gametes by mitosis. This would exclude organisms sometimes placed in the PROTISTA.

PLANT GEOGRAPHY. Study of plant distributions, past and present, and of their causes. Includes such fields as geology, palaeontology, plant genetics, evolution and taxonomy. See ZOOGEOGRAPHY.

PLANTIGRADE. Mode of walking involving the ventral surface of the whole foot, i.e. the metacarpus or metatarsus and digits. Humans are plantigrade. Compare DIGITIGRADE, UNGULIGRADE.

PLANT SOCIOLOGY (PHYTOSOCIOLOGY). Study of plant COMMUNITIES comprising the vegetation, their origins, compositions and structures.

PLANULA. Solid free-swimming ciliated larva of most classes of Cnidaria and a few of the Ctenophora. Composed of an outer ectoderm and inner endoderm, the latter formed by migration of cells into the interior.

PLAQUE. A clear area in a bacterial culture on nutrient agar caused by localized destruction of bacterial cells through BACTERIOPHAGE activity. Similarly applied to a zone of lysis caused by a virus in animal tissue culture. Counting plaques provides a simple technique for estimating virus concentration in a fluid applied to the culture.

PLASMA. See BLOOD PLASMA.

PLASMA CELL. One kind of mature B-CELL, active in antibody secretion.

PLASMAGEL. The gel-like state of the outer cytoplasm of many cells, as distinct from the more fluid inner cytoplasm (*plasmasol*) with which it is interconvertible. May be highly vacuolated in planktonic protozoans. Involved in several forms of CELL LOCOMOTION, notably *amoeboid locomotion*.

PLASMALEMMA. Outer membrane of a cell. See CELL MEMBRANES.

PLASMID. A piece of symbiotic DNA, mostly in bacteria but also in yeast, not forming part of the normal chromosomal DNA of a cell and capable of replicating independently of it. It constitutes a form of *extrachromosomal DNA*, sometimes housing genes encoding enzymes (e.g. conferring drug resistance) of critical value to the host cell or organism. Sometimes confer no recognizable phenotype on the host cell (*cryptic plasmids*). Those that may be considered to do so include the bacterial sex factor (F FACTOR) and some forms of PHAGE (*temperate phages*, e.g. *P1*, in their non-temperate phase). A segment (T-DNA) of the tumour-inducing plasmid (T_i) can be transferred from the bacterium *Agrobacterium* to plant cells at a wound site. Some plasmids, if not all, can be transferred from one cell to another by conjugation (see F FACTOR), TRANSDUCTION and TRANSFORMATION. Often act as vehicles for TRANSPOSABLE ELEMENTS such as ANTIBIOTIC RESISTANCE ELEMENTS. Plasmids which can pick up such transposable elements are termed *R plasmids*. Another type of plasmid, *Col plasmids*, are harboured by many bacterial cells and enable them to produce toxins which specifically kill cells of other bacteria; e.g. a *colicin* will kill *E. coli*. The plasmid also renders cells which contain it immune to the colicins. The implications for public health of R plasmids, particularly their mobility between bacteria of different species and genera, are serious. Antibiotic resistance (selected for in non-pathogenic intestinal bacteria by too frequent administration of antibiotics) can be transferred to pathogenic bacteria (e.g. *Salmonella* and *Shigella*). Such resistance has been recorded in sewers and polluted rivers and is a strong reason for restricting antibiotic administration to essential cases. See CYTOPLASMIC INHERITANCE.

PLASMODESMATA (sing. **PLASMODESMA**). Extremely fine cytoplasmic tubes, often a fraction of a micrometre wide, which pass through the walls of living plant cells and are lined by the plasma membranes of two adjacent cells, connecting their cytoplasms (see SYMPLASM). Through the centre of most runs a *desmotubule* composed of membranes continuous with the endoplasmic reticula of the two cells. They may be scattered, or grouped in the membranes of PITS.

PLASMODIUM. (1) Generic name of malarial parasites. See MALARIA. (2) Vegetative stage of slime-fungi (Myxomycota). Multinucleate, amoeboid mass of protoplasm bounded by a plasma membrane but lacking definite shape or size. See COENOCYTE, SYNCYTIUM.

PLASMOGAMY. Union of two or more protoplasts, typically of gametes, unaccompanied by union of any nuclei they may have. Compare KARYOGAMY.

PLASMOLYSIS. Shrinkage of a cell's protoplasm (in plants, away from cell wall) when placed in a hypertonic solution, as a result of osmosis. Compare TURGOR.

PLASTIDS. Small, variously shaped, organelles in cytoplasms of plant cells, one to many per cell. Each surrounded by an envelope of two membranes and containing a system of internal membranes, pigments and/or reserve food material, a more or less homogeneous ground substance (*stroma*) in which ribosomes of a prokaryotic type may be present. Arise either by division of existing plastids or from *proplastids*, and are classifiable when mature on the basis of their pigments into: CHLOROPLASTS (sites of photosynthesis containing chlorophylls and carotenoids), CHROMOPLASTS, and non-pigmented LEUCOPLASTS. See PLASTOGENE.

PLASTOGENE. Plastid-borne genetic factor. A plasmogene. See CYTOPLASMIC INHERITANCE, cpDNA.

PLASTRON. (1) Flat horny scutes on underside of body of a turtle or tortoise, connected to the carapace above by a bony bridge. (2) An air store forming thin film over bodies of several aquatic insects (some beetles and heteropteran bugs), communicating with the spiracles and held by hydrofuge bristles or scales. Can act as a 'physical gill' if there is adequate dissolved oxygen, enabling the insect to remain under water.

PLATELETS (THROMBOCYTES). Small cell-like blood-borne fragments (3–4 μm long) shed from megakaryocytes in vertebrate bone marrow. Discoid in shape, containing secretory granules which increase permeability of capillaries, to which they adhere in BLOOD CLOTTING, they activate COMPLEMENT and attract LEUCOCYTES. Platelets are destroyed by RETICULO-ENDOTHELIAL SYSTEM after about 10 days.

PLATYHELMINTHES (FLATWORMS). Invertebrate phylum, containing bilaterally symmetrical triploblastic acoelomates, dorso-ventrally flattened, with only one opening to the gut (when present), and without a blood system. Include planarians (Turbellaria), flukes (Trematoda) and tapeworms (Cestoda). They have bulky parenchymatous tissue derived from mesoderm, FLAME CELLS and complex hermaphroditic reproductive organs.

PLATYPUS. Duck-billed platypus (*Ornithorhynchus*). See PROTOTHERIA.

PLATYRRHINES. Members of the PRIMATE Suborder Platyrrhini (Ceboidea, New World monkeys). Probably isolated from other primates since the Eocene. Characterized by broad nasal septum; often have prehensile tails. Include marmosets, tamarins, howler monkeys. See ANTHROPOIDEA, CATARRHINE.

PLECOPTERA. Stoneflies. Small order of exopterygote insects with aquatic nymphs, long antennae, biting mouthparts and weak flight. Two pairs of wings, held flat over back at rest; hind pair usually larger. Adults tend to feed on lichens and unicellular algae. Good pollution indicators.

PLECTENCHYMA. Tissue composed of fungal hyphae; either *prosenchyma*, in which component hyphae are loosely woven and recognizable as such, or *pseudoparenchyma*, comprising closely-packed cells which can no longer be distinguished as hyphae and resemble parenchyma of higher plants.

PLECTOSTELE. A protostele, split into many plate-like units.

PLEIOMORPHISM (PLEOMORPHISM). Form of POLYMORPHISM in which there are several different forms in the LIFE CYCLE; e.g., in the life cycles of rust fungi, where there is a succession of different spore forms.

PLEIOTROPY. The capacity of allelic substitutions at one gene locus to affect more than one aspect of phenotype. For any given allele some effects may be dominant, others recessive (see DOMINANCE). A gene locus may have pleiotropic effects if it encodes an enzyme or regulatory protein whose product is involved in several biochemical pathways, as commonly occurs in development. Compare POLYGENES.

PLEISTOCENE. Geological epoch; first of Quaternary period (see GEOLOGICAL PERIODS). An upper value of duration would be 2.5 Myr–11 000 BP. Characterized by four major glacials, separated by interglacials lasting tens of thousands of years. The epoch saw rapid hominid evolution (see HOMO), and the appearance of elephants, cattle and the modern horse.

PLEROME. See APICAL MERISTEM.

PLESIOMORPHIC. In PHYLOGENETICS, describing the original pre-existing member of a pair of homologous characters, the evolutionary novelty being the *apomorphic* member of the pair. See CLADISTICS.

PLESIOSAURIA. Extinct suborder of marine euryapsid reptiles (Order Sauropterygia), prominent in the Mesozoic. Typical length 3 metres; long-necked, short-bodied, with limbs modified into powerful paddles. Derived from similar *nothosaurs*, they were not dinosaurs.

PLEURA. Serous membranes lining PLEURAL CAVITY and covering the lung surfaces in mammals. Normally only a thin fluid layer separates the pleura from body wall.

PLEURAL CAVITY. Coelomic space surrounding each lung in a mammal, separated from rest of perivisceral coelom by the diaphragm. Pleural cavities are separated from each other by the mediastinum and the pericardium.

PLEUROCARPUS. Growth form in some mosses, where the gametophore is multibranched and creeping, with the sporophyte borne upon a very short lateral branch.

PLEUROPNEUMONIA-LIKE ORGANISMS (PPLOs). See MYCOPLASMAS.

PLEXUS. See NERVE PLEXUS.

PLIOCENE. Epoch of the Tertiary period. See GEOLOGICAL PERIODS.

PLOIDY. The number of haploid chromosome sets in a nucleus, or cell. A haploid nucleus has a ploidy of 1; a diploid nucleus 2, a triploid nucleus 3, etc. See ANEUPLOID, POLYPLOID.

PLUMULE. (Bot.) Terminal bud of well-differentiated embryos in seed plants. (Zool.) Down feather of nestling birds, persisting in some adult birds between contour feathers. See FEATHER.

PLURILOCULAR. Possessing many chambers, e.g. describing gametangia and mitosporangia in brown algae (PHAEOPHYTA).

PLURIPOTENCY. Property of a *pluripotent* cell which can give rise to many cell phenotypes of the organism to which it belongs when suitably challenged, but lacks complete TOTIPOTENCY. Pluripotent cell progeny eventually become DETERMINED and show the same phenotype in different environments.

PLUTEUS. Larva of echinoid (sea urchin) or ophiuroid (brittle star). Ciliated and planktonic.

PNEUMATOCYST. Hollow region of stipes of some brown algae, containing gas helping to keep the alga afloat.

PNEUMATOPHORE. (Bot.) Negatively geotropic specialized root branch, produced in large numbers by some vascular plants growing in water of tidal swamps, e.g. mangrove; grows into the air above the water and contains well-developed intercellular system of air spaces communicating with the atmosphere through pores on the aerial portion. (Zool.) Gas sac of a siphonophoran coelenterate.

POGONOPHORA. Phylum of benthic and tubicolous worm-like animals; mostly very thin, but attaining large sizes near hydrothermal oceanic vents. Deuterostomes, the coelom in three parts (*oligomerous*), with a tentacular hydrostatic system comparable to the lophophore of phoronids. Lacking a gut, absorbing dissolved organic compounds across their tentacles.

POIKILOTHERMY (ECTOTHERMY). Condition of any animal whose body temperature fluctuates considerably with that of its environment. Characteristic of invertebrates, and vertebrates other than birds and mammals among living forms. See HOMOIOTHERMY.

POLAR BODY. One of up to three small haploid nuclei, usually with a covering of cytoplasm, extruded from primary and secondary OOCYTES during MEIOSIS in animals. The first polar body may undergo the second meiotic division, producing three polar bodies in all. They usually play no further genetic role, and get lost or broken down in early development.

POLARITY. (1) Having a difference (morphological, physiological, or both) between the two ends of an axis, e.g. shoot/root, head/tail. Characteristic of most organisms, unicellular and multicellular, and of many cells within organisms. (2) Non-random organization of the cytoplasm of an animal egg prior to fertilization, sometimes with consequences for future development of embryo. Determined largely by position of egg in ovary, its association with nutritive cells, and its attachment to ovary wall. May change at fertilization or cleavage. Sometimes marked by yolk and/or centriole distribution. Animal/vegetal axis may dictate future anterior–posterior AXIS of embryo. See ANIMAL POLE, MATERNAL EFFECT.

POLAR NUCLEUS. One of two nuclei that migrate to centre of megagametophyte of a flowering plant, fusing to form the central fusion nucleus prior to DOUBLE FERTILIZATION.

POLAR PLASM. See GERM PLASM, GERM LINE.

POLLEN ANALYSIS (PALYNOLOGY). Techniques for reconstructing past floras and climates using pollen grains (and other spores), especially those preserved in lake sediments and peat. Their resistance to decay and distinctive sculpturing enable quantitative and qualitative estimates of past species abundance, typically within a timescale of a few decades up to millennia.

POLLEN GRAIN. Microspores of seed plants, containing a mature or immature microgametophyte. Outer EXINE coat sculptured. See MICROSPORANGIUM.

POLLEN TUBE. Tube formed on germination of a pollen grain, which transports the two 'male' nuclei to the egg. In flowering plants, each penetrates tissues of the stigma and style, entering the ovary and growing towards an ovule, where it passes through the micropyle, penetrating the nucellus and rupturing at the tip of the embryo sac to set free two nuclei. See DOUBLE FERTILIZATION, MICRO-GAMETOPHYTE.

POLLEX. 'Thumb' of pentadactyl fore-limb. Often shorter than other digits. Compare HALLUX.

POLLINATION. Transfer of pollen (by wind, water, insects, birds or other animals) from the anther where they were formed to a receptive stigma. Not to be confused with any later fertilization which may result.

POLLINIUM. Coherent mass of pollen grains, as in orchids; generally transferred as a unit in pollination.

POLYADELPHOUS. (Of stamens) united by their filaments into several groups, as in St. John's wort. Contrast MONADELPHOUS, DIADELPHOUS.

POLYCHAETA. Class of ANNELIDA, including ragworms, tubeworms, fanworms and lugworms. Marine; free-swimming, errant, burrowing or tube-dwelling. Clearly segmented; typically a pair of PARAPODIA per segment. Usually a well-marked head, often with jaws and eyes. Large coelom, usually divided by septa. Separate sexes; fertilization external; larva a trochophore. Compare OLIGOCHAETA, HIR-UDINEA.

POLYCLAVE. See IDENTIFICATION KEYS.

POLYCLONE. Well-defined anatomical region (e.g. insect wing COM-PARTMENT) consisting of several entire clones of cells.

POLYEMBRYONY. Asexual formation of more than one embryo from each zygote produced. In plants, accessory embryos sometimes arise by budding from the pro-embryo. In animals, sometimes involves fission at some early stage of development, or budding in a larval stage. Common in some parasites. See CERCARIA, HYDATID CYST, MONOZYGOTIC TWINS, PROGENESIS.

POLYGENES. Genes, at more than one locus, variations in which in a particular population have a combined effect upon a particular phenotypic character (said to be determined *polygenically*, or to be a *polygenic character*). Allelic substitutions at such loci tend to have little individual effect upon phenotypic differences between individuals, the normal situation in much of developmental biology, where expression of several different genes is required for production of a character. Such *polygenic inheritance* increases the likelihood that a character will exhibit continuous VARIATION in the population. Human height is an example. Compare PLEIOTROPY, EPISTASIS.

POLYGENIC INHERITANCE. See POLYGENES.

POLYMER. A very large molecule comprising a chain of many similar or identical molecular subunits (monomers) joined together (polymerized). In biochemistry, this usually involves condensation reactions. Proteins, polysaccharides and nucleic acids are characteristic biopolymers.

POLYMERASE. Enzyme joining monomers together to form a polymer; e.g. DNA POLYMERASE, RNA POLYMERASE, starch synthase.

POLYMORPH (POLYMORPHONUCLEAR GRANULOCYTE). Any of several types of LEUCOCYTE with granular cytoplasms and lobed nuclei, produced in bone marrow; forming 60–70% of human blood leucocytes. Also found in tissue fluid and lymph (they can adhere to and penetrate capillaries). Predominantly phagocytic (esp. neutrophils), but include EOSINOPHILS, BASOPHILS and MAST CELLS. PLATELETS also have granular cytoplasms, but are not polymorphs. Neutrophils degranulate into their PHAGOSOMES; eosinophils, basophils and mast cells degranulate by exocytosis.

POLYMORPHISM. Major category of discontinuous VARIATION within species. (1) *Genetic polymorphism.* Simultaneous occurrence in a species population of two or more discontinuous forms in such a ratio that the rarest of them could not be maintained solely by recurrent mutation. This excludes reproductively non-overlapping seasonal forms of a species (e.g. some PLEIOMORPHISMS), rare mutants, geographical races and all continuous variation. These polymorphisms generally involve clear genetic alternatives between individuals in the population, as controlled by SUPERGENES, INVERSIONS or loci where allelic substitutions tend to bring about marked differences in phenotype (so-called 'major genes'). Examples include many forms of Batesian MIMICRY, human BLOOD GROUPS, the MHC system, some forms of INDUSTRIAL MELANISM and HETEROSTYLY, and many forms of SEX DETERMINATION. Enzyme and DNA polymorphisms, however, are often quite cryptic. The term *balanced polymorphism* describes polymorphism maintained by a balance of selective agencies whose effects tend to cancel each other out (e.g. see FREQUENCY-DEPENDENT SELECTION).

Where a mutant gene increases in frequency in a population it may eventually replace its alternative allele(s). Such a situation is termed *transient polymorphism*, but may never go to complete fixation of the new mutant. The spread of melanic forms is an example. Relative fitnesses of different *morphs* can sometimes be calculated using the HARDY-WEINBERG THEOREM. See HETEROZYGOUS ADVANTAGE.

(2) *Non-genetic polymorphism*, where distinct forms of a species occur but without the definitional rigour of the cases above. Environmental influences play a large part in insect CASTE polymorphisms. Genetic factors (e.g. haplo-diploidy) may be important, but nutritional influences (e.g. in hymenopterans) and pheromones (e.g. in termites) are instrumental in developmental decisions. The term polymorphic is also applied to the different ZOOIDS in siphonophore and ectoproct colonies.

POLYNUCLEOTIDE. Long-chain molecule formed from a large number of nucleotides (e.g. NUCLEIC ACID).

POLYP (HYDRANTH). Sedentary form of the CNIDARIA. Cylindrical stalk-like body attached to substratum at one end, mouth surrounded by tentacles at the other. Many bud asexually; some are sexual. Polyp-like stages in the scyphozoan life cycle (*scyphistomas*) produce ephyra larvae (which develop into sexual medusae) by strobilation. Most hydrozoans and anthozoans have (or are) polyps. See POLYPIDE.

POLYPEPTIDE. Molecule consisting of a single chain of many amino acid residues linked by PEPTIDE BONDS. See PROTEIN.

POLYPETALOUS. (Of flowers) having petals free from one another; e.g. buttercup. Compare GAMOPETALOUS.

POLYPHOSPHATE GRANULES. Phosphate storage granules found in cells of blue-green algae (CYANOBACTERIA).

POLYPHYLETIC. Term for a group of taxa (species, genera, etc.) when, despite their being classified together as one taxonomic category, it is thought that not all have descended from a common ancestor which was also a member of the group. Such a taxon forms a GRADE rather than a CLADE. If the classification is to correspond with phylogeny, the group should be split into two or more distinct taxa. See CLADISTICS, HOMOLOGY, MONOPHYLETIC.

POLYPIDE. Polyp-like individuals, forming a COLONY of ectoprocts or entoprocts.

POLYPLOIDY. Condition in which a nucleus (cell, individual, etc.) has three or more times the HAPLOID number of chromosome sets characteristic of its species (or ancestral species). The haploid number is usually represented by the letter n, so a *triploid* is commonly represented as $3n$, a *tetraploid* as $4n$, a *pentaploid* as $5n$, and so on.

Tetraploidy can arise in otherwise normal diploid somatic tissue, both plant and animal, through failure of replicated chromosomes to separate in MITOSIS. This can be induced artificially by exposure to a compound such as colchicine, which inhibits formation of the spindle apparatus in mitosis through its disruptive effect upon microtubule formation; or by heat treatment. Such AUTOPOLYPLOID cells are much more likely to become part of the germ line in plants than in animals; but there may be some reduction in fertility due to unequal segregation of chromosomes at meiosis. However, some autotetraploid plants (e.g. cocksfoot grass, *Dactylis glomerata*; purple loosestrife, *Lythrum salicana*) are fully fertile. ALLOPOLYPLOID organisms are more likely to be fully fertile since each chromosome has only one fully homologous partner to pair with. Crossing and backcrossing between these and normal (usually diploid) individuals often produces a *polyploid complex*, resulting in gene INTROGRESSION. Most instances of plant APOMIXIS involve polyploid species, giving them an 'escape from sterility'. Polyploidy is comparatively rare in animals, probably because most animals are bisexual and obligate cross-fertilizers. A tetraploid (4*n*) animal could only hybridize with diploids and then only produce sterile triploid (3*n*) offspring, if any at all. Only a few amphibians and fish among vertebrates seem to be polyploid, possibly because of their method of SEX DETERMINATION. See also HYBRIDIZATION.

In plants, the range of variation is often narrower in allotetraploids than in a related diploid because each gene is duplicated. Polyploids are often self-pollinated, even when related diploids are mainly cross-pollinated, which reinforces the decrease in variability.

Polyploids are of interest because of the examples they may provide of 'instantaneous' SPECIATION, and have been extremely important in the evolution of certain groups of plants. One of the most important polyploid (hexaploid) species is wheat, e.g. bread wheat, *Triticum aestivum*, which has 42 chromosomes. It was derived some 8000 years ago following the spontaneous hybridization of a cultivated wheat possessing 28 chromosomes with a grass of the same group with 14 chromosomes, followed by chromosome doubling.

Among plant polyploids are many of our most important crops, including, besides wheat, bananas, cotton, potatoes, sugar cane and tobacco. Many garden flowers are polyploids, e.g. chrysanthemums, day lilies. See ENDOMITOSIS, POLYTENY.

POLYSACCHARIDE (GLYCAN). Carbohydrate produced by condensation of many monosaccharide subunits to form polymers, forming long, often fibrous, molecules which are poorly soluble, osmotically inactive, and lacking a sweet taste. Extremely important structurally (e.g. CELLULOSE, CHITIN and GLYCOSAMINOGLYCANS) and as energy reserves (e.g. STARCH, GLYCOGEN and INULIN).

POLYSEPALOUS. (Of a flower) having sepals free from one another; e.g. buttercup. Compare GAMOSEPALOUS.

POLYSOME (POLYRIBOSOME). See RIBOSOME.

POLYSOMY. Condition in which one chromosome (rarely more) in a nucleus is represented more frequently than the remaining chromosomes. In a *trisomy* (see DOWN'S SYNDROME) one chromosome is present three times in the nucleus as opposed to twice for other chromosomes. The nucleus would have a ploidy of $2n + 1$ (where n = haploid number). Polysomy can occur in nuclei which are otherwise of any ploidy. See NON-DISJUNCTION, POLYPLOID.

POLYSPERMY. (Zool.) Entry of more than one sperm nucleus into an ovum at fertilization. Occurs normally only in very yolky eggs (e.g. shark, bird) and commonly in insects; only one nucleus fuses with egg nucleus, the rest taking no part in development. May upset normal development in other eggs. See FERTILIZATION, PSEUDO-GAMY.

POLYSTELIC. (Bot.) Possessing more than one STELE, comprising more than one (sometimes many) vascular bundles.

POLYTENY. Condition of some chromosomes, nuclei, cells, etc., in which many identical parallel copies of each chromosome are formed by repeated replication without separation, and lie side-by-side forming thick cable-like chromosomes. Best studied in salivary gland cells of the fruit-fly *Drosophila*, these and LAMPBRUSH CHROMOSOMES are often termed *giant chromosomes*, occurring only in cells which have lost the power to divide mitotically (permanent interphase). They tend to be longer than normal on account of being less coiled. Close pairing of sister strands may produce transverse 'banding', which enables both the detection of chromosome inversions and study of chromosome 'puffing' (see ECDYSONE). May occur in secretory tissues where increase in cell size rather than number has occurred. Compare ENDOMITOSIS, POLYPLOIDY.

POLYTHETIC. (Of classification) a system where membership of a taxon depends upon possession of a large number of characters in common.

POLYTOPIC. Designating taxa occurring in two or more separate areas.

POLYTYPIC. Designating species which occur in a variety of geographical forms, or subspecies. See INFRASPECIFIC VARIATION.

POLYZOA (BRYOZOA). See ECTOPROCTA, ENTOPROCTA.

POME. 'False fruit', the greater part of which is developed from the receptacle and not the ovary; e.g. apple, pear. The edible fleshy part represents the receptacle, and the core represents the ovary.

PONGID. Member of primate family Pongidae (see HOMINOID), including all ANTHROPOID APES and DRYOPITHECINES.

PONS. Floor of fourth ventricle of mammalian brain, lying in front of medulla oblongata and under cerebellum. Connects cerebrum to cerebellum and houses respiratory centres (see VENTILATION).

POPULATION. Group of conspecific individuals, commonly forming a breeding unit, sharing a particular habitat at a given time. See -DEME, INFRASPECIFIC VARIATION.

PORE COMPLEX. See NUCLEUS.

PORIFERA. Sole phylum of Subkingdom PARAZOA. Sponges. Have an apparently loose diploblastic organization lacking both nerve and clear muscle cells, and comprising barely discrete tissues, there being no basement membranes. Basically vase-shaped, the outer 'epithelium' of flattened contractile cells (*pinacocytes*) with amoeboid locomotion; inner layer of cells consists of CHOANOCYTES. Internal skeleton may be of either calcareous (Class Calcarea) or siliceous (Class Demospongia) spicules, and frequently contains fibres of the protein spongin. Pinacocytes may develop into *porocytes* whose intracellular lumens form the surface pores (ostia) through which water enters propelled by flagella of the food-collecting choanocytes, and after entering the central cavity (spongocoel) leaves through one or more oscula at the top. Complexity of internal chambering increases from *ascon*, via *syconoid* and *sicon*, to *leucon* levels of organization. A gelatinous MESOGLOEA between the two main layers contains amoeboid cells (see AMOEBOCYTES) whose varied functions include spicule secretion, gamete production and transport of sperm or food. Locomotion confined to flagellated larvae. See MULTICELLULARITY.

PORPHYRIN. Prosthetic group of several conjugated protein pigments; nitrogen atoms of tetrapyrrole nucleus often coordinated to metal ions (magnesium in chlorophyll, iron in haemoglobin and cytochromes).

PORTAL VEIN. Blood vessel connecting two capillary beds, as in *pituitary portal system* and *hepatic portal system*.

POSITION EFFECT. Occurrence of phenotypic change resulting not from gene mutation as such but from a change in position of a piece of genetic material as occurs in inversion, translocation or crossing-over. Some effects of ONCOGENES and TRANSPOSONS are position effects.

POSITIONAL INFORMATION. In embryological theory, indicating those signals which enable cells in a developing metazoan to respond as though they appreciated their spatial positions in the embryo. Would normally involve specification of information as to where each cell lay on both antero-posterior and dorso-ventral axes. Such information would give cells their *positional values* and ultimately determine their fates. *Positional signals* are graded and can produce multiple outcomes, unlike *inductive signals*, which produce all-or-none responses and result in a single cell state. It is likely cells retain their positional values even when differentiated, making even cells of the same type (e.g. osteocytes) non-equivalent if they lie in different body regions. A cell might pick up and 'remember' different cues at different times prior to differentiation, such as chemical gradients (see MORPHOGEN) or those arising from key cell contacts. A cell's position on the various axes of the body might be given in sequence using a form of digital framework, evidence for which is emerging from, among others, studies of the genetics of *Drosophila* development. Compare PREPATTERN. See AXIS, BICOID GENE, GAP GENES, PAIR-RULE GENES, PATTERN, SEGMENT-POLARITY GENES.

POSTERIOR. (Bot.) Of lateral flowers, part nearest main axis. See FLORAL DIAGRAM. (Zool.) Situated away from head region of an animal, or away from region foremost in locomotion (e.g. dorsal surface in bipedal animals such as humans).

PPLO. See MYCOPLASMAS.

P-PROTEIN (PHLOEM-PROTEIN). Proteinaceous substance found in cells of flowering plant phloem, particularly in sieve-tube members, where it may block or hinder TRANSLOCATION through sieve pores.

PRECAPILLARY SPHINCTER. Sphincter muscle around metarteriole where it gives rise to capillary bed. See CAPILLARY.

PRECIPITIN. Antibody combining with, and causing precipitation of, a soluble antigen. Such precipitations form the basis of *in vitro* procedures for separating and identifying antigens, often on gels (*precipitin reaction*).

PRECOCIAL. Those young of mammals and birds which hatch or are born in a partially independent condition. Compare ALTRICIAL, NIDICOLOUS, NIDIFUGOUS.

PREFORMATION. Doctrine, generally accepted in Europe in 17th and 18th centuries, that all parts of the adult are already perfectly formed at the beginning of development. 'Ovists' believed these were present only in the ovum; 'spermatists' believed they were only in the sperm. Compare EPIGENESIS.

PREMAXILLA. Dermal bone forming front part of upper jaw in most vertebrates, bearing teeth (incisors in mammals). Forms most of upper beak in birds.

PREMOLAR. One of the crushing cheek teeth of mammals, anterior to the molars and posterior to the canines (or incisors). Unlike molars, has a predecessor in the milk teeth; usually has more than one root and a pattern of ridges on biting surface. See DENTAL FORMULA.

PREPATTERN. A theoretical model of animal DEVELOPMENT in which a prior spatial deployment of molecular gene regulators is 'read' by the cells of an embryo (rather than decoded, as with gradient models), the patterns of deployment dictating the gene activities and thereby the paths of differentiation of the cells. In the simplest form of the model, the 'plan' of the embryo would be laid out at the time of fertilization. More reductionist than the POSITIONAL INFORMA-TION model, the chief difficulties of the model arise in explaining regulative development (see REGULATION) and increases in spatial complexities of embryos. It also fails to account satisfactorily for the origin of prepattern itself. See EPIGENESIS.

PRESUMPTIVE. Term indicating that a particular cell or cell group in the early animal embryo will normally differentiate in a particular way. Thus, 'presumptive epidermis' indicates that a group of cells so described will normally become epidermal cells, even prior to their becoming DETERMINED.

PRIAPULIDA. Phylum of burrowing marine worms of dubious affinity. Possibly pseudocoelomate. A distended proboscis, everted with great force, is used to anchor the worms while the body is drawn up over it.

PRIBNOW BOX. Nucleotide sequence in prokaryote PROMOTER regions, located about six bases upstream of a transcribed region. Contains the sequence TATAATG. Also called a (-10) region, because of the invariant T residue at base 10 upstream from the start of the transcribed region. However, many promoters whose expression is controlled by SIGMA FACTORS other than the normal vegetative one have different consensus sequences in different positions upstream of the actual transcription start point. See TATA BOX.

PRIMARY ENDOSPERM NUCLEUS. Nucleus resulting from fusion of one generative nucleus with the two polar nuclei in a flowering plant embryo sac. Usually triploid, its mitotic products produce the en-dosperm. See DOUBLE FERTILIZATION.

PRIMARY IMMUNE RESPONSE. See B-CELL.

PRIMARY PIT FIELD. See PITS.

PRIMARY PRODUCTION. The total organic material synthesized in a given time by autotrophs of an ecosystem. See PRODUCTIVITY.

PRIMARY STRUCTURE. See PROTEIN.

PRIMATES. Members of the mammalian Order Primates. Placentals, including today lemurs, lorises and tarsiers (Prosimii), and the ANTHROPOIDEA. Arose from arboreal insectivore-like mammals, probably similar to present-day tree shrews, in the late Palaeocene. Clavicle retained (lost in many mammalian orders), with shoulder joint permitting freedom of movement in all directions, and an elbow joint allowing rotation of forearm; five digits on all limbs, with generally enhanced mobility and usually opposability of thumb (pollex) and big toe (hallux); claws modified into flattened nails; reduced snout; binocular vision, great alertness, acuity and colour vision; brain enlarged relative to body, with expansion of cerebral cortex; two mammary glands only, and young usually born singly. Most of these features are adaptations to arboreal living (most primates are tree-dwellers). There have been independent moves towards terrestrialization, notably in the HOMINOID line. See AUSTRALOPITHECINE, HOMINID, HOMO, PONGID.

PRIMITIVE. Referring to a characteristic present in an early ancestral plant or animal and which may be present, relatively unmodified, in living forms. A term indicating that a structure, taxon, etc. is regarded as similar in appearance to a stem line. Such a line may have given rise through adaptive radiation to diverse and often more specialized modifications of form. Has temporal rather than ecological implications.

PRIMITIVE STREAK. Longitudinal thickening in disc-like early mammalian or bird embryo during gastrulation. Produced by accumulation of mesoderm and some ectoderm during movement of cells from surface to interior.

PRIMORDIAL MERISTEM. See PROMERISTEM.

PRIMORDIUM. (Bot.) An immature cell or group of cells that will eventually give rise to a specialized structure. Thus, a leaf primordium will become a leaf.

PRISERE. Primary sere. Complete natural succession of plants, from bare habitat to climax. Compare PLAGIOSERE.

PROBAND. Propositus. Index case. Individual affected by a genetic disorder, whose presentation of it enables a pedigree for the disease to be drawn up.

PROBASIDIUM. Cell in which karyogamy occurs in BASIDIO-MYCOTINA. See METABASIDIUM.

PROBE DNA. See DNA PROBE.

PROBOSCIDEA. Elephants. Placental mammal order, probably derived during early Tertiary from the CONDYLARTHRA. Today characterized by trunk (formed from nose and upper lip), large size (largest terrestrial mammals), massive legs, greatly lengthened incisors (tusks) and huge grinding molars, only two pairs of which are in use at one time. Related to hyraxes.

PROBOSCIS. Suctorial feeding apparatus of some dipteran flies and most Lepidoptera. In the former most of the mouthparts contribute to its formation, and the labella contains food channels (pseudotracheae). In the latter it is formed largely from maxillae (mandibles being absent).

PROCAMBIUM. Tissue of narrow elongated cells in vascular plants, grouped into strands, differentiating just behind growing points of stems and roots and giving rise to vascular tissue. See APICAL MERISTEM.

PROCARYOTE. See PROKARYOTE.

PROCHLOROPHYTA (PROCHLOROBACTERIA, CHLOROXYBACTERIA). Large (9-30 μm diameter) prokaryotic cells resembling chloroplasts in morphology and in releasing oxygen in photosynthesis. Contain stacked thylakoid membranes and chlorophyll *b* in addition to chlorophyll *a*, thereby resembling chloroplasts of chlorophytes and higher plants rather than blue-green algae (CYANOBACTERIA), from which they also differ in lacking phycobiliproteins. Gram-negative cell walls contain muramic acid and are sensitive to lysozyme. *Prochloron* is found in the cloacal walls of didemnid sea squirts, some of which gain nutritionally from them. Recent work suggests *Prochloron* is not as distinct from cyanobacteria as was once supposed, and may have evolved from them independently of green plant chloroplasts. New genera have been discovered recently. See ENDOSYMBIOSIS.

PROCONSUL. Alternative name for *Dryopithecus africanus*, an early Miocene DRYOPITHECINE ape.

PROCTODAEUM. Intucking of embryonic ectoderm meeting the endoderm of the posterior of the alimentary canal and forming anus or cloacal opening.

PRODUCTIVITY (PRODUCTION). Rate at which solar energy and carbon dioxide are absorbed and utilized in photosynthesis. The *gross primary production* of an ecosystem is the total amount of organic matter produced by its autotrophs, and is usually recorded in kilojoules per hectare per year. *Net primary production* is gross primary production less that used by plants in respiration and represents food potentially available to consumers in the ecosystem. See STANDING CROP.

PROEMBRYO. In seed plants, a group of cells formed by initial divisions of the zygote, which by further development differentiates into the suspensor and embryo proper.

PROGENESIS. Form of HETEROCHRONY in animals, leading to PAEDOMORPHOSIS through precocious sexual maturity and a shortening of the developmental pathway, often resulting in a 'juvenilized' morphology. Believed to have been a source of important new taxa during evolution. Most successful paedomorphs seem to have been small progenetic larvae, e.g. six-legged myriapod larva (ancestral to insects) and the urochordate tadpole (ancestral to vertebrates).

Several gall midges have progenetic larvae; wingless aphids and many parasites are progenetic. Progenesis tends to be favoured by *r*-SELECTION. Compare NEOTENY.

PROGESTERONE. Steroid hormone secreted by CORPUS LUTEUM of mammalian ovary, and made by PLACENTA from maternal and foetal precursors. For timing of secretion, and roles in human female, see MENSTRUAL CYCLE. Broken down by LIVER. See CONTRACEPTIVE PILL.

PROGESTOGEN. Any substance with progesterone-like effects in the female mammal.

PROGLOTTIS (pl. PROGLOTTIDES). Segment-like unit of the tapeworm body which, when mature, leaves the gut of the primary host in the faeces. Normally each develops male reproductive organs first and later loses these as female parts develop. Proglottides are not true segments since they are budded off from the anterior of the worm (from the scolex). This does not amount to asexual reproduction (but see HYDATID CYST). See CESTODA.

PROKARYOTE (PROCARYOTE). Prokaryotes (BACTERIA, blue-green algae (CYANOBACTERIA), PROCHLOROPHYTA, MYCOPLASMAS and ARCHAEBACTERIA) are typically either unicellular or filamentous, and small (up to 3 μm in diameter). Their DNA is not housed within a nuclear envelope (see NUCLEOID), and no prokaryotic cell is descended from such a nucleated cell. Within the plasma membrane, often folded and convoluted within cell interior, lies the rest of the cytoplasm, containing smaller ribosomes than those of EUKARYOTES, along with granular inclusions. They lack the tubulin, actin and histones diagnostic of eukaryotes and so have other methods of CELL DIVISION and CELL LOCOMOTION (see FLAGELLUM). Their GENETIC CODE is remarkably similar to that of eukaryotes. Cell division commonly lags behind chromosome replication, so that prokaryotic cells commonly contain at least two chromosomes, each consisting of DNA and non-histone proteins, often attached for a time to the plasma membrane. Mitochondria and chloroplasts are absent, but they possess structures that function similarly (MESOSOME, CHROMATOPHORE). Some authors recognize two superkingdoms, the Prokaryota and Eukaryota. The Monera would then form the one prokaryotic kingdom; but see MESOKARYOTE.

PROLACTIN (LACTOGENIC HORMONE, LUTEOTROPIC HORMONE, LTH). A protein gonadotrophic hormone secreted by the vertebrate anterior pituitary gland. In mammals it promotes secretion of progesterone by the corpus luteum and is involved in LACTATION. In pigeons stimulates crop milk production and is necessary for maintenance of incubation (which itself stimulates its secretion) by both sexes.

PROLEGS. Stumpy unjointed appendages on ventral surface of abdomen of caterpillar.

PROLIFERATION. Growth by active cell division.

PROMERISTEM. Extreme tip of an APICAL MERISTEM comprising actively dividing, but as yet undifferentiated, cells.

PROMOTER. RNA polymerase binding site in the JACOB-MONOD THEORY of gene expression, also found upstream of yeast and probably other eukaryotic transcribed regions. Some components of the promoter sequence are highly conserved (see PRIBNOW BOX, TATA BOX). See TRANSCRIPTION FACTORS.

PRONATION. Position of fore-limb, or rotation towards it, such that fore-foot (hand) is twisted through 90° relative to the elbow, the radius and ulna being crossed. In this way the fore-foot points forwards (the natural position for walking in many tetrapods). Humans and other primates can untwist the fore-arm (*supination*).

PRONEPHROS. See KIDNEY.

PRO-OESTROUS. See OESTROUS CYCLE and MENSTRUAL CYCLE.

PROPAGULE. (Bot.) A dispersive structure, such as a seed, fruit, gemma or spore, released from the parent organism.

PROPHAGE. Non-infectious phage DNA, integrated into a bacterial chromosome and multiplying with the dividing bacterium but not causing cell lysis except after excision from the chromosome, which can be *induced* by certain treatments of the cell. See BACTERIO-PHAGE.

PROPHASE. First stage of MITOSIS and MEIOSIS.

PROPLASTID. Minute, self-reproducing, immature PLASTID occurring in cells of meristematic tissues and comprising a double membrane enclosing a granular stroma.

PROPOSITUS. See PROBAND.

PROPRIOCEPTORS. (1) Receptors involved in detection of position and movement including MUSCLE SPINDLES, organs of balance in the vertebrate VESTIBULAR APPARATUS, and campaniform sensillae occurring in all parts of the cuticle of the insect body subject to stress (especially in joints, at wing and haltere bases). Do not usually show sensory ADAPTATION. (2) In a wide sense, any receptor detecting changes within the body other than those caused by substances taken into the gut and respiratory tract; e.g. deep pain receptors and BARORECEPTORS; STATOCYSTS in invertebrates.

PROP ROOT. Adventitious supportive root, arising from the stem above soil level.

PROSENCEPHALON. See FOREBRAIN.

PROSENCHYMA. See PLECTENCHYMA.

PROSIMIAN. Member of the suborder Prosimii of the PRIMATES. Primitive nocturnal primates, mostly arboreal. Relatively small-brained compared with Anthropoidea.

PROSOMA. One of the tagmata of CHELICERATA, a combined head and thorax. Comprises one pre-oral segment and five post-oral segments, the latter bearing walking limbs. See OPISTHOSOMA.

PROSTAGLANDINS (PGs). Family of 20-carbon fatty acid derivatives continuously synthesized (in mammals at least) by most nucleated cells from precursor phospholipids of the plasma membrane. Rapidly degraded by enzymes on release; but if appropriately triggered a cell will increase its output of PG, raising local levels and influencing both itself and its neighbours. Like hormones, they may be released into the blood but, unlike them, are only effective over short distances. Promote contraction of smooth muscle, platelet aggregation, inflammation and secretion. Some bind membrane receptors and exert their effects by altering intracellular cyclic AMP levels. Of great clinical potential in alteration of blood pressure, in bronchodilation and constriction, inducing labour, reducing gastric secretion, etc. Also called local or tissue hormones.

PROSTATE GLAND. Gland of male mammalian reproductive system lying below the urinary bladder and around part of the urethra. Secretes alkaline fluid comprising up to a third of semen volume. Its effect on vaginal pH assists sperm motility. Rich source of PROSTA-GLANDINS in humans.

PROSTHETIC GROUP. Non-protein group which when firmly attached to a protein results in a functional complex (a conjugated protein). Many respiratory pigments (e.g. HAEMOGLOBIN) are conjugated proteins, while many enzymes require prosthetic groups (some of them metal ions). The carbohydrates and lipids in glycoproteins and lipoproteins are prosthetic groups of their proteins. DNA is the prosthetic group of the histones in chromatin. Compare COENZYME.

PROTAMINES. Basic proteins of low molecular weight (about 5000), lacking prosthetic groups, containing many arginine and lysine residues, and found in association with DNA. See CHROMOSOME.

PROTANDRY. (1) In flowers, the situation in which anthers mature before carpels. See DICHOGAMY. (2) In animals, the condition in sequential hermaphrodites in which sperm are produced prior to eggs (e.g. in many tapeworms).

PROTEASE. Enzyme that digests protein by hydrolysis of peptide bonds. See PROTEOLYSIS.

PROTEIN. Polymer of very large or enormous molecular mass, composed of one or more polypeptide chains, and whose monomers are AMINO ACIDS, joined together (in condensation reactions) by

PEPTIDE BONDS. In addition, some have covalent 'sulphur bonds' formed by oxidization between two cysteine radicals in the polypeptide. The potential variety of polypeptides is infinite: there are 20 common amino acids, so joining any two together would give 400 (= 20 × 20) possible *dipeptides*. Biological polypeptides are often several hundred amino acids long, so few of the possible polypeptides actually occur in organisms (see PROTEIN SYNTHESIS). Several proteins (e.g. ACTIN, TUBULIN) form filaments of polypeptide molecules. The molecular mass which results (as in the protein coat of tobacco mosaic virus) may exceed 40 million daltons.

Each polypeptide has a *primary structure*: the number and sequence of its amino acids. There is an amino-terminal ($-NH_2$) end and a carboxy-terminal (–COOH) end to the molecule. During its production on a ribosome a polypeptide commonly assumes a corkscrew-like ALPHA-HELIX (α-helix) as its *secondary structure*, due to hydrogen bonding between the hydrogen atom attached to the nitrogen in one amino acid radical and the oxygen attached to a carbon atom three radicals along the chain. Other *intra*molecular hydrogen bonds may contribute to the secondary structure, as with antiparallel folding of the molecule back along itself in the same plane, hydrogen atoms of one side being linked to oxygen atoms of the side parallel to it. Such β-pleated sheets occur in many globular (spherical) protein molecules and commonly link several polypeptides of the same type (as in fibroin of silk, and some other fibrous proteins). Regions of α-helix and β-pleated sheets, along with less clearly organized (i.e. more random) stretches of the amino acid chain, may all contribute to the three-dimensional configuration (*tertiary structure*) of a polypeptide, the α-helical portions often being thrown into folds by electrostatic attraction and repulsion resulting from the charge distribution on the amino acid R-groups (see AMINO ACID). Globular proteins in particular bristle with charge. The tertiary structures of globular proteins in solution depend upon pH (since charges on R-groups depend upon pH) and upon sulphur (disulphide) bonds. ENZYMES are, typically, globular proteins. Their functions depend upon their shapes and are affected by changes in pH and temperature (see DENATURATION, ISOELECTRIC POINT). Many important proteins (*conjugated proteins*) are formed by covalent union of a non-protein radical to the protein molecule, forming a hybrid molecule (e.g. HAEMO-GLOBIN, CYTOCHROMES). The *quaternary structure* of a protein is the shape adopted when two or more polypeptide chains associate (non-covalently) to produce the functional protein molecule. Both the non-enzyme haemoglobin and the enzyme lactate dehydrogenase (see ISOENZYME) are proteins formed by quaternary association between four polypeptides of two different kinds. Indeed, functional enzyme molecules often consist of two or more different polypeptide subunits.

Fibrous proteins often have major structural roles (e.g. in CYTO-

SKELETONS, CONNECTIVE TISSUE, STRIATED MUSCLE, CHROMOSOMES). This is more obviously true in animals than in plants, although plant CELL WALLS contain structural glycoproteins. All cells depend upon catalytic activities of enzymes, and have TRANSPORT PROTEINS in their membranes. Generally, RESPIRATORY PIGMENTS are conjugated proteins. Proteins also form major components of ANTIBODIES and other GLYCOPROTEINS, and of LIPOPROTEINS. They commonly act as BUFFERS (e.g. in blood plasma), and being colloids reduce the WATER POTENTIALS of cells and intercellular fluids. Although proteins are insoluble in lipid solvents, globular proteins (but not fibrous) dissolve in water and dilute salt solutions. Proteins are digested hydrolytically by PROTEOLYTIC ENZYMES and mineral acids, and are usually separable by ELECTROPHORESIS. Only autotrophic organisms are capable of making the amino acid components of proteins from inorganic precursors.

PROTEIN KINASE. An enzyme transferring a phosphate group from ATP to an intracellular protein, often also an enzyme, thereby increasing or decreasing its activity. Acts in opposition to protein phosphatases. Of enormous importance in regulation of metabolism. Its activity is often a function of local concentrations of cyclic AMP and calcium ions. See KINASE, REGULATORY ENZYME.

PROTEIN SYNTHESIS. Proteins are manufactured by cells on RIBOSOMES, which involves joining together in the correct sequence possibly hundreds of amino acid molecules. Cells produce particular proteins either all the time (constitutively) or as and when required (see GENE EXPRESSION). Sequence of amino acids forming primary structure of a protein is encoded in the sequence of nucleotides of genetic material of the cell, usually DNA (in some viruses it is RNA). When a piece of DNA becomes involved in protein synthesis, an RNA POLYMERASE first breaks the hydrogen bonds holding the two DNA strands together, then uses one of the strands as a template on which to incorporate the nucleotides making a complementary RNA molecule, in an order dictated by BASE PAIRING rules. Once produced (the process is called *transcription*), this RNA molecule is commonly modified (see RNA PROCESSING) to form a shortened *messenger* RNA (mRNA) molecule, always with the nucleotide triplet *start codon* AUG at one end (coding for the amino acid methionine). This mRNA molecule passes into the cytoplasm (via the nuclear pore apparatus in nucleated cells) and triggers a ribosome to assemble upon it at the AUG codon nearest the 5′-end of the mRNA molecule (see START CODON).

Free amino acids are not assembled directly into protein but are first loosely bound to an *activating enzyme*, so-called because it (a) hydrolyses an ATP molecule, providing energy for (b) attachment of the activated amino acid to one of small number of specific transfer RNA (tRNA) molecules, of which there is a pool in the cell. The result of this ATP-dependent catalysis is a pool of amino acyl-tRNAs from which protein synthesis proceeds.

Each ribosome draws amino acyl-tRNAs from its surroundings in an order determined by the nucleotide sequence of the mRNA molecule to which it attaches. AUG is the CODON for the amino acid methionine, and the newly-assembling ribosome already has tRNA-methionine bound to it (see P-SITE). This tRNA base-pairs with the AUG codon of the mRNA by hydrogen bonds with a triplet of nucleotides exposed at one end of the molecule: its anticodon triplet. Then another tRNA molecule gets bound by the ribosome (see A-SITE), but only if its anticodon base-pairs with the codon next to AUG in the mRNA. It brings with it its own attached amino acid. This codon–anticodon hydrogen bonding provides the essential working principle of the GENETIC CODE.

The ribosome will continue to draw in appropriate tRNA molecules, joining each of their amino acids together to form a growing polypeptide chain. This chain elongation requires hydrolysis of two GTP molecules per hydrogen bond formed. Each of the tRNA molecules is released from the ribosome once it has donated its amino acid load.

As it draws in each amino acyl-tRNA in turn, the ribosome moves one codon further along the mRNA molecule, towards the 3′-end of the molecule. At an appropriate STOP CODON on the mRNA, the ribosome stops protein synthesis and releases the completed polypeptide. The N-terminal methionine gets cleaved off.

This ribosomal phase of protein synthesis is called *translation*. Each mRNA molecule is simultaneously the site of attachment of many ribosomes (see POLYSOME), and when each ribosome reaches the stop codon it releases another identical polypeptide molecule. Both chloroplasts and mitochondria make some of their own proteins but, as with prokaryotic systems in general, their start codon (AUG) binds N-formylmethionine rather than methionine. Protein synthesis is energy-dependent, each amino acid incorporated into the polypeptide requiring hydrolysis of three HIGH-ENERGY PHOSPHATE bonds. Several ANTIBIOTICS stop protein synthesis, during either transcription or translation.

PROTEOGLYCAN. Class of acidic GLYCOPROTEINS found in varying amounts in extracellular matrices of animal tissues, notably connective tissues. Contain more carbohydrate than protein. One forms, with collagen, the rubbery material in cartilage preventing bone ends from grating together. As with MUCINS, a carbohydrate-free area serves for cross-links between the protein chains to produce aggregation. The fibrous polysaccharide HYALURONIC ACID serves as a chain along which many of these proteoglycan molecules align themselves. See CHITIN.

PROTEOLYSIS. Hydrolysis of a protein; achieved by mineral acids, but much more widely and importantly by proteases (proteolytic enzymes).

PROTEOLYTIC ENZYME. Any enzyme taking part in breakdown of proteins, ultimately to amino acids. Occur frequently in plant and animal cells, and extracellularly in plant seeds and animal digestive juices. Include PEPSIN, TRYPSIN, CHYMOTRYPSIN, PEPTIDASES and intracellular CATHEPSINS.

PROTEROZOIC. Geological division of Earth history between end of Archaean (about 2600 Myr BP) and onset of Phanerozoic (about 600 Myr BP). In it atmospheric oxygen levels rose to about one tenth of present levels. Limestones became abundant for the first time, often containing STROMATOLITES.

PROTHALLUS. Independent gametophyte stage of ferns and related plants. Small, green, parenchymatous thallus bearing antheridia and archegonia, showing little differentiation. Usually prostrate on the soil surface, attached by rhizoids. May be subterranean and mycotrophic.

PROTHORACIC GLAND. See ECDYSONE.

PROTHROMBIN. See BLOOD CLOTTING.

PROTISTA. Term frequently used to denote a kingdom. Originally included all unicellular organisms, but since enlarged to include all eukaryotic organisms of relatively simple organization, mainly algae, fungi, slime-fungi and protozoans. The term *Protoctista* is sometimes employed as a kingdom to include all eukaryotic microorganisms excluding plants, fungi and animals. See KINGDOM.

PROTOCHORDATA. Subgroup of the CHORDATA comprising the HEMICHORDATA, UROCHORDATA and CEPHALOCHORDATA: all three invertebrate chordate subphyla. Some would include pogonophorans and/or graptolites as well.

PROTOGYNY. (1) The condition in flowers (termed *protogynous*) whose carpels mature before their anthers, as in plantains. See DICHOGAMY. (2) The condition in sequentially hermaphrodite animals in which first eggs are produced, then sperm. See PROTANDRY.

PROTONEMA. Branched, multicellular, filamentous or (less commonly) thalloid structure, produced on germination of a bryophyte spore, from which new plants develop as buds.

PROTONEPHRIDIUM. See NEPHRIDIUM.

PROTOPLASM. Cell contents within and including the plasma membrane but usually taken to exclude large vacuoles, masses of secretion or ingested material. In most eukaryotic cells it includes, besides the CYTOPLASM, one or more nuclei. Prokaryotic cells lack nuclei. Cell walls, if present, are non-protoplasmic. Each protoplasmic unit constitutes a *protoplast*.

PROTOPLAST. (Bot.) Actively metabolizing part of a cell (its PROTO-PLASM), as distinct from cell wall. Equivalent to 'cell' in zoology. See CELL FUSION.

PROTOPODITE. See BIRAMOUS APPENDAGE.

PROTOSTELE. Simplest and most primitive type of STELE, comprising a central core of xylem surrounded by a cylinder of phloem. Present in stems of some ferns and club mosses and almost universal in roots. In a *haplostele*, xylem forms a central rod; in an *actinostele*, xylem is ribbed and appears star-shaped in transverse section; in a *plectostele*, xylem is in several parallel, longitudinal strips embedded in the phloem.

PROTOSTOMIA. Those coelomate metazoans (sometimes termed an infragrade) in which the blastopore develops into the mouth of the adult, cleavage tends to be determinate, and the coelom tends to form by SCHIZOCOELY. Includes annelids, arthropods, molluscs and, usually, those phyla with LOPHOPHORES. Compare DEUTER-OSTOMIA.

PROTOTHERIA. Mammalian subclass, of which only the monotremes (six species) survive. Includes extinct orders MULTITUBERCULATA, Triconodonta and Docodonta, and the extant Order Monotremata. The latter comprise the duckbill, or platypus (*Ornithorhynchus*), of Australia and Tasmania, and the spiny anteaters (*Tachyglossus*, *Zaglossus*) of Australia, Tasmania and New Guinea. Fossil forms have been found only in Australia, and have not so far pre-dated the Pleistocene. All species have hair and mammary glands and are homoiothermic; but all have the reptilian features of egg-laying (ovipary), retention of separate coracoid and interclavicle bones in the PECTORAL GIRDLE, and epipubic bones attached to the PELVIC GIRDLE. Brain size in relation to body size is lower than in placental mammals, resembling marsupials in this respect. A CLOACA is present. Echidnas incubate in a pouch (marsupium); duckbills incubate in a nest.

PROTOXYLEM. The first elements to be differentiated from procambium; extensible. Described as *endarch* when internal to the later-formed metaxylem (as in roots), and as *mesarch* when surrounded by metaxylem (as in fern stems).

PROTOZOA. Phylum, or subkingdom, of the ANIMALIA, comprising unicellular and colonial animals of varied form. Generally subdivided into four classes: SARCODINA (amoebae, radiolarians, foraminifera), MASTIGOPHORA (flagellates), CILIATA (ciliates and suctorians) and SPOROZOA (e.g. coccidians, gregarines); but some of these at least represent grades rather than clades. The first two are sometimes united as the Sarcomastigophora. Reproduction commonly by binary or multiple FISSION, but sometimes by CONJUGATION. Ubiquitous, inhabiting aquatic and damp terrestrial habitats. Several of them are serious pests of humans and their domestic animals (e.g. see MALARIA).

PROTURA. Order of the APTERYGOTA. Minute (0.5–2.5 mm long) whitish insects lacking antennae and eyes. Inhabitants of moist soils, leaf litter, etc. Stylet-like mandibles for piercing. Metamorphosis consists of addition of three abdominal segments and development of genitalia. No evidence that the adult moults.

PROVENTRICULUS. Anterior part of the bird stomach, where digestive enzymes are secreted, posterior part being the GIZZARD. Used synonymously with gizzard in crustaceans and insects.

PROVIRUS. Viral genomes which are integrated into the host cell chromosome and most of the time remain there unexpressed (*latent*). Bacteria harbouring bacteriophages which do this are termed *lysogenic*. See BACTERIOPHAGE.

PROXIMAL. Situated relatively near to a point of attachment or origin. Compare DISTAL.

PROXIMATE FACTOR. Explanations in biology are mostly either proximate or ultimate. Former are characteristically mechanistic and indicate how some outcome or change is intelligible in terms of antecedent causes. The latter are teleological, rendering phenomena intelligible in terms of probabilities of future states of affairs. Thus, change in photoperiod may be cited as the proximate factor causing change of winter coat colour in stoats, etc.; but better camouflage may be cited as an ULTIMATE FACTOR responsible for the change, bringing an increase in individual fitness.

PRYMNESIOPHYTA. Division of ALGAE; mainly marine. Uninucleate flagellates possessing a HAPTONEMA between two smooth flagella. Cells, commonly covered in scales, contain chlorophylls a, c_1, and c_2, with fucoxanthin the major carotenoid. Outer of the two chloroplast membranes is continuous with outer nuclear envelope. Storage product is chrysolaminarin. An important part of marine nanoplankton.

PSEUDOALLELES. Two mutations in the same cistron which give rise to different phenotypes when in the *cis* and *trans* conditions respectively. In the CIS-TRANS TEST they fail to complement one another.

PSEUDOCOELOM (PSEUDOCOEL). Fluid-filled cavity between body wall and gut with, however, an entirely different origin from true COELOM. The pseudocoelom is a persistent blastocoel, lacking a definite mesoderm lining. In some cases, as in nematode worms, it may be filled with vacuolated mesenchyme cells. Pseudocoelomate invertebrates include ASCHELMINTHES, ENDOPROCTA and possibly priapulid worms. Their interrelationships are unclear, but it is likely that all pseudocoelomate animals have had progenetic origins (see PROGENESIS), and that they are polyphyletic and represent a grade rather than a clade.

PSEUDOGAMY. Phenomenon where fertilization is required for development of sexually-produced offspring which derive all their genes from their maternal parent. In the grass *Poa*, and in *Potentilla*, apomictic plants produce perfectly functional pollen and fertilization precedes seed development, but only fertilization of the endosperm nucleus occurs, the egg cell nucleus remaining unfertilized. See GYNO-GENESIS, PARTHENOGENESIS.

PSEUDOGENE. A DNA sequence which, despite being largely homologous to a transcribed sequence elsewhere in the genome, is not transcribed. Some probably arise through gene duplication (*processed* pseudogenes); others, more numerous, must have originated by reverse transcription of mRNA and insertion into the gene, since they have lost their introns and are dispersed in location. The gene cluster for human haemoglobin contains two pseudogenes (designated by the symbol ψ in front of the gene symbol).

PSEUDOPARENCHYMA. See PLECTENCHYMA.

PSEUDOPODIUM. Temporary protrusion of some cells (in some sarcodine protozoans, e.g. *Amoeba*; MACROPHAGES) involved in amoeboid forms of CELL LOCOMOTION and food capture. See PHAGOCYTE.

PSEUDOPREGNANCY. State resembling pregnancy in female mammal, but in absence of embryos. Due to hormone secretion of CORPUS LUTEUM, and occurs in species where copulation induces ovulation (e.g. rabbit, mouse), but when such copulation is sterile, or when normal oestrous cycle includes a pronounced luteal phase (e.g. bitch).

PSEUDOSCORPIONES. Arachnid order containing the pseudo-scorpions: minute scorpion-like animals lacking tail and sting and common in soil where they are predatory, using pincer-like pedipalps.

PSILOPHYTA. Whisk ferns. Two living genera (*Psilotum* and *Tmesipteris*). Sporophytes very simple. *Psilotum* lacking both roots and leaves but with an underground rhizome system with many rhizoids and a dichotomously branching aerial portion with small, scale-like outgrowths, *Tmesipteris* being epiphytic, with larger leaf-like appendages. Vascular system prostelic to siphonostelic. Homosporous, spores producing bisexual gametophytes which absorb nutrients from the sporophyte. Both are mycotrophic.

P-SITE. Binding site of ribosome for the tRNA corresponding to the START CODON of mRNA (usually AUG, but sometimes GUG) and for the tRNA linked to the growing end of the polypeptide chain (hence P for peptidyl) during PROTEIN SYNTHESIS. The P-site receives this tRNA from the A-SITE. See RIBOSOME.

PSOCOPTERA. Booklice (psocids). Order of exopterygote insects. Small, some wingless. Biting mouth parts. Feed on fragments of animal and vegetable matter and paste of book-bindings.

PSYCHROPHILIC. (Of a microorganism), with optimum temperature for growth below 20°C.

PTERIDOPHYTA. In older plant classifications, a division containing spore-bearing (as opposed to seed-bearing) tracheophytes: Psilophyta, Lycophyta, Sphenophyta and Pterophyta.

PTERODACTYLA. See PTEROSAURIA.

PTEROPHYTA. The ferns. Relatively abundant as fossils since the Carboniferous period; not known from the Devonian period. About 12 000 extant species, roughly two thirds in tropical regions, the other third in temperate regions. Display great diversity of form and habit, but are almost entirely terrestrial. The fern plant, the sporophyte, is the dominant generation; the gametophyte is the prothallus. Most temperate ferns comprise an underground siphonostelic rhizome which produces new leaves each year; roots are adventitious; leaves (fronds) are the megaphylls. Unique among seedless vascular plants in possessing megaphylls. All but a few ferns are homosporous, with sporangia variously placed and commonly borne in clusters called sori. Fern heterospory is restricted to two specialized groups of aquatic ferns. Spores of most homosporous ferns germinate to give rise to free-living gametophytes which develop into prothalli; these are monoecious, antheridia and archegonia developing on the ventral surface. Sperm are motile and require water to swim to the eggs. The sporophyte resulting from fertilization is initially dependent upon the gametophyte, but growth is rapid and it soon becomes independent. See LIFE CYCLE.

PTEROSAURIA. Pterodactyls. Extinct order of ARCHOSAURS, originating in late Triassic and disappearing at end of Cretaceous. Winged reptiles, fourth finger of the fore-limb greatly elongated and supporting a membrane. Hind-limbs feeble. Long tail tipped by a membrane and acting as a rudder. Wing spread up to 16 metres.

PTERYGOTA (METABOLA). Insect subclass including all but the APTERYGOTA. Some are secondarily wingless (e.g. fleas). Includes the ENDOPTERYGOTA and EXOPTERYGOTA.

PTYALIN. An AMYLASE present in saliva of some mammals, including humans.

PUBIC SYMPHYSIS. See PELVIC GIRDLE.

PUBIS. See PELVIC GIRDLE.

PUFF (BALBIANI RING). Swelling of giant chromosome (e.g. from dipteran salivary gland cell) normally regarded as representing regions being actively transcribed and consisting of many strands of decondensed DNA and its associated messenger RNA. Some recent work indicates some transcription of these chromosomes in regions lacking puffs, and some puff presence where transcription is lacking. See POLYTENY, ECDYSONE.

PULMONARY. (Adj.) Relating to the LUNG. Vertebrate *pulmonary arteries* (where present, derived from sixth AORTIC ARCH) carry deoxygenated blood to the lungs from the right ventricle of the heart in crocodiles, birds and mammals (or from the single ventricle of lungfish and amphibians) to the lung capillaries; in tetrapods, *pulmonary veins* return oxygenated blood to the left atrium.

PULMONATA. Order of GASTROPODA. Lung develops from mantle cavity; e.g. snails, slugs.

PULP CAVITY. Internal cavity of vertebrate tooth or denticle, opening by a channel to the tissues in which it is embedded. Contains *tooth pulp* of connective tissue, nerves and blood vessels, with odontoblasts lining dentine wall of cavity. See TOOTH.

PULSE. (Zool.) Intermittent wave of raised pressure passing rapidly (faster than rate of blood flow) from heart outwards along all arteries each time the ventricle discharges into the aorta. The increased pressure dilates the arteries, and this can be felt. Each pulse experiences resistance from the elastic walls of the arteries. See ARTERY. (Bot.) Seed of leguminous plant (Fabaceae), rich in proteins; e.g. soya bean.

PUNCTUATED EQUILIBRIUM. See EVOLUTION, CLADISTICS.

PUPA (CHRYSALIS). Stage between larva and adult in life cycle of endopterygote insects, during which rearrangement of body parts (METAMORPHOSIS) occurs, involving development of IMAGINAL DISCS. In some pupae (*exarate*) the appendages are free from the body; in many others (*obtect*) they are glued to the body by a larval secretion. Pupae of most culicine mosquitoes are active.

PUPIL. Opening in IRIS of vertebrate and cephalopod EYE, permitting entry of light.

PURE LINE. Succession of generations of organisms consistently homozygous for one or more characters under consideration. Initiated by crossing two appropriately homozygous individuals or, in plants, by selfing one such individual. Such organisms *breed true* for the characters, barring mutation. See INBREEDING.

PURINE. One of several nitrogenous bases occurring in NUCLEIC ACIDS, nucleotides and their derivatives. By far the commonest are ADENINE and GUANINE; rarer *minor purines* including methylation products of these. See PYRIMIDINE.

PURKINJE FIBRES. Modified CARDIAC MUSCLE fibres forming the bundle of His conducting impulses from PACEMAKER, and the fine network of fibres piercing the myocardium of the ventricles. See also HEART CYCLE. Not to be confused with *Purkinje cells*, which are large neurones of the cerebellar cortex.

PUROMYCINS. Antibiotics interrupting the translation phase of protein synthesis in both prokaryotes and eukaryotes by their addition to growing polypeptide chain, causing its premature release from the ribosome. Compare ACTINOMYCIN D. See ANTIBIOTIC.

PUTREFACTION. Type of largely anaerobic bacterial decomposition of proteinaceous substrates, with formation of foul-smelling amines rather than ammonia.

PYCNIDIUM. Closed sporocarp containing a cavity bearing conidia found in some members of the Ascomycotina and Deuteromycotina.

PYLORUS. Junction between vertebrate stomach and duodenum. Has a sphincter muscle within a fold of mucous membrane closing off the junction while food is digested in the stomach.

PYRAMID OF BIOMASS. Diagram, pyramidal in form, representing the dry masses in each TROPHIC LEVEL of a community or food chain. The standing crops of populations are normally summated within each trophic level. Not all biomass has the same energy content, however. See ENERGY FLOW, PYRAMID OF NUMBERS, PYRAMID OF ENERGY.

PYRAMID OF ENERGY. Diagram representing the energy contents within different TROPHIC LEVELS of a community or food chain. Generally pyramidal in form, it is difficult to represent the DECOMPOSERS, and does not easily indicate the unavailability (to the next trophic level) of storage if standing crops are used. See PYRAMID OF BIOMASS, PYRAMID OF NUMBERS.

PYRAMID OF NUMBERS. Diagram, pyramidal (sometimes inversely) in form, representing the numbers of organisms in each TROPHIC LEVEL of a community or food chain. All organisms are equated as identical units, a massive tree being equivalent to a flagellate cell. See PYRAMID OF BIOMASS.

PYRENOID. Proteinaceous region within chloroplasts of many algae, taking early products of photosynthesis and converting them to storage compounds. Occur in every eukaryotic algal group, and are considered a primitive characteristic. The pyrenoid is denser than the surrounding stroma and may or may not be traversed by thylakoids. In green algae (Chlorophyta), they are associated with starch synthesis and surrounded by starch deposits.

PYRIDOXINE. One of the water-soluble vitamins in B6; precursor of *pyridoxal phosphate*, a PROSTHETIC GROUP associated with transaminase enzymes of mitochondria and/or the cytosol. Required by, among others, yeasts, bacteria, insects, birds and mammals.

PYRIMIDINE. One of a group of nitrogenous bases occurring in NUCLEIC ACIDS, nucleotides their derivatives. The three commonest are cytosine, thymine and uracil. Methylated forms also occur; they are termed *minor pyrimidines*.

PYRROPHYTA (DINOPHYTA). Division of the ALGAE; important as freshwater and marine plankton, diverse group of biflagellate unicells (*dinoflagellates*), together with some non-motile forms (coccoid, filamentous, palmelloid and amoeboid). Possess chlorophylls *a* and c_1, β-carotene and unique xanthophylls (peridinin, etc.). Photosynthetic members have chloroplast surrounded by one membrane of chloroplast endoplasmic reticulum, not continuous with nuclear envelope. Heterotrophic nutrition not uncommon; parasitic, saprotrophic, symbiotic and holozoic forms all represented. Food reserve starch; cell wall (or theca), when present, composed largely of cellulose, often in the form of plates. In dinoflagellates (Dinophyceae), the nucleus is unusual in containing chromosomes which remain condensed at all stages of the cell cycle (*mesokaryotic*); a typical motile cell comprising two valves separated by a transverse girdle (*cingulum*) in which runs a long transverse flagellum, another running in a longitudinal sulcus perpendicular to the cingulum. Some forms produce dangerous toxins, particularly those lodged in shellfish gills.

PYXIDIUM. Type of CAPSULE.

Q

Q_{10}. TEMPERATURE COEFFICIENT. Increase in rate of a process (expressed as a multiple of initial rate) produced by raising the temperature by 10°C. Rate of most enzymatic reactions approximately doubles for each rise in temperature ($Q_{10} \simeq 2$), but this varies a little from one enzyme to another.

Q_{O_2}. OXYGEN QUOTIENT. Rate of O_2 consumption, often of whole organisms or tissues. Often expressed in $\mu l.mg^{-1}. hr^{-1}$.

QUADRAT. (1) Delineated area of vegetation (standard size is one square metre) chosen at random for study of its composition. Quadrat frames of different dimensions are often 'thrown' on to vegetation to ascertain which frame size provides most information in least time.

QUADRATE. Cartilage-bone of posterior end of vertebrate upper jaw. Develops within the PALATOQUADRATE and attaches to neuro-cranium, in most cases articulating with lower jaw. In mammals becomes the incus (see EAR OSSICLES).

QUALITATIVE INHERITANCE. See VARIATION.

QUANTASOMES. Granules occurring on inner surfaces of THYLA-KOIDS of CHLOROPLASTS; thought to be basic structural units involved in light-dependent phase of photosynthesis rather than artefacts of electron microscope preparations.

QUANTITATIVE INHERITANCE. See VARIATION.

QUATERNARY. The GEOLOGICAL PERIOD comprising both the Pleistocene and Holocene.

QUATERNARY STRUCTURE. See PROTEIN.

QUIESCENT CENTRE. Area at the tip of the root apical meristem where rate of cell division is lower than in surrounding tissue.

R

RACE. Category of INFRASPECIFIC VARIATION.

RACEME. Kind of INFLORESCENCE.

RACHIS. (1) Main axis of an INFLORESCENCE; (2) axis of a pinnately compound leaf to which leaflets are attached.

RADIAL CLEAVAGE (BILATERAL CLEAVAGE). See CLEAVAGE.

RADIAL SECTION. Longitudinal section cut parallel to the radius of a cylindrical body (e.g. a stem or root).

RADIAL SYMMETRY. Capable of bisection in two or more planes to produce halves that are approximately mirror images of each other. Characteristic of bodies of coelenterates and echinoderms; and of many flowers (see ACTINOMORPHIC) and some algae, e.g. the CENTRIC DIATOMS, when viewed in valve view. Compare BILATERAL SYMMETRY.

RADICAL. (Bot.) Arising from the root or crown of the plant.

RADICAL. Root of embryo seed plants.

RADIOACTIVE LABELLING. See LABELLING.

RADIOISOTOPE (RADIOACTIVE ISOTOPE). Unstable isotope of element, decaying spontaneously and emitting a characteristic radiation pattern. Of common use in biology, notably Carbon-14 (^{14}C), Phosphorus-32 (^{32}P) and Tritium (^{3}H). Distinguished from *heavy isotopes*, which have a higher atomic mass than the commonest isotope but are not radioactive. See LABELLING.

RADIOLARIA. Order of marine planktonic sarcodine protozoans, lacking shells but with a central protoplasm comprising chitinous capsule and siliceous spicules perforated by numerous pores through which spines project between vacuolated and jelly-like outer protoplasm. Between the spines radiate out branching pseudopodia exhibiting cytoplasmic streaming. Many house yellow symbiotic algal cells (zooxanthellae). *Radiolarian oozes* cover much of the ocean floor and are important in flint production.

RADIUS. One of two long bones (the other being the ulna) in the tetrapod fore-limb. Articulates with the side of the fore-foot bearing the thumb. See PENTADACTYL LIMB.

RADULA. 'Tongue' of molluscs; a horny strip, continually renewed, with teeth on its surface for rasping food. Found in Amphineura, Gastropoda and Cephalopoda; absent in Bivalvia, which are microphagous. May be modified for boring in some species. Pattern of teeth helpful in identification.

RAMAPITHECUS. Miocene ape (about 10 Myr BP), first discovered in Siwalik hills of India, with the hominid tendencies of large, heavily enamelled molars, small canines and arched palate. Canine-like premolar unlike hominid's. Only teeth and jaw remains so far found. Recent proposition is that ramapithecines and sivapithecines (fossil genus *Sivapithecus*) belong to a single group, *Sivapithecus*, with affinities to orang-utan clade. Compare DRYOPITHECINE.

RAMET. An independent individual of a CLONE.

RAPHE. (1) In seeds formed from anatropous ovules, a longitudinal ridge marking position of the adherent FUNICLE. (2) Elongated slit (or slit pair) through valve wall of some pennate diatoms, involved in movement of cell over the substratum.

RAPHIDE. Needle-shaped crystals of calcium oxalate in vacuoles of many plant cells.

RATITES. See PALAEOGNATHAE.

RAUNKIAER'S LIFE FORMS. System of vegetational classification based on position of perennating buds in relation to soil level, indicating how plants survive the unfavourable season of their annual life cycle. The following are recognized: *phanerophytes*: woody plants whose buds are borne more than 25 cm above soil level (many trees and shrubs); *chamaerophytes*: woody or herbaceous plants whose buds are above soil level but less than 0.25 m above; *hemicryptophytes*: herbs with buds at soil level, protected by the soil itself or by dry dead portions of the plant; *geophytes*: herbs with buds below soil surface; *heliophytes*: herbs whose buds lie in mud; *hydrophytes*: herbs with buds in water; *therophytes*: herbs surviving the unfavourable season as seeds.

RAY. Tissue initiated by CAMBIUM and extending radially in secondary xylem and phloem. Mainly parenchymatous, but may include tracheids in the xylem.

RAY FLOWER. See DISC FLOWER.

REACTION TIME. See LATENT PERIOD.

recA. CISTRON of the bacterium *E. coli* whose product, the *rec*A (or *Rec*A) protein is involved in promoting general DNA recombination in the genome, and has ATPase activity in the presence of single-stranded DNA as well as catalysing DNA base-pairing and strand annealing (assimilation). *Rec*A protein binds to single-stranded DNA and anneals it to any complementary sequence in a double-stranded (duplex) DNA in such a way as to replace one of the two

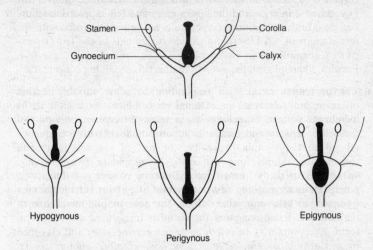

Stamen — Corolla
Gynoecium — Calyx

Hypogynous   Epigynous

Perigynous

Fig. 56. *Diagrams of different types of floral receptacle, showing the position of the gynoecium (black) relative to other flower parts.*

DNA strands of the original duplex, forming a HETERODUPLEX. Similar gene products are being found in eukaryotes.

RECAPITULATION. See BIOGENETIC LAW.

recB, recC. CISTRONS of the bacterium *E. coli* whose combined products make up the RecBC enzyme, which initiates DNA-unwinding at any free duplex end and has nuclease activities. Once bound to a free duplex end (not normally present in *E. coli*) RecBC proceeds to unwind the duplex, but wherever the enzyme encounters the DNA strand sequence 5′-GCTGGTGG-3′ (termed *Chi*), it cuts the strand, leaving it exposed to binding by RecA protein (see recA). For this reason, recombination is promoted in *Chi* regions, of which there may be one thousand in the *E. coli* genome.

RECENT. See HOLOCENE.

RECEPTACLE (THALAMUS, TORUS). Apex of flower stalk, bearing flower parts (perianth, stamens, carpels). Its relation to the gynoecium determines whether carpels are *inferior* or *superior*. When carpels are at the apex of a conical receptacle and other flower parts are inserted in turn below, the gynoecium is superior and the flower *hypogynous* (e.g. buttercup, Fig. 56). When carpels are at the apex (centre) of a concave receptacle with other flower parts borne around its margin, the gynoecium is superior and the flower *perigynous* (e.g. rose, upper and lowest diagrams in Fig. 56). When receptacle completely encloses carpels and other flower parts arise from receptacle above, the gynoecium is inferior and the flower *epigynous* (e.g. apple, dandelion, Fig. 56). In this condition, carpel walls are intimately fused with the receptacle wall. (2) Describing the shortened axis of the INFLORES-CENCE (capitulum) in Compositae. (3) In some brown algae (e.g. Fucales, Phaeophyta), the swollen thallus tip containing conceptacles.

RECEPTOR (SENSE CELL). Cell responding to some variable feature of an animal's internal or external environment by a shift in its membrane voltage. Sometimes, as in *primary receptors*, only part of the cell is sensory and generates action potentials with a frequency related to the stimulus intensity; the rest of the cell is axonal, transmitting signals long distances. In *secondary receptors* (e.g. MUSCLE SPINDLES) the altered membrane voltage initiates action potentials in a synapsing neurone by bringing about voltage changes (*receptor*, or *generator potentials*) in the postsynaptic membrane of the neurone. In all receptors the signal is transduced into electrical form. Receptors may be *interoceptors* or *exteroceptors* and classified by modality into *chemoreceptors*, *mechanoreceptors*, *photoreceptors*, etc. See SENSE ORGAN.

RECEPTOR POTENTIAL. See RECEPTOR.

RECEPTOR SITE. Part of cell membrane or molecule (*binding site*), combining specifically with a molecule or other particle (e.g. virus, hormone, repressor molecule). See CELL MEMBRANES, GATED CHANNELS.

RECESSIVE. (Of phenotypic characters) only expressed when the genes determining them are HOMOZYGOUS (complete recessivity). When heterozygous, the allele for the character is either 'silent', giving rise to no cell product, or its effect is masked by the presence of the other allele. Sometimes 'recessive' is used to describe alleles themselves, but since many gene loci are pleiotropic it would need to be made clear which aspect of phenotype was being described as recessive. Compare DOMINANCE. See PENETRANCE.

RECIPROCAL ALTRUISM. See ALTRUISM.

RECIPROCAL CROSS. (1) Cross between two hermaphrodite individuals, in which the male and female sources of the gametes used are reversed. (2) Crossing operation between stocks of two different genotypes, where each stock is used in turn as the source of male and female gametes. Employed when testing for SEX-LINKAGE (also sex-limited and sex-controlled inheritance), where one sex has a greater influence than the other in determining offspring phenotype. When reciprocal crosses give very different results (e.g. in F1 or F2) the character studied is likely to be sex-linked or under cytoplasmic control. See CYTOPLASMIC INHERITANCE.

Recombinant DNA. DNA whose nucleotide sequence has undergone alteration as a result of incorporation of, or exchange with, another DNA strand. Such DNA occurs naturally as a result of CROSSING-OVER during RECOMBINATION, and also during recombinant DNA techniques employed during GENE MANIPULATION.

Recombination. Any process, other than point mutation, by which an organism produces cells with gene combinations different from any it inherited. Offspring resulting from such recombinant cells are *recombinant offspring*. A major source of GENETIC VARIATION, its effectiveness is dependent upon mutation for initial gene differences, from which recombination events can generate further gene rearrangements.

(1) In meiosis (eukaryotes only), two kinds of recombination between chromosomes commonly occur. *Free combination* or *reassortment* (not always regarded as recombination) occurs when non-homologous chromosomes assort randomly to form the two haploid nuclei during anaphase of the first meiotic division. As a result, if N = the number of chromosome pairs in the parent cell, each chromosome pair being heterozygous at least at one locus, then the number of possible nuclear genotypes from the first division is 2^N, and this in the absence of crossing-over. *Non-random (restricted) recombination* (the most usual sense of 'recombination') results from crossing-over between homologous chromosomes during first meiotic prophase, producing recombinant DNA. This is non-random in the sense that it only occurs between homologous sequences of DNA (and non-randomly then; see CROSSING-OVER, SYNAPTONEMAL COMPLEX). (2) The process involved in most exchanges of DNA between chromosomes, including prokaryotic (see SEXUAL REPRODUCTION). Termed *reciprocal recombination* when equivalent lengths of DNA are reciprocally exchanged between duplexes (i.e. between double helices), and *non-reciprocal recombination* (see GENE CONVERSION) when only one duplex retains its original length, as often happens in the immediate vicinity of crossovers (hence restricted, since crossovers themselves are of limited occurrence). In one model, homologous duplex DNA sequences first align themselves side-by-side; one strand of each duplex is then cut (nicked) by an enzyme and its broken ends joined up with their opposite partners by a DNA ligase (i.e. not merely rejoined again as before), to form two homologous duplexes whose nucleotide sequences have been altered. See recA, recB and recC.

In bacterial TRANSFORMATION and TRANSDUCTION (examples of *homologous* recombination) and some eukaryotic gene transfers, homologous DNA duplexes first align, the donor duplex undergoes DENATURATION (separates into its two strands), and one strand invades the host duplex, aligning with the host strand having the greater base-pairing conformity. It is then nicked while the host strand without a partner is nicked at two places, donor DNA getting inserted by ligases in its place. The evidence for this comes from electron microscopy and CHROMOSOME MAPPING. See GENE MANIPULATION.

During the generation of ANTIBODY DIVERSITY during B-CELL maturation, genes from different parts of a chromosome are brought together in such a way that an RNA transcript is produced which effectively 'omits' the intervening DNA sequences.

RECOMBINATION NODULE. See SYNAPTONEMAL COMPLEX.

RECTUM. Terminal part of intestine, opening via anus or cloaca and commonly storing faeces. In insects, often reabsorbs water, salts and amino acids from the 'urine' (see MALPIGHIAN TUBULES); some insect larvae have tracheal gills in the rectum, while larval dragonflies also eject water forcibly from the rectum for propulsion. Ectodermal in origin (see PROCTODAEUM).

RED BLOOD CELL (RED BLOOD CORPUSCLE, ERYTHROCYTE). Most abundant vertebrate blood cell; generated in bone marrow, usually from RETICULOCYTES. Contains many molecules of HAEMO-GLOBIN loading and unloading molecular oxygen (and carbon dioxide to a much lesser extent) and serving as a blood BUFFER. Mammalian erythrocytes are flattened, circular, biconcave discs (about 8 μm diameter in humans), lacking nuclei, mitochondria and most internal membranes. Tend to be larger and oval in shape in other vertebrates, retaining a nucleus. Damaged by passage through capillaries, they last about four months in humans (judged by radioactive tracers) before being destroyed by the liver's RETICULO-ENDOTHELIAL SYSTEM. Their surface antigens specify BLOOD GROUP. Their membrane SODIUM PUMPS regulate cell volume, but hypotonic solutions cause osmotic swelling and rupture, leaving erythrocyte membranes as *ghosts*. The important enzyme CARBONIC ANHYDRASE catalyses the reversible reaction:

$$H_2O + CO_2 \underset{\longleftarrow \text{lungs}}{\overset{\longrightarrow \text{tissues}}{\longleftrightarrow}} H^+ + HCO_3^-$$

enabling rapid exchanges of gases in the lungs and body tissues. Role of erythrocytes in CO_2 transport is primarily to generate HCO_3^- ions for carriage in the plasma and to reconvert them back to CO_2 molecules in the lungs, where exhalation occurs. However, about 23% of CO_2 carried in human blood is in the form of *erythrocytic carbaminohaemoglobin*, which breaks down in the lungs to release CO_2 again. See HAEMOPOIESIS.

REDIA. One of the larval types in endoparasitic Trematoda, developing asexually from the sporocyst and from other rediae (see POLY-EMBRYONY). Often parasitic in snails, developing into cercariae.

REDOX REACTIONS. Oxidation-reduction reactions, in biology generally catalysed by enzymes. Involve transfer of electrons from an electron donor (reducing agent) to an electron acceptor (oxidizing agent). Sometimes hydrogen atoms are transferred, equivalent to electrons, so that dehydrogenation is equivalent to oxidation. Respiration involves many *redox pairs*, one member donating, the other accepting, electrons, determined by their relative *standard oxidation-reduction potentials*. See ELECTRON TRANSPORT SYSTEM.

RED TIDE. Water in which very large numbers of dinoflagellates or other organisms colour it red. Sometimes associated with neurotoxin production.

REDUCING SUGAR. Sugar capable of acting as a reducing agent in solution, as indicated by a positive BENEDICT'S TEST and ability to decolourize potassium permanganate solution. Depends upon presence of potentially free aldehyde or ketone group. Most monosaccharides are reducing sugars (and all are in weakly acid solution), as are most disaccharides except sucrose.

REDUCTION DIVISION. First division of MEIOSIS, the chromosome numbers of the daughter cells produced being half that of the parent cell.

REDUCTIONISM. See EXPLANATION.

REFLEX. Innate (inherent) and often invariant neuromuscular animal response to an internal or external stimulus, usually with little delay involved. In its simplest form, mediated by a *reflex arc* involving input along a sensory neurone of a spinal nerve to the CENTRAL NERVOUS SYSTEM (CNS) and output along a motor neurone of the same or a different spinal nerve to an effector (muscle, gland). Frequently in vertebrates an association (relay) neurone intervenes in the CNS, acting to transmit impulses via white fibres to the brain. This may enable perception and learning, as in CONDITIONED REFLEXES. Vertebrate reflexes which involve only the spinal component of the CNS are termed *spinal reflexes*. Examples of innate reflexes in humans include blinking, sneezing, coughing, ventilation, regulation of heart rate and pupil diameter (the iris reflex), the thirst reflex (see HYPOTHALAMUS) and complex postural reflexes involving the CEREBELLUM during walking and running. See NERVOUS INTEGRATION, NERVOUS SYSTEM.

REFRACTORY PERIOD. Time taken for nerve and muscle membranes to recover their resting ionic imbalance after passage of an impulse. See IMPULSE, MUSCLE CONTRACTION.

REFUGIUM. Locality (e.g. a tableland or mountain *nunatak*) which has escaped drastic alteration following climatic change (in this case glaciation), in contrast to the region as a whole. Refugia usually form centres for RELIC species populations or communities.

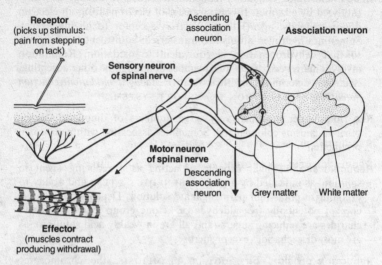

Fig. 57. *Diagram of section of spinal cord showing nervous pathways involved in a spinal reflex on one side. Ascending and descending association neurones permit involvement of higher and lower body regions.*

REGENERATION. Restoration by regrowth of parts of body which have been removed, as by injury or AUTOTOMY. Commoner in lower than higher animals, but extensive in some planarians (where new individuals can be regenerated from small body fragments), in the polychaete *Chaetopterus* (which can regenerate a whole animal from one segment) and in crustaceans (which can replace limbs) and echinoderms (which replace arms). Involves the re-establishment of local tissue differentiation. Some examples require long-range interactions and extensive cell movement (*morphallaxis*), while others involve short-range interactions and extensive growth (*epimorphosis*). Vertebrate embryonic limb regeneration involves production of a BLASTEMA. If a large part of an organ is removed (e.g. of liver, pancreas) the remaining organ commonly returns to normal size. This is *compensatory hypertrophy*. Removal of a mammalian kidney usually results in compensatory hypertrophy of the one remaining. Compare REGULATION. Very common in plants, occurring e.g. in higher plants by growth of dormant buds, formation of secondary meristems and production of adventitious buds and roots. Exploited on a large scale in plant propagation.

REGULATION. Ability of animal embryo to compensate for disturbance (e.g. involving removal, addition or rearrangement of cells) and to produce an apparently normal individual. Eggs may be *regulative*, meaning that removal of some early CLEAVAGE products has no effect on the eventual animal, other than decreasing its size. Embryonic regulation is a broad term, including *twinning* (production of two complete animals from bisected embryo), *fusion* (fusion of more than one embryo to give a single giant embryo), *defect regulation* (removal of part of the early embryo does not disturb development), and *inductive reprogramming* (see INDUCTION, where a grafted inducer causes surrounding parts to pursue a developmental route different from that expected from the FATE MAP). Compare MOSAIC DEVELOPMENT.

REGULATIVE DEVELOPMENT. See REGULATION.

REGULATOR GENE and PROTEIN. See JACOB-MONOD THEORY.

REGULATORY ENZYMES. Enzymes specifically involved in switching metabolic pathways on or off. Include a) ALLOSTERIC enzymes, where activity is modulated through the non-covalent binding of a specific molecule (e.g. see ATP, ADP, GLYCOLYSIS) at a site other than the active site; b) *covalently-modulated enzymes*, which alternate between active and inactive forms as the result of other enzyme activity (e.g. see KINASE). Regulatory enzymes are exquisitely responsive to alterations in the metabolic needs of the cell.

REINFORCEMENT. Process whereby presentation of a stimulus (the *reinforcer*) to an animal just after it has performed some act alters the probability or intensity of future performances of that act when repeated either in isolation or in the presence of the stimulus. *Positive reinforcers* increase the probability and/or intensity of response; *negative reinforcers* decrease its probability and/or intensity. See CONDITIONING.

RELATEDNESS. Incestuous and non-diploid matings excluded, the degree of relatedness between two conspecific individuals, male A and female B, is found by first locating their nearest common ancestor(s) and then counting the number of generations passed by moving from B to the common ancestor(s) and then on to A (the *generation distance*). This number is the power to which $\frac{1}{2}$ is first raised and then multiplied by the number of nearest common ancestors. For two second cousins, the calculation is as shown on p. 49. Generation distance is given by numbers in parentheses from B to A, i.e. 6. There are two nearest common ancestors of A and B, namely 1 and 1′, so degree of relatedness, $R = 2 \times (\frac{1}{2})^6 = \frac{1}{32}$, meaning that two second cousins (such as A and B) are expected on average to share 1 in 32 of their genes. See HAMILTON'S RULE.

RELEASER. A standard external stimulus evoking a stereotyped response; commonly applied in the context of INSTINCT. A *social releaser* emanates from a member of the same species as the respondent. Releasers include the *sign stimuli* (e.g. claw-waving in fiddler

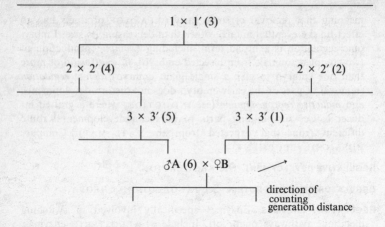

$$1 \times 1' \ (3)$$

$$2 \times 2' \ (4) \qquad\qquad 2 \times 2' \ (2)$$

$$3 \times 3' \ (5) \qquad 3 \times 3' \ (1)$$

$$\male A \ (6) \times \female B$$

direction of
counting
generation distance

crabs, postures adopted by courting birds, etc.) which have been so valuable in ETHOLOGY.

RELEASING FACTOR (RF, RELEASING HORMONE, RH). Substance stimulating release of hormone into blood. The *locus classicus* for their production is the vertebrate HYPOTHALAMUS, several of whose neurosecretions initiate hormone release from the anterior pituitary.

RELIC. (1) Surviving organism, population or community characteristic of an earlier time. (2) A *relic* (*relect*) *distribution* of fauna or flora is one representing the localized remains of an originally much wider-ranging distribution such as those organisms, now confined to mountain tops, which were far more widespread in glacial times (see REFUGIUM).

RENAL PORTAL SYSTEM. In jawed fish, amphibians, and variably in reptiles and birds, blood is brought from capillaries of posterior part of body (hind limbs, tail) to the kidneys by a pair of renal portal veins (originally part of the posterior cardinal veins), blood being diverted into capillaries around the kidney tubules (but not joining the glomeruli). Blood is recollected either into posterior cardinal veins (fish) or renal veins and inferior vena cava (amphibia, etc.). See PORTAL VEIN.

RENIN. Enzyme produced by juxtaglomerular cells of kidney when blood pressure falls. See ANGIOTENSINS.

RENNIN. Enzyme secreted by stomachs of young mammals; converts soluble caseinogen to the insoluble protein casein, which coagulates and therefore takes longer to leave the stomach. *Rennet* is impure rennin.

REPETITIVE DNA. Apart from the repeated nucleotide sequences of

SATELLITE DNA, there exist in human and rodent genomes (at least) transposable sequences of about 300 nucleotides (*Alu* sequences) which are individually repeated and inserted throughout the chromosomes, making up about 5% of the genome mass. They have changed far more slowly in mammalian evolution than have satellite DNAs and appear to have arisen from a 7SL RNA gene. Much longer dispersed and transposable repetitive sequences occur in primates, notably the 'LINE' sequences, which comprise about 4% of the human genome mass and move by an RNA-requiring process which also involves a reverse transcriptase (which one sequence may encode). They are probably 'junk' (parasitic) sequences, but may have had significant effects on expressions of nearby genes (see ONCOGENE). See GENE DUPLICATION, INVERTED REPEAT SEQUENCE.

REPLACING BONE. Synonym for cartilage bone. See OSSIFICATION.

REPLICA PLATING. Method employed to isolate mutant microorganisms sensitive to a component of, or deficiency in, a growing medium and which, by failing to grow on it, are difficult to detect. The diluted organism is plated on to a solidified medium suitable for growth and colony formation. An adsorbent material (e.g. filter paper) is then pressed lightly on to the medium surface to pick up cells from each colony, and then pressed on to fresh solidified medium with the additive or deficiency. The distribution of resulting colonies on this new medium can be compared with the original distribution, any gaps indicating sensitive strains.

REPLICATION. See DNA REPLICATION.

REPLICON. Nucleic acid molecule containing a nucleotide sequence forming a replication origin, at which replication is initiated. Usually one per bacterial or viral genome, but often several per eukaryotic chromosome.

REPRESSION. (1) (Of enzyme) see ENZYME feedback inhibition. (2) (Of gene) see JACOB-MONOD THEORY.

REPRESSOR MOLECULE. See JACOB-MONOD THEORY.

REPTILIA. Reptiles. Vertebrate class including the first fully terrestrial tetrapods. Probably a MONOPHYLETIC group, whose members lay AMNIOTIC EGGS, have horny or scaly skin and metanephric kidneys. Stem reptiles (Cotylosaurs, Subclass Anaspida) evolved in the Lower Carboniferous, or even earlier, and it has been generally held that they had affinities with the fossil amphibian group, the anthracosaurs. They had solid skulls lacking fenestrations (modern turtles and tortoises are representatives). Synapsida had a single fenestration low on side of skull and included pelycosaurs. Therapsida (MAMMAL-LIKE REPTILES) appeared from the mid-Permian onwards; some were large and herbivorous, others smaller and carnivorous. They were dominant in the Permian fauna and had varied cranial morphology. By the close of the Triassic, therapsids had become very mammal-like, as indicated by the *cynodonts* (dog-sized), which gave rise to mammals. The great Mesozoic reptilian radiation produced three subclasses: Euryapsida (marine reptiles such as ichthyosaurs and plesiosaurs), Archosauria (ARCHOSAURS), including dinosaurs and crocodiles, and Lepidosauria, including extinct crocodile-like forms as well as lizards and snakes (see SQUAMATA). Modern forms are poikilothermic (ectothermic), but some extinct archosaurs were probably homoiothermic (endothermic). For *Sphenodon*, see RHYNCHOCEPHALIA.

RESIN DUCT. Relatively large intercellular spaces lined by thin-walled parenchyma cells secreting resin into the duct. Wounding, pressure, frost and wind damage can stimulate their formation. Resin apparently protects the plant from attack by decay fungi and bark beetles. Found in conifers.

RESOLUTION, RESOLVING POWER. See MICROSCOPE.

RESPIRATION. (1) (*Internal, tissue* or *cellular* respiration.) Enzymatic release of energy from organic compounds (esp. carbohydrates and fats) which either requires oxygen (*aerobic respiration*) or does not (*anaerobic respiration*). Anaerobic respiration is sometimes used as a synonym of FERMENTATION. Among eukaryotic cells only those with mitochondria, and among prokaryotes only those with mesosomes, can respire aerobically. All cells can respire anaerobically, using enzymes in their cytosol; but not all obtain sufficient energy release for their needs this way. A high proportion of the energy released during respiration is coupled to ATP synthesis (50% during anaerobic respiration in erythrocytes; 42% in aerobic respiration involving metochondria). The major anaerobic pathway in cells is GLYCOLYSIS. This is a metabolic 'funnel' into which compounds from a variety of original sources may be fed (e.g. from proteins, fats, polysaccharides). The energy-rich products are ATP (from ADP) and NADH (from NAD); pyruvate formed is either catabolized further in mitochondria (see KREBS CYCLE) or converted to lactate or alcohol. The actual ATP-generating steps of aerobic respiration occur during passage of electrons along the ELECTRON TRANSPORT SYSTEM. (2) (*External respiration.*) See VENTILATION.

RESPIRATORY CENTRES. See VENTILATION.

RESPIRATORY CHAIN. Alternative for ELECTRON TRANSPORT SYSTEM.

RESPIRATORY MOVEMENT. See VENTILATION.

RESPIRATORY ORGAN. Animal organ specialized for gaseous exchanges of oxygen and carbon dioxide. Include LUNGS, LUNG BOOKS, GILLS, GILL BOOKS and the arthropod TRACHEAL SYSTEM. See VENTILATION.

RESPIRATORY PIGMENT. (1) Substance found in animal blood or other tissue, involved in uptake, transport and unloading of oxygen (and/or carbon dioxide) in solution, always by weak and reversible bonds. Some visibly colour the blood (HAEMOGLOBIN, HAEMERYTHRIN, CHLOROCRUORIN); but MYOGLOBIN is not blood-borne, and does not noticeably colour the tissues where it is found. (2) Substances (e.g. CYTOCHROMES, FLAVOPROTEINS) found in all aerobic cells (in insufficient concentrations to colour them), and involved in the ELECTRON TRANSPORT SYSTEM of aerobic respiration.

RESPIRATORY QUOTIENT (RQ). Ratio of volume of carbon dioxide produced to oxygen consumed by an organism during aerobic respiration. The theoretical RQ value for oxidation of carbohydrates is 1; for fats 0.7; for protein 0.8. Cells can oxidize more than one of these at a time, making *in vivo* results problematic, although data on levels of nitrogen excreted can be helpful. Interconversion of carbohydrates to fats can also affect RQ values. With people on a normal diet an RQ of 0.82 is expected.

RESPONSE. Change in organism (or its parts) produced by change in its environment. Usually adaptive. See IRRITABILITY.

RESTING NUCLEUS (RESTING CELL). See INTERPHASE.

RESTING POTENTIAL (MEMBRANE POTENTIAL). Electrical potential across a cell membrane when not propagating an impulse. All plasma membranes exhibit such voltage gradients (inside negative with respect to the outside) generated by TRANSPORT PROTEINS, the two most important being the Na^+/K^+ pump (SODIUM PUMP) and the K^+ leak channel. The former pumps sodium out and potassium in, and is ATP-dependent. The latter lets potassium ions flow back out of the cell along their electrochemical gradient until internal negativity of -75 mV (nerve axons) is achieved – the potential which retards loss of potassium to the extent that an *equilibrium potential* for K^+ results. Any sodium that enters via this route is pumped out by the Na^+/K^+ pump. Membrane potentials of 20–200 mV internally negative can be generated, depending on species and cell type. See IMPULSE.

RESTITUTION NUCLEUS (RESTITUTION MEIOSIS, RESTITUTION DIVISION). Inclusion within the same nuclear membrane of all the chromatin after meiotic chromosome or chromatid separation has occurred, giving an unreduced egg which may develop by automictic PARTHENOGENESIS (e.g. in dandelions, *Taraxacum*). See AUTOMIXIS.

RESTRICTION ENDONUCLEASES (RESTRICTION ENZYMES). Class of nucleases originally extracted from the bacterium *E. coli* (where it digests phage DNA but leaves the cell DNA intact). Type I restriction enzymes bind to a recognition site of duplex DNA, travel along the molecule and cleave one strand only, about 75 nucleotides long, apparently randomly. Type II are more valuable in GENE MANIPULATION and cleave the duplex at specific target sites at or near the binding site. Their naming uses the first letter of the genus and the first two letters of the specific name of the organism (host) from which they were derived, with host strain identified by a full or subscript letter. Roman numerals then identify the particular host enzyme if there are more than one. Thus *Eco*RI is the first restriction enzyme from *E. coli* strain RY13, and *Hind*II comes from *Haemophilus influenzae* Rd. Target sites for these nucleases are very specific. See MODIFICATION, PHAGE RESTRICTION, RESTRICTION MAPPING.

RESTRICTION MAPPING. Method employed in CHROMOSOME MAPPING, usually of organelle and viral genomes. Complete nucleotide sequencing may result.

RETE MIRABILIS (pl. RETIA MIRABILIA). Network of arterioles running towards an organ and breaking up into capillaries adjacent to, but in the opposite direction from, a similar set of capillaries returning blood to the venous system. Exchanges between the two networks form a COUNTERCURRENT SYSTEM, passively increasing local value of some variable (e.g. metabolite concentration, temperature). They occur for example in teleost gills and GAS BLADDERS and in feet of birds which stand in cold water or on ice.

RETICULAR FIBRES. Fine branching protein fibres (*reticulin*) forming an extracellular network in many vertebrate connective tissues and holding tissues and organs together. The protein is collagen-like and may develop into collagen in wound tissue and embryo. Stains selectively with silver preparations. Abundant in BASEMENT MEMBRANES under many epithelia, sarcolemma around muscle and around fat cells. Forms framework for lymph corpuscles in LYMPHOID TISSUE.

RETICULAR FORMATION (RETICULAR ACTIVATING SYSTEM). Nerve cells scattered throughout the vertebrate BRAINSTEM, some forming nuclei, receiving impulses from the spinal cord, cerebellum and cerebral hemispheres and returning impulses to them. Stimulation of some of these cells produces AROUSAL in unanaesthetized animals, the formation playing an important role in waking, attentiveness and, in higher vertebrates, consciousness. A *decerebrate* animal, whose brain is sectioned between basal ganglia and reticular formation, exhibits the effects of overactive stimulatory and underactive inhibitory mechanisms of the reticular formation, producing *decerebrate rigidity*.

RETICULATE THICKENING. Internal thickening of a wall of a xylem vessel or tracheid in the form of a network.

RETICULIN. See RETICULAR FIBRES.

RETICULOCYTE. Immature, non-nucleated, mammalian RED BLOOD CELL. Develops from nucleated *myeloblasts* and is released into the blood during very active HAEMOPOIESIS. Cytoplasm basophilic, due to RNA.

RETICULO-ENDOTHELIAL SYSTEM (RES). Tissue macrophages either circulating in blood, dispersed in connective tissue or attached to capillary endothelium. Of MYELOID origin, these cells carry (F_c) receptors for IgG antibodies, complement and lymphokines. Attack foreign or tumour cells either by ingesting them or by lysis after adherence to them. Abundant in liver sinusoids (*Kupffer cells*), where e.g. they ingest worn erythrocytes, in kidney glomeruli (*mesangium cells*), alveoli of lung, attached to capillaries of brain (*microglial cells*), spleen and lymph node sinuses.

RETICULUM. (1) Second compartment of the RUMINANT stomach, receiving partially-digested material which has been in the rumen, rechewed, and swallowed a second time. Water is here pressed out before the food passes to the abomassum. (2) See ENDOPLASMIC RETICULUM.

RETINA. Photosensitive layer of vertebrate and cephalopod EYES, non-sensory in region of ciliary body and iris. Contains ROD CELLS and CONES, both of which synapse with intermediary neurones before impulses leave via the optic nerve. Also present are blood vessels and glial cells; in vertebrates light must traverse these tissues (the ganglion layer) before impinging upon photoreceptors, an arrangement termed an *inverted retina*. By contrast in cephalopod retinas, derived from external ectoderm, light impinges first upon photoreceptors. Vertebrate retinas arise as outpushings of the brain, photoreceptors having a pigmented layer behind them abutting the choroid. See FOVEA, BLIND SPOT.

RETINOIC ACID (RA, VITAMIN A ACID) Naturally occurring oxidation product of vitamin A, considered by some to be an endogenous vertebrate MORPHOGEN, in addition to playing a part in the growth and maintenance of epithelia and inhibiting growth of some malignant cells. In chick limb buds, where retinol can serve as a precursor, a gradient of RA levels provides the *positional information* specifying anterior–posterior digit pattern in the wing. An RA receptor (RAR) protein, once bound to its ligand, then binds to specific DNA regions within the nucleus (see NUCLEAR RECEPTORS).

RETINOIDS. Diterpenoids derived from a monocyclic parent compound and containing five C–C double bonds and a functional group at the terminus of the acylic portion VITAMIN A (retinol), retinene (retinal), RETINOIC ACID and 3,4 didehydroretinoic acid (ddRA) are important animal examples; BACTERIORHODOPSIN is a prokaryote retinoid.

RETINOL (VITAMIN A). See RHODOPSIN, VITAMIN A.

RETROINHIBITION. See END-PRODUCT INHIBITION.

RETROVIRUS. See VIRUS, REVERSE TRANSCRIPTASE.

REVERSE TRANSCRIPTASE. A DNA POLYMERASE of retroviruses (initially Rous sarcoma virus), synthesizing complementary single-strained DNA (cDNA) on a single-stranded RNA template. Its inhibition could prevent retroviral infection. cDNA formed may be cloned and screened to detect differences in mRNAs between cells, mRNA acting as template for reverse transcriptase. See GENE MANIPULATION, REPETITIVE DNA, VIRUS.

RHESUS SYSTEM. See BLOOD GROUP.

RHIPIDISTIA. Order or suborder (probably a clade) of crossopterygians, appearing in early Devonian and extinct by end of Palaeozoic. Probably ancestral to coelacanths and amphibians. Up to 4 m long. Cranium divided into anterior and posterior parts. Many (osteolepids) were well covered in thick scales, and adapted to life at the water's edge.

RHIZOID. Single- or several-celled hair-like structure serving as a root. Present at bases of moss stems and on undersurfaces of liverworts and fern prothalli, as well as in some algae and fungi.

RHIZOME. More or less horizontal underground stem bearing buds in axils of reduced scale-like leaves. Serves in perennation and vegetative propagation; e.g. mint, couch grass, *Scirpus*.

RHIZOMORPH. Root-like strand of hyphae produced by some fungi, transporting food materials from one part of the thallus to another, and increasing in length by apical growth. May be internally differentiated and complex.

RHIZOPLANE. Root-surface component of the RHIZOSPHERE.

RHIZOPODA. Protozoan class, or subclass of the SARCODINA.

RHIZOSPHERE. Zone of soil immediately surrounding root, and modified by it. Characterized by enhanced microbial activity and by changes in the ratios of organisms compared with surrounding soil.

RHODOPHYTA. Division of the ALGAE (red algae). Seaweeds characterized by complete absence of flagellated cells (including the male gametes), by presence of chlorophylls *a*, *d* and the accessory pigments *phycoerythrin* and *phycocyanin*, by non-aggregated thylakoids within chloroplasts and by the storage product *floridean starch*. Cell walls generally contain cellulose (xylan in some) and amorphous mucilages including commercial agars and carrageenan. Sexual reproduction oogamous, if present, involving specialized female cells (carpogonia) and male cells (spermatia). Red algae are generally smaller than brown algae, but as diverse. The thallus in a few is unicellular, in others filamentous (often branched), or flattened and membranous. Most are marine and benthic, some epiphytic, some partial parasites, others colourless parasites. Coralline forms are important in reef-building.

RHODOPSIN (VISUAL PURPLE). Light-sensitive pigment of ROD CELLS, whose molecules contain a protein component (*opsin*) whose prosthetic group, the chromophore 11-*cis* retinal (vitamin A, a component of CONE pigments), undergoes isomerization to *trans* retinal in light, becoming free of opsin in the process. Opsin seems to undergo an allosteric change, closing sodium channels in the rod cell membrane. An intermediary (perhaps Ca^{++}) probably links these events. Photons may initiate calcium release from the discs in the outer segment of the cell, while rhodopsin's shape change possibly starts a CASCADE of reactions reducing intracellular cyclic GMP levels. Conversion of *trans* retinal to *cis* retinal occurs in absence of light and involves reduction by reduced NAD, *retinol* being an intermediary in the regeneration. See BACTERIORHODOPSIN.

RHO FACTOR (rho FACTOR, ρ). Protein binding to the *termination site* of a cistron and, during transcription, causing release of both RNA POLYMERASE and newly-synthesized RNA. See SIGMA FACTOR.

RHOMBENCEPHALON. See HINDBRAIN.

RHYNCHOCEPHALIA. Order of lepidosaurian reptiles, related to eosuchians. Now represented solely by the tuatara (*Sphenodon*), restricted to New Zealand (a RELIC from GONDWANALAND), where it was in danger of extinction. Appeared in the Mesozoic, but never a dominant group. Lizard-like with many primitive features, e.g. lack of external ear opening, a pineal eye (see PINEAL GLAND) and lack of an intromittant organ. See REPTILIA.

RHYNCOTA. See HEMIPTERA.

RHYNIOPHYTA. Fossil plant division containing earliest-known vascular plants, dated to 400 Myr BP (Silurian). Seedless, they comprised a simple dichotomously branched axis bearing terminal heterosporous sporangia. Differentiation into stems, leaves or roots not apparent. *Rhynia* is the best-known example; leafless, but possessing a rhizome with water-absorbing rhizoids. Internal structure resembled extant vascular plants, having an epidermis surrounding the photosynthetic cortex while centre of axis contained the xylem. See TRIMEROPHYTA.

RHYTIDOME. See BARK.

RIBOFLAVIN (VITAMIN B$_2$). Precursor of FMN and FAD (prosthetic groups of CYTOCHROMES). Insufficient intake in humans causes inflammatory lesions in corners of mouth and blocking of sebaceous glands of nose and face. See VITAMIN B COMPLEX.

RIBONUCLEASE. See RNase.

RIBONUCLEIC ACID. See RNA.

RIBOSE. A pentose (5-carbon) monosaccharide; $C_5H_{10}O_5$. Component of RNA and its nucleotide triphosphate precursors, etc.

RIBOSOMAL. See RIBOSOMES, RNA.

RIBOSOMES. Non-membranous, but often membrane-bound, organelles of both prokaryotic and eukaryotic cells, of chloroplasts and mitochondria. Sites of PROTEIN SYNTHESIS, each is a complex composed of roughly equal ratios of *ribosomal RNA* (rRNA) and 40 or more different types of protein. Prokaryotic ribosomes are slightly smaller than eukaryotic, with a sedimentation rate (Svedberg number, S) of 70S as opposed to 80S. Each is composed of one large (50S or 60S) and one smaller (30S or 40S) sub-particle, forming ribosomes with diameters of approx. 29 nm in prokaryotes and 32 nm in eukaryotes respectively. Each sub-particle is formed from one to three species of rRNA plus associated proteins. A functional ribosome has one conformational groove to fit the growing protein chain and another for the messenger RNA (mRNA) molecule, whose presence initiates formation of composite ribosomes from their sub-particles (see A-SITE, P-SITE), several ribosomes on an mRNA strand forming a *polysome*.

Ribosomes are not attached to membranes in prokaryotes, chloroplasts and mitochondria, but are commonly membrane-bound in eukaryotic cells (esp. actively secreting ones), forming rough ENDOPLASMIC RETICULUM, as well as being attached to outer surface of outer nuclear membrane. The origin of rRNA is the NUCLEOLUS. See CELL FRACTIONATION.

RICKETTSIAE. Group of bacterium-like prokaryotes occurring naturally in tissues of arthropods (e.g. fleas, lice, ticks), which can transmit them to mammals where they can produce fatal diseases (e.g. typhus in humans). Some can be transmitted via food as well, notably those causing Q fever and a form of encephalitis, both in untreated goat's milk.

RINGER'S SOLUTION. Physiological saline; aqueous solution, containing sodium, potassium and calcium chlorides, employed in maintaining cells or organs alive *in vitro*. Always appropriately buffered, usually with bicarbonate or phosphate (see BUFFER).

RING SPECIES. Species characterized by circular or looped geogra-

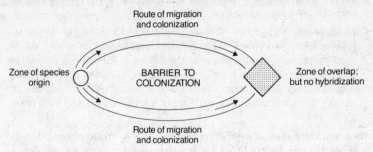

Fig. 58. Ring species.

phical distributions (see Fig. 58), adjacent populations interbreeding on the two arms of the loop but not where arms overlap, i.e. those populations presumably furthest from the original area of spread. See SPECIES, TRANSITIVITY.

RITUALIZED BEHAVIOUR. Behaviour which has become stereotyped in evolution through its role in communication. Many display and threat sequences are ritualized and some DISPLACEMENT ACTIVITIES have become so through selection as information-bearers.

RNA (RIBONUCLEIC ACID). Nucleic acid class, differing from DNA in being usually either single-stranded or looped, in containing ribose not deoxyribose, and in that uracil replaces thymine. Synthesized by RNA POLYMERASES from nucleoside triphosphates (ATP, GTP, CTP, UTP); with 5'- and 3'-ends to the molecule, as in DNA (see DNA).

The three RNA types, *messenger RNA* (mRNA), *transfer RNA* (tRNA) and *ribosomal RNA* (rRNA), are all involved in PROTEIN SYNTHESIS, in both prokaryotic and eukaryotic cells, but in eukaryotes especially, RNA PROCESSING accompanies mRNA production. In RNA viruses, RNA is sometimes double-stranded, serving as genetic material (see REVERSE TRANSCRIPTASE, CENTRAL DOGMA); in some RNA viruses the RNA is transcribed into RNA and in others (*retroviruses*) it is reverse-transcribed into DNA. tRNA molecules fold back upon themselves by complementary base-pairing to form double-stranded 'stems' and single-stranded 'loops'. A loop at one end bears a specific nucleotide triplet (the *anticodon*) while the 3'-end of the molecule carries a tRNA-specific amino acid – both essential for protein synthesis to proceed by means of a GENETIC CODE. Ribosomal RNA subunits associate with protein molecules to form RIBOSOMES. All tRNA and rRNA molecule types are encoded by DNA (see GENE), and there are many more of these molecules per cell than there are of mRNA. Some

RNA molecules have catalytic activity (see RNA PROCESSING, ORIGIN OF LIFE, RIBOZYMES).

RNA CAPPING. Eukaryotic RNAs are processed in nuclei prior to release into the cytosol. This involves attachment of a *cap* of 7-methylguanosine triphosphate to their 5'-end. Ribosomes recognize this cap and commence translation at the AUG codon nearest to the cap, finishing at the first stop codon, ensuring that translation is usually monocistronic. See CODON, RNA PROCESSING, PROTEIN SYNTHESIS.

RNA POLYMERASES. Enzymes producing RNA from ribonucleoside triphosphates. Unlike DNA polymerases they do not require a polynucleotide primer. Three types occur in eukaryotic cells, *polymerase I* making large ribosomal RNAs, *polymerase II* transcribing structural genes (introns and exons), *polymerase III* making small RNAs such as tRNAs and rRNAs. See RNA PROCESSING.

RNA PROCESSING. mRNA transcription within the nucleus produces RNAs of various sizes (*heterogeneous RNA, hnRNA*) which are modified (processed) before passage to the cytosol for translation on ribosomes. The 5'-end of the molecule is first capped (see RNA CAPPING) and then has a long poly-AMP sequence bound to the 3'-end, which may facilitate the rest of processing and passage to the cytosol. Major feature of nuclear processing is the excision from hnRNA of non-coding INTRON sequences. This is achieved by cutting these sections out using a PHOSPHODIESTERASE, and then SPLICING the transcript. This may rejoin one encoding region (exon) to another that is not its official nearest neighbour in the hnRNA. Alternatively, an exon may get cut out. This provides flexibility in eventual protein production and is important to lymphocytes in generating ANTIBODY DIVERSITY. Eukaryotic ribosomal RNA is processed in the nucleus prior to assembly into ribosomes. Both RNA processing and 'gene splicing' are involved in the production of ANTIBODY DIVERSITY by different mature B-cell clones.

RNase (RIBONUCLEASE). Any of several enzymes which hydrolyse RNA by breaking their phosphodiester bonds.

ROD CELL. Highly light-sensitive secondary receptor of vertebrate RETINA, providing monochromatic vision even in very dim light. Its outer segment, closest to the choroid, contains parallel-stacked membranous discs with embedded molecules of *cis-retinal* (see RHODOPSIN), responding even to single photons. Attached to this by a cilium-containing restriction is the mitochondrion-rich *inner segment* which leads to the nuclear region (cell body) with its synaptic processes. Much integration of rod cell output is achieved by horizontal, bipolar and AMACRINE CELLS before ganglion cells lead off via the optic nerve. Each primate retina contains about 120 million rods, few of them in the fovea. See SCOTOPIC VISION, CONE.

RODENTIA. Most widespread and numerous of all mammalian orders. Placentals, they include rats, mice, capybaras, squirrels, beavers and porcupines. Herbivores or omnivores; one pair of large, continually growing, chisel-like incisors in both upper and lower jaws (see DENTITION, LAGOMORPHA). Large diastema; grinding molars.

ROGUE. See INFRASPECIFIC VARIATION.

ROOT. The vascular plant organ that usually grows downwards into the soil, anchoring the plant and absorbing water and nutrient mineral salts. Roots cannot be distinguished from stems on the basis of their position with respect to the soil; some plants have roots wholly above ground and others have underground stems (e.g. RHIZOMES). Externally, roots principally differ from stems in not bearing leaves or buds; they also possess at their tips a protective layer of cells, the ROOT CAP. Internally, roots differ from stems in having their vascular tissue in a central region and with protoxylem exarch. See FIBROUS ROOT, TAP ROOT, ADVENTITIOUS.

ROOT CAP. Cap of loosely arranged cells covering apex of growing point of a root, protecting it from damage when forced through the soil. Formed from promeristem, dermatogen, or from a meristematic layer external to the dermatogen known as calyptrogen.

ROOT HAIR (PILUS). Tubular outgrowth of root epidermal cell, possessing a thin, delicate wall in intimate contact with soil particles. Root hairs are produced in large numbers behind the region of active cell division at root tip, forming the *piliferous* layer. They enormously increase the root's absorbing surface area, taking up some ions actively, and are continually replaced from new root tissue formed by the apical meristem.

ROOT NODULES. Small swellings on roots of leguminous plants (e.g. pea, bean, clover) produced through infection by nitrogen-fixing bacteria. See NITROGEN FIXATION.

ROOT PRESSURE. Pressure under which water passes from living root cells into the xylem; demonstrated by exudation of liquid from the cut end of a decapitated plant. May continue for long periods, arising through maintenance of water potential gradients by active transport of solutes. See CAVITATION, GUTTATION, TRANSLOCATION.

ROTIFERA. Abundant and widespread phylum (or class of ASCHELMINTHES); microscopic unsegmented pseudocoelomates, all aquatic and many capable of producing resistant sexual eggs. Parthenogenesis common; one group (bdelloids) thelytokous. Feed usually by ciliated trochal disc; gut entire, commonly with a muscular pharynx and chitinous 'jaws'. Excretion involves flame cells. Nervous system very simple. Elastic cuticle covers most of body. Extraordinary in being of approx. protozoan size, but at the 'organ' level of organization.

ROUNDWORMS. See NEMATODA.

r-**SELECTION.** Selection tending to operate in scattered and transient habitats, favouring organisms with ability to colonize rapidly, make use of short-lived resources, and complete their reproduction rapidly. Such organisms are selected for high *r*-values (r = intrinsic rate of increase); are clearly exploiters of their environments, have a tendency to PROGENESIS, and are at opposite end of range of strategies from organisms subject to *K*-SELECTION.

RUDERAL. (Of a plant) living in waste places near human habitations.

RUFFLED MEMBRANE. See LAMELLIPODIUM.

RUMEN. Diverticulum of RUMINANT oesophagus; the chamber of the 'stomach' in which storage and initial digestion of food occur.

RUMINANT. Artiodactyl (Suborder Ruminantia) characterized by very complex herbivorous stomach and feeding method. Grazed food is swallowed into the RUMEN and mixed with mucus, undergoing partial and anaerobic digestion by cellulase from a symbiotic bacterial flora. Some products are absorbed here; but 'pulp' can be returned to the mouth as 'cud' for chewing. After this (often done at leisure), the bolus is returned to the RETICULUM and omassum, where water is removed. The *abomassum* is the true stomach, in which further digestion occurs. In some ruminants, food can be seen regularly travelling up and down the oesophagus in the neck. Include deer, antelopes, giraffes, cattle, sheep and goats.

RUNNER. A STOLON that roots at its tip and forms a new plant which is eventually freed from connection with the parent through decay of the runner. Also used horticulturally of the daughter plant itself.

S

SACCHAROMYCES. Genus of YEAST fungi (ASCOMYCOTINA, Endomycetales), including organisms used in the production of bread and alcoholic drinks.

SACCULUS. See VESTIBULAR APPARATUS.

SACRAL VERTEBRAE. Vertebrae of tetrapod lower back, articulating by rudimentary ribs with pelvic girdle. Just one occurs in most amphibia; two or more are fused in other vertebrae. See SACRUM.

SACRUM. Group of fused sacral vertebrae, ilia of the PELVIC GIRDLE being united to some or all of them. See SYNSACRUM.

SALIENTIA. See ANURA.

SALIVA. Fluid secretion of salivary glands (labial glands in insects). In terrestrial vertebrates contains mucus and, in some of these and insects generally, contains amylases (e.g. *ptyalin* in mammals). *Invertases* occur in some insects (e.g. worker bees). Anticoagulants are present in salivas of blood-sucking leeches, insects and vampire bats, while those of the last two groups may contain pathogens.

SALIVARY GLAND CHROMOSOMES. See POLYTENY.

SALTATORY CONDUCTION. See NODE OF RANVIER.

SAMARA. Simple, dry, one- or two-seeded indehiscent fruit, the pericarp bearing wing-like outgrowths (e.g. sycamore).

SAPROPHYTE. See SAPROTROPH.

SAPROTROPH (SAPROPHYTE, SAPROBE). Organism obtaining organic matter in solution from dead and decaying organisms. Usually involves secretion of extracellular enzymes on to the material, followed by absorption of the digestion products. Common in bacteria and fungi engaged in decay; of enormous importance in recycling and making available carbon dioxide, nitrates, phosphates and other nutrients via the DECOMPOSER food chain.

SAPROZOIC. (Of animal) taking up nutrients from the environment in solution; e.g. gut parasites utilizing the host's digestion products, some free-living flagellate protozoans.

SAPWOOD. Outer region of xylem of tree trunks, containing living cells (e.g. xylem parenchyma, medullary ray cells); functions in water conduction and food storage, as well as providing mechanical support. Usually distinguished from HEARTWOOD by its lighter colour.

SARCODINA. Protozoan class characterized by PSEUDOPODIA. Some bear flagella in part of life cycle, supporting unification with flagellate protozoans in the single class Sarcomastigophora. Most are free-living, but some amoebae (e.g. *Entamoeba*) are parasitic. Many secrete shells. Includes subclasses Rhizopoda (amoebae, FORAMINIFERA) and Actinopoda (RADIOLARIA, HELIOZOA).

SARCOMA. See CANCER CELL.

SARCOPLASMIC RETICULUM. Endoplasmic reticulum of muscle fibres; especially important in STRIATED MUSCLE in regulating calcium ion level in myofibril environment. Forms sac-like *terminal cisternae* around transverse tubule system, giving triad appearance in section. See MUSCLE CONTRACTION and Fig. 61a.

SARCOPTERYGII. See CHOANICHTHYES.

SATELLITE DNA. Large repeat sequences of chromosomal DNA forming about 30% of the eukaryotic genome; recondensing rapidly during DNA HYBRIDIZATION. Consists of constitutive HETEROCHROMATIN. The three major repeat sequences in humans are subject to rapid evolutionary change. See REPETITIVE DNA.

SAURISCHIA. Extinct order of lizard-hipped reptiles forming (with ORNITHISCHIA) the dinosaur archosaurs. Distinguished by pubes pointing downwards and forwards and ischia pointing downwards and backwards (see PELVIC GIRDLE). Included *theropods* (bipedal carnivores, e.g. *Tyrannosaurus, Allosaurus*) and herbivorous *sauropodomorphs* (e.g. *Apatosaurus* (= *Brontosaurus*), *Diplodocus*). The last two were the largest land animals yet discovered, a semi-aquatic life supporting their bulk.

SAVANNA. Tropical or subtropical grassland containing scattered trees; transitional regions between evergreen tropical rain forests and deserts.

SCALARIFORM THICKENING. Internal thickening of cell wall of xylem vessel or tracheid, taking the form of more or less transverse bars, suggestive of the rungs of a ladder.

SCALE. (Bot.) In some algae, an external element covering the cell. May be organic or inorganic in composition, with intricate ornamentation (e.g. some members of the CHRYSOPHYTA, PRYMNESIOPHYTA); in vascular plants, a platelike outgrowth. (Zool.) (1) An epidermal structure characteristic of most vertebrate groups, usually plate-like or tooth-like, and serving for protection. Employed in fish classification (see PLACOID, COSMOID and GANOID SCALES). Some fish, notably some eels and trout, have scales so small and/or so deeply embedded that they appear to be scaleless. Early AMPHIBIA were scaled, as are many reptiles (e.g. SQUAMATA), in some of which bony as well as horny scales may be present. Birds have scaly legs;

some mammals (e.g. rodents) have scaly tails. The production and deployment of vertebrate scales is termed *squamation*. (2) In some insects (e.g. butterflies and moths), one of numerous hollow sac-like ectodermal secretions attached to wing membranes, containing air or pigment. Their distributions may produce interference patterns, providing colouration.

SCALE INSECTS (MEALY BUGS). Hemipterans (Superfamily Coccoidea); females wingless and scale-like, gall-like or covered in waxy secretion. Males usually with one pair of wings and vestigial mouthparts. Important economic pests; cottony cushion scale (*Icerya purchasi*) damages citrus fruits, others coconuts. Control by coccinellid beetles (ladybirds) is usually successful.

SCAPE. Leafless flowering stem arising from ground level (e.g. dandelion, daffodil).

SCAPHOPODA. Small class of bilaterally symmetrical marine molluscs, (tusk shells, e.g. *Dentalium*) somewhat intermediate between bivalves and gastropods. The tubular shell opens at both ends, while the reduced foot is used in burrowing. Ctenidia absent; larva a trochosphere.

SCAPULA. Dorsal component of vertebrate PECTORAL GIRDLE; the shoulder-blade of mammals.

SCHISTOSOMA. Blood fluke (see TREMATODA) of Africa (esp. Egypt), S. America and China causing *bilharzia* (= *schistosomiasis*) in man. Smaller male lives attached to female within a groove of her body; eggs laid in abdominal blood vessels penetrate walls (causing most damage) and pass to bladder. Larval miracidia hatch in urine diluted on contact with freshwater and burrow into snail, in which rediae develop and from which cercariae emerge and penetrate skin of bathing/rice-planting people, debilitating over 200 million worldwide. Three common species infect humans. Education, support of irrigation engineers, molluscicides and medically prescribed drugs (e.g. antimony-containing) are all involved in prevention/cure.

SCHIZOCARP. Dry fruit formed from a syncarpous ovary that splits at maturity into its constituent carpels forming several partial fruits, which are usually single-seeded, resemble ACHENES, and are called *mericarps*; e.g. hollyhock, mallow, geranium.

SCHIZOCOELY. Mode of formation of the COELOM in most annelids, in arthropods, molluscs and higher chordates. The coelom arises *de novo* as a cavity in the embryonic mesoderm. Compare ENTEROCOELY.

SCHIZOGENOUS. (Bot.) (Of structure) formed by splitting due to separation of cells; e.g. oil-containing cavities in leaves of St. John's wort. See LYSIGENOUS.

SCHIZOMYCOPHYTA (SCHIZOMYCETES). In old classifications, denoting BACTERIA when they were regarded as fungi.

SCHWANN CELL. Specialized vertebrate GLIAL CELL, derived from NEURAL CREST, ensheathing axons of peripheral nervous system and responsible for formation there of the MYELIN SHEATH. See NODES OF RANVIER.

SCION. Twig or portion of twig of one plant grafted on to the stock of another.

SCLERA (SCLEROTIC). The white of the vertebrate EYE. Coat of fibrous or cartilaginous tissue covering whole eyeball except for cornea.

SCLEREID. A SCLERENCHYMA cell possessing a thickened, lignified secondary wall having many pits. Variable in form but not typically very long. They may or may not be alive at maturity.

SCLERENCHYMA. Tissue giving mechanical support to plants. Cells are thick-walled, usually with lignin deposited so thickly in and on their primary cell walls as to leave only a very small lumen. Usually without living protoplasts at maturity. Two major types: *fibres* and *sclereids*. Fibres are very elongated cells with tapering ends, occurring singly or variously grouped into strands, forming caps of vascular bundles in herbaceous dicots, and cylinders around those of many monocots; they are used in manufacture of rope, linen, paper, etc. (being largely cellulose in flax, lignified in hemp). Sclereids occur singly or in groups and are common in fruits such as pear and nuts and in seed coats. See COLLENCHYMA.

SCLERITE. Region of arthropod CUTICLE in which the exocuticle is fully differentiated and *sclerotized*. Between two such sclerites there is usually a region of flexible membranous (unsclerotized) cuticle where exocuticle is undeveloped, allowing for articulation at joints. The body wall of a typical arthropod segment is divisible into four sclerotized regions: dorsal tergum, ventral sternum and a lateral pleuron on each side. Sclerites of the tergum are *tergites*, those of the sternum are *sternites* and those of each pleuron are *pleurites*.

SCLEROPROTEINS. Group of insoluble proteins forming major components of CONNECTIVE TISSUES. Include COLLAGEN, KERATIN and ELASTIN.

SCLEROSIS. (1) Hardening, as occurs in vertebrate tissues after injury. Involves deposition of COLLAGEN. (2) *Atherosclerosis* involves deposition of CHOLESTEROL and triglycerides in arterial walls, often in response to damage to their endothelial walls. See LIPOPROTEIN and references there.

SCLEROTIC. See SCLERA.

SCLEROTIUM. (Of fungi) compact tissue-like mass of fungal hyphae, often possessing a thickened rind, varying in size from that of a pin's head to that of a man's head; capable of remaining dormant for long, perhaps unfavourable, periods, and commonly giving rise to fruiting bodies; e.g. ERGOT.

SCLEROTIZATION. Tanning of arthropod CUTICLE. Involves cross-linking of protein by quinones (tyrosine derivatives). In some insects (e.g. *Calliphora* larvae), a quinone precursor is released onto the cuticle surface via pore canals and subsequently enzymatically oxidized by a phenol oxidase, tanning the protein cuticulin in the epicuticle and diffusing inwards to tan the outer procuticle to produce exocuticle. This is a major cause of cuticles becoming hard and brittle.

SCLEROTOME. See MESODERM.

SCOLEX. Part of tapeworm attached by suckers and hooks to gut wall of host; sometimes called the head. Proglottides are budded off behind it.

SCORPIONES. Order of ARACHNIDA containing scorpions. Pedipalps form large pincers, chelicerae small ones. End part of abdomen a segmented flexible tail bearing sting. Viviparous; terrestrial. See PSEUDOSCORPIONES.

SCORPION FLIES. See MECOPTERA.

SCOTOPHILE. See SKOTOPHILE.

SCOTOPIC VISION. Dark-adaptation of eye. Decrease in threshold of sensitivity of the ROD CELLS with increasing length of time in the dark. Involves enzymatic resynthesis of RHODOPSIN, increase in sensitivity of cells proceeding faster than the resynthesis. See EYE.

SCROTUM. Pouch of skin of perineal region of many male mammals, containing testes (at least during breeding season) and keeping them cooler (by 2 °C in humans) than body temperature, without which sperm formation is impaired. Female spotted hyaenas have a pseudo-scrotum, important in formation of dominance hierarchy.

SCUTELLUM. Single cotyledon of a grass embryo, specialized for absorption of the endosperm.

SCYPHISTOMA. Polyp stage in life cycles of SCYPHOZOA, during which the ephyra larvae are budded off by a kind of strobilation.

SCYPHOZOA. Class of CNIDARIA containing jellyfish. Life cycle exhibits ALTERNATION OF GENERATIONS. Medusoid stage the rather complex adult jellyfish (gonads endodermal; enteron of four pouches, with complex connecting canals). Fertilized eggs develop into ciliated planula larvae which settle to become SCYPHISTOMAS.

SEA ANEMONES. See ACTINOZOA.

Sea cucumbers. See HOLOTHUROIDEA.

Seals. See PINNIPEDIA.

Seaweeds. Red, brown or green algae living in or by the sea.

Sebaceous gland. Mammalian holocrine skin gland opening into hair follicle. Secretes oily lipid-containing *sebum* that helps to waterproof fur and epidermis. Epidermal in origin but projecting into dermis.

Secondary growth (secondary thickening). In plants, growth derived from secondary or lateral meristems, the vascular or cork cambia. Results in increase in diameter of gymnophyte and dicotyledonous stems and roots, providing additional conducting and supporting tissues for growing plants and, in most cases, makes up greater part of mature structure. Rare in monocotyledons.

Secondary meristem. (Bot.) Region of active cell division that has arisen from permanent tissues, e.g. cork cambium (phellogen), wound cambium. See PRIMARY MERISTEM.

Secondary sexual characters. Characters in which the two sexes of an animal species differ, excluding gonads, their ducts and associated glands. In humans include mammary glands, subcutaneous fat deposition, shape of pelvic girdle, voice pitch, mean body temperature, mass and extent of muscle development. See OESTROGENS, TESTOSTERONE.

Secondary wall. Innermost layer of plant CELL WALL, formed by some cells after elongation has finished; possesses highly organized microfibrillar structure.

Second messenger. Organic molecules and sometimes metal ions, acting as intracellular signals, whose production or release usually amplifies a signal such as a hormone, received at the cell surface. Some hormones bind to the cell membrane and activate an enzyme there to generate the second messenger. Alternatively, the ligand may be a non-hormone which opens or closes a GATED CHANNEL affecting membrane permeability to an ion. Calcium ion (Ca^{++}) concentration is extremely important in control of many cell functions (see CALCIUM PUMP, CALMODULIN). First organic molecule hailed as a second messenger was cyclic AMP (see AMP) but others have been discovered. See CASCADE, INOSITOL 1,4,5-TRIPHOSPHATE.

Secretin. Polypeptide hormone of intestinal mucosa, produced there in response to acid chyme from stomach. Inhibits gastric secretion, reduces gut mobility but stimulates secretion of alkaline pancreatic juice, bile production by liver and intestinal secretion. See CHOLECYSTOKININ, GASTRIN.

Secretion. (1) Production and release from a cell of material useful either to it or to the organism of which it is a part. Material is commonly packaged into *secretory vesicles* which bud off from the GOLGI APPARATUS and fuse with the cell's apical plasmalemma (see MEROCRINE GLAND); other methods are employed by APOCRINE and HOLOCRINE GLANDS. Plant cell walls are secreted. (2) *Active secretion*. Process whereby a substance (often an ion) is actively pumped out of a cell against its concentration gradient, as in the ascending loop of Henle in the vertebrate kidney, and in fish gills. See ACTIVE TRANSPORT.

SEDIMENTATION COEFFICIENT. See SVEDBERG UNIT.

SEED. Product of fertilized ovule, comprising an embryo enclosed by protective seed coat(s) derived from the integument(s). Some seeds (castor oil, pine) are provided with food material in the form of ENDOSPERM tissue surrounding the embryo, while in other (*nonendospermic*) seeds food material is stored in the cotyledons (e.g. pea). The seed habit is the culmination of an evolutionary development involving, in sequence, heterospory, reduction of a free-living female gametophyte generation dependent on water for fertilization, and its retention within the tissues of the sporophyte by which it is protected and supplied with food. It occurred early in geological history, and in more than one group of plants. Of equal importance was the reduction of the male gametophyte to the pollen grain and its pollen tube, again avoiding the need for water in fertilization. Independence from water in sexual reproduction increased immensely the range of environments open for colonization. Biological advantages of the seed habit include: continued protection and nutrition of the embryonic plant during its development, provision of dispersal mechanisms, provision of food to tide over critical periods of growth of the embryo, and its establishment as an independent plant after the seed is shed. See LIFE CYCLE.

SEGMENTATION. (1) Alternative term for CLEAVAGE of an egg. (2) In zoology, commonly synonym for *metameric segmentation* or *metamerism*, to indicate production of a body plan of repeating organizational units, variably distinguishable, along the antero–posterior body axis. Metamerism most marked in annelids and arthropods (see TAGMOSIS) but vertebrates exhibit it in the segmental organization of nervous system and muscles, most clearly in the embryo (see MESODERM). Has often become reduced or wholly lost in evolution, notably in molluscs and echinoderms. Debate surrounds role of the HOMOEOBOX in control of segmentation. See COMPARTMENT.

SEGMENT-POLARITY GENES. A class of segmentation genes, studied particularly in *Drosophila*, mutants in which cause not only a loss of the same part in each segment, but also a mirror-image duplication of the remaining part to be produced in its place. In *Drosophila* at least, domains of mRNA transcript expression of segment-polarity genes are dependent largely upon prior distributions of GAP GENE expression domains. For instance, *engrailed* (*en*) and *wingless* (*wg*) express as stripes of transcript in the early embryo, *wg* stripes representing the posterior limits of PARASEGMENTS, while *en* stripes represent anterior limits. It is likely that some at least of these genes encode transmembrane proteins involved in translating extracellular signals into changes in cell fate. Compare PAIR-RULE GENES.

SEGREGATION. (1) See MENDEL'S LAWS. (2) Process whereby alleles usually present together in somatic cells of an organism separate into different cells during meiosis in the germ line. Such alleles are said to segregate.

SEGREGATION DISTORTER (SD). See ABERRANT CHROMOSOME BEHAVIOUR (4).

SEISMONASTY. (Bot.) Response to non-directional shock stimulus; e.g. rapid folding of leaflets and drooping of leaves in *Mimosa pudica* when lightly struck or shaken.

SELACHII. Elasmobranch Order containing modern sharks. Members have five to seven gill openings on each side of head and an upper jaw not fused to skull. Group appeared in the Jurassic and is characterized by the *rostrum* (snout) that hangs over the mouth. Includes largest living fishes (over 20 metres in whale sharks, *Rhinedon*). Rays and skates (Batoidea) are sometimes also included. See ELASMOBRANCHII.

SELECTION. See ARTIFICIAL SELECTION, NATURAL SELECTION.

SELECTION COEFFICIENT (s). See COEFFICIENT OF SELECTION, FITNESS.

SELECTION PRESSURE. See NATURAL SELECTION.

SELF-ASSEMBLY. Usually reserved for biological structures which become assembled from components without help of enzymes or 'scaffolds' which do not form part of the functional structure. Some VIRUSES (e.g. tobacco mosaic virus, TMV) assemble within the host cell from separate nucleic acid and capsomere protein molecules. Several other structures are either fully self-assembling (e.g. bacterial ribosome) or require surprisingly few enzyme steps or accessory proteins acting as 'jigs' in assembly. Some viruses, membranes, cilia, mitochondria and myofibrils are in this second category. It is thought that some steps in these more complex cases of partial self-assembly require appropriate timing, and are irreversible if disassembly is imposed upon them.

SELF-FERTILIZATION. Fusion of micro- and macrogametes from the same individual. See INBREEDING.

SELF-POLLINATION. Transfer of pollen from anther to stigma of same flower, or to stigma of another flower of same plant. See INBREEDING.

SELF-STERILITY. (Of some HERMAPHRODITES) inability to form viable offspring by self-fertilization. See INCOMPATIBILITY.

SEMEN. Sperm-bearing fluid produced by the testes and accessory glands (e.g. PROSTATE GLAND, SEMINAL VESICLES and Cowper's gland in mammals), particularly of animals with internal fertilization.

SEMICIRCULAR CANALS. Component of vertebrate VESTIBULAR APP-ARATUS detecting (directional) acceleration of the head.

SEMINAL VESICLE. (1) Organ of lower vertebrates and of some invertebrates (e.g. earthworms, insects) that stores sperm. (2) Diverticulum of the *vas deferens* (sperm duct) of male mammals, whose alkaline secretions lower the pH of the semen and counteract vaginal/uterine acidity.

SEMINIFEROUS TUBULES. See TESTIS.

SEMISPECIES. Populations in the process of acquiring mechanisms isolating one from the other reproductively. Regarded therefore as *incipient species* in the sense of the biological species concept (see SPECIES). Semispecies are generally expected to exhibit greater genetic differences than do geographical subspecies (see INTRASPECIFIC VARIATION).

SENESCENCE. See AGEING.

SENSE ORGAN. Group of sensory RECEPTORS and associated non-sensory tissues specialized for detection of one sensory modality (e.g. light in the EYES, sound in the EARS, etc.).

SENSITIZATION. Process rendering an organism or cell more reactive to a specific ANTIGEN or antigenic determinant. See IMMUNITY, ALLERGIC REACTION.

SEPAL. Component member of the calyx of dicotyledonous flowers; usually green and leaf-like. See FLOWER.

SEPTICIDAL. (Bot.) Describing the dehiscence of multilocular capsules by longitudinal splitting along septa between the carpels, separating the carpels from one another, e.g. St John's wort. See LOCULICIDAL.

SEPTUM. Partition or wall. The structure concerned is said to be *septate*.

SERE. Particular example of plant succession. Seres originating in water are referred to as HYDROSERES; those arising under dry conditions as XEROSERES, of which those developing upon exposed rock surfaces are known as LITHOSERES.

SERIAL HOMOLOGY. See HOMOLOGY.

SERIAL SECTIONS. Series of successive microtome sections, from which three-dimensional structure can be built up.

SERINE PROTEASES. Family of homologous proteolytic enzymes (see GENE DUPLICATION) including several involved in digestion (e.g. chymotrypsin, trypsin) and BLOOD CLOTTING (e.g. thrombin). Implicated in the genesis of POSITIONAL INFORMATION during *Drosophila* development. See CASCADE.

SEROLOGY. The study of ANTIGEN-ANTIBODY REACTIONS in vitro.

SEROSA. Outermost layer of most parts of vertebrate GUT. A SEROUS MEMBRANE.

SEROTONIN (5-HYDROXYTRYPTAMINE, 5-HT). Tryptophan-derived NEUROTRANSMITTER of vertebrate brain, especially of PONS, and of PINEAL GLAND (a precursor of *melatonin*). Hyperpolarizes postsynaptic membranes, and activates *phosphofructokinase* in liver (see GLYCOLYSIS), both mediated via cyclic AMP. See also PLATELETS, BLOOD CLOTTING.

SEROUS MEMBRANES. Mesothelial layers overlying deeper connective tissue and lining the vertebrate coelomic spaces (pericardial, perivisceral, pleural, peritoneal cavities).

SERTOLI CELLS. See TESTIS, MATURATION OF GERM CELLS.

SERUM. See BLOOD SERUM.

SESAMOID BONE. Bone (e.g. patella) developing within tendon of vertebrate, particularly where tendon operates over ridge of underlying bone.

SESSILE. (1) (Of animals) living fixed to the substratum, e.g. sponges, corals, barnacles, limpets, tunicates. (2) Lacking stalks, e.g. eyes of some crustaceans. Opposite of *pedunculate* in plants.

SETA. (Bot.) Stalk of sporagonium in mosses and liverworts; in algae, a stiff hair, bristle or bristle-like process – an elongated hollow cell extension. (Zool.) Invertebrate epidermal bristle, consisting solely of cuticular material (CHAETA), or of a hollow projection of cuticle enclosing epidermal cell or its part (e.g. in insects).

SEWALL WRIGHT EFFECT. See GENETIC DRIFT.

SEX. (1) Often used as synonym of SEXUAL REPRODUCTION. (2) Used in several senses. *Germ cell sex* distinguishes individuals by their abilities to produce gametes of particular morphological types, namely, microgametes (sperm, generative nuclei, etc.), or macrogametes (eggs, egg cells, etc.). Males (with 'male' sex organs) produce the former; females (with 'female' sex organs) produce the latter. HERMAPHRODITES produce both, either simultaneously or sequentially. *Genetic sex* concerns an individual's genotype in so far as it bears on SEX DETERMINATION; *phenotypic sex* relates to anatomical appearances normally associated with one or other sex (e.g.

SECONDARY SEXUAL CHARACTERS), sometimes distinguished from *behavioural sex*, where the two sexes behave in distinctive ways. *Hormonal sex*, identified by the particular hormonal production from an individual's sex organs, is determined by sex organ appearance and physiology rather than by genotype (see SEX-REVERSAL GENE). *Brain sex* refers to distinctive anatomical differences between the brains of the two sexes, itself sometimes causally related to behavioural sex.

The *origin and evolution of sex* are problematic, but the processes common to all forms of genetic recombination from cellular DNA REPAIR MECHANISMS. Some hold that since the only advantages of sexual reproduction over asexual seem to be long-term ones, some kind of GROUP SELECTION is necessary to account for prevalence of the former. Others argue that natural selection alone can account for eukaryotic sexual reproduction, despite the COST OF MEIOSIS. See RECOMBINATION.

Separation of sexes (*dioecism*) has the effect of reducing the potential number of individuals between which fertilization may occur (see INCOMPATIBILITY).

SEX CHROMATIN. See BARR BODY.

SEX CHROMOSOME. Chromosome having strong causal role in SEX DETERMINATION, usually present as a homologous pair in nuclei of one sex (the HOMOGAMETIC SEX) but occurring either singly (or with a partial homologue) in those of the other sex (the HETERO-GAMETIC SEX). In cases of partial homology one chromosome (Y) is often much smaller than the other (X), and both may resemble autosomes. In birds, where females are heterogametic, males are sometimes given the genotype ZZ and females ZW. Sometimes sex-determining loci are situated on such a short region of a single chromosome pair as to make sex chromosomes indistinguishable in appearance. In species with an XX/XY system (e.g. humans) sex chromosomes usually pair up at meiosis and form a bivalent; in mammals, there may be crossing-over and chiasma formation between homologous (often sub-terminal) regions. Organisms with both types of sex organ combined in one individual (e.g. monoecious plants, hermaphrodites) lack specialized sex chromosomes. See AUTO-SOME, HEMIZYGOUS, SEX LINKAGE.

SEX-CONTROLLED CHARACTER. See SEX-LIMITED CHARACTER.

SEX DETERMINATION. Control of occurrence of, and differences between, sexes (see SEX). Where male and female differentiation occurs within a single individual (e.g. a HERMAPHRODITE), it is commonly restricted to the sex organs. See GYNANDROMORPH, FREEMARTIN.

Bisexual (dioecious) species often have genetic sex-determining mechanisms. These are occasionally cytoplasmic (see CYTOPLASMIC

INHERITANCE), sometimes depend upon the individual's ploidy (see MALE HAPLOIDY), but most commonly involve a pair of SEX CHROMOSOMES.

In some cases (e.g. many gonochorist species of fish, amphibians and reptiles) one or a few loci on an unspecialized chromosome pair may be responsible for sex determination through segregation of a pair or a few pairs of alleles (resembling some INCOMPATIBILITY mechanisms). In cases of sex organ differentiation within a single hermaphrodite individual (e.g. some moss and fern gametophytes) the mechanism involved resembles normal cytological DIFFERENTIATION. Sex-determining loci may not be restricted to sex chromosomes, even where these are differentiated (e.g. see SEX REVERSAL GENE). But where sex is determined by a differentiated sex chromosome pair it may be absence of a second X or presence of a Y that results in sex differentiation. In humans it is the latter. Individuals with KLINEFELTER'S SYNDROME (XXY) are male while those with TURNER'S SYNDROME (XO) are female. In mammals, the Y-chromosome encodes a testis-determining factor (TDF in humans) which causes embryos with a Y-chromosome to develop testes and become males, while those lacking a Y-chromosome develop ovaries and become females. But in the fruit fly *Drosophila* the X-chromosome is female-determining while the autosomes collectively are male-determining, an individual's sex resulting from the balance, or ratio, between the number of sets of autosomes and X-chromosomes. This was established by artificial production of INTERSEXES which are triploid for their autosomes but diploid (XX) for their sex chromosomes. 'Superfemales' were diploid for their autosomes but triploid for sex chromosomes (XXX); 'supermales' were triploid for their autosomes but hemizygous for an X-chromosome (XO). So an XY/XX sex chromosome system may in itself tell us little about the sex-determining mechanism involved. The worm *Bonellia* (Echiuroidea) has environmental sex-determination, albeit with genetic involvement. If a larva develops independently it becomes female; if it is influenced by pheromones produced by the adult female's proboscis it develops into a male. In some reptiles (e.g. turtles, crocodilians) sex is strongly determined by the temperature at which the embryo develops.

SEXDUCTION (F-DUCTION). See TRANSDUCTION, F FACTOR.

SEX FACTOR. See F FACTOR.

SEX HORMONES. See OESTROGENS, TESTOSTERONE.

SEX-LIMITED CHARACTER (SEX-CONTROLLED CHARACTER). Character determined by a genetic element expressed differently between the two sexes, commonly as a result of hormonal differences. *Pattern baldness* in humans is a character expressed in men who are either homozygous or heterozygous for the allele, but only in those women who are homozygous for it.

SEX LINKAGE. A character is said to be sex-linked if it is determined by a genetic element occurring either (a) on an X-chromosome where the method of SEX DETERMINATION involves an XX/XO system, or (b) on any non-homologous region of a pair of SEX CHROMOSOMES where the method involves an XX/XY system. Most readily detected by an appropriate RECIPROCAL CROSS. In an XX/XY system, if the character in question is recessive it will appear in the heterogametic sex with the same frequency (say, n) as the chromosome bearing the element; hence it will appear in the homogametic sex with the square of that frequency (i.e. n^2). In humans, red-green colour blindness and haemophilia are such examples. *X-linked* characters cannot be transmitted from father to son since a father contributes a Y-chromosome to his male offspring. A mother may be a carrier, in which case half her male offspring would tend to be affected. Dominant sex-linked characters would be expressed in hemizygous, heterozygous and homozygous conditions.

SEX PILI. See PILI, F FACTOR.

SEX RATIO. Proportion of males to females in a species population, usually expressed as the number of males per 100 females. *Primary sex ratio* is assessed immediately after fertilization; *secondary sex ratio* is assessed at birth (or hatching); *tertiary sex ratio* is assessed at maturity. The sex ratio among the offspring of a particular female may be subject to fluctuations, as with queen bees and other female eusocial insects. This is the subject of much theorizing and investigation.

SEX-REVERSAL GENE (*Sxr*). Dominant autosomally determined character of mice causing the somatic gonadal tissues of females to develop as testis. This then secretes testosterone and brings about male phenotype despite the presence of female genotype. See TESTICULAR FEMINIZATION.

SEX-ROLE REVERSAL. Occurrence, notably in some bird species (e.g. some arctic wading birds), where females are more brightly coloured than males and display for mates, sometimes leaving them to incubate the offspring alone.

SEXUAL DIMORPHISM. Occurrence of populations where individuals differ in respect of two distinct sets of phenotypic characters, sex itself being one of them. Some cases may be attributable to SEXUAL SELECTION.

SEXUAL REPRODUCTION. In its broadest sense, any process by which genetic material is transferred from one cell, and/or individual, to another (see CONJUGATION); or, any process involving genetic RECOMBINATION. Such a view of sexual reproduction would have to include PLASMID transfer, viral infection, antibody production and even some forms of GENE MANIPULATION. One possible benefit of throwing the concept wide open this way is that it may reveal clues as to the possible origin of SEX.

The processes of *sex* (recombination and/or genetic transfer) and *reproduction* (production of new individuals) are sometimes only tenuously linked (e.g. in PARTHENOGENESIS in *Paramecium*), and may even be uncoupled. When sexual events are accompanied by production of new individuals, only very rarely will any offspring produced be genetic copies of any previous individuals.

Excision of PLASMIDS and their incorporation within the genome of an unrelated recipient species raise questions about the validity of SPECIES definitions which rely on the concept of a shared GENE POOL. In prokaryotes, lacking MEIOSIS and FERTILIZATION, sexual processes commonly involve a specialized form of conjugation (e.g. see F FACTOR). In eukaryotes, the distinctive features of sexual LIFE CYCLES are meiosis and fertilization; but gametes are not invariably produced by meiosis, and fertilization is not always followed by mitosis. The genetic recombination normally associated with sexual processes tends to promote GENETIC VARIATION among offspring (but see INBREEDING). This is generally considered adaptive in unstable or patchy environments. However, meiosis has a potentially disruptive effect upon coadapted gene complexes.

SEXUAL SELECTION. Form of selection, generally contrasted with NATURAL SELECTION although also proposed by DARWIN. Results from the exercise of mating preferences (by either sex, but most commonly by females) in favour of individuals expressing certain genetically determined characters. As a result, genes for these characters tend to spread through the population, being expressed as SEX-LIMITED CHARACTERS. These could be defined as 'attractive' to the choosing sex.

This is a form of natural selection, a male's fitness being enhanced by his expression of 'attractive' characters, a female's fitness being improved by contributing genetically both to male offspring likely to have these characters and to female offspring likely to be genetic carriers of them. Sexual selection is often invoked to explain cases of extreme SEXUAL DIMORPHISM (e.g. where males are huge compared with females), and such dimorphisms do tend to evolve where males are polygynous. Sexual selection may enhance the directional component of selection, but its role in the spread of genes which have undramatic phenotypic effects is the subject of debate and experimentation. See ARMS RACE.

SHARED HOMOLOGUE. A character found in two or more taxa whose most recent common ancestor also had (or can be inferred to have had) it. See SYMPLESIOMORPHY, SYNAPOMORPHY.

SHARKS. See SELACHII.

SHELL. No clear-cut definition; includes many hardened animal secretions having protective and/or skeletal roles, often attached only to part of body surface. Structures termed shells are found in many animal groups, e.g. the SARCODINA, MOLLUSCA, BRACHIOPODA, in CRUSTACEA (a *carapace*) and in echinoid echinoderms (a *corona*).

SHORT-DAY PLANTS. See PHOTOPERIODISM.

SHOULDER GIRDLE. See PECTORAL GIRDLE.

SIBLINGS (sibs). Brothers and/or sisters; offspring of the same parents.

SIBLING SPECIES. Very closely related SPECIES differing only in minor respects, or appearing identical, but in fact reproductively isolated. Separation is often important where one of the species is vector to an economically important parasite (as in *Simulium* and *Anopheles* spp.), when DNA probes may be used to distinguish them. See SUPERSPECIES.

SICKLE-CELL ANAEMIA. A hereditary, genetically determined disorder affecting many newborn African and American negroes and others of tropical climates where malaria is, or has recently been, endemic. Caused by homozygosity for allele Hb^S, producing a single amino acid substitution in the β-chain of the normal haemoglobin molecule determined by allele Hb^A. Individuals heterozygous for the allele ($Hb^A Hb^S$) are resistant to the most serious form of MALARIA (subtertian malaria caused by *Plasmodium falciparum*; see HETEROZYGOUS ADVANTAGE), and allele Hb^S therefore attains a high frequency in malarial areas. However the homozygote $Hb^S Hb^S$ suffers from sickle-cell anaemia, so-called because red blood cells have a sickle shape even at high blood oxygen levels and block capillaries, becoming phagocytozed and causing anaemia and other symptoms. An example of NATURAL SELECTION in humans.

SIEVE AREA. Area of sieve element wall containing clusters of pores through which protoplasts of adjacent sieve elements are connected. See PHLOEM, SIEVE PLATE.

SIEVE CELL. See PHLOEM.

SIEVE PLATE. End-wall of SIEVE TUBE MEMBER, where pores are larger than in lateral sieve area; highly differentiated.

SIEVE TUBE. Series of SIEVE TUBE MEMBERS, arranged end-to-end and interconnected by SIEVE PLATES. Functions in transport of food materials (e.g. sucrose, amino acids). Lacking nuclei at maturity. See TRANSLOCATION.

SIEVE TUBE MEMBER. Component of a SIEVE TUBE. Elongated, unlignified tube-like cells found primarily in flowering plants, but also in some brown algae (Laminariales); typically associated with a COMPANION CELL. See PHLOEM.

SIGMA FACTOR (σ FACTOR). Protein component of prokaryotic RNA POLYMERASES binding loosely to the core enzyme and restricting mRNA transcription to one of the two DNA strands and appropriate promoter region (see JACOB-MONOD THEORY). Core RNA polymerases (lacking sigma factor) tend to start transcription randomly, on either DNA strand. The σ factor tends to dissociate itself from the RNA polymerase after a few RNA nucleotides have been incorporated, when its place may be taken by elongation factors involved in chain elongation and termination. Eukaryotic RNA polymerases may have sigma factors too. See RHO FACTOR, PRIBNOW BOX, TRANSCRIPTION FACTORS.

SILICOFLAGELLATES. Members of the CHRYSOPHYTA (DICTYCHALES) possessing cells with an external silicified skeleton. Originated in the Cretaceous. Fossil forms often found in calcareous chalks, with members of the PRYMNESIOPHYTA. Today, these algae form an important part of the phytoplankton of colder seas.

SILICULA. See SILIQUA.

SILIQUA (SILIQUE). Special type of capsule found in cabbages and related plants (fam. Cruciferae). Dry, elongated fruit, formed from an ovary of two united carpels and divided by central false septum into two compartments (locules). Dehiscing by separation of carpel walls from below upwards, leaving the septum bounded by persistent placentas (replum), with seeds adhering to it. The *silicula* has a similar structure, but is short and broad; e.g. honesty, shepherd's purse.

SILK. Fibroin (a β-keratin protein) produced by modified salivary glands of silkworms (silkmoth larvae) and by spiders, among other arthropods. Has small amino acid R-groups, while anti-parallel polypeptide molecules hydrogen-bond to form very stable β-pleated sheets.

SILURIAN. Period of Palaeozoic era (from about 440–400 Myr BP). Its deposits have produced scales of ostracoderm fish; common fossils include corals, crinoids, trilobites, brachiopods and graptolites. See GEOLOGICAL PERIODS.

SINANTHROPUS (PEKING MAN). See HOMO.

SINOATRIAL NODE. See PACEMAKER.

SINUS. (Bot.) Space or recess between two lobes of leaf or other expanded organ. (Zool.) A *blood sinus* is an expanded vein, particularly of selachian fish; sometimes also used of the HAEMOCOEL. *Nasal sinuses* are air-filled spaces within some facial bones of mammals lined by mucous membrane and communicating with the nasal cavity.

SINUSOID. Type of CAPILLARY.

SINUS VENOSUS. Chamber of vertebrate HEART, between veins and auricle(s). Thin-walled; absent from adult birds and mammals. See PACEMAKER.

SIPHONACEOUS (SIPHONOUS). Coenocytic; e.g. members of several orders of algae, which are filamentous, sac-like or tubular, without cross-walls.

SIPHONAPTERA. Order of endopterygote insects. Fleas. Secondarily wingless, with legless detritus-feeding larvae; exarate pupa enclosed within a cocoon. Mouthparts consist of long serrated mandibles, short palped maxillae and reduced palped labium. Laterally compressed, with legs adapted for running between body hair and for jumping. Include human flea *Pulex irritans* and rat flea (transmitter of bacterial plague) *Xenopsylla cheopis*.

SIPHONOPHORA. Order of CNIDARIA containing complex and polymorphic colonial animals. Colonies often form by budding from original medusoid individual.

SIPHONOSTELE (SOLENOSTELE). Type of STELE containing hollow cylinder of vascular tissue surrounding a pith. See ECTOPHLOIC and AMPHIPHLOIC.

SIPHUNCULATA (ANOPLURA). Sucking lice. Wingless exopterygotan insects, ectoparasitic on mammals. Eyes (and ocelli) reduced or absent; mouthparts modified for piercing and sucking. Thorax of fused segments. The human louse *Pediculus humanus* can carry *Rickettsia prowazeki*, distributed via the insect's faeces and causing epidemic typhoid fever.

SIRENIA. Order of placental mammals. Manatees, dugongs (seacows). Aquatic (coastal and river-dwelling) and herbivorous, with transversely expanded tail, front legs modified as flippers and vestigial hindlegs. One pair of mammary glands. Not closely related to cetaceans.

SIVAPITHECUS. See RAMAPITHECUS.

SKATES. See ELASMOBRANCHII.

SKELETAL MUSCLE. See STRIATED MUSCLE.

SKELETON. See ENDOSKELETON, EXOSKELETON.

SKIN (CUTIS). Vertebrate organ covering most of body surface, comprising *epidermis* (ectoderm) above, produced by the MALPIGHIAN LAYER, with varying amounts of underlying connective tissue of *dermis* (mesoderm) and *subcutaneous fat* (adipose tissue). Often impervious to water, but permitting gaseous exchange in modern amphibia. Produces, variously, SCALES, HAIR, FEATHERS, NAIL and sometimes bone (see DERMAL BONE). Glands located here include mucus glands (fish, amphibia), and sebaceous and sweat glands (mammals). May be variously pigmented (see CHROMATOPHORE) and contain several kinds of receptors in the dermis (e.g. pain, pressure, temperature). Attachment to underlying organs by loose connective tissue helps it move over them and return elastically. In amphibia the attachment is particularly loose. See CORNIFICATION, KERATIN.

SKOTOPHILE. Literally, dark-loving; dark-receptive phase of circadian rhythms, lasting about 12 hours. See PHOTOPHILE.

SLIDING FILAMENT HYPOTHESIS. Hypothesis that the apparently diverse activities of MUSCLE CONTRACTION, CYCLOSIS, CYTO-PLASMIC STREAMING, and various kinds of CELL LOCATION all depend fundamentally upon energy-dependent sliding of micro-filaments of ACTIN over MYOSIN. Similar mechanism, involving MICROTUBULES of tubulin, explains the beating of eukaryotic flagella (see CILIUM), and organelle and chromosome movement.

SLIME FUNGI. See MYXOMYCOTA.

SMALL INTESTINE. See ILEUM, JEJUNUM.

SMOOTH MUSCLE (INVOLUNTARY MUSCLE). Contractile tissue (generally called *visceral muscle* in vertebrates) comprising numbers of elongated spindle-shaped cells (sometimes syncytial) lacking transverse striations or other obvious ultrastructure. Position of the one nucleus per cell varies widely. Responsible among much else for peristaltic movement of food along gut, of blood along some contractile vessels (esp. invertebrates) and in regulation of vertebrate blood pressure (see VASOMOTOR CENTRE). Often arranged in *muscle coats* as bundles of cells averaging 100 μm in diameter, frequently within vertebrate connective tissue but occurring widely in both invertebrates and ver-tebrates. Among the latter principally in visceral, vascular and other locations where hollow organs occur (e.g. uterus). Cell lengths vary from 30–450 μm; diameters from 2–6 μm. Cells in the muscle bundle are separated by a basement membrane of glycosaminoglycans, glyco-proteins, collagen and elastic fibres about 60 nm wide; cell contacts include tight junctions and gap junctions. Proteins involved in its contraction are basically those of striated and cardiac muscle; but the myosin filaments are not as regularly arranged, while the actin filaments are more randomly distributed, lying along the longitudinal axis of the cell parallel to the myosin filaments; some of them terminate via accessory proteins at the cell membrane.

In vertebrates, almost always under control of AUTONOMIC NERVOUS SYSTEM, often with intrinsic PACEMAKERS. Hormones may alter the threshold for contraction. The sympathetic nervous system is generally excitatory on vascular smooth muscle but in-hibitory on gut muscle; the parasympathetic system may be either

excitatory or inhibitory, depending on the organ. See MUSCLE CON-
TRACTION.

SOCIETY. (Bot.) Minor climax community within a consociation,
arising as a result of local variations in conditions and dominated by
species other than the consociation dominant.

SODIUM PUMP. ACTIVE TRANSPORT mechanism present in plasma
membranes of most animal cells, consuming an estimated third of a
cell's ATP production in pumping sodium ions (Na^+) out of the cell and
potassium ions (K^+) into it in the ratio 3:2. As a result of its
(electrogenic) role in the establishment of a cell's RESTING POTENTIAL
this pump serves to regulate cell volume by casting out Na^+ which
would tend to enter along its electrochemical gradient, adding to the
negative osmotic potential of the cell and drawing water in. The cell's
internal electrical negativity prevents chloride ions (Cl^-) from entering
and having the same effect. The pump is blocked by external *ouabain*,
and animal cells may therefore swell or burst if this or other inhibitors of
ATP synthesis or hydrolysis are added. It is indirectly responsible for
glucose and amino acid uptake by cells since it creates a sodium gradient
necessary for Na^+-based symports (see TRANSPORT PROTEINS).

SOIL PROFILE. Series of recognizably distinct layers or horizons
visible in a vertical section through the soil down to the parent
material. Study of soil profiles yields valuable information on the
character of soils. See Fig. 59.

SOLAR PLEXUS. See AUTONOMIC NERVOUS SYSTEM, NERVE PLEXUS.

SOLONOSTELE. See SIPHONOSTELE.

SOMA. See CELL BODY, SOMATIC CELL.

SOMATIC CELL. Body cell; any cell of multicellular organism other
than gametes. Mutations in somatic cells do not generally play
significant role in evolution, being unlikely to be passed to further
generations in gametes. However, asexual budding may produce new
individuals with copies of somatic mutations, and some plants may
generate polyploid microspores and megaspores from polyploid
somatic precursor cells. See GERM PLASM.

SOMATOMEDINS. Liver-produced growth-promoting (anabolic) peptides of vertebrate blood serum whose synthesis and activity are
dependent upon presence of GROWTH HORMONE and whose activities they *mediate*. Cause instantaneous release of hypothalamic
SOMATOSTATINS. Enhance wound-healing.

SOMATOSTATIN. Hypothalamic *release-inhibiting factor* (RIF) preventing release of growth hormone releasing factor (GHRF) and hence
of GROWTH HORMONE. Some pancreatic cells (delta cells) also
release somatostatin, here inhibiting release of pancreatic hormones.

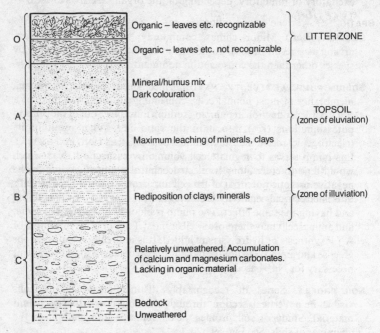

Fig. 59. Generalized soil profile indicating the major horizons. Not all soils have all these and their relative thicknesses vary greatly.

SOMATOTROPHIN. See GROWTH HORMONE.

SOMITE. See MESODERM.

SOREDIA. Organs of vegetative reproduction in LICHENS. Minute clusters of algal cells surrounded by fungal hyphae, formed in large numbers over lichen surface and dispersed by various agencies (e.g. wind, water, insects).

SOUTHERN BLOT TECHNIQUE. Very sensitive method for detecting presence among restriction fragments of DNA sequences complementary to a radiolabelled DNA or RNA probe sequence. After initial separation by agarose gel electrophoresis, restriction fragments are denatured to form single-stranded chains and then trapped in a cellulose nitrate filter on to which the probe suspension is poured. Hybridized fragments are detected by autoradiography, after washing off excess probe.

SPACER DNA. DNA separating one gene from another, often not transcribed itself. Common where there is GENE AMPLIFICATION (e.g. between ribosomal RNA genes).

SPADIX. Kind of INFLORESCENCE.

SPATHE. Bract enclosing inflorescence of some monocotyledons (e.g. *Arum*).

SPATIAL SUMMATION. See ALL-OR-NONE RESPONSE.

SPECIAL CREATIONISM. The view that each *kind* of organism (present and past) owes its existence to a unique creative act of God. Many influential scientists of the 17th and 18th centuries (Ray, Burnet, Whiston) believed science would vindicate a literal interpretation of the biblical creation story, as did many in the 19th (Buckland, Sedgwick, Agassiz). Darwin, whose views oppose special creationism, thought he had refuted this theory, holding that an intelligent Creator would have distributed organisms according to their physiological requirements rather than their taxonomy.

- **SPECIALIZED**. Having special adaptations to a particular ecological niche which often result in wide divergence from the presumed ancestral form. Such *specializations* evolve and may result in niche restriction.

SPECIATION. The origin of SPECIES. If species are not real ('objective') entities, speciation cannot be regarded as a real process. If, however, species are lineages with temporal continuity, then speciation is whatever generates independence of lineages from one other.

The major models of speciation are the *allopatric* and *sympatric* models. In the former a parent species becomes physically separated into daughter populations by geography, restricting (or eliminating) gene flow between non-overlapping populations. In sympatric models, a parent species differentiates into lineages in the absence of any physical restriction on gene flow.

Some phenotypic characters bear directly on speciation. Shape of genital aperture, timing of breeding, degree of assortative mating, compatibility of pollen and stigma, degree of developmental homeostasis – all are aspects of phenotype subject to genetic variation. They are therefore responsive to disruptive or directional selection, and can alter cohesion of the population as an evolutionary unit or lineage. Debate surrounds whether speciation is adaptive, or merely a stochastic process which perhaps selection cannot prevent. Allopatric speciation, at least, seems to involve non-selective splitting of lineages. In time, the daughter populations may simply gain sufficient distinctive genetic and phenetic characteristics for taxonomists to recognize separate species. Sympatric speciation models more often invoke the adaptiveness of species formation. But some insist that in either model speciation is *de jure* incomplete until sufficient overlapping of daughter populations has occurred for the biological species concept to be applied. It is often said that after overlap, selection against hybrids either reinforces lineage independence or is too weak to prevent collapse of population identities; but much will hinge on the species concept, phenetic or biological, being employed.

Major sympatric models involve disruptive selection on a deme already polymorphic for an ecological requirement (e.g. food plant, oviposition site), and assortative mating in favour of individuals sharing that requirement. Its occurrence in the wild is slowly gaining acceptance.

Stasipatric speciation postulates that a widespread species may generate internal daughter species whose chromosomal rearrangements play a primary role in speciation (through reduced fecundity or viability of individuals heterozygous for the rearrangements). Daughter species might then extend their ranges at the expense of the parent species and might hybridize where ranges abut, although resulting offspring would be less fertile.

Polyploidy can be a form of 'instantaneous' speciation, since it may result in the complete reproductive isolation of an individual from other gene pools. Allopolyploids are more significant here than autopolyploids, permitting regular bivalent formation of meiosis as well as originating through combination of genomes from different species. They breed true for their hybrid character. Moreover, crosses of allotetraploids to their diploid progenitors give sterile triploids, preventing backcrossing and gene flow. They are widespread in plants, where vegetative propagation of sterile hybrids can enhance colonization prior to allopolyploidy, successful meiosis and improved fertility.

Debate surrounds the role of speciation in phyletic (macro-)evolution. Apart from stasipatric and polyploid speciation, there are no clear accounts of the genetics of speciation.

SPECIES. Term used both of a formal taxonomic category ('the species') and of taxa exemplifying it (particular species). In the system of BINOMIAL NOMENCLATURE, taxa with species status are denoted by Latin binomials, each species being a member of a genus. The naturalist John Ray, writing of plants in 1672, did not entirely avoid circularity in stating that the true criterion of species as taxa is that they are 'never born from the seed of another species and reciprocally': the cross-sterility criterion. To a special creationist like Ray such cross-sterility was to be expected (compare the nominalism of BUFFON). For Darwin, all degrees of sterility should be discoverable if sterility *barriers* (the quite different, evolutionary, concept of ISOLATING MECHANISMS) take time to evolve. See SPECIATION.

Darwin's writings sometimes reveal nominalism on the species question, to be expected on the view that species evolve gradually. At other times he drew a clear distinction between *taxa* and *categories*, and although doubtful about the possibility of defining the category 'species' (i.e. the taxonomic unit) he was in no doubt that taxa with specific status actually existed. He did not espouse ESSENTIALISM with regard to species.

Zoologists find the criterion of reproductive isolation especially valuable in demarcating species in the wild. As such, the *biological species concept* includes as species groups of populations which are phenotypically similar and reproductively isolated from other such groups, but which are actually or potentially capable of interbreeding among themselves. Problems in its application arise (a) when reproductive isolation from other populations admits of degrees, being incomplete over part of a species range, (b) with RING SPECIES, (c) with obligately asexual species (*agamospecies*), (d) with animals which although sexual lack males (obligate THELYTOKY), and (e) with ANAGENESIS. The biological species concept also fails to incorporate a historical dynamic into its account, thereby ignoring that species are genealogically unique. The majority of fungi, plants and marine invertebrates broadcast their gametes widely, making it impossible to establish reproductive isolation in the wild. So complexes of phenetic characters (the stock-in-trade of museum taxonomists) usually serve to identify such species. But SIBLING SPECIES pose problems, and classification rests here upon cytological techniques (e.g. use of DNA probes) rather than external morphology, although reproductive isolation is often the original clue to their existence.

Recently favour has developed for ecological and evolutionary species concepts. The former fix a classification to independently existing environmental states or niches (difficult to isolate independently of the organisms which occupy them), equating speciation with niche change but leaving open the extent of change required; the latter stress the genealogical uniqueness of species (useful in asexual and thelytokous forms), but emphasize the cladistic (branching) nature of speciation at the expense of gradual anagenesis. Most species probably comprise two or more *subspecies* or *races* (see INFRASPECIFIC VARIATION, SEMISPECIES) and are said to be *polytypic*.

Inability to find a unified species concept is no disgrace, reflecting the variety of reproductive systems and the dynamic state of biological material. It is probably no accident that the species concept does not figure prominently in biological theory: species may be best regarded not as NATURAL KINDS (such as elements in chemistry) but as individuals, each historically unique and irreplaceable once extinct. If species are individuals then species names are proper names, so that properties of species would describe but not define them. Definitions of taxa would then be *necessary* in philosophical terms, even though there is no list of necessary properties for any biological taxon. See TRANSITIVITY.

SPECIES FLOCK. A species-rich, narrowly endemic and ecologically diverse group of organisms, usually considered to have radiated rapidly in geological terms from one, or more than one, very closely related stem species. They are sometimes quoted as *prima facie* candidates for sympatric speciation, but insufficient evidence is usually available to ascertain the mode of speciation which produced them. Examples include cichlid fish in African lakes, Galapagos finches, honeycreepers and fruitflies on Hawaii.

SPECIES GROUP. Informal taxonomic category used in preference to such formal categories as subgenus or infragenus.

SPECTRIN. Important protein of the erythrocytic cytoskeleton, apparently forming a network just under the plasmalemma and associated with actin and another globular protein, *ankyrin*. See CYTO-SKELETON.

SPERM. See SPERMATOZOID, SPERMATOZOON.

SPERMATELIOSIS. See MATURATION OF GERM CELLS.

SPERMATHECA (SEMINAL RECEPTACLE). Organ in some female or hermaphrodite animals which receives and stores sperm from the other mating individual.

SPERMATID. Haploid animal cell resulting from the meiotic division of a secondary spermatocyte. Undergoes extensive cytoplasm loss and condensation of its nucleus during spermiogenesis (see MATURA-TION OF GERM CELLS).

SPERMATIUM. Non-motile 'male' sex cell present in red algae and some members of the ASCOMYCOTINA and BASIDIOMYCOTINA fungi.

SPERMATOCYTE. (Bot.) Cell which becomes converted into a sperma-tozoid (without intervention of cell division). See SPERMATID. (Zool.) Cell undergoing meiosis during sperm formation. *Primary spermatocytes* undergo first meiotic division; *secondary spermatocytes* undergo second meiotic division. See MATURATION OF GERM CELLS.

SPERMATOGENESIS. Formation of spermatozoa. See MATURATION OF GERM CELLS.

SPERMATOGONIUM. Cell within testis, commonly lining seminiferous tubules, which either divides mitotically to produce further sperma-togonia or else gives rise to primary spermatocytes. See TESTIS, MATURATION OF GERM CELLS.

SPERMATOPHORE. Small packet of sperm produced by some species of animals with internal fertilization, e.g. many crustaceans, snails, mites, scorpions, *Peripatus*, newts, etc. Those of cephalopod molluscs are very complex and pump seminal fluid into the female by syringe-like structures.

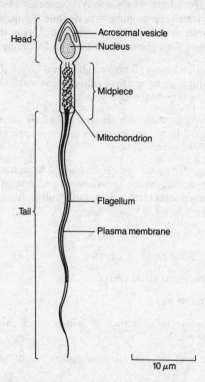

Head {
— Acrosomal vesicle
— Nucleus

— Midpiece

— Mitochondrion

Tail {
— Flagellum

— Plasma membrane

10 μm

Fig. 60. *Diagrammatic illustration of a spermatozoon.*

SPERMATOPHYTA. In some classifications, the Division containing all seed-bearing vascular plants. See ANTHOPHYTA, GNETOPHYTA, GINKGOPHYTA, CYCADOPHYTA, CONIFEROPHYTA.

SPERMATOZOID (ANTHEROZOID). (Bot.) Small, motile, flagellated microgamete.

SPERMATOZOON. (Zool.) Microgamete (see Fig. 60), usually motile, produced by testes; commonly flagellated but amoeboid in nematodes and aschelminthes. Sperm of *Drosophila* are about 2 mm in length. Most use lipids as fuel; complex sperm store glycogen. See ACROSOME, MATURATION OF GERM CELLS.

SPERMIOGENESIS. See MATURATION OF GERM CELLS.

SPERMOGONIUM (SPERMAGONIUM). Flask-shaped or flattened, hollow structure in which spermatia are formed.

SPHENODON. See RHYNCHOCEPHALIA.

SPHENOPHYTA. Horsetails. Homosporous vascular plants, possessing jointed stems with very prominent nodes and elevated siliceous ribs. Sporangia borne in a strobilus at stem apex. Have scale-like leaves and motile sperm. One genus (*Equisetum*) and 15 living species.

SPHEROPLAST. Cell whose wall material has been largely but not entirely removed (protoplasts have theirs entirely removed). Often employed in DNA cloning (e.g. bacterium, yeast).

SPHEROSOME. Single, membrane-bound, spherical structure in plant cell cytoplasms. Many contain lipids and are apparently centres of lipid synthesis and accumulation.

SPHINCTER. Ring of smooth muscle in wall of tubular organ, opening of hollow organ, etc., whose contractions and relaxations close and open the tube or aperture; e.g. pyloric and cardiac sphincters of stomach, anal sphincter.

SPHINGOLIPIDS. See PHOSPHOLIPIDS.

SPHINGOMYELIN. See PHOSPHOLIPIDS.

SPHINGOSINE. See PHOSPHOLIPIDS.

SPIDERS. See ARANEAE.

SPIKE. Indeterminate inflorescence in which main axis is elongated and flowers are sessile.

SPIKELET. Basic unit of grass inflorescences; small group of grass flowers.

SPINAL COLUMN. See VERTEBRAL COLUMN.

SPINAL CORD. The part of the vertebrate CENTRAL NERVOUS SYSTEM lying within the vertebral canal, protected by the vertebral column, and consisting of a hollow cylinder of mixed nervous tissue (derived from the NEURAL TUBE) with walls of relatively uneven thickness. Paired and segmental spinal nerves leave it on each side between vertebrae. Contains both grey and white matter, former usually H-shaped in cross-section with the *cerebrospinal canal* running through the 'cross-bar', the latter surrounding the grey matter, the whole covered by the MENINGES. Carries sensory and motor information via ascending and descending tracts of white fibres; also provides for reflex arcs (intra- and intersegmental, ipsi- and contralateral). Continuous with the medulla oblongata of BRAINSTEM. See REFLEX for diagram (Fig. 57).

SPINAL NERVES. Peripheral nerves arising from the vertebrate spinal cord between vertebrae; typically one on each side per segment. Each has a dorsal (afferent) root and a ventral (efferent) root, which typically fuse on emergence from the vertebral column. Compare CRANIAL NERVES.

SPINDLE. (1) MICROTUBULE complex appearing during mitoses and meioses. Has two sources: regions around the CENTRIOLES, and the KINETOCHORES. Microtubules from the former extend to the equator of cell where they apparently overlap and generate sliding forces which often push the poles of the cell apart; kinetochores somehow pull the centromeres of sister chromatids towards opposite poles. Microtubules (*polar*) developed from the centriolar region have their fast-growing (+) ends away from the pole; those from the kinetochore have their (+) ends attached to the kinetochore. While a cilium-like dynein has been implicated in the sliding of polar microtubules, the origin of the force pulling chromatids to the poles is not clear. (2) See MUSCLE SPINDLE.

SPINDLE ATTACHMENT. See CENTROMERE.

SPINNERETS. Maximum of three pairs of silk-spinning organs on the posterior opisthosomas of spiders (ARANEAE); most probably modified legs. The silk, which hardens on emergence, may be used in construction of egg cocoon, feeding web, and cords wrapped around prey trapped in the web. When released as a long line it may also provide sufficient wind resistance to lift the spider into the air for its dispersal.

SPIRACLE. (1) Reduced first gill slit of many fish. Dorsally situated, its small size results from the connection formed between mandibular and hyoid arches for firm attachment of jaws, the spiracle lying between these arches. In most living bony fish the spiracle is closed up; the gill pouch of embryo tetrapods representing the spiracle develops into the cavity of the middle ear and Eustachian tube. See VISCERAL ARCHES. (2) External opening (stigma) of insect and other TRACHEAE. Often contains valves to regulate water loss and gaseous exchange.

SPIRAL CLEAVAGE. See CLEAVAGE.

SPIRAL THICKENING. Internal thickening of wall of xylem vessel or tracheid, in form of a spiral. Occurs in cells of protoxylem and, while providing mechanical support, permits longitudinal stretching as neighbouring cells grow.

SPIRAL VALVE. Spiral fold of mucous membrane projecting into the intestine of some fish, notably elasmobranchs, ganoids and Dipnoi. Probably serves to increase surface area for absorption.

SPIRILLUM. Long, coiled or spiral bacterium.

SPIROCHAETES. Elongated, spirally twisted, unicellular bacteria with thin, delicate walls; up to 500 μm long (large for a bacterium); motility by a helical wave along the cell. Some are free-living, some parasitic and pathogenic (e.g. *Treponema pallidum*, causing syphilis).

Spleen. Largest mass of LYMPHOID TISSUE, lying in mesentery of stomach or intestine of jawed vertebrates. Unlike LYMPH NODES, perfused by blood rather than lymph. Important lymphocyte and PLASMA CELL reservoir, and component of the RETICULO-ENDOTHELIAL SYSTEM, its cells phagocytozing worn red blood cells and platelets. A store of red blood cells, sympathetic contraction of smooth muscle squeezing them into the circulation (often giving a 'stitch') in emergency. See HAEMOPOIESIS.

Splicing. Of INTRONS, their combined enzymatic removal from a primary RNA transcript (hnRNA) followed by joining together (splicing proper) of the exons on either side to generate messenger RNA (mRNA). It is now known that some intron RNA, particularly in mitochondrial transcripts of fungi, can splice itself out catalytically (see RNA PROCESSING). The ability of genetic elements to insert themselves into chromosomes, plasmids, etc., is employed in GENE MANIPULATION. See INSERTION SEQUENCE, HETERO-CHROMATIN, TRANSPOSABLE ELEMENT, ONCOGENE.

Split gene. Gene with at least one INTRON sequence embedded within it.

Sponge. See PORIFERA.

Spongy mesophyll. See MESOPHYLL, LEAF.

Spontaneous generation. The view that life can arise from non-life, independently of any parent. In the sense of the ORIGIN OF LIFE, this may be regarded as a scientifically respectable view but it has been held in the past that many individual organisms arise abiogenically from e.g. fermenting broth, rotting meat, etc. (see BIO-GENESIS). In the 19th century, endorsement of spontaneous generation by Lamarck and Geoffroy St. Hilaire resulted in its commonly being associated in France with any evolutionary theory. See LAM-ARCK, PASTEUR, SPECIAL CREATIONISM.

Sporangiole. (Of fungi) small sporangium containing only one or a few spores (sporangiospores), and lacking a columella.

Sporangiophore. (Of fungi) hypha bearing one or more sporangia; sometimes morphologically distinct from vegetative hyphae.

Sporangium. Organ within which are produced asexual spores; typically in fungi and plants.

Spore. A single- or several-celled reproductive body (propagule) that detaches from the parent and gives rise, directly or indirectly, to a new individual; a general term. Spores are usually microscopic, of many different types, produced in various ways. Thin- or thick-walled, they often serve for very rapid increase in the population, as when produced in enormous numbers and distributed far and wide by wind, water, animals, etc. Others are resting spores, enabling survival through unfavourable periods. Spores occur in all plant groups, in fungi, bacteria and protozoans. The term sexual spore usually indicates a spore that can engage directly in fertlization; less commonly it indicates a spore produced by meiosis or fertilization. Asexual spores therefore either do not engage in fertilization, or are not produced by meiosis. See MICROSPORE, MEGASPORE.

SPORE MOTHER CELL. (Bot.) Diploid cell giving rise by meiosis to four haploid spores or nuclei.

SPORIC MEIOSIS. Pattern of ALTERNATION OF GENERATIONS where the spores are produced by meiosis and develop into multi-cellular gametophytes before the gametes are produced; occurs in many algae, all bryophytes and vascular plants.

SPOROCYST. (1) Cyst of some SPOROZOA. (2) Stage in life cycles of many flukes (see TREMATODA); lacks mouth and gut; can produce daughter sporocysts or rediae (see POLYEMBRYONY).

SPORODOCHIUM. (Of fungi) a cluster of conidiophores arising from a stroma or mass of hyphae.

SPOROGONIUM. Spore-producing structure of liverworts and mosses that develops after fertilization; the sporophyte generation of these plants.

SPOROPHORE. (Of fungi) general term for a structure producing and bearing spores; e.g. a sporangiophore, conidiophore (simple sporophore), or mushroom (complex sporophore).

SPOROPHYLL. Leaf, bearing sporangia. In some plants, indistinguishable from ordinary leaves except by presence of sporangia, e.g. in bracken fern; in others, much modified and superficially quite unlike ordinary leaves, e.g. stamens and carpels of flowering plants.

SPOROPHYTE. Spore-producing diploid phase in LIFE CYCLE of a plant (see ALTERNATION OF GENERATIONS). Arises by union of sex-cells produced by haploid gametophytes.

SPOROPOLLENIN. Tough substance of which the exine (outer wall) of spores and pollen grains is composed. One of the most resistant organic substances known, not affected by hot hydrofluoric acid or concentrated alkali. It consists of complex polymers with an empirical formula $[C_{90}H_{142}O_{36}]$. Formed by oxidative polymerization of carotenoids and their esters.

SPOROZOA. Class of parasitic protozoans, many intracellular, some with alternate hosts. Mature stages lack locomotor organelles, but young may be amoeboid or flagellated. Large numbers of young are produced, either naked or in spores, by multiple fission after syngamy, when transmission to another host may occur. Includes *Plasmodium* (see MALARIA).

SPORT (ROGUE). Individual exhibiting the effect of a mutation, often an unusual or rare one.

SPOT DESMOSOME. See DESMOSOMES.

SPRINGTAIL. See COLLEMBOLA.

SQUAMATA. Order of the REPTILIA (Subclass Lepidosauria) containing lizards (Lacertilia), snakes (Ophidia), amphisbaenids (Amphisbaenia) and the tuatara (RHYNCHOCEPHALIA). Males generally have unique paired copulatory organs. In *lizards*, mandibles are joined at a symphysis but, as in snakes, there is a freely movable quadrate bone to which lower jaw is hinged. *Snakes* lack eardrums and movable eyelids and their eyes are covered by transparent eyelids. The jaws are exceptionally mobile and only loosely attached to the skull; mandibles are only loosely attached anteriorly by elastic ligaments and laterally by skin and muscles, with great relative mobility. Snakes feeding on prey large enough to struggle are usually constrictors (e.g. boas, pythons, anacondas) or venomous (e.g. cobras, mambas, coral and sea snakes, pit vipers and true vipers). The group is apparently undergoing rapid speciation. *Amphisbaenids* are legless burrowing forms, mostly under 1.5 m in length, with a single tooth in midline of upper jaw fitting into a space in lower jaw (effective nippers). Head shape blunt or flattened. Eardrums absent and eyes rudimentary. The Squamata have bodies covered in horny epidermal scales.

SQUAMOSAL. A MEMBRANE BONE of skull, in mammals taking over from the quadrate the articulation of lower jaw (dentary). See EAR OSSICLES.

SQUAMOUS CELLS. Animal cells, commonly forming an epithelium, which are flattened and resemble paving stones in shape. Cells of the inside of the cheek form a squamous epithelium; those lining blood vessels are squamous endothelium. See ENDOTHELIUM, EPITHELIUM.

SQUAMULOSE. Lichen growth form which is similar to the FOLIOSE form, but with numerous small, loosely-attached thallus lobes or *squamules*.

STAINING. Treatment of biological material with chemicals (dyes, stains) that only colour specific organelles or parts of a structure, thus providing contrast, as between nucleus and cytoplasm, mitochondria and other organelles, or between cell wall and cytoplasm. For use in light MICROSCOPY, most stains are organic compounds (dyes) comprising a negative and positive ion. In acid stains, the colour arises from the organic anion; in basic stains it arises from the organic cation, while neutral stains are mixtures of both acid and basic stains. Stains are applied to a biological material by its immersion in a stain solution. *Vital staining* refers to staining of living tissue when no damage to the tissue occurs (e.g. neutral red stains vacuoles and granules in many cells; Janus green stains mitochondria specifically). *Non-vital staining* refers to the staining of dead tissue. Various modified techniques are required for staining different types of tissues for light microscopy; e.g. *counterstaining* (double staining), where two stains are used in sequence so as to stain different parts of the specimen. See DEHYDRATION, FIXATION.

Staining techniques may give quantitative data, as when the amount of stain taken up is proportional to the amount of stained component, and methods exist (microspectrophotometry) to measure through the microscope the amount of stain present at a given site within a cell.

Stains in electron microscopy are not the coloured stains of light microscopy; rather they contain heavy metal atoms (e.g. lead, uranium) in forms that combine with chemical groups characteristic of specific structures in the cell. Presence of such atoms permits fewer electrons to pass through the specimen, so that an image is produced. 'Stained' (electron-dense) structures appear darker than their surroundings. *Negative staining* for electron microscopy is used to examine three-dimensional and surface aspects of cell structure. Specimens are not sectioned; rather they are placed directly on a thin plastic film, covered with a drop of solution containing heavy metal atoms and allowed to dry, leaving the specimen with a thin layer of electron-dense material.

STAMEN. Organ of FLOWER which forms microspores (shed after development as pollen grains); a microsporophyll; comprises stalk or filament bearing anther at apex. Anther comprises two lobes united by a prolongation of the filament connective, and in each lobe there are two pollen sacs (microsporangia) producing pollen.

STAMINATE. FLOWER which has stamens but no functional carpels, i.e. is male. See PISTILLATE.

STAMINODE. Sterile stamen; one that does not produce pollen.

STANDARD FREE ENERGY. Measure of the difference between the free energies of products and reactants of a reaction, where reactants and products are all present at the same initial concentration of 1 mole/dm³. See THERMODYNAMICS.

STANDING CROP. Biomass per unit area (or per unit volume) at any one time. Not equivalent to biomass productivity, which is a *rate* measure. Standing crop values are often given in terms of energy content and commonly relate to populations or trophic levels.

STAPES. Stirrup-shaped mammalian EAR OSSICLE, representing columella auris of other tetrapods and the hyomandibular of fish. See COLUMELLA.

Starch. A complex insoluble polysaccharide carbohydrate of green plants, one of their principal energy ('food') reserve materials. Formed by polymerization (condensation) of several hundred glucose subunits $(C_6H_{10}O_5)$ and easily broken down enzymatically into glucose monomers (see AMYLASE). Comprises two main components: *amylose* and *amylopectin*. The former consists of straight chains of α[1,4]-linked glycosyl residues and stains blue-black with iodine/KI solution; the latter contains in addition some α[1,6] branches in its molecules and stains red with iodine/KI solution. Found in colourless plastids (LEUCOPLASTS) in storage tissue and in the stroma of CHLOROPLASTS in many plants. Formed into grains, laid down in a series of concentric layers. See DEXTRIN, STATOLITH.

Starch sheath. Innermost layer of cells of cortex of young flowering plant stems, containing abundant and large starch grains; considered homologous with ENDODERMIS, it may sometimes lose starch and become thickened as an endodermis at later stage.

Starfish. See ASTEROIDEA.

Start codon. The AUG mRNA codon closest to the 5'-end of the molecule, acting as an initiation codon for translation by ribosomes. AUG codons to the 3'-end of the start codon do not initiate. See PROTEIN SYNTHESIS, STOP CODON.

Statoblasts. Resistant internal buds with chitinous shells produced by some ECTOPROCTA asexually and capable of withstanding unfavourable conditions, as during winter. They break open in spring to produce new colonies.

Statocyst(otocyst). Mechanoreceptor and/or position receptor evolved independently by several invertebrate groups (Cnidaria, Platyhelminthes, Crustacea) and vertebrates (see MACULA). Typically a fluid-filled vesicle containing granules of lime, sand, etc. (*statoliths*), which impinge upon specialized setae (*statolith hairs*) and stimulate sensory cells as the animal moves. Resulting nerve impulses initiate reflexes which often serve to right the animal after it has been turned upside down. See STATOLITH.

Statocyte. Plant cell containing one or more STATOLITHS.

Statolith. (1) A solid inclusion of a plant cell, commonly a starch grain, free to move under influence of gravity; thought to provide stimulus for GEOTROPISM and some other gravity responses. (2) See STATOCYST.

Statospore. Resting spore produced by some members of the CHRYSOPHYTA and XANTHOPHYTA. Ornamentation is species-specific. At germination, the protoplast emerges in an amoeboid manner.

Stele (vascular cylinder). (Bot.) Cylinder or core of vascular tissue in centre of roots and stems, comprising xylem, phloem, pericycle, and in some steles, pith and medullary rays; surrounded by endodermis. Structure of the stele differs in different groups of plants. See DICTYOSTELE, PROTOSTELE, SIPHONOSTELE, VASCULAR BUNDLE.

STEM. Normally aerial part of axis of vascular plants, bearing leaves and buds at definite positions (nodes), and reproductive structures, e.g. flowers. Some are subterranean (e.g. rhizomes) but these, like all stems, are distinguished externally from roots by the occurrence of leaves (scale leaves on rhizomes) with buds in their axils, and internally by having vascular bundles arranged in a ring forming a hollow cylinder, or scattered throughout tissue of the stem, with the protoxylem most commonly endarch.

STENOHALINE. Unable to tolerate wide variations in environmental salinity.

STENOPODIUM. See BIRAMOUS APPENDAGE.

STENOTHERMOUS (-THERMIC). Unable to tolerate wide variations in environmental temperature. Compare EURYTHERMOUS.

STEREOCILIUM. Specialized microvillus. See HAIR CELL.

STERIGMA. (Of fungi) a minute stalk bearing a spore or chain of spores.

STERILE. (1) Unable to produce viable gametes and/or sexual offspring, unlike normal individuals. (2) Free from microorganisms. See ANTISEPTIC, AUTOCLAVE, DISINFECTANT.

STERNUM. (1) Breast bone. Tetrapod bone lying ventrally in midchest to which ventral ends of most ribs are attached. Attached anteriorly to PECTORAL GIRDLE. (2) Cuticle on ventral side of each segment of an arthropod, often forming a thickened plate. See TERGUM.

STEROIDS. Chemically similar but biologically diverse group of LIPIDS originating from *squalene*. Include bile acids, vitamin D, adrenal cortex and gonadal hormones, active components of toad poisons and digitalis. Saturated hydrocarbons, with 17 carbon atoms in a system of rings, three 6-membered and one 5-membered, condensed together. *Sterols* (e.g. CHOLESTEROL and ergosterol) form a large steroid subgroup, having a hydroxyl group at C_3 and an aliphatic chain of 8 or more carbon atoms at C_{17}.

STEROLS. See STEROIDS.

STH. Somatotropic hormone. See GROWTH HORMONE.

STICK INSECTS. See PHASMIDA.

STICKY ENDS. See DNA LIGASE, TELEOMERE.

STIGMA. (Bot.) (1) Terminal portion of the style; surface of the carpel which receives the pollen. (2) See EYESPOT. (Zool.) Rare alternative name for insect SPIRACLE.

STIMULUS. Any change in the internal or external environment of an organism intense enough to evoke a response from it without providing the energy for that response. See IRRITABILITY, ADAPTATION, HABITUATION.

STIPE. Stalk. (1) Of fruit bodies of certain higher fungi, e.g. Basidiomycotina. (2) Of thallus of seaweeds (e.g. *Laminaria*), the organ between the holdfast and the blade.

STIPULE. Small, usually leaf-like, appendage found one on either side of leaf stalk in many plants, protecting axillary bud; often photosynthetic.

STOCK. Part of plant, usually comprising the root system together with a larger or smaller part of the stem, on to which is grafted a part of another plant (the *scion*). See GRAFT.

STOLON. Stem growing horizontally along the ground, rooting at nodes (e.g. strawberry runner).

STOMA (pl. STOMATA). (Bot.) (1) Pore in plant epidermis, present in large numbers, particularly in leaves, through which gaseous exchange occurs. Each surrounded by two specialized crescent-shaped cells (GUARD CELLS), whose movements, due to changes in turgidity, govern opening and closing of pore. (2) Includes both pore and guard cells. See TRANSPIRATION.

STOMACH. Enlargement of anterior region of ALIMENTARY CANAL. In vertebrates it follows oesophagus and usually has thick walls of SMOOTH MUSCLE to churn food and lining mucosal cells secreting mucus, pepsinogen and hydrochloric acid (see GASTRIC). Cardiac and pyloric sphincters can close the ends during churning. See GASTRIN, RENNIN, RUMINANT.

STOMIUM. Place in wall of fern sporangium where rupture occurs at maturity, releasing spores.

STOMODAEUM. Intucking of ectoderm meeting endoderm of anterior part of ALIMENTARY CANAL, forming mouth.

STONE CELL. Type of sclereid. See SCLERENCHYMA.

STONEFLIES. See PLECOPTERA.

STOP CODON. mRNA codons signifying chain-termination during PROTEIN SYNTHESIS. Also sometimes called *nonsense codons*. See CODON.

STRATUM CORNEUM. Outer layer of epidermis of vertebrate skin. Cells undergo CORNIFICATION (keratinization) and die, becoming worn off.

STREPSIPTERA (STYLOPIDS). Small endopterygote insects; females endoparasitic; males free-living, short-lived, with large metathorax, anterior wings haltere-like, hind wings large and fan-shaped. Females degenerate, apodous and larviform, enclosed in persistent larval cuticle. Several forms parasitize hymenopterans. Larvae emerge from host and probably find new hosts by waiting on flowers or through contact in nest. Probable affinities with the Coleoptera.

STREPTOCOCCUS. Genus of non-spore-producing, Gram-positive bacterium, forming long chains. Many are harmless colonizers of milk; some commensal in the vertebrate gut; but the *pyogenic* group are human pathogens, some producing haemolysins, destroying erythrocytes. The *viridans* streptococcal group lives usually non-pathogenically in the upper respiratory tract, but can cause serious infections (some chronic), and may cause bacterial arthritis.

STREPTOMYCIN. ANTIBIOTIC inhibiting translation of mRNA on prokaryotic, but not eukaryotic, ribosomes and can be used to distinguish these translation sites. Streptomycin resistance is conferred upon prokaryotes by plasmid-borne transposon. See ANTIBIOTIC RESISTANCE ELEMENT, CHLORAMPHENICOL.

STRETCH RECEPTOR. See MUSCLE SPINDLE, PROPRIOCEPTOR.

STRIATED MUSCLE (SKELETAL/VOLUNTARY/STRIPED MUSCLE). Contractile tissue, consisting in vertebrates of large elongated muscle fibres formed by fusion of MYOBLASTS to form syncytia. The cytoplasm of each fibre is highly organized, producing conspicuous striations at right angles to its long axis, and contains numerous longitudinal fibrils (*myofibrils*), each with alternating bands (A, anisotropic, I isotropic), H zones and Z discs caused by distributions of ACTIN and MYOSIN myofilaments and of α-ACTININ (see Fig. 61). The cross-striations of a whole fibre result from similar bands lying side by side. Each fibre is bounded by a *sarcolemma* (plasmalemma and basement membrane), the plasmalemma of which is deeply invaginated into the fibre forming *transverse tubules* (T-system), generally between the Z discs and H zones in insects but over the Z discs in vertebrates. These bring membrane depolarizations right into the fibre, ensuring uniform contraction. Mitochondria abound between myofibrils. The endoplasmic reticulum is modified to form a confluent system of sacs (*sarcoplasmic reticulum*) controlling calcium ion concentration.

On stimulation, a striated muscle fibre contracts by shortening and thickening. Fibres are bound together by connective tissue to form muscle tissue, and bring about locomotion by moving the skeleton, to which they are attached in vertebrates by tendons. See MUSCLE

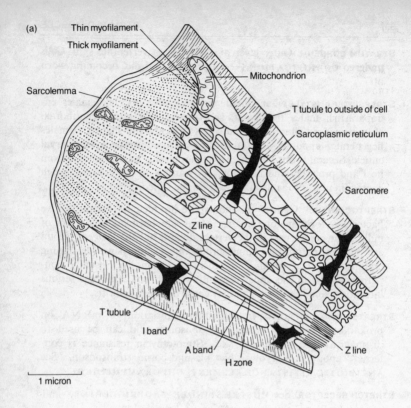

(a)

Thin myofilament

Thick myofilament

Mitochondrion

Sarcolemma

T tubule to outside of cell

Sarcoplasmic reticulum

Sarcomere

Z line

T tubule

I band

A band

H zone

Z line

1 micron

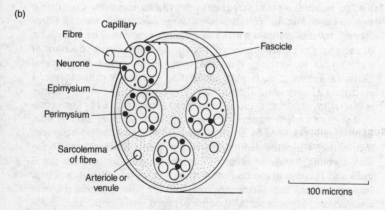

(b)

Fibre

Capillary

Neurone

Fascicle

Epimysium

Perimysium

Sarcolemma
of fibre

Arteriole or
venule

100 microns

*Fig. 61 (a) Diagram of the ultrastructure of a vertebrate striated muscle
fibre showing a complete sarcomere and two adjacent parts of sarcomeres.
Several such fibres together, with appropriate connective tissue, form a
muscle fascicle several of which in turn comprise a striated muscle as
shown in (b).*
*Fig. 61 (b) Four striated muscle fascicles and surrounding connective tissue
forming a small muscle.*

CONTRACTION, CARDIAC MUSCLE, SMOOTH MUSCLE, NEURO-MUSCULAR JUNCTION.

STROBILATION. Process of transverse fission which produces proglottides from behind the scolex of a tapeworm and ephyra larvae from the jellyfish scyphistoma (Scyphozoa). Regarded as a method of asexual reproduction in the latter, but less commonly in the former. The whole ribbon-like chain of tapeworm proglottides may be referred to as a strobila.

STROBILUS. Cone. Reproductive structure comprising several modified leaves (sporophylls), or ovule-bearing scales, grouped terminally on a stem.

STROMA. (Bot.) (1) Tissue-like mass of fungal hyphae, in or from which fruit bodies are produced. (2) Colourless matrix of the CHLOROPLAST, in which grana are embedded. (Zool.) Intercellular material (matrix), or connective tissue component of an animal organ.

STROMATOLITES. Macroscopic structures produced by certain blue-green algae (CYANOBACTERIA) where there is deposition of carbonates along with trapping and binding of sediments. Predominantly hemispherical in shape, they possess fine concentric laminations produced by growth responses to regular (often daily) environmental change. Fossil stromatolites occur from early Precambrian (more than 3000 Myr BP) to the Recent period.

STYLE. Slender column of tissue arising from top of ovary and through which pollen tube grows.

SUBARACHNOID SPACE. Area between the arachnoid and the pia mater, filled with cerebrospinal fluid. See MENINGES.

SUBCUTANEOUS. Immediately below dermis of vertebrate skin (i.e. the *hypodermis*). Such tissue is usually loose connective tissue, blood vessels and nerves and generally contains fat cells (see ADIPOSE TISSUE). In many tetrapods, also includes a sheet of striated muscle (*panniculus carnosus*) to move skin or scales.

SUBERIN. Complex mixture of fatty acid oxidation and condensation products present in walls of cork and most endodermis cells, rendering them impervious to water.

SUBERIZATION. Deposition of SUBERIN.

SUBSPECIES. Formal taxonomic category used to denote the various forms (types), usually geographically restricted, of a polytypic species. See INFRASPECIFIC VARIATION.

SUBSTRATE. (1) Substance upon which an ENZYME acts. (2) Ground or other solid object on which animals walk or to which they are attached. (3) Material on which a microorganism is growing, or solid surface to which cells in tissue culture attach.

Subtidal. Zone in sea or ocean extending from low-tide mark to edge of continental shelf.

Succession. Progressive change in composition of a community of organisms, e.g. from initial colonization of a bare area (*primary succession*), or of an already established community (*secondary succession*), towards a largely stable climax. See SERE.

Succulent. Type of xerophytic plant which stores water within its tissues and has a fleshy appearance (e.g. cacti).

Succus entericus (intestinal juice). Digestive juice (pH about 7.6 in humans) containing enterokinase, peptidases, nucleases, sucrase, etc., all secreted by the glandular *crypts of Lieberkuhn* between intestinal villi. Completes hydrolysis of food molecules begun higher in the gut. About 2–3 litres per day secreted in humans. See DIGESTION.

Sucker. Sprout produced by roots of some plants, giving rise to a new plant.

Sucrase (invertase). See SUCROSE.

Sucrose (cane sugar). Non-reducing disaccharide, comprising one glucose and one fructose moiety, linked between C_1 of glucose and C_2 of fructose. Abundant transport sugar in plants. Digested by the enzyme *sucrase* (*invertase*) and dilute mineral acids to glucose and fructose. See INVERT SUGAR.

Suctoria. Predatory ciliates, ciliated only in larval stage. See CILIATA.

Sulcus. (1) Longitudinal furrow, as in groove containing trailing flagellum in dinoflagellates; (2) thin furrowed area of pollen wall, notably in cycads and *Ginkgo* pollen.

Summation. Additive effect at synapses when arrival of one or a few presynaptic impulses is insufficient to evoke a propagated response but a train of impulses can do so. Termed *temporal summation* when impulses arrive at the same synapse, *spatial summation* when at different synapses of the same cell. In conjunction with nervous INHIBITION and FACILITATION it enables fine control over an animal's effector responses. See NERVOUS INTEGRATION, SYNERGISM, TRANSMITTER.

Supergene. Group of gene loci with mutually reinforcing effects upon phenotype that have come (through selection) to lie on the same chromosome, increasingly tightly linked so as to be inherited as a block (e.g. supergenes for the heterostyle or homostyle system in primroses (*Primula*) and shell colour and banding pattern in the snail *Cepaea*). See POLYMORPHISM.

SUPERIOR OVARY. See RECEPTACLE.

SUPERNUMERARY CHROMOSOME (ACCESSORY CHROMOSOME). Chromosome additional to normal karyotype of the species (B chromosome in botany). Either not homologous, or only partially homologous, with members of normal karyotype. In some populations of a species most of the individuals carry such chromosomes; in others, frequency is low. Most are heterochromatic. Their presence does not seem to affect markedly the individual's appearance. Geographical distribution often non-random.

SUPERSPECIES. Informal taxonomic category usually applied to allopatric arrays of species where evidence suggests common ancestry and where the species are sufficiently similar. Such species complexes tend to be discoverable with difficulty due to cryptic distinguishing characteristics between the species (usually sibling species), such as *Anopheles gambiae* and *Simulium damnosum* complexes. See DNA PROBE.

SUPINATION. See PRONATION.

SUPPRESSOR MUTATION. (1) (*Intragenic.*) A mutation (nucleotide addition or deletion) at a site in a chromosome sufficiently close to a prior nucleotide deletion or addition to restore the reading frame and thus suppress the effect of the original mutation. (2) (*Intergenic.*) A mutation at one locus which prevents the expression of a mutation at another locus. (3) (Uncommonly) a mutation preventing local or complete crossing-over in meiotic cells.

Some *suppressor genes* seem to be responsible for preventing oncogenic phenotype in normal cells; their loss may result in ONCOGENE expression. See ABERRANT CHROMOSOME BEHAVIOUR (1), MUTATION, HYPOSTASIS.

SUPRARENAL GLAND. See ADRENAL GLAND.

SURVIVAL VALUE. Characters and genes are said to have positive survival value if they increase an individual's FITNESS, and (less commonly) negative survival value if they decrease it. Neutral survival value is attributable to characters and genes with no effect on fitness. See NATURAL SELECTION.

SUTURE. Area of fusion of two adjacent structures. (1) (In flowering plants) the line of fusion of edges of carpel is known as a ventral suture. Mid-rib of carpel is known as the dorsal suture, not implying any fusion of parts to form it but to distinguish it from the ventral (true) suture. (2) Junction between the irregular interlocking edges of adjacent skull bones, or between plates of hardened cuticle of exoskeleton. Suture-lines occur on shells of ammonites, marking edge of a septum with side-wall of shell. (3) (Surgical) to sew a wound together.

SV40. See VIRUS.

SVEDBERG UNIT. The unit (S), related to sedimentation coefficient, used to indicate the time taken for large molecules, small organelles, etc., to reach an equilibrium level when ultracentrifuged in water at 20°C. The unit (S) is defined as a sedimentation coefficient of 1×10^{-13} seconds. Sedimentation coefficient is given by the equation:

$$s = \frac{dx/dt}{\omega^2 x}$$

where x = the distance of the sedimenting boundary from the centre of rotation (cm), t = time (s) and ω = angular velocity (radian s^{-1}). The S-value of an object is termed its *Svedberg number*. See RIBOSOMES.

SWARM SPORE. See ZOOSPORE.

SWEAT. Aqueous secretion of mammalian SWEAT GLANDS containing solutes in lower concentrations than in blood plasma, to which it is hypotonic. In humans, contains 0.1–0.4% sodium chloride, sodium lactate and urea; about one litre is lost per day in temperate climates (up to 12 litres in hot dry conditions when salt and water supply permit). The urea and lactate may be regarded as excretory, but may also reduce colonization by skin microflora. Its production, along with vasodilation of skin capillaries, removes body heat as latent heat of vapourization of water. See HOMOIOTHERMY.

SWEAT GLANDS. Epidermal glands, projecting into mammalian dermis, releasing SWEAT when stimulated by sympathetic nervous system. (1) *Apocrine sweat glands* are simple, branched tubular glands of (mainly) armpits, pubic region and areolae of breasts. Secretion more viscous than from eccrine sweat glands. (2) *Eccrine sweat glands* are simple, coiled tubular glands. Widely distributed; rows of them open via pores between the papillary ridges (responsible for finger prints) on hands and feet of many mammals. Absent from cats and dogs except between toes. See MAMMARY GLANDS.

SWIM BLADDER. See GAS BLADDER.

SYMBIONT. A symbiotic organism. See SYMBIOSIS, MYCOBIONT, PHYCOBIONT.

SYMBIOSIS. The living together in permanent or prolonged close association of members (*symbionts*) of usually two different species, with beneficial or deleterious consequences for at least one of the parties. Included here are: *commensalism*, where one party (the commensal) gains some benefit (often surplus food) while the other (the host) suffers no serious disadvantage; *inquilinism*, where one party shares the nest or home of the other, without significant disadvantage to the 'owner'; *mutualism*, where members of two different species benefit and neither suffers (symbiosis in a restricted sense) and *parasitism*, where one party gains considerably at the other's expense (see PARA-SITE). Some include intraspecific relationships within symbiosis.

SYMMETRODONTA. Order of extinct mammals (Infraclass Trituber-culata). Small Mesozoic forms; insectivorous. Possibly ancestral to PANTOTHERIA.

SYMPATHETIC NERVOUS SYSTEM. See AUTONOMIC NERVOUS SYSTEM.

SYMPATRIC. Of populations of two or more species, whose geographi-cal ranges or distributions coincide or overlap. From an ecological and genetic viewpoint, often more valuable to distinguish between sympatry as defined above and *effective* sympatry. Where adults are settled in different geographical regions but gametes, developmental stages, etc., are widely dispersed, there may be effective sympatry despite adult distribution. In lice, several species may inhabit the same host but settle in well-defined and species-specific anatomical regions. They are sympatric, yet ALLOTOPIC. See ALLOPATRIC, SYNTOPIC, SPECIATION.

SYMPETALOUS. See GAMOPETALOUS.

SYMPHYSIS. Type of joint allowing only slight movement, in which surfaces of two articulating bones (both covered by layer of smooth cartilage) are closely tied by collagen fibres. Symphyses occur between centra of vertebral column. For *pubic symphysis*, see PELVIC GIRDLE.

SYMPLAST. Interconnected protoplasts and their plasmodesmata which effectively result in the cells of different plant organs forming a continuum (e.g. from root hairs to stele). See TRANSLOCATION.

SYMPLESIOMORPHY. Any character which is a SHARED HOM-OLOGUE of two or more taxa, but which is thought to have occurred as an evolutionary novelty in an ancestor earlier than their earliest common ancestor. Compare SYNAPOMORPHY. See PLESIOMOR-PHIC, CLADISTICS.

SYMPODIUM. Composite axis produced, and increasing in length, by successive development of lateral buds just behind the apex. Compare MONOPODIUM.

SYMPORT. See TRANSPORT PROTEINS.

SYNANGIUM. Compound structure formed by lateral union of spor-angia; present in some ferns.

SYNAPOMORPHY. Term used in phylogenetics (see CLADISTICS) to denote what are otherwise termed *homologous* characters in many writings on evolution. Any character which is a SHARED HOM-

OLOGUE of two or more taxa and is thought to have originated in their closest common ancestor and not in an earlier one. Synapomorphy has been used in a more inclusive sense than has homology, characters involved including, e.g., geographical ranges. See HOMOLOGY, SYMPLESIOMORPHY.

SYNAPSE. Region of functional contact between one neurone and another, or between it and its effector. Commonly a gap (*synaptic cleft*) of at least 15 nm occurs between the apposed plasma membranes, the direction of impulses defining *presynaptic* and *postsynaptic* membranes. An individual neurone commonly receives 1000–10 000 synaptic contacts with 1000 or more other neurones, most synapses being between axon terminals of the stimulating neurone and dendrites of the receiving neurone, but axon–axon and dendrite–dendrite synapses occur. Most synapses are chemical (humoral), involving release of neurotransmitter via *synaptic vesicles* (see ACETYLCHOLINE, IMPULSE); others are electrical, impulses passing without delay from one neurone to another via gap junctions (see INTERCELLULAR JUNCTION). Chemical synapses, although encountering greater resistance due to diffusion time of transmitter, do permit NERVOUS INTEGRATION. See SUMMATION, NEUROMUSCULAR JUNCTION.

SYNAPSIS. Pairing-up of chromosomes. See MEIOSIS.

SYNAPTIC VESICLE. Vesicle produced by the Golgi apparatus of a nerve axon, in which neurotransmitter is stored prior to release into synaptic cleft. Bud off endocytotically again and re-fuse with the Golgi apparatus after release of transmitter. See COATED VESICLES.

SYNAPTONEMAL COMPLEX. Ladder-like (zip-like) protein complex holding chromosomes together during synapsis of MEIOSIS, chromosome loops pointing away from it. Within it lie *recombination nodules*, apparently determining where CHIASMA formation occurs. The complex dissolves at diplotene.

SYNCARPOUS. (Of the gynoecium of a flowering plant) with united carpels (e.g. tulip). See FLOWER.

SYNCHRONOUS CULTURE. Culture (of microorganisms or tissue cells) in which, through suitable treatment, all cells are at any one time at approximately the same stage of development, or of the CELL CYCLE. In mammalian tissue culture, cells in mitosis (M phase) round up and can be separated from others by gentle agitation to start a synchronous culture. They will rapidly enter G_1.

SYNCYTIUM. Animal tissue formed by fusion of cells, commonly during embryogenesis, to form multinucleate masses of protoplasm. May form a sheet (as in mammalian TROPHOBLAST) or cylinder (STRIATED MUSCLE), or network of more or less discrete cells linked by intercellular bridges (as in mammalian spermatogenesis; see MATURATION OF GERM CELLS). Smooth muscle is occasionally syncytial. See ACELLULAR, SYMPLAST.

SYNECOLOGY. Ecology of communities as opposed to individual species (autecology).

SYNERGIDAE (SYNERGIDS). Two short-lived cells lying close to the egg in mature EMBRYO SAC of flowering plant ovule.

SYNERGISM. Interaction of two or more agencies (e.g. hormones, drugs) each influencing a process in the same direction. Their combined effects are either greater than their separate effects added (*potentiation*) or roughly the sum of their separate effects (*summation*). Compare ANTAGONISM.

SYNGAMY. See FERTILIZATION.

SYNGENESIOUS. (Of stamens) united by their anthers; e.g. Compositae.

SYNGRAFT. See ISOGRAFT.

SYNOVIAL MEMBRANE. Connective tissue membrane forming a bag (*synovial sac*) enclosing a freely movable joint, e.g. elbow joint, being attached to the bones on either side of the joint. Bag is filled with viscous fluid (*synovial fluid*) containing glycoprotein, lubricating the smooth cartilage surfaces making contact between the two bones.

SYNTOPIC. Two or more organisms are syntopic if they share the same habitats (or microhabitats) within the same geographical range. They are *microsympatric*. This may be the expression of different phenotypes within a single species. See SYMPATRIC.

SYNTYPE. Each of several specimens used to describe a new species when a single type specimen was not chosen. See HOLOTYPE.

SYRINX. Sound-producing organ of birds, containing typically a resonating chamber with elastic vibrating membranes of connective tissue (VOCAL CORDS); situated at point where trachea splits into bronchi. Compare very different LARYNX of mammals, which in birds lacks vocal cords.

SYSTEM (ORGAN SYSTEM). Integrated group of ORGANS, performing one or more unified functions, e.g. NERVOUS SYSTEM, VASCULAR SYSTEM, ENDOCRINE SYSTEM.

SYSTEMATICS. Term often used as synonym of TAXONOMY, but sometimes used more widely to include also identification, practice of classification, nomenclature. See BIOSYSTEMATICS, CLASSIFICATION.

SYSTEMIC. Generally distributed throughout an organism.

SYSTEMIC ARCH. Fourth AORTIC ARCH of tetrapod embryo, becoming in adult main blood supply for the body other than the head. In Amphibia both left and right arches persist in the adult; in birds only the right; in mammals only the left (the aorta).

SYSTOLE. (1) Phase of HEART CYCLE when heart muscle contracts. (2) Phase of contraction of contractile vacuole.

T

TACHYCARDIA. Abnormally rapid heart rate.

TACTILE. (Adj.) Referring to the sense of touch. Sense organs include many surface pressure receptors.

TADPOLE. Term given to aquatic larval stages of tunicates (URO-CHORDATA) and of most modern amphibians. Metamorphoses into the adult.

TAGMA. (pl. TAGMATA). Functional and anatomical regionalizations of the arthropod body associated with division of labour; thus appendages with similar functions tend to be grouped on adjacent segments. Tagmata include *head*, *thorax* and *abdomen*, but each class within the phylum has a characteristic pattern of *tagmosis*. Insects have distinct head, thorax and abdomen; crustaceans generally lack a clear head–thorax distinction; arachnids have PROSOMA and OPISTHOSOMA, etc. Tagmosis in post-head region is closely related to mode of locomotion. See ARTHROPODA.

TAGMOSIS. See TAGMA.

TAIGA. Vegetation extending over vast areas of the USSR and N. America. Northern coniferous (boreal) forest, characterized by severe winters and persistent snow cover. Flanked by either grassland or deciduous forest to the south, grading unevenly into tundra to the north. Compare TUNDRA.

TANGENTIAL SECTION. Longitudinal section cut at right angles to radius of cylindrical structure (e.g. root or stem).

TANNINS. Group of complex astringent substances occurring widely in plants, dissolved in cell sap. Particularly common in tree bark, unripe fruits, leaves and galls; also found in animals (e.g. insect cuticles). Include phenols and hydroxy acids or glycosides. Some deter herbivores. Employed in the tanning process, which is usually enzymatic (see CUTICLE). Commercial leather production involves cross-linking collagen molecules in skins by treatment with benzoquinone, and a similar process occurs naturally.

TAPETUM (TAPETUM LUCIDUM). (Zool.) Reflecting layer of vertebrate choroid (retina in some teleosts), especially in nocturnal forms and deepwater fish. Generally contains guanine crystals reflecting light back through the retina, increasing its stimulatory effect but reducing visual acuity. The pigment melanin is commonly expanded over the crystals in bright light. See CHROMATOPHORE. (Bot.) In vascular plants, a layer of cells, rich in reserve food, surrounding a group of spore mother cells; e.g. in fern sporangium, pollen sacs of the anther. Gradually disintegrate and liberate their contents, which are absorbed by developing spores.

TAPEWORM. See CESTODA.

TAP ROOT. Primary root of a plant, formed in direct continuation with root tip or radicle and forming prominent main root, directed vertically downward, bearing smaller lateral roots; e.g. dandelion. Sometimes swollen with food reserves; e.g. carrot, parsnip. See FIBROUS ROOT.

TARDIGRADA. Order of minute aquatic arthropods of uncertain status, but unique among these in having a terminal mouth. The four pairs of stumpy appendages are simple and lobopod (resembling ONYCHOPHORA), ending in claws. Clear circulatory and respiratory systems lacking. Most pierce and suck plant juices. Resistant to desiccation, living among mosses, etc.

TARSAL BONES (TARSALS). Bones of proximal part of tetrapod hind-foot (roughly the ankle). Primitively 10–12 bones in a compact group, reduced during evolution by fusion and/or loss. There are seven in humans, one (calcaneum) forming the heel. They articulate proximally with tibia and fibula, distally with metatarsals. See PEN-TADACTYL LIMB.

TARSUS. (1) Region of tetrapod hind-foot containing TARSAL BONES (approx. the ankle). Compare CARPUS. (2) A segment (fifth from the base) of an insect leg.

TASTE BUD. Vertebrate taste receptor, consisting of group of sensory cells, usually located on a papilla of the TONGUE, but in aquatic forms often widely scattered (especially in barbels around fish mouths). Buds are often (e.g. in humans) localized in regions sensitive to salt, sweet, bitter and acid (sour) tastes; other taste modalities are detected in fish.

TATA BOX. Short nucleotide consensus sequence in the PROMOTER regions of eukaryotes, about 25–30 base pairs upstream of the sequence to be transcribed (hence -25 to -30). An AT-rich region, commonly including the interrupted sequence TATAT . . . AAT . . . A. The sequence is recognized and bound by transcription factor IID (also known as TATA factor), which complex is in turn bound by RNA polymerase II. See PRIBNOW BOX, TRANSCRIPTION FACTORS.

TAXIS. Directional locomotory response, as opposed to growth, to external stimulus, such as temperature gradient (thermotaxis), light (phototaxis), aerial or dissolved substances (chemotaxis). Movement into increasing stimulus intensity is a *positive* response; movement into decreasing stimulus intensity is a *negative* response. Compare TROPISM.

TAXON. The organisms comprising a particular taxonomic entity, such as a particular class, family or genus. The members of a *particular* species form a taxon, but the TAXONOMIC CATEGORY species does not.

TAXONOMIC CATEGORY. A category, formal or informal, used in CLASSIFICATION. Formal categories include Kingdom, Phylum (Division), Class, Order, Family, Genus and Species. Informal categories include superspecies and race. Instances of these (e.g. Phylum Arthropoda) are taxa, not taxonomic categories. See INFRA-SPECIFIC VARIATION.

TAXONOMY. Theory and practice of CLASSIFICATION. Classical taxonomy is concerned with morphology (including cytological, biochemical, behavioural), and may involve weighting phenetic characters in some scheme of relative taxonomic importance or value. *Numerical taxonomy* dispenses with such weighting and involves computerized analysis of data obtained from observing whether or not organisms being compared have or do not have any of the 'unit characters' involved in the comparison. Unit characters have an 'all-or-none' nature, being either present or absent. The data tend to arrange themselves into sets or phenons, which can then be organized into a DENDROGRAM. Relationships between organisms which can be thus evaluated are termed *phenetic* if determined by overall morphological similarity between organisms, or *cladistic* if they depend upon community of descent. In *experimental taxonomy*, breeding work and field experiments may be used to clarify the taxa to which organisms belong.

T-CELL (T-LYMPHOCYTE). Lymphocytes which travel from the bone marrow via the blood and enter the thymus, after which they enter the circulation again and settle in spleen and lymph nodes. While in the thymus a T-cell acquires ability to recognize a specific antigen (*T-cell maturation*), which involves ANTIGEN-PRESENTING CELLS. T-cells do not produce antibody, but antibody production by B-cells often requires T-cell help. A *T-helper cell* (T_H) recognizes a specific antigen on an antigen-presenting cell, binds to it, and then assists a B-cell binding the same antigen to proliferate into specific antibody-secreting cells. *Cytotoxic T-cells* (T_C) recognize tumour or virus-infected cells by their surface antigens in combination with their MHC markers, and will kill them. Other T-cells (*macrophage-activating cells*) produce LYMPHOKINES which promote macrophage activity. *Suppressor T-cells* (T_S) specifically suppress the immune response, probably through affecting antigen-presenting cells and/or through more direct interactions with T_H or B-cells. See Fig. 41.

T-DNA. A DNA segment becomes transferred from a Ti plasmid and integrated into host (usually plant) DNA. See PLASMID.

TECTUM (OPTIC TECTUM). Dorsal region of vertebrate midbrain (see BRAIN). A correlating region integrating vestibular, tactile and visual information and initiating or modifying reflex motor responses. Except in mammals, it is the primary visual centre.

TELEOLOGY. Biological discourse becomes teleological when reference is made to function, purpose and design, as when structure is related to function. In recent history, vitalists, organismic biologists and others have stressed that biological systems demand a functional interpretation, but tended to explain this goal-directedness or adaptedness by a retroactive (teleological) causation peculiar to living systems 'pulling' towards perfection. As Aristotle probably intimated, and we tend to believe, the goal-directedness of the vast bulk of biological structure and behaviour is more easily interpreted as the outcome, or expression, of an organizational complexity far greater than that of normal inert objects. It is largely attributable to DARWIN that, since the mid 19th century, an account of adaptation has been available which trivializes classical teleology and interprets contexts in which teleological terms are used as inevitable consequences of biological reproduction, accompanied as that is by selective transmission of heritable variation from one generation to the next as organisms compete for finite environmental resources. The concept of teleology is difficult to explicate without circularity. See FUNCTION, PROXIMATE FACTOR, ULTIMATE FACTOR, ADAPTATION.

TELEOSTEI. Higher bony fishes (about 20 000 living species). A fish taxon of uncertain category (e.g. suborder, series) but within Subclass ACTINOPTERYGII of Class OSTEICHTHYES (body fishes). Probably arose from a holostean stock in the Mesozoic, but became abundant in the Cretaceous. Vertebral axis turns upwards in the tail, which is superficially symmetrical; paired fins are small; scales are usually rounded, without ganoid covering, thin and bony. Among living teleosts soft-rayed fish (e.g. salmon, herring, carp) are regarded as relatively primitive, spiny-rayed fish, which include the ACANTHOPTERYGII (e.g. perches), as more progressive. See FINS.

TELOCENTRIC. Of chromosomes, with the CENTROMERE at one end (terminal).

TELOLECITHAL. Of eggs (e.g. amphibian) with yolk more concentrated at one end and a marked POLARITY along the axis of yolk distribution.

TELOMERE. Originally, term used by cytogeneticists for an end of a eukaryotic chromosome. Such ends lacked any tendency to fuse together spontaneously (i.e. they were 'non-sticky'). Telomeres appeared never to become incorporated within the chromosome (e.g. by terminal inversion). It is now known that they are examples of INVERTED REPEAT SEQUENCES of DNA, replicated by a specific *telomerase* enzyme (in a 5′ to 3′ end direction), which is a specialized reverse transcriptase providing the template for new telomeres. Telomerase is an enzyme composed of both RNA and protein, and as such may be an evolutionary link between systems using RIBOZYMES only and those using purely protein enzymes (see ORIGIN OF LIFE).

TELOPHASE. Terminal stage of MITOSIS or MEIOSIS, when nuclei return to interphase.

TELSON. Hindmost segment of arthropod abdomen, forming sting of scorpions and part of tail fan of lobsters and their allies. In insects present only in the embryo.

TEMPERATE. Of BACTERIOPHAGE that can insert its genome into that of its host so that it is replicated with it. See also EPISOME.

TEMPERATURE COEFFICIENT. See Q_{10}.

TEMPERATURE-SENSITIVE MUTANT. Mutant form expressed only under certain temperature conditions, the wild-type phenotype being expressed at others. Commonly due to heat-labile gene product.

TEMPLATE. In nucleic acids, the strand used by a polymerase to build a new and complementary polynucleotide strand.

TENDON. Cord or band of relatively inextensible vertebrate connective tissue attaching muscle tissue to another structure, often bone. Consists almost entirely of closely packed collagen fibres with rows of fibroblasts between. With high tensile strength and coefficient of elasticity.

TENDRIL. Stem, or part of leaf, modified as a slender branched or unbranched thread-like structure; used by many climbing plants for attachment to a support, either by twining around it (e.g. pea, grape vine) or by sticking to it by an adhesive disc at the tip (e.g. virginia creeper). See HAPTOTROPISM.

TERATOGEN. Any factor or agent causing malformation in embryos (e.g. X-rays, certain chemicals).

TERATOMA. Growth of cell mass from an unfertilized egg within mammalian ovary (or of germ cells in the male's testis) which becomes disorganized and uncontrolled. Many differentiated cell types may be represented, all mixed up. Teratomas may become malignant (*teratocarcinomas*). See PARTHENOGENESIS.

TERGITE. A SCLERITE composing an arthropod TERGUM.

TERGUM. Thickened plate of cuticle on dorsal side of arthropod segment.

TERMITE. See ISOPTERA.

TERPENE. Unsaturated LIPID consisting of multiples of *isoprene* (a 5-carbon hydrocarbon), sesquiterpenes having three such units. Includes vitamins A, E and K, carotenoids and many odorous substances in plants. *Squalene* (see CHOLESTEROL) is another. Rubber and gutta-percha are polyterpenes.

TERRITORY. Area or volume of habitat occupied and defended by an animal, or group of animals of same species. Territories commonly arise during or prior to breeding, with an important spacing role in many vertebrate and arthropod populations. Compare HOME RANGE, INDIVIDUAL DISTANCE.

TERTIARY. GEOLOGICAL PERIOD, lasting from about 70–1 Myr BP. With Quaternary it comprises the Cenozoic era. Often called the Age of Mammals on account of the radiation which occurred during it. Birds dominated the air throughout, almost all major modern groups being present early in the period.

TESTA. Seed coat. Protective covering of embryo of seed plants, formed from the integument(s); usually hard and dry.

TEST CROSS. Cross involving mating an individual of unknown genotype, or one heterozygous at a number of loci, with another homozygous for recessive characters at loci of interest. Offspring ratios are then used to deduce the genotype of the unknown, and/or the linkage relationships of its heterozygous loci. Not necessarily a BACKCROSS. See CHROMOSOME MAPPING.

TESTICULAR FEMINIZATION. Syndrome of some mammals, including humans. Due to presence of the *Tfm* allele at an X-linked locus, individuals chromosomally male (XY) are phenotypically female since male genital (Wolffian) ducts fail to develop, lacking cell membrane receptors capable of binding androgens (testosterone and its derivatives). Uterus absent or rudimentary; gonads usually abdominal or inguinal testes.

TESTIS. Sperm-producing organ of male animal; in vertebrates producing also 'male' sex hormones (androgens) and derived in part from *genital ridges* of coelomic epithelium in dorsal abdomen adjacent to the mesonephros, forming the testis cortex. Amoeboid *primordial germ cells* invade the ridges from the endoderm of the yolk sac and as the ridges hollow out, primitive sex cords develop from mesenchyme to form *seminiferous tubules* (see Fig. 62), to which primordial germ cells attach. Tubules are separated by septa and discharge sperm into the EPIDIDYMIS via ducts (*vasa efferentia*). Each testis is attached to the abdominal wall by a ligament (the *gubernaculum*) and under testosterone influence is drawn into the scrotum through a canal (the inguinal canal) as body elongation occurs. Endocrine cells (*interstitial cells*, or *Leydig cells*) between the tubules secrete androgens, most potent being TESTOSTERONE. See MATURATION OF GERM CELLS, WOLFFIAN DUCT, OVARY.

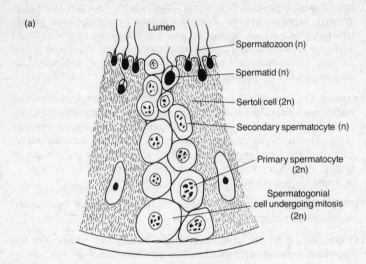

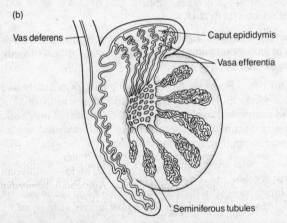

Fig. 62. (a) *Section through a vertebrate seminiferous tubule indicating the various cell stages involved in the production of spermatozoa.* (b) *Section through mammalian testis.*

TESTOSTERONE. The main vertebrate androgen ('male' sex hormone). Anabolic steroid produced largely by the Leydig cells of the testes from cholesterol or acetyl coenzyme A, and in smaller quantities by the adrenal cortex. Responsible for maintaining testes and for male growth spurt at puberty by stimulation of longitudinal bone growth and deposition of calcium. Closes epiphyses a few years after puberty, arresting growth. Promotes protein synthesis (hence muscle development), sexual behaviour and spermateliosis (see MATURATION OF GERM CELLS). In humans promotes hair growth in pubic, axillary, facial and chest regions and enlargement of laryngeal cartilage and voice deepening.

TETANUS. (1) Disease caused by toxin from anaerobic spore-forming bacterium *Clostridium tetani*. Increasing muscle spasms make opening of jaw difficult (hence lock-jaw). Progressive convulsions may be fatal, often by asphyxia or exhaustion. (2) Sustained muscle contraction resulting from nervous stimulation at a rate too great to allow muscle relaxation. Results from overabundance of calcium ions in sacroplasm. See MUSCLE CONTRACTION.

TETRAD. Group of four haploid cells or nuclei produced by meiosis, while they are adjacent.

TETRAPLOID. Of nuclei, cells, individuals, having four times (4n) the haploid number of chromosomes. See POLYPLOID.

TETRAPOD. Four-limbed vertebrate (see Fig. 63). Includes amphibians, reptiles, birds and mammals. Secondary loss of one or both pairs of limbs, or modification into wings, flippers, etc., has occurred in some taxa. See PENTADACTYL LIMB.

TETRASPORES. In some red algae, the four spores formed through meiosis in a tetrasporangium. Borne usually on a free-living and diploid sporophyte.

THALAMUS. (Bot.) Receptacle of flower. (Zool.) Part of telencephalon of vertebrate forebrain, forming roof and/or lateral walls of third ventricle and composed largely of grey matter. Organized into nuclei relaying sensory information from spinal cord, brainstem and cerebellum to cerebral cortex. See HYPOTHALAMUS.

THALLOPHYTA. In older classifications, a Division of plant kingdom housing prokaryotes and simple plant-like eukaryotes possessing simple vegetative bodies (thalli). Included bacteria and cyanobacteria, algae, fungi, lichens and slime fungi. Term now largely abandoned.

THALLUS. Simple, vegetative plant body, lacking differentiation into root, stem and leaf. Unicellular or multicellular, comprising branched or unbranched filaments; or more or less flattened and ribbon-shaped.

THECODONTIA. 'Stem' order of ARCHOSAURS, arising in early Triassic and extinct by its close, giving rise to DINOSAURS and probably birds (see AVES). South African fossil *Euparkeria* is representative, having several DIASPID characteristics, but with numerous small

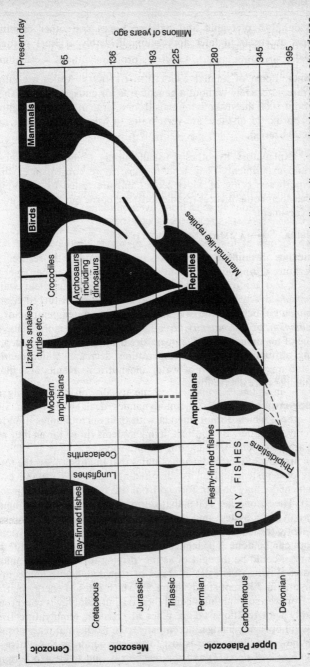

Fig. 63. Diagram indicating approximate tetrapod phylogenies. The width of a group's entry indicates its approximate species abundance.

teeth on both pterygoid and palatine bones and other primitive features absent from later archosaurs. Probably at least partially bipedal.

THELYTOKY. Form of animal PARTHENOGENESIS. Males are either very rare, effectively without a genetic role, or entirely absent. There are about 1000 thelytokous animals from a large number of taxa, but only approx. 25 of these are vertebrates (4 fish, 2 salamanders and about 20 lizards).

THEORY. Explanatory hypothesis, usually firmly founded in observation and experiment. They tend to have more consequences than do hypotheses, being of wider scope, and are tested by examining whether their consequences (predictions) are borne out by observation and experiment.

THERAPSIDA. See MAMMAL-LIKE REPTILES.

THERMOCLINE. Stratification of lakes and oceans with respect to temperature during summer months, characterized by upper layer of more or less uniformly warm, circulating, fairly turbulent water (*epilimnion*) overlying deeper, cold and relatively undisturbed region (*hypolimnion*). Between the two is a region of steep temperature drop (*metalimnion*, or *discontinuity layer*). The thermocline is the plane or surface of maximum rate of temperature drop with respect to depth. During autumn, epilimnetic temperatures decrease, water density rises and mixing (*overturn*) of water and nutrients results in a more even temperature distribution.

THERMODYNAMICS. Classical thermodynamics deals with *closed systems* – those which do not interact with their surroundings in terms of energy or matter. By contrast, living systems (in so far as they are alive) are *open systems*, interacting in both ways.

The first law of thermodynamics states that in any process the total energy of a system and of its surroundings remains constant, even though energy may be transformed from one form to another (transduced). The second law states that during any process the combined *entropy* (S) of a system and its surroundings (its disorder, randomness) tends to increase until equilibrium is attained, at which point no work can be done. The tendency for entropy in the universe to be maximized could be thought of as the 'driving force' of all chemical processes. Living cells exist in states of thermodynamic non-equilibrium, and in different steady states, in which rates of energy/matter input from the surroundings equal their output. Biology is concerned primarily with reactions taking place at constant temperature (*isothermally*) but heat changes accompany even isothermal reactions. If under these conditions heat is lost to the surroundings the system is said to lose *enthalpy* (H) and the reaction is said to be exothermic. Absorption of heat characterizes endothermic reactions. The form of energy capable of doing work in a system under constant temperature and pressure is its *free energy* (G), whose value is the key to predicting the direction of a chemical reaction. These relationships may be summarized:

$$\Delta G = \Delta H - T\Delta S$$

where ΔG = the free energy change of the system, ΔH = its change in enthalpy, T = absolute temperature and ΔS = its change in entropy. Reactions are *exothermic* if ΔH is negative, *endothermic* if ΔH is positive; they are *exergonic* (and may do work) if ΔG is negative, and *endergonic* (not doing work) if ΔG is positive. Spontaneous reactions are characterized by a loss of free energy (exergonic). Endergonic reactions cannot proceed spontaneously.

The free energy change of any reaction at constant temperature and pressure involves a fixed constant for that reaction known as the *standard free energy change* of the reaction (ΔG_o), giving the loss of free energy when the reaction is allowed to go to equilibrium starting with certain standard conditions – in particular, when all reactants and products are present at 1.0 molar concentrations. It is a measure of the difference between the sum of free energies of products and of reactants and is related to the equilibrium constant K (since $\Delta G_o = -RTlnK$). Free energy of a reaction varies with concentrations of reactants and products, affecting its probability of occurrence. A chemical reaction only occurs if its ΔG is negative in sign, and the maximum amount of work it then does equals this decrease in free energy.

THERMONASTY. (Bot.) Response to a general, non-directional, temperature stimulus; e.g. opening of crocus and tulip flowers with temperature increase.

THERMOPHILIC. (Of a microorganism) with optimum temperature for growth above 45°C.

THEROPHYTES. Class of RAUNKIAER'S LIFE FORMS.

THIAMINE (VITAMIN B₁). Vitamin precursor of coenzyme *thiamine pyrophosphate* (TPP), serving enzymes transferring aldehyde groups during decarboxylation of α-keto acids (such as pyruvate in mitochondria), and formation of α-ketols. Deficiency causes beri-beri in man, and polyneuritis in birds. See VITAMIN B COMPLEX.

THIGMOTROPISM. See HAPTOTROPISM.

THORACIC DUCT. Main mammalian lymph vessel, receiving lymph from trunk (including lacteals) and hindlimbs and running up the thorax close to vertebral column, discharging into left subclavian vein (humans), or another major anterior vein. Often paired in fish, reptiles and birds.

THORAX. (1) In terrestrial vertebrates, part of body cavity containing heart and lungs (i.e. the chest); in mammals, clearly separated from abdominal cavity by diaphragm. (2) In arthropods, body region between head and abdomen; often not clearly separable, or tagmatized. In adult insects comprises three segments, each typically with a pair of walking legs and one or two of them commonly bearing a pair of wings. See TAGMA.

THREAD CELL. See CNIDOBLAST.

THREADWORMS. See NEMATODA.

THRESHOLD. Critical intensity of a stimulus below which there is no response by a tissue (e.g. nerve, muscle).

THROMBIN. Proteolytic enzyme derived from prothrombin during sequence of reactions involved in BLOOD CLOTTING, converting fibrinogen to fibrin.

THROMBOCYTES. See PLATELETS.

THROMBOPLASTIN (THROMBOKINASE). Phospholipid and protein, liberated from PLATELETS and damaged tissues on wounding, initiating the CASCADE of reactions resulting in BLOOD CLOTTING.

THYLAKOID. Flattened membranous vesicle containing CHLOROPHYLL pigments; site of photochemical reactions in PHOTOSYNTHESIS. Vary in form and arrangement between different groups of organisms.

THYMIDINE. Nucleoside comprising the base THYMINE linked to ribose by a glycosidic bond.

THYMINE. Pyrimidine base of the NUCLEOTIDE thymidine monophosphate, a monomer of DNA but not RNA. Also a component of thymidine di- and triphosphates.

THYMUS GLAND. Bilobed vertebrate organ containing primary LYMPHOID TISSUE, usually situated in the pharyngeal or tracheal region. Originates from gill pouches (in mammals, the third pouches) or gill clefts. In mammals lies in thorax (mediastinum), the gill pouch cells mixing with lymphocytes. Attains maximum size at puberty, slowly diminishing afterwards. Responsible for maturation of T-CELLS, and possibly with an endocrine function.

THYROCALCITONIN. See CALCITONIN.

THYROGLOBULIN (TGB). Transport protein for thyroid hormones produced by thyroid gland. Stored there and carried in blood plasma.

THYROID GLAND. Vertebrate endocrine gland, unpaired and in neck region. Derived from endoderm of pharyngeal floor; probably homologous with ENDOSTYLE of cephalochordates, etc. Comprises saclike *thyroid follicles* in which THYROID HORMONES (thyroxine and triiodothyronine) are stored, colloidally complexed with THYRO-GLOBULIN, surrounded by parafollicular cells (secreting CAL-CITONIN). Inadequate dietary iodine causes enlargement of the thyroid (goitre) due to raised TSH secretion (see THYROID STIMULATING HORMONE).

THYROID HORMONES. In addition to the hormone CALCITONIN, most vertebrate thyroids produce small iodine-containing hormones as derivatives of the amino acid tyrosine. These seem to act like steroid hormones, passing through cell membranes and binding to NUCLEAR RECEPTORS, the complex then binding specific chromosomal DNA sites to bring about GENE EXPRESSION.

Thyroxine (T_4) is produced in greater quantities than *triiodothyronine* (T_3), but is about a quarter as potent. About one third of circulating T_4 is converted to T_3 in lungs and liver. Both are homeostatically controlled, and if circulating hormone levels or blood temperature are too low, the hypothalamus releases thyrotropin releasing factor (TRF) into the anterior pituitary portal system, stimulating release of TSH (THYROID STIMULATING HORMONE), which, in turn, releases bound and unbound hormones from the thyroid. These increase BASAL METABOLIC RATE (especially in the liver) and oxygen consumption, uncoupling electron transport from ATP synthesis in liver mitochondria. Respiration then releases heat into the blood (raising its temperature) rather than generating ATP.

They have general catabolic effects upon fat and carbohydrate, promoting gluconeogenesis; but they stimulate protein synthesis. Are important in tissue growth and development, and in metamorphosis of amphibian tadpoles. Deficiency in growing vertebrates causes reduced nerve and organ development; presence promotes neural activity, heart rate, blood pressure. Dietary intake of iodine (in humans 100–400 μg day^{-1}) is required for their normal production.

THYROID STIMULATING HORMONE (TSH, THYROTROPIN). Protein product of anterior pituitary gland, stimulating thyroid production and release of thyroxine (T_4) and triiodothyronine (T_3). Secretion influenced by TSH-releasing factor (TRF) from the hypothalamus. Negative feedback occurs between levels of (a) (T_4) and (T_3) and (b) TSH and TRF in blood. Fall in T_3 and T_4 raises outputs of TRF and TSH; rise of the last two shuts off TRF and TSH release.

THYROXINE (T_4). See THYROID HORMONES.

THYSANOPTERA. Thrips. Order of minute exopterygote insects, most with piercing mouthparts for sucking plant juices (and occasionally aphids). Have prepupal and pupal stages despite being exopterygote. Of economic importance as transmitters of disease, and ecologically in studies of population regulation.

THYSANURA. Silverfish. See APTERYGOTA.

TIBIA. (1) Shinbone; the more anterior of the two long bones below knee of tetrapod hindlimb, the other being the fibula. See PEN-TADACTYL LIMB. (2) Fourth segment from base of an insect's leg.

TICKS. See ACARI.

TIGHT JUNCTION. See INTERCELLULAR JUNCTION.

TILLER. (Of grasses) a side shoot arising at ground level.

TISSUE. Association of cells of multicellular organism, with a common embryological origin or pathway and similar structure and function. Often, cells of a tissue are contiguous at cell walls (plants) or cell membranes (animals) but occasionally the tissue may be fluid (e.g. blood). Cells may be all of one type (a *simple tissue*, e.g. squamous epithelium, plant parenchyma) or of more than one type (a *mixed tissue*, e.g. connective tissue, xylem, phloem). Tissues aggregate to form organs. See ACELLULAR.

TISSUE CULTURE (EXPLANTATION). Technique for maintaining fragments of animal or plant tissue or separated cells alive after their removal from the organism. A sterile bathing medium (commonly RINGER'S SOLUTION) surrounds the cells, while appropriate temperature, pH and nutrient levels are maintained and waste products removed. Usually the cells form a monolayer, one cell thick. Similar techniques are employed in organ culture. See CONTACT INHIBITION.

TISSUE FLUID (INTERSTITIAL FLUID). Fluid derived from blood plasma by filtration through capillaries in the tissues. Differs from blood chiefly in containing no suspended blood cells and in having lower protein levels. Bathes tissue cells in appropriate salinity and pH and acts as route for reciprocal diffusion of dissolved metabolites between them and blood. Most of the water filtered from capillaries is re-absorbed by them osmotically; that which remains (along with some solutes) enters the LYMPHATIC SYSTEM, when it is termed *lymph*. See OEDEMA.

TITRE. Usually, a relative measure of the amount of antibody in a fluid, usually serum. The greatest dilution capable of producing a particular detectable antibody–antigen reaction is its titre. This assay technique can be used to compare titres in different samples. See PRE-CIPITIN.

TMV. Tobacco mosaic virus. See VIRUS.

TOADSTOOL. Common name for fruiting bodies of fungi (other than mushrooms) belonging to the Agaricaceae (BASIDIOMYCOTINA). Often wrongly assumed that all are poisonous, although some are.

TOCOPHEROL. Vitamin E. See VITAMINS.

TOLERANCE. See IMMUNE TOLERANCE.

TONE (TONUS). See MUSCLE CONTRACTION.

TONOPLAST. (Bot.) Cytoplasmic membrane surrounding vacuole in plant cells.

TONSILS. Masses of non-encapsulated LYMPHOID TISSUE in mouth or pharynx of tetrapods, lying close underneath mucous membrane. Humans have a pair of palatine tonsils at the junction of mouth and pharynx, and a single pair of pharyngeal tonsils (adenoids) at the back of the nose. Sites of lymphocyte production. See PEYER'S PATCHES.

TONUS (TONE). See MUSCLE CONTRACTION.

TORNARIA. Ciliated planktonic larval stage of the HEMICHORDATA. Its development by radial cleavage and ENTEROCOELY suggests links with echinoderms.

TORUS. (1) Receptacle of a FLOWER; (2) central thickened portion of pit membrane in bordered PITS of conifers and some other gymnophytes.

TOTIPOTENCY. A cell (e.g. a zygote) is said to be *totipotent* if it can give rise to all the cell phenotypes of the organism to which it belongs if suitably challenged by different environments. Compare PLURIPOTENCY.

TOXIN. Any poisonous substance produced by an organism; commonly injurious to potential competitors/predators. May be of microbial, plant or animal origin; some are among the most potent poisons known. Some antibodies are *antitoxins*. See ENDOTOXIN, EXOTOXIN, TOXOID.

TOXOID. Toxin modified so as to retain its antigenicity (its epitopic domain) but to lose its toxicity.

TRABECULAE. Slender bars of tissue lying across cavities. (Zool.) They form the mesh-like interiors of spongy BONE, commonly growing along stress lines. (Bot.) Slender rod-shaped supporting structures found in various plant organs: (1) radially elongated endodermal cells, each becoming a row of cells by division, that support the stele in stems of the lycopod *Selaginella*, linking stele to cortex; (2) membranes, comprising rows of cells, traversing interior of sporangia in the lycopod *Isoetes*; (3) rod-like outgrowths of cell wall lying across lumen of tracheids.

TRACE ELEMENT. Element required in minute amounts for an organism's normal healthy growth. A *micronutrient*. Often a component or activator of an ENZYME. *Essential* plant trace elements (without which death eventually ensues) include zinc, boron, manganese, copper and molybdenum. Fluorine is essential for hardening tooth enamel. Iodine is a component of thyroxine and triiodothyronine. For cobalt, see CYANOCOBALAMIN. Compare VITAMIN.

TRACER. Term sometimes used for an isotopic label. See LABELLING.

TRACHEA. (1) Vertebrate windpipe. A single tube, covered and kept open by incomplete rings of cartilage; with smooth muscle in its wall, and a ciliated lining. Leads from glottis down neck, branching into two bronchi. (2) In most insects and some other arthropods (e.g. woodlice), hollow tubes of epidermis and cuticle conducting air from spiracles directly to tissues. The finest branches (*tracheoles*, diameter about 0.1 μm) are intracellular. Both tracheae and tracheoles may be permeable but tracheoles are more important in gaseous exchanges. See VENTILATION, ECDYSIS.

TRACHEAL GILL. See GILL.

TRACHEARY ELEMENT. General term for the water-conducting cell in vascular plants. Comprise TRACHEIDS and VESSEL MEMBERS.

TRACHEID. Non-living XYLEM element, characteristic of vascular plants other than flowering plants. Formed from a single cell, it is elongated with tapering ends and thick, lignified and pitted walls: an empty firm-walled tube running parallel to long axis of organ in which it lies, overlapping and in communication with adjacent tracheids by means of pits. Functions in water conduction and mechanical support. Primitive compared with a VESSEL.

TRACHEOLE. See TRACHEA.

TRACHEOPHYTA. A group (Division) in classifications regarding possession of vascular tissue as of greater taxonomic significance than the seed habit. Includes all vascular plants, thus comprising the majority of the world's terrestrial vegetation.

TRAIT. A particular phenotypic character, as opposed to character mode. Thus eye colour in *Drosophila* is a character mode, whereas red eye colour is a character trait. Traits can be *autosomal* or *sex-linked*, and determined either by a single locus or polygenically. Sometimes, in heritable disorders, *trait symptoms* are contrasted with *disease symptoms*, in which case the former are usually milder and occur in heterozygotes while the latter are more severe and occur in homozygotes (see SICKLE-CELL ANAEMIA).

TRANSAMINASE. Enzyme carrying out TRANSAMINATION.

TRANSAMINATION. Transfer by a *transaminase* of an amino group from an amino acid to a keto acid (e.g. pyruvate, ketoglutarate), producing respectively a keto acid and an amino acid. See PYRIDOXINE.

Trans-configuration. (Of mutations) see CIS-TRANS TEST.

Transcriptase. An RNA POLYMERASE. Compare REVERSE TRANS-CRIPTASE.

Transcription. Production of an RNA molecule off a DNA template by an RNA polymerase. See PROTEIN SYNTHESIS.

Transcription factors. Regulatory proteins determining the efficiency with which RNA polymerases bind to DNA promoter regions during transcription. *Transcription activators* are involved in positive, and *transcription repressors* in negative control of gene expression. The DNA-binding ability of these proteins often comprises the common structural motif of an α-helix, a short turn and a second α-helix (the 'helix-turn-helix'). The first helix interacts with the DNA backbone while the second lies in a major groove of the DNA molecule. Many eukaryotic transcription factors contain in addition tandemly repeated amino acid sequences each producing a metal-binding (often Zn^{++}) finger-like protrusion in the functional molecule. These finger sequences often include identical or near-identical (*consensus*) sub-sequences which are thought to be conserved and homologous. Several eukaryotic transcription activators have been shown to have one nucleic acid-binding region and a second 'activating region' binding additional transcription factors or perhaps RNA polymerase itself. See HOMOEOBOX.

Transdetermination. Change brought about in groups of cells (e.g. insect imaginal disc cells) which when subsequently cultured differentiate into a structure normally produced by cells from a different lineage within the organism (e.g. a different disc). The effect resembles mutation in a stem cell, but occurs to a group of cells which shifts from one heritable state to another as a result of environmental influences. See HOMOEOTIC MUTATION.

Transduction. Process in which usually a BACTERIOPHAGE picks up DNA from one bacterial cell and carries it to another, when the DNA fragment may become incorporated into the bacterial host's genome. Two basic types: (a) *generalized transduction*, where the phage DNA-packaging mechanism picks up 'by mistake' any phage-sized fragment of chromosomal DNA, which can be integrated by homologous RECOMBINATION into the recipient genome after injection into the cell by the phage apparatus; (b) *specialized transduction*, in which, on induction, integrated phage DNA (e.g. λ) genome is imprecisely excised from the chromosome, carrying adjacent chromosomal DNA with it; since the phage is generally integrated at a specific site in the chromosome, only a few bacterial genes can be transduced this way. A similar process, not usually regarded as transduction, occurs in *F-mediated sexduction* (see F FACTOR) in which an F factor (a plasmid, not a phage) carries bacterial DNA from one cell to another. All three types are employed in CHROMO-SOME MAPPING.

TRANSECT. Line or belt of vegetation selected for charting plants; designed to study changes in composition of vegetation across a particular area.

TRANSFECTION. In prokaryotes, reserved for the uptake by transformation of naked phage DNA to produce a bacteriophage infection. See TRANSFORMATION for use in eukaryotes.

TRANSFERASE. Enzyme (e.g. acyl transferases, transaminases) transferring a non-hydrogen functional group from one molecule to another. Coenzymes are often involved.

TRANSFER CELLS. Specialized parenchyma cells possessing ingrowths of the cell wall and producing unusually high surface-to-volume protoplast ratios. Believed to play an important role in transfer of solutes between cells over short distances. Exceedingly common; probably serving a similar role throughout the plant. Occur in association with XYLEM and PHLOEM of small veins in cotyledons and leaves of many herbaceous dicotyledons; in xylem and phloem of leaf traces, at nodes in both dicots and monocots; in various tissues of reproductive structures, e.g. embryo sacs, endosperm; in various glandular structures, e.g. nectaries and glands of carnivorous plants, and in absorbing cells of haustoria of dodder (*Cuscuta*).

TRANSFERRIN. Iron-binding β-globulin protein transferring iron from reticulo-endothelial cells in bone marrow to immature red blood cells, where it binds and passes iron (Fe^{++}) into the cell. See FERRITIN.

TRANSFER RNA. See RNA, GENETIC CODE, PROTEIN SYNTHESIS.

TRANSFORMATION. (1) Change in certain bacteria (occasionally other cells) which, when grown in presence of killed cells, culture filtrates or extracts from related strains, take up foreign DNA and acquire characters encoded by it. Transformed cells retaining this DNA may pass it to their offspring. When cultured mammalian or other animal cells take up foreign DNA the term *transfection* is often used. Discovery implicated DNA as a genetic material. (2) Event producing CANCER CELL properties in an otherwise normal animal cell. May involve viral infection. See CARCINOGEN, TRANSGENIC.

TRANSFORMATION SERIES. See EVOLUTIONARY TRANSFORMATION SERIES.

TRANSFUSION. Transfer directly into the bloodstream of whole blood or blood components. A form of TRANSPLANTATION.

TRANSGENIC. Describing an organism whose normal genome has been altered by introduction of a gene by a manipulative technique (microinjection of DNA into egg; use of plasmid or virus-based DNA vector), often involving introduction of DNA from a different species. The foreign DNA may then integrate into the host genome. Such organisms are potentially valuable in plant and animal husbandry. The technique is yielding results in the genetics of animal development. See TRANSDUCTION.

TRANSITIVITY. A logical relation of importance in the context of SPECIES definitions. The relational property of populations '*interbreeds with . . .*' is transitive if, when population A interbreeds with population B and population A interbreeds with population C, then population B interbreeds with population C. RING SPECIES indicate that '*interbreeds with . . .*' is not transitive at the population level and so cannot do the job which the biological species concept demands of it.

TRANSLATION. Ribosomal phase of PROTEIN SYNTHESIS.

TRANSLOCATION. (1) (Bot.) Long-distance transport of materials (e.g. water, minerals, photosynthates) within a plant. Absorption of mineral ions from the soil solution by roots may be under metabolic control, often taking place against concentration gradients. Most inorganic ions are taken up by ACTIVE TRANSPORT; others flow in passively along electrochemical gradients. Most ions travel via the SYMPLAST from root epidermis to xylem, and then upwards in the TRANSPIRATION STREAM. Movement of organic molecules (e.g. sugars, amino acids) and some inorganic ions occurs upwards and downwards in the phloem from leaves or storage organs to regions of active growth (or storage), the pattern changing with the season and stage of development. Rates of movement in phloem ($60–120$ cm h^{-1}) greatly exceed normal sucrose diffusion rates in water. According to the *pressure-flow hypothesis*, assimilates move from source to sink along an osmotic turgor pressure gradient. In leaves (the source), sugars are actively secreted into sieve tubes and absorbed there by companion cells, while at the sink (e.g. roots) they are absorbed by parenchyma cells, all of which causes MASS FLOW in the intervening sieve-tube, which in this hypothesis plays a purely passive role throughout. Physical models suggest that pressure-flow could produce the flow rates observed with radio-labelled sugars, and at the pressures detected within sieve-tubes, but only on the improbable view that most of the pores in sieve-plates are unblocked during translocation. Some support the view that phloem is involved actively in translocation, maybe through a form of cytoplasmic streaming. See TRANSPIRATION. (2) Form of chromosomal MUTATION in which an excised chromosome piece either rejoins the end of the same chromosome or is transferred to another.

TRANSMITTER. See NEUROTRANSMITTER.

TRANSPIRATION. Loss of water vapour from plant surfaces. More than 20% of the water taken up by roots is given off to the air as water vapour, most of that transpired by higher plants coming from the leaves. Differs from simple evaporation in taking place from living tissue and in being influenced by the plant's physiology. Most occurs through STOMATA, and to a much lesser extent through the leaf cuticle. Transpiration rates are affected by levels of CO_2, light, temperature, air currents, humidity and availability of soil water. Most of these affect stomatal behaviour, whose closing and opening are controlled by changes in the turgor pressure of guard cells, in turn correlated with the potassium ion (K^+) levels inside them. So long as stomata are open (during exchange of gases between leaf and atmosphere), loss of water vapour to the atmosphere must occur and, although for a healthy plant this is inevitable, it is harmful when excessive and causes wilting and even death.

TRANSPIRATION STREAM. The TRANSLOCATION of water and dissolved inorganic ions from roots to leaves via xylem, caused by TRANSPIRATION. Water enters plant through root hairs, much of it moving along a WATER POTENTIAL gradient to the leaves via the XYLEM. This gradient is via both the SYMPLAST and APOPLAST, the latter route offering much less resistance. According to the *cohesion-tension theory*, water within xylem vessels is under considerable tension because polar water molecules cling together in continuous columns pulled by evaporation above. This tension is experimentally confirmed, and if the water column breaks CAVITATION occurs. See TRANSLOCATION.

TRANSPLANTATION. (1) Artificial transfer of part of an organism to a new position either in the same or a different organism. Practically synonymous with *grafting*; but no close union with tissues in new position is implied. See GRAFT. (2) For nuclear transplantation, see NUCLEUS.

TRANSPORT PROTEINS (CARRIER PROTEINS). Proteins intrinsic to some cell membrane (not just plasmalemma), mediating either passive diffusion, FACILITATED DIFFUSION or ACTIVE TRANSPORT of a solute across it. All appear either to traverse the membrane or be part of a structure which does. Similar to enzymes in having a specific binding site for transported solute, which may be saturated similarly to an active site. Reversible allosteric changes in the protein are probably responsible for carriage, and for actively transported molecules these conformational changes are in part linked to ATP hydrolysis. Transport proteins are either *uniports* (taking one solute type one way), *symports* (transport of one solute depends upon simultaneous carriage of another in the same direction) or *antiports* (carriage of one solute into the cell depends upon simultaneous

transport of another out of the cell, e.g. SODIUM PUMP). See BAC-
TERIORHODOPSIN, GATED CHANNELS, IMPULSE, IONOPHORE.

TRANSPORT VESICLES. Membrane-bound vesicles, many originating
either from plasmalemma or Golgi apparatus, involved in endocytosis
or exocytosis.

TRANSPOSABLE ELEMENTS. Prokaryotic INSERTION SEQUENCES and
ANTIBIOTIC RESISTANCE ELEMENTS, and several eukaryotic
DNA sequences (e.g. the COPIA elements among over thirty elements
in *Drosophila* making $\simeq$ 30% of its total genome, Ty elements in
yeast, IAP and VL30 in rodents, *Alu* sequences in humans, Ac, Spm
and other elements from maize, and many others), all being incapable
of replication independently of the host cell genome, and inherited
only when physically integrated into that genome or that of a
PLASMID or BACTERIOPHAGE. They mediate a wide range of
genetic effects in their host, notably inactivation or change of expression of genes, and sometimes chromosomal rearrangements (mutations) such as inversions, translocations and deletions. Their insertion
into, or adjacent to, another gene often blocks its expression, but
insertion into a gene promoter region may increase that gene's
expression. If recombination occurs between copies of a transposable
element at different sites in a chromosome, then *non-homologous
crossing-over* sometimes occurs, possibly a source of reciprocal translocations, deletions, duplications or GENE AMPLIFICATION. It is
now thought that CROSSING-OVER is often associated with previously unsuspected transposable elements. One class of elements, e.g.
Drosophila copia, yeast Ty (sometimes called *Class I elements* in
eukaryotes), have structural affinities with retroviruses and seem to
transpose via RNA intermediates, being flanked by direct (i.e. not
inverted) repeat DNA sequences. Another class, e.g. *Drosophila* P
ELEMENT, maize Ac and Spm (sometimes called *Class II elements* in
eukaryotes), seem to transpose directly from DNA to DNA and are
flanked by either long or short INVERTED REPEAT SEQUENCES.
Copying and insertion of these elements into a genome may involve
processes similar to mRNA SPLICING. See REPETITIVE DNA,
TRANSPOSITION, TRANSPOSON.

TRANSPOSITIONS. Phenomena in which genetic elements insert and
excise themselves from chromosomal and plasmid DNA. Those that
can replicate independently of the host genome or plasmid are
termed episomal (e.g. prophages and temperate viruses); those without their own replication origin are termed TRANSPOSABLE
ELEMENTS.

TRANSPOSON. A genetic element, varying from 750 base pairs to 40
kilobase (kb) pairs in length, having at least the genes necessary for
its own transposition (movement from a site in one genome to
another site in the same or in a different genome). *Simple* transposons

(INSERTION SEQUENCES) carry this information alone; *complex* transposons house additional genes, such as ANTIBIOTIC RESIST-ANCE ELEMENTS. Many if not all encode the enzyme *transposase* which facilitates their insertion, although its level may be kept low in the host cell by repression from a (*resolvase*) gene also carried on the transposon. In simple transposition, the transposon is moved to the new DNA site but leaves a lethal gap in the old DNA; in replicative transposition the transposon is replicated so that one copy is left *in situ* while the other is moved to a new DNA site. Replicative transposition often requires the enzyme resolvase. See TRANSPOS-ABLE ELEMENTS.

TRANSVERSE PROCESS. Lateral projection, one on each side, of neural arch of tetrapod vertebra, with which the head of the rib articulates.

TRANSVERSE SECTION. Section cut perpendicular to longitudinal axis of an organism.

TREMATODA. Flukes. Class of parasitic PLATYHELMINTHES. Ectoparasites (Monogenea) or endoparasites (Digenea) of vertebrates. Mouth and gut are retained, adults having a thick epidermal cuticle. All have a sucker around mouth. Adult *digeneans* live in liver, gut, lung or blood vessel of primary host and may cause serious disease (see SCHISTOSOMA). Ciliated larvae (miracidia) pass out of the host with egesta and parasitize snails (secondary host) in which they reproduce asexually (see POLYEMBRYONY). Sporocysts then give rise to REDIAE and these to CERCARIAE. *Monogeneans* normally have a large multiple central sucker for attachment, *Polystomun* alone among them being endoparasitic (frog bladder).

TRIASSIC (TRIAS). GEOLOGICAL PERIOD, lasting from around 225–180 Myr BP. Permian and Triassic are sometimes united as the *Permo–Trias*. See EXTINCTION.

TRIBE. Minor taxon of plant classification, used within large families for groups of closely-related genera. Tribe names end in *-eae*.

TRICARBOXYLIC ACID CYCLE (TCA CYCLE). See KREBS CYCLE.

TRICHOGYNE. Unicellular or multicellular projection from female sex organ of some green (CHLOROPHYTA) and red (RHODOPHYTA) algae, fungi (ASCOMYCOTINA, BASIDIOMYCOTINA) and lichens; receives male gamete or male nuclei prior to fertilization.

TRICHOME. In blue-green algae (CYANOBACTERIA), a filament comprising a uniserate or multiserate chain of cells. See also HAIR.

TRICHOPTERA. Caddis flies. Order of moth-like, weakly-flying endopterygote insects. Two pairs of membranous wings covered in bristles (hairs) rather than scales. Mandibles reduced or absent in adult. Larvae aquatic, often encased, with biting mandibles and important as herbivores, carnivores and as prey items. Some are indicators of unpolluted water.

TRICONODONTA. Extinct mammalian order (Lower Cretaceous–Upper Triassic), apparently lacking descendants. Up to four incisors in each half-jaw; molars typically with three sharp conical cusps in a row. Cat-sized or smaller; possibly derived from therapsids. Brain small.

TRICUSPID VALVE. Valve between atrium and ventricle on right side of bird or mammalian heart. Consists of three membranous flaps preventing backflow of blood into the atrium on ventricular contraction. Compare MITRAL VALVE.

TRIGEMINAL NERVE. Fifth vertebrate CRANIAL NERVE.

TRIGLYCERIDE. Ester formed by condensation of three fatty acid molecules to the trihydric alcohol glycerol. See FAT, GLYCEROL.

TRILOBITA. Extinct class of marine arthropods, abundant from Cambrian–Silurian, surviving until Permian. The conservative body, oval and depressed, comprised the following tagmata: shield-like *cephalon* with one pair of antennae and paired sessile compound eyes, a *trunk* of freely movable segments terminating in the united segments of the *pygidium*. Along length of body, longitudinal grooves separated large pleural plates (covering the biramous limbs) from a higher middle region on each trunk segment. Average length 5 cm, but some reached 0.6 m. Probably benthic, feeding on mud and suspended matter. See ARTHROPODA, MEROSTOMATA.

TRIMEROPHYTA. Primitive fossil vascular plants, thought to have evolved directly from the RHYNIOPHYTA and to represent ancestors of ferns, progymnophytes, and even horsetails. Appeared 360 Myr BP, during the mid-Devonian. The plants lacked leaves, but the main axis formed lateral branch systems that dichotomized several times, some of the smaller branches bearing sporangia in which one type of spore was formed. Other branches remained vegetative. Cortex was wide, with cells having thick walls; in the centre was the vascular strand.

TRIPLOBLASTIC. Animals with a body organization derived from three GERM LAYERS (ectoderm, endoderm, mesoderm). Includes all metazoans except coelenterates, which are diploblastic.

TRIPLOID. Nuclei, cells or organisms with three times the haploid number of chromosomes. A form of POLYPLOIDY. Triploid individuals are usually sterile due to failure of chromosomes to pair up during meiosis. See ENDOSPERM.

TRISOMY. Nucleus, cell or organism where one chromosome pair is represented by three chromosomes, giving a chromosome complement of $2n + 1$. May result from NON-DISJUNCTION. See DOWN'S SYNDROME.

TROCHANTER. (1) Any knob for muscle attachment on the vertebrate femur. Three occur on mammalian femurs, largest in humans being conspicuous at hip joint. (2) Segment lying second from base of an insect leg.

TROCHLEAR NERVE. Fourth vertebrate CRANIAL NERVE.

TROCHOPHORE (TROCHOSPHERE). Ciliated larval stage of polychaetes, molluscs and rotifers. Usually planktonic, developing by spiral cleavage. The main ciliary band (*prototroch*) encircles the body in front of mouth (preorally). Commonly a second ring (*telotroch*) surrounds the anus. Coelom arises by schizocoely, not enterocoely. Two protonephridia may be present.

TROPHIC LEVEL. Theoretical term in ecology. One of a succession of steps in the transfer of matter and energy through a COMMUNITY, as may be brought about by such events as grazing, predation, parasitism, decomposition, etc. (see FOOD CHAIN). For theoretical and heuristic purposes, organisms are often treated as occupying the same trophic level when the matter and energy they contain have passed through the same number of steps (i.e. organisms) since their fixation in photosynthesis. Primary producers, herbivores, primary, secondary and tertiary carnivores and decomposers all commonly figure as trophic levels in the analysis of ecosystems. Different developmental stages and/or sexes within a species may occur in more than one trophic level. The number of trophic levels in a community is thought to be limited by inefficiency in ENERGY FLOW from one trophic level to the next; however, food chains are no longer in tropical communities, where energy input is high, than they are in Arctic communities, where energy input is low.

TROPHOBLAST. Epithelium surrounding the mammalian BLASTOCYST, forming outer layer of chorion and becoming part of the embryonic component of the PLACENTA or of the EXTRA-EMBRYONIC MEMBRANES. In humans, produces HUMAN CHORIONIC GONADOTROPHIN and HUMAN PLACENTAL LACTOGEN.

TROPICAL RAIN FOREST. Forests found in three major regions of the world. Largest is in the Amazon basin of South America, with extensions into coastal Brazil, Central America, eastern Mexico and some islands in the West Indies; the second is in the Zaire basin of Africa, with an extension along the Liberian coast; the third extends from Sri Lanka and eastern India to Thailand, the Philippines, the large islands of Malaysia and a narrow strip along the north-east coast of Queensland in Australia. Neither water nor low temperatures are limiting factors for photosynthesis, and rainfall averages from

200–400 cm. yr^{-1}. It is the richest biome in terms of both plant and animal diversity. The trees are evergreen, characterized by large leathery leaves. There is a poorly developed herbaceous layer on the forest floor because of low light levels; but there are many vines and epiphytes at higher levels. Many of the animals inhabit tree tops. Little accumulation of organic debris occurs because decomposers rapidly break it down, released nutrients being quickly absorbed by mycorrhizal roots or leached from the soil by rain (very little in undamaged forest). Tropical rain forest now forms about half the forested area of the planet, but is in process of being destroyed by human activities. Many of the soils of tropical rain forests are conditioned by high and constant temperatures and abundant rainfall and are relatively infertile (*latosols* – red clays largely leached of their nutrients). When the forest is cleared, this leaching process accelerates and the soils either erode or form thick, impenetrable crusts on which cultivation is impossible. It is estimated that by the end of this century most of the tropical rain forests (with huge untapped genetic resources) will have disappeared to produce fields which will be completely useless to agriculture within a very few years and then turn to desert, with considerable effect on global climate.

TROPISM. Response to stimulus (e.g. gravity, light) in plants and sedentary animals by growth curvature; direction of curvature determined by direction of origin of the stimulus. See NASTIC MOVEMENT.

TROPOMYOSIN. See MUSCLE CONTRACTION.

TROPONIN. See MUSCLE CONTRACTION.

TRUFFLE. Subterranean fruiting body of an ascomycotinan fungus (Order Tuberales); prized as a gastronomic delicacy.

TRUMPET HYPHAE. Drawn-out sieve cells, wider at the cross-walls than in the middle, in some brown algae (Order Laminariales). Active transport of photosynthates (mostly mannitol) through sieve cells occurs in this Order.

TRYPANOSOMES. Flagellated protozoans containing a kinetoplast. Parasites in blood of vertebrates and their blood-sucking invertebrates (e.g. insects, leeches) which act as vectors. The genus *Leishmania* parasitizes vertebrate lymphoid/macrophage cells, causing visceral leishmaniasis (kala-azar) in humans; *Trypanosoma* is a vertebrate blood parasite, different species causing sleeping sickness and Chagas' disease in humans and nagana in domesticated cattle. Notorious for ANTIGENIC VARIATION, making search for vaccines against them almost pointless.

TRYPSIN. Protease enzyme secreted in its inactive form (trypsinogen) by vertebrate pancreas and converted to active form by EN-TEROKINASE. Converts proteins to peptides at optimum pH 7–8.

TRYPSINOGEN. Inactive precursor of TRYPSIN.

TSETSE FLY. Genus (*Glossina*) of dipteran fly, sucking vertebrate blood and acting as vector for trypanosomes (spreading sleeping sickness).

TSH. See THYROID STIMULATING HORMONE.

TUATARA. See RHYNCHOCEPHALIA.

TUBE-FEET (PODIA). Soft, hollow, extensile and retractile echinoderm appendages, responding to pressure changes within WATER VASCULAR SYSTEM. In some groups (starfish, sea urchins) they end in suckers and are locomotory; in other groups (brittle stars, crinoids) suckers are absent. Those around mouth may be used for feeding (sea cucumbers, brittle stars). Ciliated in crinoids, forming part of food-collecting mechanism.

TUBER. Swollen end of RHIZOME, bearing buds in axils of scale-like rudimentary leaves (stem tuber), e.g. potato; or swollen root (root tuber), e.g. dahlia. Tubers contain stored food materials and are organs of vegetative propagation.

TUBULIDENTATA. Aardvarks. Order of fossorial termite-eating African mammals with long snouts and tongues; once regarded as edentates but now considered related to the extinct CONDYLARTHRA. One family (Orycteropidae) and species (*Orycteropus afer*). Large digitigrade mammal, with large ears and strong, digging claws. Skin nearly naked. No incisors or canines in the permanent dentition; cheek teeth rootless, growing throughout life. Arrangement of dentine around tubular pulp cavities gives Order its name.

TUBULIN. A characteristic eukaryotic protein. Filamentous (F) tubulin is composed of globular (G) monomers, 13 of these filaments adopting the composite cylindrical (tubular) form characteristic of MICROTUBULES.

TUMOUR (NEOPLASM). Swelling caused by uncontrolled growth of cells; may be malignant (see CANCER CELLS) or benign.

TUNDRA. Treeless region to the farthest limits of plant growth. A huge biome, occupying about one fifth of the planet's surface, and best developed in northern hemisphere, being found mostly north of Arctic circle. In general, permanent ice (permafrost) lies within a metre of the surface and ground conditions in tundra are usually moist as water cannot drain through the soil due to ice. A wide variety of perennial plants is found, of which vegetative propagation is characteristic, but few annuals or woody plants. Bryophytes are an important plant group of tundra, while lichens dominate in the drier regions.

TUNICA CORPUS. Organization of shoot of most flowering plants and a few gymnophytes, comprising one or more peripheral layers of cells (*tunica layers*) and an interior (*corpus*). Tunica layers undergo surface growth and corpus undergoes volume growth.

TUNICATA. See UROCHORDATA.

TURBELLARIA. Class of PLATYHELMINTHES. Mostly aquatic and free-living. Epidermis ciliated, especially ventrolaterally, for locomotion. Gut a simple or branched enteron, with ventral mouth and protrusible pharynx.

TURBIDITY. Level of cloudiness of a suspension, commonly of cells. Often used as indicator of density of cells in suspension.

TURGOR. State of plant cell in which cell wall is rigid, stretched by an increase in volume of vacuole and protoplasm during absorption of water. The cell is described as *turgid*. An essential feature of mechanical support of the plant tissues, loss of turgor (when water loss exceeds absorption) being followed by wilting. Aids expansion of young cells in growth. Compare PLASMOLYSIS. See WATER POTENTIAL.

TURION. Detached winter bud by means of which many aquatic plants survive the winter; e.g. *Myriophyllum*.

TURNER'S SYNDROME. Human genetic disorder caused by absence of a sex chromosome, leaving one X chromosome per cell (XO). NONDISJUNCTION is often the cause. Individuals usually have webbing of the neck, narrowing of the aorta, and reduced height. Gonads undifferentiated secreting no hormones; so no menstrual cycle or pubic hair occur. Abnormal XO conditions also occur in mice, horses and rhesus monkeys. See SEX DETERMINATION.

TURNOVER NUMBER. See MOLECULAR ACTIVITY.

TYLOSE (TYLOSIS). Balloon-like enlargement of membrane of a PIT in wall between a xylem parenchyma cell and a vessel or tracheid, protruding into the cavity of the latter and blocking it; wall may remain thin or become thickened and lignified. Tyloses occur in wood of various plants, often abundantly in heartwood of trees, and may be induced to form by wounding. Often occur in the xylem below developing abscission layer before leaf-fall.

TYMPANIC CAVITY. See EAR, MIDDLE.

TYMPANIC MEMBRANE (TYMPANUM). See EAR DRUM.

TYNDALLIZATION. Fractional sterilization devised by the physicist John Tyndall (1820–93). Method involved lengthy immersion of the material in boiling brine (over 100°C), killing vegetative microorganisms, followed by a period in which unkilled spores were allowed to germinate, followed by further immersion in boiling brine.

This could be repeated two or three times, gradually eliminating all bacterial sphores. Compare PASTEURIZATION.

TYPE SPECIMEN. See HOLOTYPE.

U

UBIQUINONE. See COENZYME Q.

UBIQUITIN. Highly conserved protein (uniform amino acid sequence), apparently present in all eukaryotic cells. Free or covalently bound to a variety of nuclear and cytoplasmic proteins, some integral to membranes. Binding to a protein may initiate the protein's selective degradation. May act as a protein modulator, possibly of certain histones in chromatin.

ULNA. Bone of tetrapod fore-limb, lateral to radius. Articulates distally with carpus and proximally with humerus, above which it may protrude as the olecranon process ('funny bone'). Serves in attachment of main extensor muscle of fore-limb. May fuse with radius or, particularly in mammals, be lost altogether. See PENTADACTYL LIMB.

ULTIMATE FACTOR. Factor considered to be a cause of some biological structure or phenomenon, whose presence it serves to explain in terms of future states of affairs rather than (mechanistic) antecedent conditions. ADAPTATION and ONTOGENY require consideration of future conditions. Compare PROXIMATE FACTOR. See TELEOLOGY.

ULTRACENTRIFUGE. High-speed centrifuge, devised in 1925 by Svedberg, capable of sedimenting (and separating) particles as small as protein or nucleic acid molecules. See CELL FRACTIONATION, SVEDBERG UNIT.

ULTRAMICROTOME. See MICROTOME.

ULTRASTRUCTURE. Fine structure of a cell or tissue, as obtained by instruments with higher resolving power than the light MICROSCOPE.

UMBILICAL CORD. Connection between ventral surface of embryo of placental mammal and placenta. Consists mainly of allantoic mesoderm and blood vessels (umbilical artery and veins), covered by amniotic epithelium. Surrounded by amniotic fluid, it usually breaks or is bitten through at birth. See EXTRAEMBRYONIC MEMBRANES.

UNGULATE. Non-formal term for a hoofed mammal; usually herding and adapted for grazing. Run on firm, open ground on tips of digits (*unguligrade* locomotion). Includes ARTIODACTYLA and PERISSODACTYLA.

UNGULIGRADE. See UNGULATE. Compare DIGITIGRADE, PLANTI-GRADE.

UNIAXIAL. (Of algae) having an axis composed of a single filament.

UNICELLULAR. (Of organisms or their parts) consisting of one cell only. Compare ACELLULAR, MULTICELLULARITY.

UNIPAROUS. Producing one offspring at birth.

UNIPORT. See TRANSPORT PROTEINS.

UNIRAMOUS APPENDAGE. Appendage of myriapods and insects, consisting of a single series of segments (podomeres) as opposed to two series as in a BIRAMOUS APPENDAGE.

UNISEXUAL. (Bot.) (1) (Of flowering plants and flowers) having stamens and carpels in separate flowers. Can be either monoecious or dioecious. (2) (Of algae) having only one type of gametangium formed on a plant. (Zool.) (Of an individual animal) not producing sperm or eggs but both. Compare HERMAPHRODITE.

UNIT MEMBRANE. See CELL MEMBRANES.

UNIT OF SELECTION. Whatever it is that NATURAL SELECTION discriminates between, and whose frequencies in populations may thereby be altered. Controversy surrounds this issue. Strong initial contenders for such units are individual organisms themselves. However, since selection discriminates between these on the grounds of phenotype, and since total phenotype is not what sexual reproduction reproduces, phenotypic characters themselves might better qualify as units of selection, individual organisms being composites of character traits. Strictly, however, it is usually not traits but their material causes (physical encodings, GENES) which pass from parent to offspring in reproduction. Yet if selection acts at the level of classical Mendelian genes, many of which are pleiotropic, then as dominance relations between genetically determined characters are subject to background genetic modifiers, genetic backgrounds as much as individual alleles (alternative genes at a locus) may qualify as units of selection. However, alleles themselves may not be durable enough to be the units, since crossing-over, when it occurs, normally does so within genes. Instead, the genetic unit favoured by selection may simply be any length of DNA which persists intact through a greater than average number of meioses. But a mutant molecular 'trick' (e.g. GENE CONVERSION) producing such better-than-average persistence would tend to be favoured automatically by selection, thereby disrupting the normal control selection exerts on genotype via expression of mutants in the phenotype: for such a mutant might have no phenotypic effect at all. In conclusion, a plurality of entities may qualify as units of selection, producing a range between neo-classical Mendelian genes and 'selfish' DNA. See FITNESS, ALTRUISM.

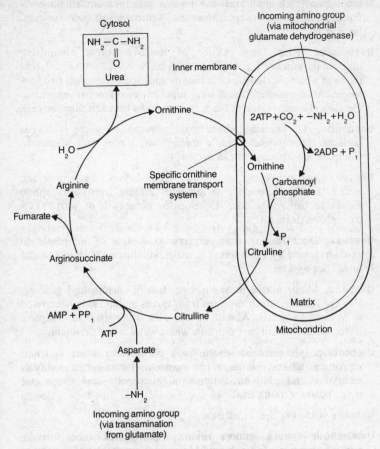

Fig. 64. *Diagram to show the involvement of mitochondria in the urea cycle.*

UNIVOLTINE. (Of animals, typically insects) completing one generation cycle in a year and then diapausing through winter, the adults dying.

URACIL. Pyrimidine base occurring in the nucleotide uridine monophosphate (UMP), whose radical is an integral component of RNA but not of DNA. The UMP radical, derived from UTP, base pairs with AMP and is incorporated into RNA by RNA POLYMERASE.

UREA. Main nitrogenous excretory product, $CO(NH_2)_2$, of ureotelic animals, produced by liver cells from deaminated excess amino acids via the *urea cycle*, part of which occurs in mitochondria. Degradation product of both pyrimidines and, via uric acid, purines (especially fish, amphibians). Elasmobranch fish have a high urea content in body fluid, enabling water to enter osmotically through soft body surfaces. See Fig. 64.

UREDINALES. Rust fungi. Order of Basidiomycotina comprising obligate parasites of higher plants, with succession of several different forms in a complicated life cycle that in some species involves two host species. Many rust fungi are of great importance as plant pathogens; e.g. *Puccinia graminis*, cause of black stem rust of wheat and other cereals.

UREOTELIC. (Of animals) whose main nitrogenous waste is UREA, such as elasmobranchs, adult amphibians, turtles and mammals. Compare AMMONOTELIC, URICOTELIC.

URETER. Duct conveying urea away from kidney. In vertebrates, usually restricted to duct of amniotes leading from metanephric kidney to urinary bladder. Develops as outgrowth of WOLFFIAN DUCT. See CLOACA.

URETHRA. Duct leading from urinary BLADDER of mammals to exterior; joined by vas deferens in males, conducting both semen and urine. See PENIS.

URIC ACID. Major purine breakdown product of adenine and guanine; also the form in which nitrogen from excess amino acids is excreted in *uricotelic* animals. Almost insoluble in water, it is an adaptive nitrogenous waste in environments where water is at a premium.

URICOTELIC. (Of animals) whose main nitrogenous waste is URIC ACID; e.g. snakes, lizards, a few mammals (some desert rodents), embryonic and adult birds, terrestrial gastropods, and insects (see MALPIGHIAN TUBULES).

URINARY BLADDER. See BLADDER.

URINIFEROUS TUBULE (KIDNEY TUBULE). Convoluted tube forming bulk of each *nephron* (excluding Malpighian corpuscle) of vertebrate KIDNEYS.

URKARYOTE. Term reserved for proto-eukaryotes, organisms (presumably single-celled) from which eukaryotes evolved after endosymbiotic incorporation of whichever prokaryotes gave rise to mitochondria, chloroplasts and perhaps some other membranous structures now included within the category of organelles. See ENDOSYMBIOSIS.

UROCHORDATA (TUNICATA). Subphylum of the CHORDATA (see PROTOCHORDATA). Sea squirts, etc. Adults sedentary (some free-swimming forms, the Larvaceae, are probably progenetic), feeding microphagously by a ciliary-mucus method employing an ENDOSTYLE, with pharyngeal gill slits, reduced nervous system, no notochord. Larvae active and tadpole-like, with well-developed nervous system and notochord. See PROGENESIS.

URODELA. Order of AMPHIBIA. Newts and salamanders; with elongated bodies, tail and short limbs. Mostly confined to moist terrestrial habitats and/or ponds and lakes. Tadpoles resemble adults more than do those of frogs and toads. *Ambystoma*, a salamander, is an example of NEOTENY.

UROSTYLE. Rod-shaped bone (representing fused caudal vertebrae) in anuran vertebral column; important in rigidifying posterior part of the animal during jumping. Flanked by two ilia, with which it articulates.

UTERUS (WOMB). Muscular expansion of MÜLLERIAN DUCT of female mammals (except monotremes) in which embryo develops after implantation. Usually paired, each connecting to a Fallopian tube; single in humans, through fusion of lower part. Connected by vagina to exterior, controlled by sex hormones (see MENSTRUAL CYCLE), oxytocin and prostaglandins; enlarging at sexual maturity or in breeding season. See CERVIX. Glandular lining membrane (endometrium, decidua) nourishes early embryo. Smooth muscle in its wall greatly increases during pregnancy; its contraction ultimately expels embryos and their placentae (PARTURITION).

UTRICULUS. See VESTIBULAR APPARATUS.

V

VACCINE. Material producing in an animal an immune reaction and an acquired IMMUNITY to a natural microorganism (*immunization*; compare INOCULATION). Commonly a suspension of killed or attenuated microorganisms, their surface antigens or toxins. Immunity to poliomyelitis may be achieved using attenuated virus. So many viral influenzas exist (none bringing cross-immunity) that a general vaccine for all forms has not been found (see PASTEUR). Almroth Wright (1861–1947) first showed that dead rather than living (typhus) bacteria could produce immunity. BCG vaccine (against tuberculosis) contains live, attenuated, tubercle bacilli. Vaccination against smallpox has eliminated the disease.

VACUOLE. Fluid-filled space or cavity within cytoplasm, bounded by membrane (*tonoplast* in plant cells). A single vacuole, occupying most of cell volume, is present in many plant cells, containing a solution of sugars and salts (cell sap) isotonic to rest of cytoplasm. Absent from cells of bacteria and blue-green algae. See CONTRACTILE VACUOLE, FOOD VACUOLE, WATER POTENTIAL.

VAGILITY. Distance in a straight line between either (a) an individual's birth place and the place where it dies, or (b) its site of conception and the place where it gives rise to a new zygote. Populations or demes of low mean vagility have more probability of recruiting rare homozygous mutants.

VAGINA. Duct of female mammal connecting uteri with the exterior via short vestibule. Usually single and median due to fusion of lower part of MÜLLERIAN DUCT in the embryo. Receives the male's penis during copulation. Lined with stratified non-glandular epithelium, which may undergo cyclical changes during oestrous cycle under the influence of sex hormones.

VAGUS NERVE. Large, mixed (sensory and motor), vertebrate CRANIAL NERVE (CN X) innervating much of the gut, ventilatory system (including fish gill muscles) and heart. Each emerges within skull from the medulla, running back on each side as an abdominal ramus, the main parasympathetic route. ACCESSORY NERVE is basically a motor element of the vagus, emerging further back along the spinal cord. Vagi tend to terminate in plexi, such as in the sympathetic ganglia of the solar plexus. See AUTONOMIC NERVOUS SYSTEM.

VALVE. (Bot.) (1) One of several parts into which a CAPSULE separates after dehiscence by longitudinal splitting. (2) One of two halves of diatom cell wall (frustrule). (Zool.) (1) (Of heart) flap or pocket of vertebrate heart wall at entry and exit of each chamber, preventing back-flow of blood (e.g. MITRAL VALVE, TRICUSPID VALVE). (2) (Of veins, lymphatics) pocket-like flap in birds and mammals, preventing back-flow.

VARIATION. Phenotypic and/or genotypic differences between individuals of a population. *Continuous* (*qualitative*) characters are those exemplified throughout the phenotypic range, tending to be determined by POLYGENES. *Discontinuous* (*quantitative*) characters are those unrepresented in all parts of the phenotypic range (a lack of intermediates), tending to be *polymorphic*, determined by genetic 'switch-mechanisms' (see POLYMORPHISM). Controversy over the relative influence of heredity and environment in producing phenotypic differences fuels the Nature-Nurture debate. Problems arise in obtaining acceptable control populations to test hypotheses. By starting with genetically uniform material (e.g. by cloning or repeated inbreeding) it is often possible to compare phenotypes produced under different environmental regimes and to estimate HERITABILITY. See GENETIC VARIATION, PHENOTYPIC PLASTICITY.

VARIEGATION. Irregular variation in colour of plant organs (e.g. leaves, flowers) through suppression of normal pigment development in certain areas. Commonly due to TRANSPOSABLE ELEMENTS, or to chloroplast- (esp. cpDNA-)based mosaicism, or maybe a disease symptom (e.g. mosaic diseases due to viral infection).

VARIETY. Taxonomic category below subspecies. Often used loosely to mean a variation of any kind within a species. See INFRASPECIFIC VARIATION.

VASCULAR BUNDLE. Longitudinal strand of conducting (vascular) tissue comprising essentially XYLEM and PHLOEM; unit of stelar structure in stems of gymnophytes and flowering plants and occurring in appendages of leaves (e.g. veins in leaves). In gymnophyte stems, arranged in a ring surrounding the pith; in monocotyledons, scattered throughout stem tissue. May be (a) *collateral*, with the phloem on same radius as xylem and external to it (typical condition in flowering plants and gymnophytes); (b) *bicollateral*, with two phloem groups, external and internal to xylem, on same radius (uncommon, occurring in some dicotyledons, e.g. marrow); (c) *concentric*, with one tissue surrounding the other. The latter are *amphicribral* when phloem surrounds the xylem, as in some ferns, and *amphivasal* when xylem surrounds phloem, as in rhizomes of certain monocotyledons. Vascular bundles are further described as *open* when cambium is present, as in dicotyledons, and *closed* when cambium is absent, as in monocotyledons.

VASCULAR CYLINDER. See STELE.

VASCULAR PLANT. Any plant containing a VASCULAR SYSTEM. See TRACHEOPHYTA.

VASCULAR SYSTEM. (Bot.) See STELE. (Zool.) System of fluid-filled vessels or spaces in animals where transport and/or hydrostatics is involved; e.g. BLOOD SYSTEM, LYMPHATIC SYSTEM, WATER VASCULAR SYSTEM.

VAS DEFERENS (pl. VASA DEFERENTIA). One of a pair of muscular tubes (one on each side) conveying sperm from testis to exterior. In male amniotes the vasa lead from the epididymis to the cloaca or urethra. See WOLFFIAN DUCT, VAS EFFERENS.

VAS EFFERENS (pl. VASA EFFERENTIA). Tube (many on each side), developing from a mesonephric tubule and conveying sperm from the testis to the epididymis in male vertebrates. See KIDNEY, TESTIS, WOLFFIAN DUCT.

VASOCONSTRICTION. Narrowing of blood vessel diameter (commonly arteriole) through smooth muscle contraction. Brought about in skin and abdomen by adrenaline, but more generally by vasoconstrictor nerves of sympathetic nervous stimulation. ANTIDIURETIC HORMONE and ANGIOTENSINS may also be involved. See VASOMOTOR CENTRE.

VASODILATION. Increase in blood vessel diameter (commonly arteriole), often through reduction in its sympathetic nervous stimulation, by cholinergic nervous stimulation, or (in heart and skeletal tissue) by adrenaline. HISTAMINE and KININS are vasodilators associated with inflammatory response. See CAPILLARY, VASOMOTOR CENTRE.

VASOMOTOR CENTRE. Group of neurones in vertebrate medulla (see BRAINSTEM) whose reflex outputs control muscle tone of arteriole walls, producing normal VASOCONSTRICTION (vasomotor tone), but increasing this tone when required by raised sympathetic output. Inputs include those from BARORECEPTORS influencing cardiac centres.

VASOPRESSIN. See ANTIDIURETIC HORMONE.

VECTOR. (1) Organism (animal, fungus) housing parasites and transmitting them from one host to another, commonly acting as a host itself. Insects, ticks and mites often act as vectors, many transmitting parasites to man, crops and domestic animals. See MALARIA, FILARIAL WORMS. (2) In GENE MANIPULATION, any REPLICON (e.g. small plasmids, bacteriophages) to which fragments of DNA may be attached and become replicated.

VEGETAL POLE (VEGETATIVE POLE). Point on surface of an animal

egg furthest from nucleus; usually in the yolkiest end. Compare
ANIMAL POLE.

VEGETATIVE REPRODUCTION (V. PROPAGATION). (Bot.) Asexual reproduction in plants through detachment of some part of the plant
body other than a spore, e.g. gemmae, rhizomes, bulbs, corms,
tubers, and its subsequent development into a complete plant. See
ASEXUAL REPRODUCTION. (Zool.) See ASEXUAL REPRODUC-
TION.

VEIN. (Bot.) Vascular bundle forming part of the conducting and
supporting tissue of a leaf or other expanded organ. (Zool.) (1) Blood
vessel (smaller diameter than venous sinus) carrying blood back from
capillaries to heart. In vertebrates, has smooth interior of EN-
DOTHELIUM surrounded by variable degrees of smooth muscle and
connective tissue, but always much less than a corresponding artery
and with a larger internal diameter. Non-return valves ensure unidirectional blood flow, low pressure being provided mainly by skeletal
muscle contractions. (2) Fine tubes of toughened cuticle within insect
wings, providing support. May contain tracheae, nerves or blood.

VELAMEN. Multiple epidermis covering aerial roots of certain plants
(e.g. epiphytic orchids), comprising layers of dead cells (often with
spirally thickened, perforated, sponge-like walls) absorbing water
running over it.

VELIGER. Planktonic molluscan larva, developing from trochophore
and with larger ciliated bands; adult organs present include foot,
mantle and shell.

VENA CAVA. Either of two major vertebrate veins, returning blood
directly to the heart. The usually paired superior (anterior) vena cava
returns blood from fore-limbs and head (only the right persists in
many mammals, including humans). The posterior (inferior) vena
cava is single and median in lungfish and tetrapods, returning blood
from the main body posterior to fore-limbs. Supplants renal portal
veins in phylogeny, shunting much blood formerly entering kidney
glomeruli directly to the heart, bypassing much or all of the renal
filtration and the liver. See CUVIERIAN DUCT.

VENATION. (Bot.) Arrangement of veins in the leaf mesophyll. Termed
netted venation where arranged in branching pattern (largest dicot
vein often extending along main leaf axis as midrib); or *parallel
venation* (most monocots) where veins are of similar size and parallel
along the leaf, much smaller veins forming a complex network.
(Zool.) Arrangement of veins in an insect wing; often used for
identification and classification.

VENTER. (Bot.) Swollen basal region of archegonium, containing
egg cell.

VENTILATION. Rhythmic breathing movements of an animal's muscles and skeleton, increasing gaseous exchange across respiratory surfaces. Often termed *external respiration*. In large active insects the abdomen commonly vibrates and undergoes reversible 'telescoping' as the sternum and tergum of each segment move, causing compression and expansion of the tracheal system, pumping air in and out of the spiracles. Thoracic pumping may also be involved during flight. In fish, gill ventilation is achieved using a pressure pump in front and a suction pump behind. Amphibians at rest on land employ throat movements to renew air in the mouth, where gaseous exchange occurs; when active, a buccal force-pump ventilates the lungs, possibly assisted by flank movements. Bird ventilation is very complex (see AIR SACS). Only mammals have a DIAPHRAGM (stimulated by phrenic nerves from the cervical spinal cord) which, with INTERCOSTALS, ventilates the lungs by changing thorax volume during inspiration and expiration under the control of the medulla (see BRAINSTEM). Here an *inspiratory centre* receives inputs from stretch receptors in the lungs (via the vagus), from CAROTID BODIES and from an inhibitory *expiratory centre* in the pons. Both centres are directly sensitive to gaseous tensions in the blood.

VENTRAL. Generally the surface resting on, or facing, the substratum. (Bot.) See ADAXIAL. (Zool.) In chordates, body surface furthest from nerve cord.

VENTRAL AORTA. See AORTA, VENTRAL.

VENTRICLE. (1) Chamber of vertebrate heart; either single or paired (tendency to separate into two in phylogeny). Thicker-walled than ATRIUM. See HEART, HEART CYCLE. (2) Chamber of molluscan heart responsible for blood flow to tissues. (3) One of the cavities of vertebrate brain filled with CEREBROSPINAL FLUID. A pair in the cerebrum, another in the rest of the forebrain, and one in the medulla.

VENULE. Small vertebrate blood vessel, intermediate in structure and position between capillary and vein. Most have non-return valves. Highly permeable, comprising layer of endothelium with coat of collagen fibres. Larger venules also have smooth muscle in their walls.

VERNALIZATION. (Bot.) Promotion of flowering by application of cold treatment; hormones, cold and daylength interact to modify the response. First studied in cereals, some of which if sown in spring will not flower in the same year but continue to grow vegetatively. Such plants (winter varieties) need to be sown in autumn of the year preceding that in which they are to flower, and contrast with spring varieties which, planted in the spring, flower in the same year. By vernalization, winter cereals can be sown and brought to flower in one season. Seed is moistened sufficiently to allow germination to begin, but not enough to encourage rapid growth, and when tips of radicles are just emerging the seed is exposed to a temperature just above 0°C for a few weeks. Seed thus vernalized acquires properties of seed of winter varieties; sown in the spring, it produces a crop in summer of same year. Even after vernalization, plant must be subjected to suitable photoperiod for flowering to occur. In some plants, gibberellin treatment can substitute for cold exposure, while vernalization effects can be reversed in some plants by exposure to high temperature in anaerobic conditions.

VERNATION. Way in which leaves are arranged in relation to one another in the bud.

VERTEBRA. See VERTEBRAL COLUMN.

VERTEBRAL COLUMN (SPINAL COLUMN). Segmentally arranged chain of bones or cartilages near vertebrate dorsal surface, surrounding and protecting the spinal cord. In most vertebrates, vertebrae form a hollow rod attached anteriorly to skull, largely replacing the notochord. Each vertebra straddles the junction of two embryonic somites, a central mass (*centrum*) replacing the notochord, and an arch above (*neural arch*) enclosing the spinal cord. A similar arch below (*haemal arch*), or ventral projections, often encloses major axial blood vessels (aorta, posterior cardinal veins/vena cava). Projections from vertebrae are for muscle attachment. Vertebrae join at their centra (see SYMPHYSIS), often also by projections of neural arch; only restricted movement is possible between any two. Fish vertebrae are very similar from skull to tail; but in tetrapods there are usually these regional differences: atlas and axis; *cervical vertebrae*, with very short ribs; *thoracic vertebrae*, bearing main ribs; *lumbar vertebrae*, without ribs; *sacral vertebrae*, attached by rudimentary ribs to pelvic girdle; and tail (*caudal*) vertebrae. See BONE, CARTILAGE.

VERTEBRATA (CRANIATA). Major subphylum of CHORDATA. Cyclostomes, fish, amphibians, reptiles, birds and mammals. Differ from non-vertebrate chordates in having a skull, vertebral column and well-developed brain. All but agnathans have paired jaws. Classes include AGNATHA, PLACODERMI, ACANTHODII, CHONDRICHTHYES, OSTEICHTHYES, AMPHIBIA, REPTILIA, AVES and MAMMALIA.

VESSEL (TRACHEA). Non-living element of xylem comprising tube-like series of cells arranged end-to-end, running parallel with long axis of the organ in which it lies and in communication with adjacent elements by means of numerous pits in side-walls. Components of a vessel, *vessel members*, are cylindrical, sometimes broader than long, with large perforations in end walls, and function in TRANSLOCATION of water and mineral salts, and in mechanical support. They

have evolved from tracheids, principal features of this evolution
being elimination of multiperforate end walls to form a single large
perforation, reduction or disappearance of taper of end walls and a
change from a long, narrow element to one relatively short and wide.
Generally thought to be more efficient conductors of water than
tracheids because water can flow relatively unimpeded from one
vessel member to another through the perforations. With few exceptions, confined to flowering plants.

VESTIBULAR APPARATUS (V. ORGAN). Complex part of the *membranous
labyrinth* forming part of inner ear of most vertebrates. Lying on
either side within otic region of the skull, each comprises a closed
system of cavities forming sacs and canals lined by epithelium and
filled with fluid *endolymph* resembling tissue fluid. Major sac-like
parts are the *utriculus* (utricle) and the *sacculus* (saccule), the latter
normally lower, each containing a MACULA with HAIR CELLS and
otoliths. These respond to head tilt and acceleratory movements; but
turning movements are detected by the *semicircular canals*, three of
which are generally present on each side. Two canals lie in vertical
planes at right angles, the third horizontal. Each has a swollen
ampulla containing a sensory *crista* with hair cells embedded in a
jelly-like *cupula* which swings to and fro under influence of endolymph
in the canals, pulling the hair cells whose combined outputs along the
AUDITORY NERVE register turning movements in all spatial planes.
The COCHLEA is the other component of the membranous labyrinth.

VESTIBULOCOCHLEAR NERVE (AUDITORY NERVE, ACOUSTIC NERVE).
Eighth vertebrate cranial nerve, innervating the inner ear. Essentially
a dorsal root of facial nerve, relaying impulses from hair cells of the
cochlea (via cochlear branch) and from vestibular apparatus (via
vestibular branch).

VESTIGIAL ORGAN. An organ whose size and structure have diminished over evolutionary time through reduced selection pressure. Only
traces may remain, comparative anatomy providing evidence of phylogeny. Remains of pelvic girdles in snakes and whales indicate tetrapod
ancestries. See APPENDIX (not vestigial in humans), EVOLUTION.

VIBRISSAE. Stiff hairs or feathers, usually projecting from face. Tactile.

VILLUS. (1) Finger-like projection from the lining of the small intestines of many vertebrates. Their great number (up to 4–5 million,
each up to 1 mm long in humans) gives the mucosa a velvet-like
appearance. Each is covered by epithelium whose brush borders
increase the surface area for absorption and digestion 25-fold (see
MICROVILLUS, DIGESTION). Each contains blood vessels and a
lacteal, extending and retracting by means of smooth muscle. (2) For
chorionic (trophoblastic) villi, see PLACENTA.

VIRCHOW, RUDOLPH. German medical microscopist (1821–1902) who did most to dispel the notion, current from the time of Hippocrates, that disease resulted from imbalance in body 'humours'. Replaced it by a cell-based theory, arguing that cells are derived only from other cells (*omnis cellula a cellula*), broadening and deepening CELL THEORY.

VIRION. See VIRUS.

VIROID. Small naked RNA loops, 300–400 nucleotides long. Replicated by host enzymes as their genomes do not encode any. Lacking capsids, only able to pass from one damaged cell to another. Analogous to PLASMIDS.

VIRUS. One of a group of minute infectious agents (20–300 nm long and/or wide), unable to multiply except inside living cell of a host, of which they are obligate parasites and outside of which they are inert. Pass through filters which trap bacteria, and resolvable only by electron microscopy. Not normally regarded as 'living' since none has any enzyme activity away from its host. When inside a cell, may pack together in a crystalline condition through symmetry and regularity of their capsids. They are biological systems, since each contains molecular information in the form of nucleic acid (but unlike cells, never both DNA and RNA), transcribed and replicated within the host cell. The fully formed virus particle (*virion*) contains its nucleic acid or nucleoprotein core either within a naked coat of protein (*capsid*) consisting of protein subunits (*capsomeres*) or within a capsid enveloped by one or more host cell membranes acquired during exit from the cell and often modified, as by addition of specific glycoproteins. Capsid symmetry may be *helical* (capsomeres forming a rod-shaped helix around the nucleic acid core, e.g. tobacco mosaic virus, TMV); or *icosahedral* (*spherical*, capsomeres forming a 20-sided structure), capsomere number ranging from 12 (φX174) to 252 (adenovirus). Some viruses (*phages*) infect fungi (*mycophages*) or bacteria (BACTERIOPHAGES).

Viruses may be classified in terms of the type of nucleic acid core they contain. In RNA viruses this may be either single-stranded (e.g. picornaviruses, causing polio and the common cold) or double-stranded (e.g. reovirus, causing diarrhoea); likewise DNA viruses (e.g. single-stranded in φX174 phage and parvovirus; double-stranded in adenovirus, herpesvirus and poxvirus). RNA viruses are of three main types: + *strand RNA viruses*, whose genome serves as mRNA in the host cell and serves as template for a minus (−) strand RNA intermediate; − *strand RNA viruses*, which cannot serve directly as mRNA but rather as templates for mRNA synthesis via a virion transcriptase; and *retroviruses*, which are + strand and can serve as mRNAs but on infection immediately act as templates for double-stranded DNA synthesis (which immediately integrates into the host chromosome) via a contained or encoded REVERSE TRANSCRIPTASE. Human T-lymphotrophic viruses (HTLVs) are single-stranded retroviruses and can cause acquired immune deficiency syndrome (AIDS).

While bacteriophages inject their genomes into the host cell, animal viruses are engulfed by endocytosis or, if encapsulated by a membrane, fuse with the host plasmalemma to release the nucleoprotein core into the cell. Some viruses (e.g. poliovirus) have receptor sites on specific host cells which enable entry. Once inside, the genome is commonly initially transcribed by host enzymes; but virally encoded enzymes usually then take over. Host cell synthesis is generally shut down, the viral genome replicated, and capsomeres synthesized prior to assembly into mature virions (see SELF-ASSEMBLY). The virus commonly encodes a late-produced enzyme, rupturing the host plasma membrane (the *lytic phase*) to release the infective progeny; but a viral genome may become integrated into the host chromosome and be replicated along with it (a *provirus*). Many eukaryotic genomes have proviral components. Sometimes this results in *neoplastic transformation* of the cell (see ONCOGENE, CANCER CELL) through synthesis of proteins normally produced only during viral multiplication. DNA tumour viruses include *adenoviruses* and *papovaviruses*; RNA tumour viruses are capsulated and include some retroviruses (e.g. Rous sarcoma virus). Viruses may be used as genetic tools in TRANSDUCTION. They may have evolved from PLASMIDS that came to encode capsid proteins.

VISCERAL ARCH. (1) One of a series of partitions in fish and tetrapod embryos on each side of the pharynx. In fish, lying between mouth and spiracle and between this and adjacent gill slits; in tetrapod embryos, between corresponding gill pouches; (2) skeletal bars (cartilage or bone) lying in these partitions in fish (see GILL BAR). First (mandibular) arch is modified to form the jaw skeleton (PALATOQUADRATE and MECKEL'S CARTILAGE); the second (HYOID ARCH), behind the spiracle, commonly attaches jaws to the skull, each remaining (branchial) arch lying behind a functional gill slit.

VISCERAL POUCH. See GILL POUCH.

VISCERAL PURPLE. See RHODOPSIN.

VITALISM. Metaphysical doctrine with early roots, popular in a variety of forms during 19th century. Opposed to alternative extreme of scientific materialism. Underlying most vitalisms was the conviction that life was more than mere complex chemistry, otherwise science would subject even human activity to deterministic explanations. Ignorance of biochemical principles made such phenomena as growth and development mystifying, employing terminology beyond physics or chemistry. Such causal agencies as 'entelechies' and 'vital forces' were invoked for really baffling phenomena but added nothing to understanding. Vitalists, often religiously inspired, eschewed Darwin's mechanistic philosophy.

VITAL STAINING. See STAINING.

VITAMIN. Organic substance not normally synthesized by an organism, which it must obtain from its environment in minute amounts (a micronutrient). ESSENTIAL AMINO ACIDS are required in larger amounts, and are not vitamins. *Provitamins*, closely related precursors of some vitamins, may occur in the diet. Absence of a vitamin from the diet for sufficient time gives symptoms of a resulting *deficiency disease*, although it is often unclear why particular symptoms occur. A vitamin for one organism may not be so for another.

Classified into fat-soluble (lipid) and water-soluble forms. For humans the only lipid vitamins are A, D, E and K, all stored in the liver, while all water-soluble vitamins are converted to coenzymes, accounting for the small amounts needed. See particular vitamin entries.

VITAMIN A. 11-*cis*-retinal, the lipid prosthetic group of the protein opsin in visual purple. Its deficiency affects all tissues, but the eyes are most readily affected. Young animals, lacking a liver store of the vitamin, are most affected by deficiency, which causes *xeropthalmia* ('dry eyes') in human infants and young children. See RHODOPSIN.

VITAMIN B COMPLEX. Several water-soluble vitamins (currently 12). B_1(aneurine, or thiamine) prevents beri-beri. Group includes BIOTIN, CYANOBALAMIN (B_{12}), FOLIC ACID, NICOTINIC ACID, PANTO-THENIC ACID, PYRIDOXINE (B_6) and RIBOFLAVIN (B_2).

VITAMIN C. See ASCORBIC ACID.

VITAMIN D. Small group of fat-soluble (lipid) vitamins of humans and some other animals, deficiency causing rickets. Some dietary in origin, but also synthesized in the skin under ultraviolet light. One form (D_2, *ergocalciferol*) derives from provitamin ergosterol; another (D_3, *cholecalciferol*) derives from provitamin dehydrocholesterol. People receiving insufficient sunlight can supplement their vitamin D levels by eating liver, particularly fish liver oils. The main circulating form of vitamin D in animals, *25-dihydroxycholecalciferol*, is very active and derived by modification of cholecalciferol in the liver. Like other forms of the vitamin it promotes uptake of calcium ions in the ileum and is required for calcification of bones and teeth. See CALCITONIN.

VITAMIN E (TOCOPHEROL). Group of fat-soluble vitamins obtained principally from plant material (seed oils, wheat germ oil) but also found in dairy produce. Deficiency produces infertility in male and female rats, and probably in most vertebrates, kidney degeneration and other general wasting symptoms. They seem to prevent oxidation of highly unsaturated fatty acids (and their polymerization in cell membranes) in presence of molecular oxygen.

VITAMIN K. Fat-soluble vitamins (K_1 and K_2), required for liver synthesis of a substance (proconvertin) required for prothrombin production, and so for conversion of fibrinogen to fibrin during BLOOD CLOTTING. Produced by many plants (K_1), and by microorganisms (K_2), including those in the gut.

VITELLINE MEMBRANE. See EGG MEMBRANES.

VITREOUS HUMOUR. Jelly-like semitransparent substance filling cavity of the vertebrate eye behind the lens.

VIVIPARITY (VIVIPARY). (Zool.) Reproduction in animals whose embryos develop within the female parent and derive nourishment by close contact with her tissues. Involves PLACENTA in placental mammals; occurs also in some reptiles, amphibians, elasmobranchs, waterfleas (cladocerans), aphids and tsetse flies. See r-SELECTION. (Bot.) Having seeds germinating within the fruit, e.g. mangrove; or producing shoots (e.g. bulbils) for vegetative reproduction, instead of inflorescences; e.g. some grasses.

VOCAL CORDS. See LARYNX, SYRINX.

VOLKMANN'S CANALS. See HAVERSIAN SYSTEM.

VOLTAGE CLAMP. Technique whereby the potential difference across a membrane is kept steady so that movements of ions across it can be measured. Employed a great deal in nerve and muscle research.

VOLUNTARY MUSCLE. Alternative term for STRIATED MUSCLE.

VOLVA. Cup-like fragment of the universal veil at base of stipe in some members of the BASIDIOMYCOTINA (Order Agaricales).

W

WALLACE, ALFRED RUSSEL. British naturalist and traveller (1823–1913). According to T. H. Huxley, his short essay on the mechanism of evolution, received by Charles DARWIN from the Moluccas in June 1858, 'seems to have set Darwin going in earnest' on writing his *The Origin of Species . . .* Their joint paper of 1858, read at the Linnean Society, was the first publicized account of the theory of natural selection. At first, Wallace held that human evolution could be explained by this theory, but later departed from Darwin on this, believing a guiding spiritual force necessary to account for the human soul. Wallace considered SEXUAL SELECTION to be less important in evolution than did Darwin, holding that, unlike Darwin, it had no role in the evolution of human intellect. His contributions to zoogeography were of great importance (see WALLACE'S LINE), and his book *Darwinism* (1889) attributes a wider role for natural selection in evolution than did Darwin's works.

WALLACE'S LINE. Boundary drawn across the Malay Archipelago, separating Arctogea and Notogea (formerly between Australian and Oriental ZOOGEOGRAPHICAL REGIONS). Not a precise demarcation line, lying along the edge of the south-east Asian continental shelf, between Bali and Lombok, Borneo and Celebes, the Phillipines and the Sangai and Talaud Islands. Named by T. H. Huxley after the zoogeographical work of WALLACE (see DARWIN).

WALL PRESSURE. (Bot.) Pressure exerted against the protoplast by plant cell wall; opposed to turgor pressure. See WATER POTENTIAL.

WARM-BLOODED. See HOMOIOTHERMY.

WARNING COLOURATION. See APOSEMATIC, MIMICRY.

WATER POTENTIAL. Term indicating the net tendency of any system to donate water to its surroundings. The water potential, ψ_ω, of a plant cell is the algebraic sum of its wall pressure (pressure potential), ψ_p, and of its osmotic (solute) potential, ψ_o;

$$\psi_\omega = \psi_p + \psi_o.$$

Since the water potential of pure water at atmospheric pressure is zero pressure units by definition, any addition of solute to pure water reduces its water potential and makes its value negative. Water movement in nature is therefore always from a system with higher (less negative) water potential to one with lower (more negative) water potential. The units normally employed are *megapascals* (MPa;

1 Pa = 1 Newton/m²), or *millibars* (1 mbar = 10^2 Pa). Standard atmospheric pressure, equal to the average pressure of air at sea level, is 1.01 bar.

The term has wide applicability. The water potential of any cell is its net tendency to donate water to its surroundings, one with a higher water potential than an adjacent cell tending to donate water to it. Nor is its use restricted to cells. Water potentials of non-cellular solutions separated by a selectively permeable membrane dictate the net direction in which water moves by osmosis, since net movement of water is always from higher to lower water potential values. Osmotic effects can therefore be subsumed under the wider category of water potential effects, for membranes may or may not be involved in phenomena explained in terms of water potential. It is therefore valuable in predicting water movement between cells, between cells and soil solutions and between cells and the atmosphere. It reduces to one the number of units employed in analysing the TRANSPIRATION STREAM in plants. Compare OSMOSIS.

WATER VASCULAR SYSTEM. System of water-filled canals forming tubular component of echinoderm coelom, usually with direct connection (the *madreporite*) with the exterior. A circumoesophageal water ring internal to the skeleton usually gives off radial water canals which pass through pores in the skeleton wherever *tube-feet* protrude. Each tube-foot is expanded by a muscular ampulla internal to the skeleton and is cooordinated with others in feeding and locomotion. See TUBE-FEET.

WAXES. Fatty acid esters of alcohols with high molecular weights. Occur as protective coatings of arthropod CUTICLE, skin, fur, feathers, leaves and fruits, often reducing water loss. Also found in beeswax and lanolin.

W-CHROMOSOME. An alternative notation for the Y-chromosome in organisms where the heterogametic sex is female. See SEX CHROMOSOME.

WEEDS. Plants growing where they are not wanted; compete for space, light, water and nutrients with garden or crop plants. Many adapted to exploit disturbed land, often being the first to colonize waste ground (when they may be beneficial by preventing soil erosion). Tend to produce vast numbers of seeds/fruits per plant, these being easily and widely disseminated, often remaining dormant until rapid growth and further seed production occur. Annuals, biennials and perennials, they often provide habitats for insect and fungal pests. Leguminous weeds can add nitrogen to the soil through N-fixing bacteria, and all increase soil fibre content on ploughing in. See PEST CONTROL, ARTIFICIAL SELECTION.

WEISMANN, AUGUST. Theoretical biologist (1834–1914); one of the first to appreciate that chromosomes are the physical bearers of hereditary determinants and to forge a comprehensive theory of heredity. With Wilhelm Roux, he advocated the theory of MOSAIC DEVELOPMENT. Best remembered for his theory of the continuity of GERM PLASM. Prolific coiner of terms and entities.

WHALEBONE. See BALEEN.

WHALES. See CETACEA.

WHITE BLOOD CELL. See LEUCOCYTE.

WHITE MATTER. Myelinated nerves forming through-conduction pathways of vertebrate BRAIN and SPINAL CORD, with associated glia and blood vessels. Compare GREY MATTER, to which it normally lies superficially.

WHORL. Three or more leaves, flowers, sporangia or other plant parts at one point on an axis or node.

WILD TYPE. Originally denoting the type (genotype or phenotype) most commonly encountered in wild populations of a species, but commonly applied to laboratory stock from which mutants are derived. As in natural populations, the laboratory wild type is subject to change. In genetics, term is usually employed in contexts of alleles and particular phenotypic characters. Such alleles are normally represented by the symbol + rather than by a letter, a genotype homozygous for a wild type allele being given by either + + or +/+. Wild type characters are by no means always genetically dominant (see DOMINANCE).

WILTING. Condition of plants in which cells lose turgidity and leaves, young stems, etc. droop. Results from excess of transpiratory water loss over absorption. Excessive wilting may be stressful, even irreversible.

WISHBONE (FURCULA). Fused clavicles and interclavicles, diagnostic of birds. Present in ARCHAEOPTERYX.

WOBBLE HYPOTHESIS. Hypothesis, due to F. H. C. Crick, that the nucleotide at the 5'-end of a tRNA anticodon plays a less important role than the other two in determining which tRNA molecule will align by base-pairing with a codon triplet in mRNA. Helps to explain pattern of degeneracy observed in the GENETIC CODE, which is restricted to the 3'-ends of codons.

WOLFFIAN DUCT (ARCHINEPHRIC DUCT). Vertebrate kidney duct; one on each side. Develops in all vertebrates (initially in both sexes) from region of pronephros, becoming the mesonephric duct. In male anamniotes, conveys both urine and semen to the exterior (opening into the cloaca). In male amniotes it becomes the vas deferens (vasa efferentia and epididymis representing mesonephric tubules which now lead into it from the testis). In all adult amniotes an outgrowth from lower end of the Wolffian duct develops into the URETER, invading the metanephric kidney. Hormonal influences from the gonads determine whether the Wolffian duct or the MÜLLER- IAN DUCT will persist and differentiate in the adult.

WOMB. See UTERUS.

WOOD. See XYLEM.

WOODLICE. See ISOPODA.

WORKER. See CASTE.

X

XANTHOPHYLL. CAROTENOID pigment; each an oxygenated derivative of carotenes.

XANTHOPHYTA. Division of mainly freshwater eukaryotic ALGAE; the yellow-green algae. Flagellated, coccoid, filamentous and siphonaceous forms; motile cells with one forward-pointing *tinsel* and a shorter *whiplash flagellum*. Chloroplasts contain chlorophylls *a* and *c*, lacking fucoxanthin, and coloured yellowish-green. 'Eyespot' always in the chloroplast, outer membrane of chloroplast endoplasmic reticulum usually continuous with outer nuclear membrane. Storage products include mannitol and a paramylon-like glucan. Asexual reproduction by fragmentation or zoospores/aplanospores. Sexual processes only recorded in three genera.

X CHROMOSOME. See SEX CHROMOSOME.

XENOGRAFT. Graft between individuals of different species (*xenogeneic* individuals). See ALLOGRAFT.

XEROMORPHY. (Bot.) Possession of morphological characters associated with XEROPHYTES.

XEROPHYTE. Plant of arid habitat, able to endure conditions of prolonged drought (e.g. in DESERTS), due either to a capacity during the brief rainy season for internal water storage, used when none can be obtained from the soil, plus low daytime transpiration rates associated with stomatal closure (e.g. in such succulents as cacti); or ability to recover from partial desiccation (e.g. such desert shrubs as the creosote bush). As long as water is freely available, it seems that non-succulent xerophytes transpire as much as, or more freely than, mesophytes but unlike the latter they can endure periods of permanent wilting during which transpiration is reduced to minimal levels. Features associated with this reduction, in different xerophytes, include dieback of leaves covering and protecting perennating buds at soil surface, shedding of leaves when water supply is exhausted, heavily cutinized or waxy leaves, closure or plugging of the stomata (which are sunken or protected), orientation or folding of leaves to reduce insolation, and a microphyllous habit reducing the possibility of drought necrosis of the mesophyll. See HYDROPHYTE, MESOPHYTE.

XEROSERE. SERE commencing on a dry site.

XIPHOSURA. See MEROSTOMATA.

X-RAY DIFFRACTION (X-RAY CRYSTALLOGRAPHY). Technique where a narrow beam of X-rays is fired through a crystalline source (e.g. of DNA or protein), the arrangement of atoms within the crystal being interpreted from the X-ray diffraction pattern produced (usually by computer reconstruction). Loose quasi-crystalline structures such as cell walls and cell membranes are sometimes analysed this way.

XYLEM. Wood. Mixed vascular tissue, conducting water and mineral salts taken in by roots throughout the plant, which it provides with mechanical support. Of two kinds: *primary*, formed by differentiation from procambium and comprising protoxylem and metaxylem, and *secondary*, additional xylem produced by activity of the cambium. Characterized by presence of tracheids and/or vessels, fibres and parenchyma. In mature woody plants, makes up bulk of vascular tissue itself and of entire structure of stems and roots. See TRANS-PIRATION STREAM.

Y

Y-CHROMOSOME. See SEX CHROMOSOME.

YEASTS. Widely distributed unicellular fungi, mainly belonging to the ASCOMYCOTINA, which multiply typically by a budding process. Of great economic importance. Brewing and baking industries depend upon capacity of yeasts, usually *Saccharomyces cerevisiae*, to secrete enzymes converting sugars into alcohol and carbon dioxide. In brewing, fermentation to alcohol is the important process; in baking, escaping carbon dioxide causes the dough to rise. Yeasts are also used commercially as a source of proteins and vitamins, and as hosts in some DNA CLONING techniques of BIOTECHNOLOGY. Some yeast-like fungi (notably the experimentally important *Schizosaccharomyces pombe*) multiply by binary fission rather than budding. Not all ascomycotine, budding and fission yeasts are at all closely related.

YOLK. Store of food material, mainly protein and fat, in eggs of most animals. See POLARITY, CENTROLECITHAL.

YOLK SAC. Vertebrate EXTRAEMBRYONIC MEMBRANE, conspicuous in elasmobranchs, teleosts, reptiles and birds. Contains yolk and hangs from ventral surface of the embryo. Has outer layer of ectoderm, inner layer (usually absorptive) of endoderm, with mesoderm containing coelom and blood vessels between. A gut diverticulum; yolk usually communicating with the intestine. In mammals it is normally devoid of yolk, forming part of the CHORION; in marsupials, it forms an integral component of the PLACENTA. As yolk is absorbed, the yolk sac is withdrawn, eventually merging into the embryo.

Y-ORGAN. Pair of epithelial endocrine glands in the heads of malacrostracan crustaceans, anterior to the brain. Appear to secrete an ECDYSONE-like hormone involved in moulting.

Z

ZEATIN. A CYTOKININ.

Z-FORM HELIX. 'Left-handed' DNA double helix, containing the alternating nucleotide sequence dG-dC. It zigs and zags, hence the name.

ZINJANTHROPUS. Genus of hominid containing the species *Z. boisei*, now renamed *Australopithecus boisei*. See AUSTRALOPITHECINE.

Z LINES. See STRIATED MUSCLE.

ZONA PELLUCIDA. Glycoprotein membrane around mammalian ovum, disappearing before implantation. Secreted by cells of the GRAAFIAN FOLLICLE.

ZONULA ADHAERENS. One kind of INTERCELLULAR JUNCTION.

ZOOCHLORELLA. Symbiotic green algae assignable to the CHLOROPHYTA (e.g. *Chlorella*, *Oocystis*); both freshwater and marine. See ZOOXANTHELLAE.

ZOOGEOGRAPHICAL REGIONS (FAUNAL PROVINCES). Subdivisions of world surface into regions identified by differences in their dominant animals. They are: (a) Arctogea (Europe, Asia, Africa, Indochina and North America). It includes the Palaearctic (Asia, Europe and North Africa), Oriental (Indochina) and Nearctic (Greenland and North America south to Mexico) regions; (b) Neogea (Central and South America), equivalent to the Neotropical region; (c) Notogea (Australasia), normally taken to include Australia, Tasmania, New Zealand, eastern Indonesia and Polynesia. The island of Madagascar is often considered sufficiently distinct faunally to be regarded as a minor region: the Malagasy region. Botanical equivalents are known as *floral regions*. See WALLACE'S LINE, ZOOGEOGRAPHY.

ZOOGEOGRAPHY. Study of, and attempt to interpret, global distributions of animal taxa. Takes into account CONTINENTAL DRIFT and other major geological processes (e.g. orogenies, ice ages and other factors affecting sea level), as well as aspects of ecology and evolution (e.g. adaptive radiation, competition, speciation, geographical isolation, dispersal and vagility). It is largely due, for example, to isolation since the Cretaceous of the fauna of Notogea (see ZOOGEOGRAPHICAL REGIONS) on the Australian plate that its present fauna is so distinctive. Likewise, separation of Malagasy region from the rest of Africa prior to higher primate radiation results in a relict lemur population persisting there today. Island faunas, as Darwin and Wallace knew, have high levels of endemism, particularly if there is considerable separation from mainland faunas. See GONDWANA-LAND, LAURASIA.

ZOOID. Member of a colony of animals (chiefly ectoprocts and entoprocts) in which individuals are physically united by living material. See COLONY.

ZOOLOGY. Branch of biology dealing specifically with animals. There is considerable overlap, however, with botany.

ZOOPLANKTON. Animal members of plankton.

ZOOSPORANGIUM. (Bot.) Sporangium producing zoospores; present in certain fungi and algae.

ZOOSPORE (SWARM SPORE). Naked spore produced within a sporangium (zoosporangium); motile, with one, two or many flagella; present in certain fungi and algae.

ZOOXANTHELLAE. Algae varying in colour from golden, yellowish, brownish to reddish, living symbiotically in a variety of aquatic animals; includes algae assignable to the Bacillariophyta, Chrysophyta, Dinophyta, Cryptophyta; especially important in coelenterates of coral reefs. See ZOOCHLORELLA.

ZORAPTERA. Small order of very small exopterygote insects (metamorphosis slight) of Subclass Orthopteroidea. Occur under bark, in humus, etc. Many species are dimorphic, apterous or winged, although wings can be shed.

ZWITTERION. Ion which has both positively and negatively charged regions. All AMINO ACIDS are zwitterions, although their charge distributions are much affected by pH.

ZYGOMORPHIC. BILATERAL SYMMETRY of flowers.

ZYGOMYCOTINA. Fungi with sexual reproduction in which two gametangia fuse through CONJUGATION, a zygospore resulting. Asexual sporangiospores non-motile, borne within a sporangium. Motile stages lacking. Include large number of saprotrophs and parasites. Commercially important in production of organic acids, pigments, fermented oriental foods, alcohols and modified steroids. Members of one class (Trichomycetes) are commonly commensals of arthropod guts. Some are common mould fungi (e.g. *Mucor*, *Rhizopus*) causing spoilage of stored grain, bread and vegetables.

ZYGOSPORE. Thick-walled resting spore; product of conjugation in the ZYGOMYCOTINA and some green algae. Also used for product of fertilization in the isogamous *Chlamydomonas* and its relatives.

ZYGOTE. Cellular product of gametic union. Usually diploid.

ZYGOTENE. Stage in the first prophase of MEIOSIS.

ZYGOTIC MEIOSIS. Meiosis occurring during maturation or germination of a zygote.

ZYMOGEN. Any inactive enzyme precursor (e.g. pepsinogen, trypsinogen or prothrombin). Activation to the functional enzyme generally involves excision of a portion of polypeptide. The several zymogens in secretory vesicles of pancreatic ACINAR CELLS are termed *zymogen granules*, although some active enzyme is contained there too. Zymogen granules fuse with the cell apex under influence of acetylcholine or cholecystokinin.

FOR THE BEST IN PAPERBACKS, LOOK FOR THE

In every corner of the world, on every subject under the sun, Penguin represents quality and variety – the very best in publishing today.

For complete information about books available from Penguin – including Puffins, Penguin Classics and Arkana – and how to order them, write to us at the appropriate address below. Please note that for copyright reasons the selection of books varies from country to country.

In the United Kingdom: Please write to *Dept E.P., Penguin Books Ltd, Harmondsworth, Middlesex, UB7 0DA.*

If you have any difficulty in obtaining a title, please send your order with the correct money, plus ten per cent for postage and packaging, to *PO Box No 11, West Drayton, Middlesex*

In the United States: Please write to *Dept BA, Penguin, 299 Murray Hill Parkway, East Rutherford, New Jersey 07073*

In Canada: Please write to *Penguin Books Canada Ltd, 2801 John Street, Markham, Ontario L3R 1B4*

In Australia: Please write to the *Marketing Department, Penguin Books Australia Ltd, P.O. Box 257, Ringwood, Victoria 3134*

In New Zealand: Please write to the *Marketing Department, Penguin Books (NZ) Ltd, Private Bag, Takapuna, Auckland 9*

In India: Please write to *Penguin Overseas Ltd, 706 Eros Apartments, 56 Nehru Place, New Delhi, 110019*

In the Netherlands: Please write to *Penguin Books Netherlands B.V., Postbus 195, NL–1380AD Weesp*

In West Germany: Please write to *Penguin Books Ltd, Friedrichstrasse 10–12, D–6000 Frankfurt/Main 1*

In Spain: Please write to *Longman Penguin España, Calle San Nicolas 15, E–28013 Madrid*

In Italy: Please write to *Penguin Italia s.r.l., Via Como 4, I-20096 Pioltello (Milano)*

In France: Please write to *Penguin Books Ltd, 39 Rue de Montmorency, F-75003 Paris*

In Japan: Please write to *Longman Penguin Japan Co Ltd, Yamaguchi Building, 2–12–9 Kanda Jimbocho, Chiyoda-Ku, Tokyo 101*

Political Ideas   David Thomson (ed.)

From Machiavelli to Marx – a stimulating and informative introduction to the last 500 years of European political thinkers and political thought.

On Revolution   Hannah Arendt

Arendt's classic analysis of a relatively recent political phenomenon examines the underlying principles common to all revolutions, and the evolution of revolutionary theory and practice. 'Never dull, enormously erudite, always imaginative' – *Sunday Times*

The Apartheid Handbook   Roger Omond

The facts behind the headlines: the essential hard information about how apartheid actually works from day to day.

The Social Construction of Reality   Peter Berger and Thomas Luckmann

Concerned with the sociology of 'everything that passes for knowledge in society' and particularly with that which passes for common sense, this is 'a serious, open-minded book, upon a serious subject' – *Listener*

The Care of the Self   Michel Foucault
The History of Sexuality Vol 3

Foucault examines the transformation of sexual discourse from the Hellenistic to the Roman world in an inquiry which 'bristles with provocative insights into the tangled liaison of sex and self' – *The Times Higher Educational Supplement*

A Fate Worse than Debt   Susan George

How did Third World countries accumulate a staggering trillion dollars' worth of debt? Who really shoulders the burden of reimbursement? How should we deal with the debt crisis? Susan George answers these questions with the solid evidence and verve familiar to readers of *How the Other Half Dies*.

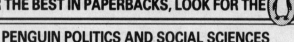

FOR THE BEST IN PAPERBACKS, LOOK FOR THE

PENGUIN POLITICS AND SOCIAL SCIENCES

Comparative Government S. E. Finer

'A considerable *tour de force* ... few teachers of politics in Britain would fail to learn a great deal from it ... Above all, it is the work of a great teacher who breathes into every page his own enthusiasm for the discipline' – Anthony King in *New Society*

Karl Marx: Selected Writings in Sociology and Social Philosophy
T. B. Bottomore and Maximilien Rubel (eds.)

'It makes available, in coherent form and lucid English, some of Marx's most important ideas. As an introduction to Marx's thought, it has very few rivals indeed' – *British Journal of Sociology*

Post-War Britain A Political History Alan Sked and Chris Cook

Major political figures from Attlee to Thatcher, the aims and achievements of governments and the changing fortunes of Britain in the period since 1945 are thoroughly scrutinized in this readable history.

Inside the Third World Paul Harrison

From climate and colonialism to land hunger, exploding cities and illiteracy, this comprehensive book brings home a wealth of facts and analysis on the often tragic realities of life for the poor people and communities of Asia, Africa and Latin America.

Housewife Ann Oakley

'A fresh and challenging account' – *Economist*. 'Informative and rational enough to deserve a serious place in any discussion on the position of women in modern society' – *The Times Educational Supplement*

The Raw and the Cooked Claude Lévi-Strauss

Deliberately, brilliantly and inimitably challenging, Lévi-Strauss's seminal work of structural anthropology cuts wide and deep into the mind of mankind, as he finds in the myths of the South American Indians a comprehensible psychological pattern.

PENGUIN PHILOSOPHY

I: The Philosophy and Psychology of Personal Identity   Jonathan Glover

From cases of split brains and multiple personalities to the importance of memory and recognition by others, the author of *Causing Death and Saving Lives* tackles the vexed questions of personal identity. 'Fascinating ... the ideas which Glover pours forth in profusion deserve more detailed consideration' – Anthony Storr

Minds, Brains and Science   John Searle

Based on Professor Searle's acclaimed series of Reith Lectures, *Minds, Brains and Science* is 'punchy and engaging ... a timely exposé of those woolly-minded computer-lovers who believe that computers can think, and indeed that the human mind is just a biological computer' – *The Times Literary Supplement*

Ethics   Inventing Right and Wrong   J. L. Mackie

Widely used as a text, Mackie's complete and clear treatise on moral theory deals with the status and content of ethics, sketches a practical moral system and examines the frontiers at which ethics touches psychology, theology, law and politics.

The Penguin History of Western Philosophy   D. W. Hamlyn

'Well-crafted and readable ... neither laden with footnotes nor weighed down with technical language ... a general guide to three millennia of philosophizing in the West' – *The Times Literary Supplement*

Science and Philosophy: Past and Present   Derek Gjertsen

Philosophy and science, once intimately connected, are today often seen as widely different disciplines. Ranging from Aristotle to Einstein, from quantum theory to renaissance magic, Confucius and parapsychology, this penetrating and original study shows such a view to be both naive and ill-informed.

The Problem of Knowledge   A. J. Ayer

How do you *know* that this is a book? How do you *know* that you know? In *The Problem of Knowledge* A. J. Ayer presented the sceptic's arguments as forcefully as possible, investigating the extent to which they can be met. 'Thorough ... penetrating, vigorous ... readable and manageable' – *Spectator*

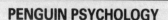
PENGUIN PSYCHOLOGY

Introduction to Jung's Psychology Frieda Fordham

'She has delivered a fair and simple account of the main aspects of my psychological work. I am indebted to her for this admirable piece of work' – C. G. Jung in the Foreword

Child Care and the Growth of Love John Bowlby

His classic 'summary of evidence of the effects upon children of lack of personal attention ... it presents to administrators, social workers, teachers and doctors a reminder of the significance of the family' – *The Times*

The Anatomy of Human Destructiveness Erich Fromm

What makes men kill? How can we explain man's lust for cruelty and destruction? 'If any single book could bring mankind to its senses, this book might qualify for that miracle' – Lewis Mumford

Sanity, Madness and the Family R. D. Laing and A. Esterson

Schizophrenia: fact or fiction? Certainly not fact, according to the authors of this controversial book. Suggesting that some forms of madness may be largely social creations, *Sanity, Madness and the Family* demands to be taken very seriously indeed.

The Social Psychology of Work Michael Argyle

Both popular and scholarly, Michael Argyle's classic account of the social factors influencing our experience of work examines every area of working life – and throws constructive light on potential problems.

Check Your Own I.Q. H. J. Eysenck

The sequel to his controversial bestseller, containing five new standard (omnibus) tests and three specifically designed tests for verbal, numerical and visual–spatial ability.

FOR THE BEST IN PAPERBACKS, LOOK FOR THE 🐧

PENGUIN PSYCHOLOGY

Psychoanalysis and Feminism Juliet Mitchell

'Juliet Mitchell has risked accusations of apostasy from her fellow feminists. Her book not only challenges orthodox feminism, however; it defies the conventions of social thought in the English-speaking countries ... a brave and important book' – *New York Review of Books*

Helping Troubled Children Michael Rutter

Written by a leading practitioner and researcher in child psychiatry, a full and clear account of the many problems encountered by young school-age children – development, emotional disorders, underachievement – and how they can be given help.

The Divided Self R. D. Laing

'A study that makes all other works I have read on schizophrenia seem fragmentary ... The author brings, through his vision and perception, that particular touch of genius which causes one to say "Yes, I have always known that, why have I never thought of it before?"' – *Journal of Analytical Psychology*

The Origins of Religion Sigmund Freud

The thirteenth volume in the *Penguin Freud Library* contains Freud's views on the subject of religious belief – including *Totem and Taboo*, regarded by Freud as his best-written work.

The Informed Heart Bruno Bettelheim

Bettelheim draws on his experience in concentration camps to illuminate the dangers inherent in all mass societies in this profound and moving masterpiece.

Introducing Social Psychology Henri Tajfel and Colin Fraser (eds.)

From evolutionary changes to the social influence processes in a given group, a distinguished team of contributors demonstrate how our interaction with others and our views of the social world shape and modify much of what we do.

PENGUIN REFERENCE BOOKS

The Penguin Guide to the Law

This acclaimed reference book is designed for everyday use and forms the most comprehensive handbook ever published on the law as it affects the individual.

The Penguin Medical Encyclopedia

Covers the body and mind in sickness and in health, including drugs, surgery, medical history, medical vocabulary and many other aspects. 'Highly commendable' – *Journal of the Institute of Health Education*

The Slang Thesaurus

Do you make the public bar sound like a gentleman's club? Do you need help in understanding *Minder*? The miraculous *Slang Thesaurus* will liven up your language in no time. You won't Adam and Eve it! A mine of funny, witty, acid and vulgar synonyms for the words you use every day.

The Penguin Dictionary of Troublesome Words   Bill Bryson

Why should you avoid discussing the *weather conditions*? Can a married woman be *celibate*? Why is it eccentric to talk about the *aroma* of a cowshed? A straightforward guide to the pitfalls and hotly disputed issues in standard written English.

A Dictionary of Literary Terms

Defines over 2,000 literary terms (including lesser known, foreign language and technical terms), explained with illustrations from literature past and present.

The Concise Cambridge Italian Dictionary

Compiled by Barbara Reynolds, this work is notable for the range of examples provided to illustrate the exact meaning of Italian words and phrases. It also contains a pronunciation guide and a reference grammar.

FOR THE BEST IN PAPERBACKS, LOOK FOR THE 🐧

PENGUIN SCIENCE AND MATHEMATICS

Facts from Figures   M. J. Moroney

Starting from the very first principles of the laws of chance, this authoritative 'conducted tour of the statistician's workshop' provides an essential introduction to the major techniques and concepts used in statistics today.

God and the New Physics   Paul Davies

Can science, now come of age, offer a surer path to God than religion? This 'very interesting' (*New Scientist*) book suggests it can.

Descartes' Dream   Philip J. Davis and Reuben Hersh

All of us are 'drowning in digits' and depend constantly on mathematics for our high-tech lifestyle. But is so much mathematics really good for us? This major book takes a sharp look at the ethical issues raised by our computerized society.

The Blind Watchmaker   Richard Dawkins

'An enchantingly witty and persuasive neo-Darwinist attack on the anti-evolutionists, pleasurably intelligible to the scientifically illiterate' – Hermione Lee in the *Observer* Books of the Year

Microbes and Man   John Postgate

From mining to wine-making, microbes play a crucial role in human life. This clear, non-specialist book introduces us to microbes in all their astounding versatility – and to the latest and most exciting developments in microbiology and immunology.

Asimov's New Guide to Science   Isaac Asimov

A classic work brought up to date – far and away the best one-volume survey of all the physical and biological sciences.

PENGUIN SCIENCE AND MATHEMATICS

The Panda's Thumb   Stephen Jay Gould

More reflections on natural history from the author of *Ever Since Darwin*. 'A quirky and provocative exploration of the nature of evolution ... wonderfully entertaining' – *Sunday Telegraph*

Genetic Engineering for Almost Everybody   William Bains

Now that the genetic engineering revolution has most certainly arrived, we all need to understand its ethical and practical implications. This book sets them out in accessible language.

The Double Helix   James D. Watson

Watson's vivid and outspoken account of how he and Crick discovered the structure of DNA (and won themselves a Nobel Prize) – one of the greatest scientific achievements of the century.

The Quantum World   J. C. Polkinghorne

Quantum mechanics has revolutionized our views about the structure of the physical world – yet after more than fifty years it remains controversial. This 'delightful book' (*The Times Educational Supplement*) succeeds superbly in rendering an important and complex debate both clear and fascinating.

Einstein's Universe   Nigel Calder

'A valuable contribution to the demystification of relativity' – *Nature*

Mathematical Circus   Martin Gardner

A mind-bending collection of puzzles and paradoxes, games and diversions from the undisputed master of recreational mathematics.

FOR THE BEST IN PAPERBACKS, LOOK FOR THE 🐧

PENGUIN DICTIONARIES

Abbreviations
Archaeology
Architecture
Art and Artists
Biology
Botany
Building
Business
Chemistry
Civil Engineering
Computers
Curious and Interesting
   Words
Curious and Interesting
   Numbers
Design and Designers
Economics
Electronics
English and European
   History
English Idioms
French
Geography
German

Historical Slang
Human Geography
Literary Terms
Mathematics
Modern History 1789–1945
Modern Quotations
Music
Physical Geography
Physics
Politics
Proverbs
Psychology
Quotations
Religions
Rhyming Dictionary
Saints
Science
Sociology
Spanish
Surnames
Telecommunications
Troublesome Words
Twentieth-Century History